Harald H. Roth Günter Merz (Eds.)

# Wildlife Resources

Springer

*Berlin*
*Heidelberg*
*New York*
*Barcelona*
*Budapest*
*Hong Kong*
*London*
*Milan*
*Paris*
*Santa Clara*
*Singapur*
*Tokyo*

Harald H. Roth Günter Merz (Eds.)

# Wildlife Resources

## A Global Account of Economic Use

With 21 Figures, 56 Photos and 56 Tables

Springer

Prof. Dr. Dr. Harald H. Roth
Tewsstr. 18
D-14129 Berlin
Germany

Dr. Günter Merz
WWF-Umweltstiftung
Hedderichstr. 110
D-60596 Frankfurt/Main
Germany

ISBN 3-540-61357-9 Springer-Verlag Berlin Heidelberg New York

Library of Congress Cataloging-in-Publication Data
Wildlife resources : a global account of economic use / Harald H.
Roth Günter Merz (eds.).
p. cm.
Includes bibliographical references and index
ISBN 3-540-61357-9 (hardcover)
1. Wildlife utilization. I. Roth, Harald H., 1926-
II. Merz, Günter
SF84.6.W54 1996
333.95´417 –DC20 96-23063



Cover Design: Erich Kirchner, Heidelberg;
Cover Illustrations: I.B. Leeds, G. Thompson
Camera ready by Pia Hofmann
SPIN 10088555 31/3137 5 4 3 2 1 0 - Printed on acid free paper -

## PREFACE

The general interest and attention paid to the use of the renewable natural resources of the world have increased greatly during the last decades. This is due to the environmental dilemma into which mankind has got itself by the total disregard of ecological facts and laws, the underprizing of natural resources and the overemphasis on economic development, coupled with unimpeded rapid population growth and the preponderance of materialistic consumption-oriented attitudes.

The management and use of natural resources such as forests, grasslands, rivers and lakes were formerly considered purely in a financial context, whereas latterly, a consciousness of their social function and the indirect economic benefits which can be derived from them has developed. Thus, as regards these traditional resources, multi-use concepts comprising economically oriented utilisation as well as recreational, educational and social use have become widely accepted. Conservation of natural resources for sustainable consumptive as well as non-consumptive uses has been recognised as a key element for maintaining economic development all over the world. Fortunately, the industrialised countries have started to apply this principle themselves and in their technical and financial aid to the developing Third World countries. This is manifest from policy documents such as the *World Bank Policy on Development of Wild Lands*, issued in 1987. While it was previously unthinkable that any Development Bank would invest in the conservation of natural resources or the management of natural forests, wetlands and other "wildlands" for long-term sustainable utilisation, this has now become a legitimate area for loan consideration.

However, despite the acceptance of wildlife as part of wildland resources, it is rarely considered as an exploitable natural resource as such. It is often not understood or not readily accepted that the same ecological and economic principles which govern the management and utilisation of trees, grasses and fish, apply equally to game animals. Everybody accepts that forests are timbered and lakes fished for economic gain, but the hunting of wild animals was and still is the subject of very diverse emotions and is obscured by traditional rituals. Forestry and fishery have always been taught and treated as special academic disciplines, whereas wildlife management has still not gained world-wide recognition as a special field of professional application.

This is largely due to the linkage of game in Europe to forestry, and the evolution of special hunting traditions and privileges in the European feudal system. Game management and conservation in Europe thus became early on a co-responsibility of the state, represented by the foresters, and the land owners. Only in the USA has the management of wildlife developed without bias into a special biological discipline and professional activity. From there this was then introduced to universities in Canada, Australia and South Africa.

Societies in Third World countries had their own traditional pragmatic relationships with wildlife, often depending on this resource for their livelihood and well-being. The European wildlife concept was forced upon them during the colonial era and became to some extent an obstacle to the development of rational wildlife management and utilisation in many of these countries, particularly in Africa. Only more recently has wildlife management been introduced in some of them as an academically taught subject.

Professional wildlife workers find it amazing that laymen, sports hunters, naturalists and animal lovers seem to know everything about wild animals and press their point of view often in a most subjective and emotional manner. This is contrary to the situation pertaining to other natural resource fields. In fact, this phenomenon, resulting from a special psychological relationship which exists between man and animals, has often prevented rational approaches to conservation and the economic use of the wildlife resources, particularly in the formerly colonised, now developing countries of the Third World.

This book sets out to describe, without bias, the principles which should govern the use of wild animals as an important natural resource under varying ecological and sociological circumstances around the world. It intends to inform on the great diversity and complexity of wildlife utilisation and its economic and other importance as a primary or secondary form of land use. Finally it discusses the uses and the processing and marketing of some of the more important wildlife products. Considering the very broad scope of the book, it can but give an overview leaving it to the reader to follow up any specific subjects in the listed literature.

In making this effort it is fully accepted that there is a strong emotional man-animal relationship and that love of animals is definitely a positive aspect of humanity. It is also recognised that the economic utilisation of wild animals, like that of other natural resources, may easily lead to over-exploitation and consequent depletion or even extermination of species, be it locally or globally. However, this does not invalidate the basic premises of the book: Wildlife is an integral natural resource of great ecological and economic importance in many parts of the world that needs to be managed and utilised to the best long-term advantage of the local population, the concerned nations and the international community as a whole. Considering the ever-increasing pressures on all natural resources, the only hope for long-term maintenance of the greatest possible diversity of wildlife is to ensure that local populations and the nations concerned derive optimum economic benefit from their wildlife in one form or another. Only this will induce and enable them to conserve their wildlife heritage.

# Contents

## PART I: CONCEPTS AND PRINCIPLES OF WILDLIFE UTILISATION

## PART II: UTILISATION OF SPECIFIC ANIMAL GROUPS



## PART III: WILDLIFE PRODUCTS

## LIST OF CONTRIBUTORS

| NAMES | CONTRIBUTIONS |
|---|---|
| **Nicholas J. Aebischer**<br>The Game Conservancy Trust<br>United Kingdom | *Utilisation of fowl-like birds* |
| **Leonid M. Baskin**<br>Institute of Ecology and Evolution<br>Russian Academy of Sciences, Moscow<br>Russia | *Utilisation of reindeer, domestication of moose and eland, and production of furskins in the former USSR* |
| **Jonathan I. Barnes**<br>WWF (US) LIFE Programme,<br>Directorate of Environmental Affairs<br>Namibia | *Economics of wildlife utilisation* |
| **Javier Bellati**<br>National Institute of Livestock<br>Technology, Bariloche<br>Argentina | *Introduction and utilisation of game animals in Argentina* |
| **W. Nigel Bonner (†)**<br>United Kingdom | *Utilisation of seals and walruses, use of animal oils and fats* |
| **Tom J. Cade**<br>The Peregrine Fund<br>World Centre for Birds of Prey<br>USA | *Utilisation of birds of prey* |
| **Ray Gambell**<br>International Whaling Commission<br>United Kingdom | *Utilisation of cetaceans* |
| **Wolfgang Gewalt**<br>formerly Zoo Duisburg<br>Germany | *Use of small whales and dolphins for research and show purposes* |
| **Eckhard W. Heymann**<br>German Primate Centre, Göttingen<br>Germany | *Utilisation of primates* |
| **Karl-Otto Honikel**<br>Institute of Chemistry and Physics<br>Federal Centre for Meat Research<br>Germany | *Composition and nutritional value of game meat* |
| **Wolfgang Krostitz**<br>formery Food and Agriculture Organisation<br>of the United Nations, Rome<br>Germany | *International trade in game meats* |
| **Iain B. Leach**<br>United Kingdom | *Processing of and trade in gameskins* |

| | |
|---|---|
| **Thomas Lenhart**<br>R & L Rosenberg & Lenhart<br>German Fur Trading Co., Frankfurt<br>Germany | *Furfarming, processing of and trade in furskins* |
| **Richard A. Luxmoore**<br>**John Caldwell**<br>World Conservation Monitoring Centre<br>United Kingdom | *Trade in wildlife products* |
| **Esmond Bradley Martin**<br>**Lucy Vigne**<br>Rhino Help International<br>Kenya | *Use and trade in ivory and rhinohorn* |
| **Günter Merz (co-editor)**<br>WWF-Umweltstiftung<br>Germany | *Wildlife-based tourism* |
| **Nicholas Mrosovsky**<br>Department of Zoology<br>University of Toronto<br>Canada | *Utilisation of sea turtles* |
| **Kay Muir-Leresche**<br>Agricultural Economics and Extension<br>University of Zimbabwe<br>Zimbabwe | *Natural resources economics and marketing of wildlife products* |
| **Heinz-Sigurd Raethel**<br>Germany | *Historical aspects of the use of fowl-like birds* |
| **Michael Rands**<br>BirdLife International<br>United Kingdom | *Utilisation of various groups of birds* |
| **Günter Reinken**<br>formerly Landwirtschaftskammer Rheinland<br>Germany | *Farming of red and fallow deer* |
| **Christian Ring**<br>Institute of Meat and Food Hygiene<br>Veterinary College Hannover<br>Germany | *Hygiene of game meat production* |
| **Harald H. Roth (editor)**<br>formerly Southern African Development Community (SADC) Coordinator for Forestry, Fisheries and Wildlife, Lilongwe<br>Zimbabwe | *History, forms and importance of wildlife utilisation; utilisation of crocodilians, ostriches, rodents, sirenians, elephants, odd-toed and suiform ungulates, camelids, deer, antelope, and buffalo; use and trade in antlers and musks, feathers and animal trophies* |
| **Walter Sachsse**<br>University of Mainz<br>Germany | *Utilisation of tortoises and freshwater turtles* |

**Sigrid Schwenk**
Research Centre for Cultural Aspects of Hunting
University of Bamberg
Germany

*Cultural aspects and history of hunting*

**Wendell G. Swank**
formerly Texas A & M University
USA

*Management of wildlife resources; game birds in North America*

**James G. Teer**
Welder Wildlife Foundation, Texas
USA

*Utilisation of exotic animals on ranches in the USA*

**Romulus Whitaker**
Centre for Herpetology,
Madras Crocodile Bank Trust
India

*Utilisation of lizards and snakes*

**Gerry A. Wild**
Environmental Training Centre, Alberta
Canada

*Management and utilisation of wild furbearers in North America*

**Keith Woodford**
**John G. Dingle**
University of Queensland, Gatton College
Australia

*Utilisation of emus*

**Michael Woodford**
formerly Food and Agriculture Organisation
of the United Nations, Rome
USA

*Veterinary aspects of trade in wildlife products*

**Antoon de Vos**
formerly Food and Agriculture Organisation
of the United Nations, Rome
Australia

*Utilisation of marsupials*

**Lir Zhirnov**
Institute of Ecology and Evolution
Russian Academy of Sciences, Moscow
Russia

*Utilisation of saiga antelope*

## ACKNOWLEDGEMENTS

The editors wish to thank most sincerely all those colleagues who are not specifically mentioned as contributors but have nevertheless furnished verbal information on certain subjects of the book or have commented on parts of the manuscript. Particular thanks are due to FIONNA WILLMOT who has spent much time in reading and improving the manuscript. PADDY WILSON has kindly produced the animal silhouette drawings. The assistance of NAOMI CHUDY and PIA HOFFMANN in the processing and layout of the text and tables is also gratefully acknowledged.

# PART I: CONCEPTS AND PRINCIPLES OF WILDLIFE UTILISATION

The use of animals for material or immaterial purposes is as old as mankind but a very complex subject. In Part I it is reviewed against the historical background of man-animal relationships in the widest sense, the taming and domestication of wild animals, the evolution of hunting and of land tenure systems and alternative uses of natural resources, but also in relation to the changing demands and preferences for products from wild animals, cultural perceptions and the recreational needs of the modern society. For thousands of years the use of wildlife was empirical and therefore haphazard. Only with the development of the natural sciences, biology and most recently ecology, has man started to understand the underlying natural processes and biological productivity. This should enable him to transform his use of wildlife resources into a scientifically planned management activity with the ultimate goal of maintaining all the earth's living resources for continued sustainable benefit to mankind. To this end it is becoming increasingly important to apply also the emerging knowledge and principles from other scientific disciplines, such as economics.

## 1.1 TERMINOLOGY AND SCOPE

As in many other emerging areas of biological endeavour there is often confusion in respect of the precise meaning of specific technical terms referring to wildlife management. This book is concerned with "*wildlife resources*" and their "*economic use*". Both of these terms are complex and require careful definition in order to make the contents of the book well understandable.

"*Wildlife*" is an Anglo-American term which is difficult to translate into other languages. The Wildlife Management Policy adopted in 1988 by the states of the Southern African Development Community (SADC) defines wildlife resources

*"to include all species of wild animals with their supporting habitat; therefore, the Wildlife Policy overlaps with forestry and fisheries policies; it refers both to problem as well as desirable species, and those that may be considered insignificant. Wildlife is recognised to be a complex natural resource that has positive as well as negative effects in relation to human needs, and that needs to be managed in a professional and scientific manner for the benefit of the people of the Region".*

Although "*wildlife*" in the wider sense is considered to also include fish, these have become the subject of specialist management and research by fisheries biologists. Therefore, the term is usually considered to comprise only terrestrial and larger aquatic vertebrates, except fish. It signifies definitely more than just "*game*" or "*wild animals*" and has also a different connotation than "*fauna*", because it refers to the total spectrum of wild animals in an ecological context. The term "*wildlife*" includes the specific environmental features on which each species depends. Such an ecological definition is important for understanding the term "*wildlife management*". This refers to any type of organised, purposeful human engagement with wildlife. Its major aim may simply be to protect, maintain or promote it: "*wildlife conservation*"; or to reduce certain species: "*wildlife control*"; or to use some species for material or recreational benefit: "*wildlife utilisation*". All of these activities are complementary and must be considered in an ecological framework. "*Wildlife conservation*" is thus a more comprehensive and enhanced concept than the formerly used term "*game protection*" which is aimed mainly at protecting certain wild animal species against excessive hunting. Management and conservation of wildlife in this broadened definition consists to a large extent of protection, manipulation or control of the habitat of certain animal species in addition to protection and manipulation of the population of the animals themselves. This elucidates how closely wildlife management is related to range management, forestry and fisheries. All of these disciplines are concerned with specific components of the renewable resources of the earth; i.e. range management with grassland resources, forestry with forest resources and fisheries with the fish resources. Wildlife constitutes another distinct component of natural resources and the term "*wildlife resources*" emphasises the importance of wild animals in the context of the management, conservation and sustainable use of all natural resources.

As will be shown below there are many different ways in which wild animals are being used by man. Wildlife utilisation may be "*consumptive*" or "*non-consumptive*" depending on the main objective of the management activity. "*Economic use*" is usually associated with consumptive wildlife utilisation, i.e. the use of wild animals for meat and other specific wildlife products, be it for commercial purposes or non-commercially for subsistence only. However, there is also an important socio-economic aspect inherent in non-consumptive utilisation of natural resources: the considerable indirect benefits which may be derived from out-door tourism, recreational or educational uses. Non-consumptive use is also an economic activity and, for example wildlife photography, in certain areas may be economically more important than consumptive use. Therefore the meaning of "*economic use*" must be broadened to include, in part, non-consumptive wildlife utilisation.

It is essential to distinguish between "*economic use*" and the "*exploitation*" of natural resources. Whereas the former should be strictly defined as *sustainable* utilisation of a renewable natural resource, the latter is usually *non-sustainable*. This implies that economic use of wildlife species may very well be compatible with their conservation and propagation, if controlled in terms of long-term wildlife management programmes. Exploitation, however, will result sooner or later in the decline or even extirpation of the species concerned.

The scope of this book must necessarily be limited. Whereas the background and the principles of wildlife utilisation (as outlined in Part I) apply to any wildlife, in any climatic or ecological setting, the specifics of wildlife utilisation (Part II) are described for each of the most important groups of reptiles, birds and mammals, and the discussion of wildlife products (Part III) covers only the more important commodities. The specific parts of the book are mainly concerned with consumptive economic use, mostly commercially oriented, of these groups of animals. The utilisation of wildlife resources in the form of non-commercial subsistence hunting and as wildlife-based tourism which comprise any wildlife is only discussed in the general context of Part I.

Literature consulted in respect of the general background and principles of wildlife utilisation is listed at the end of Part I and includes selected references up to 1993. Specific references and statistical material concerning the utilisation of the different animal groups or species are cited separately for each chapter.

## 1.2 HISTORY OF WILDLIFE UTILISATION

### Early relationships between man and animal

Man emerged during the Pleistocene about 500,000 years ago. Although a primate, he evolved as a carnivorous predator depending on his ability and skill to kill animals for food and raw materials. His original association with animals may thus be described as a **predator-prey relationship**. For hundreds of thousands of years little has changed in this basic relationship apart from man developing greater efficiency in his hunting skills. Man's association with animals has continued throughout his history and has influenced human society in many different ways. Although this association became more refined and complex in the course of his socio-cultural development, some features of the original predator-prey relationship are still evident in human behaviour patterns.

For example, despite their estrangement from nature and wild animals, many men still love to hunt "for sport", and "killing" is a thrill to some of them. In taming and domesticating wild animals man has continued to seek dominance, and in human society "leadership" of a few is still an important social feature. References to the "fierceness" of wild animals and to "wilderness" have an emotional effect on many people in industrialised countries, whilst the consumption of game meat has an almost magical significance in many developing countries.

When man learnt to manipulate fire and develop tools these great achievements had important applications in his ability to utilise wild animals. The control of fire enabled him to burn grass and forests and thus to hunt more efficiently, just as hunters in Africa still do. By melting ore he could eventually produce iron weaponry and this enabled him to kill much larger animals than before such as the mammoth. With the aid of fire he could also preserve part of the meat and was less dependent on constant pursuit of the prey. With these achievements man could expand his range from tropical zones into more inhospitable regions and survive the Ice Ages.

Prehistoric man lived a fairly nomadic life in small family bands with a division of work between men and women. The men hunted in groups, providing meat as the most important food item, while women would gather nuts, berries and herbs, and tend the children. The **economy** of this **early hunter-gatherer society** is fairly well established from osteological evidence in many archaeological deposits and from rock paintings. Although probably any animal species would be eaten, similar to the present-day nutrition of many peoples in developing countries, there were already preferences for certain species. These are reflected by the predominance of the remains of certain species in the archaeological deposits in different regions and during different periods. This might have been mainly due to the geographical distribution of the different species and their local abundance, but to some extent it also reflects differences in hunting practices and attitudes. Thus, in economic terms, some species became more important to man than others as far back as the late Pleistocene and beginning of the Holocene.

Fig. 1 shows that in more than 95% of 165 studied late palaeolithic and mesolithic sites in Europe, remains of *Red deer* were found. This species did not only supply meat, but also hides, bones, sinews and antlers as raw materials of great value to early cultures. In Western Asia *gazelles*, in some places *wild goats* and *onager*, and in North America the *bison*, were of greatest importance at that time.

**Figure 1:** The presence of mammal remains at 165 late palaeolithic and mesolithic archaeological sites throughout Europe

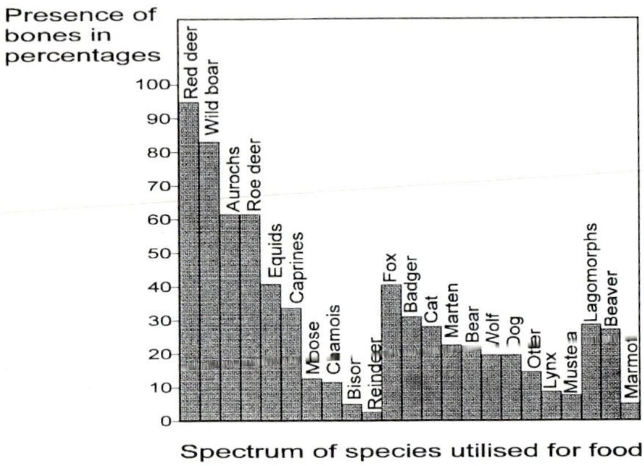


Source: JARMAN, M.R. (1972): European deer economies and the advent of the Neolithic (in HIGGS (ed.) (1972) - see selected literature references)

It may be assumed that even at this early stage, about 12,000 to 9,000 years ago, man started to select those species which offered the greatest benefit to him. It is contended by some authors that man started at the end of the Ice Ages, i.e. about 10,000 years ago, to actively manage *Red deer*, culling males and husbanding the herds. There is archaeological evidence from the late Palaeolithicum in Northern Europe that suggests that some groups of hunters had already associated themselves with *reindeer*, not managing them, but following and living off them on their seasonal migrations all through Europe. In Western Asia, after the use of *gazelles* had gained greatest importance over the initial deer utilisation, it is assumed that this species was also actively managed by man in pre-neolithic times, as a source of food.

The association of man with animals has thus evolved from a general predator-prey relationship into much more complex and differentiated forms of interaction.

Apart from the original use of wildlife, the indiscriminate killing of any available animal, which has continued in many parts of the world to-date, a deliberate **selective use of specific animal species** for different purposes emerged towards the end of the Ice Ages. It is probable that this development occurred before man started to cultivate plants and become sedentary. The phase of agricultural development which followed the Ice Ages in various regions, such as the Nile valley and Western Asia, spreading rapidly into Southern and Northern Europe, was probably the most important in the history of mankind and is referred to as the "neolithic revolution". A major feature of this was, no doubt, the capability of man to use wild animals selectively. This ability was developed in two different ways. On the one hand man started to keep and to breed wild animals under his control with a view to exploiting them more efficiently; and on the other hand he learnt to hunt them selectively.

Much of this initial selective use of herded or kept animals must have been incidental, because man saw only the immediate benefit of having them always readily available for slaughter. Only later did he realise the potential inherent in these and other species for various uses and start to breed them purposefully for special products, such as meat, fat, skins, hides, feathers, wool, medicine, and so on, or to obtain desired services, such as hunting aid, protection and herding, ploughing, transport, power for warfare, and so on. Not all of these early efforts led to the eventual domestication of selected species, as is discussed below. Throughout man's history many other species have been tamed and kept for specific purposes in various different cultures, in Eastern Asia, India, Persia, Mesopotamia, Ancient Egypt, and the Roman Empire. Only a few of these emerged as domestic animals; others were simply exploited without selective breeding.

After the initial early domestications, industrialisation in the 19th century brought about a new surge of selection of animals for economic utilisation in one way or another. This was partly due to the gradual decline of exploited species, but also a response to the emerging diverse needs of the industrialised society for luxury commodities such as special foods, furs and leathers, and for pets and laboratory animals. Throughout this century whims of fashion and economics have induced the selection of ever more species for various forms of intensive use, as discussed in Chapt. 1.4. Some more recent selection has resulted from the urgent need to improve land use through locally adapted and more productive animal species, in the light of increasing biological and ecological knowledge.

Concurrently with the domestication efforts, man also developed his ability to seek out and to kill only those wild-living species which would be of greatest use to him or would provide specially needed raw materials. Methods of hunting and trapping were constantly improved, resulting in an ever increasing efficiency in the exploitation of specific wild animals. This has lead in many parts of the world, to drastic reduction, extirpation or even extermination of the selected species. With the socio-cultural changes at the beginning of the Middle Ages, hunting in Europe became the privilege of the noble classes who were no longer concerned with the economic exploitation of wildlife; they hunted selectively for sport and the procurement of prestigious trophies. In other parts of the world, however, selective use of wild-living animals as food and raw material, through hunting or trapping, has continued to-date and retained its original superior importance.

Apart from the above-mentioned material aspect of the man-animal association, wild animals and later domestic ones as well, also had an important immaterial **mythological function**, featuring prominently in the spiritual world of man. It is easy to imagine how the intimate association with animals, their dominating importance for survival and well-being,

Rock painting from Lion's Head, Msana Reserve in Mashonaland, Zimbabwe, dating back to the later Stone Age and depicting a hunting scene

Source: SUMMERS R. (ed.) (1959): Prehistoric Rock Art of the Federation of Rhodesia and Nyasaland. Chatto and Windus, London; on behalf of the National Publications Trust, Salisbury, Rhodesia

their menace and also their beauty, has influenced man throughout his evolution. Wild animals have stimulated his imagination, have inspired him since early times to depict them and have thus contributed to the development of man's artistic capabilities. They have become part of his philosophies and religious beliefs and many societies have and still are worshipping animals.

Although all this does not necessarily constitute a benefit to man he has used wild animals since the earliest times for his spiritual needs as much as for his physical ones. This function of wild animals may also be seen as a form of "selective use" which continued throughout history in many societies; it is still present even in the "developed" countries in the form of superstitions.

The role of wild animals in mythology is a special and wide subject which surpasses the scope of this book. However, the desire to obtain certain parts of wild animals for ritual or medical purposes has been and still is one of the strongest incentives for hunting them. The need to have certain species readily available for sacrifice and the deification of others, has been as much the original motive for keeping them as material consumptive considerations. Therefore, the mythological aspect cannot be ignored altogether in the context of socio-cultural and economic use of wildlife resources dealt with in this book.

## Domestication of wild animals

As described above, the development of agriculture during the Neolithicum brought about important changes in the use of the wildlife resources. Hunter-gatherer communities, particularly in Western Asia, but also in South-East and Eastern Asia and later in Northern Europe, started to herd or contain certain species of wild animals, thus facilitating their utilisation. This led eventually to the domestication of a variety of species, as discussed above. The **reasons for the domestication** of wild species, however, were certainly more complex than previously assumed.

Two aspects are important in this context:

1. Domestication could not have been planned for any particular purpose as the early neolithic man neither knew the genetic potential inherent in the wild animals kept or herded, nor did he have any notion of selective breeding. In fact, the initial reason for herding or captive keeping of wild animals could have been only practical: they became more readily available for slaughter whenever meat was required. Archaeological and socio-cultural evidence shows that even species which later on were used for other purposes, such as *dogs, horses* and *asses*, served initially as a source of food. The *aurochs* may have been kept initially for ritual purposes as was the case with many other species and as still applies to the *gayal* in Assam. Only after taming the powerful aurochs was it also used for ploughing and pulling carts, long before the horse and ass were available for this purpose. Thus, only in the course of prolonged captive breeding did man realise the different uses to which particular domesticated animals could be put, and commence selecting empirically for desired properties; *dogs* for herding and protection, *sheep* for wool, *cattle* for beef and milk, *horses* for riding and warfare, *asses* for transport and so on. It took thousands of years to fully achieve these uses and efforts at improving the properties of domestic animals for specific purposes have continued to-date.

   Only under particularly harsh environmental conditions would man associate very closely with some animal species which would enable him to survive in that particular environment; *Camels* and *dromedaries* in very arid regions, *llamas* and *alpacas* in barren high altitudes, *reindeer* in very cold tundra. In these cases, a kind of symbiosis developed between man and animal in which both depended on each other for survival. Consequently there was very little artificial selection in the course of this domestication process. These species were so well adapted to the environment that there was no need to try to improve them.

   In other cases, wild animals were tamed and immediately kept for specific purposes which were obvious: *elephants* for power, *cheetahs* for hunting speed, *ferrets* for rabbit hunting, *birds of prey* for hunting birds, *cormorants* for fishing and so on. Although these uses all have a long history, most of the species concerned were not bred in captivity. Stocks were replenished from the wild and thus these species never became fully domesticated.

2. A change from hunting-gathering to a sedentary agricultural life, i.e. from a procurement to a production economy, did not always and everywhere improve or facilitate life in the early human communities. Previously it was thought that this fundamental change occurred voluntarily and constituted evolutionary and socio-cultural progress. This notion is certainly doubtful as can be learnt from existing primitive hunter-gatherer or nomadic societies. As long as natural resources, on which such societies depend, are plentiful and ecologically balanced, there is normally no incentive for adopting a production oriented way of life other than the allurements of modern technology which did not exist previously. To abandon his roving exploitation of the renewable natural resources in favour of

a sedentary life-style associated with the hard physical work of clearing forest, ploughing the ground, tending and husbanding domestic stock could not have been an easy voluntary choice for prehistoric man. It certainly did not come about spontaneously and everywhere; it occurred probably only under severe pressure of food scarcity as a result of ecological changes combined with local population growth and perhaps already resource mismanagement, such as over-exploitation. These are likely to have occurred in most centres of the so-called neolithic revolution, in the Nile valley, and in Western Asia.

Wherever man did embark upon the new laborious production oriented life-style, he learned to utilise selected animal species most efficiently through captive breeding and domestication. There was no way back in this evolutionary process and it has successfully sustained the continuous growth and expansion of mankind to its present-day civilisation. In fact the new life-style has probably significantly accelerated the growth of the human population by reducing mortality in the youngest and oldest age classes, which were most affected by the previous nomadic way of life.

Only during the last century has man learned to plan the use of wild animals for specific purposes through genetically controlled captive breeding. Nowadays enhanced knowledge of genetics and behaviour patterns might enable him to transform a wild species into a domestic animal serving specific needs much more quickly than the process of domestication of wild animals has taken in the past. Many fur animals, such as *chinchilla, nutria, fox,* and *mink* have been subjected to selective breeding for almost a hundred years. *Rats* and *mice* have been bred in very large numbers with known genetic make-ups for biomedical purposes for many decades. But some species with particular promise for meat production, such as the *cane rat, deer* species, and *various antelope* have also been experimented with as discussed in Chapt. 1.4.

Apart from the initiative of man, domestication also depends on **conditions inherent in the animals**. It is possible to tame almost any animal, if removed early enough from its natural mother to be behaviourly imprinted by man. However, even at the earliest stage of keeping wild animals there are great physiological differences of hardiness and nutritional requirements between species which make it more difficult to rear some animal species than others. To be able to rear the young easily and with a minimum of mortality is the first important condition for any attempt at domestication of wild animals. If the nutritional needs are met and the species is kept according to its social requirements most wild animals will breed in captivity, as the records of Zoological Gardens show. However, even if a species is bred for many generations in captivity this does not necessarily imply that it accepts man in terms of social hierarchy and behaviour, which is an absolute prerequisite for ultimate successful domestication.

It has been fairly well established archaeologically that already mesolithic man in Western Asia probably kept *gazelles* in enclosures for food supply and that also the *onager* was tamed and bred in captivity. Nevertheless, it was the *Bezoar goat*, the *Wild horse* and the *African ass* that were eventually domesticated and replaced them. In more northerly regions *deer, boar* and *fox* were the most important food animals during the palaeolithic time, but only *boar* and *wolf* were domesticated. The predominant *deer*, however, were replaced by *sheep* and *goats*, and later on by *cattle,* although deer remained plentiful.

Hand feeding of tamed Oryx antelopes (*Oryx dammah*) in Ancient Egypt, depicted in tomb 3, Beni Hasan, abut 1900 BC.

Source: NEWBERRY, P.E. (1893): Beni Hasan, part 1. In: Archaeological survey of Egypt, ed. GRIFFETH, F.L.; Egypt Exploration Fund; Kegan Paul, Trench and Trübner, London (by courtesy of the Natural History Museum London)

There are many cases in historical times in which wild animal species were kept by man without becoming domesticated. The ancient Egyptians kept and tamed a wide variety of wild animals as depicted in many of the houses of nobles, temples and tombs. The reasons for such captive keeping are not always understood. It might have been part of the ritual use of wild animals which was particularly prominent in that culture, or it may have served them otherwise. It is highly unlikely that any of the depicted species - with the exeption of the ass - were truly domesticated for utilitarian purposes as in such cases the domestic forms would have survived into later periods.

Also the Sumerians and Assyrians are known to have kept game animals in large enclosures, probably for hunting and pleasure. A particular peak in the keeping of wild animals in captivity occurred during the Roman era when many wild species were imported from other countries to be kept in menageries and special holding facilities for different purposes. *Hares, rabbits* and *dormice*, amongst others, were specially kept to provide favoured food, but only the rabbit emerged eventually as a domestic animal.

Only now that our knowledge of animal behaviour and sociology has developed can this be partly explained. Such an explanation should enable us also to judge the possible success of the most recently initiated domestication efforts. True domestication is only possible if a definite social relationship between man and animal, either as competing predator or as prey, is established. This presupposes a means of non-verbal communication and social integration. Although man has evolved from primates, he is a carnivorous predator with a social hierarchical system in which the dominance of a leader is important. He has still, to some extent, the capability of non-verbal expression by gestures, postures and facial expressions with which he can communicate to receptive animals. He will be most readily accepted by animal species of a similar social organisation, for example *wolves, Wild horses, sheep* or *goats*. The wild species should be gregarious and not solitary. Males should not be territorial, but be leaders of mobile, social family or herd units. Such species can be easily flocked together into large herds which can be directed and managed by man and dog. In order to eliminate competitive males, castration became an important aspect early on in the domestication process, as for example, with *reindeer, sheep* and *cattle*. Most *deer* and *antelope species*, though gregarious, are territorial and thus not amenable to full domestication. However, they may nevertheless be utilised in a semi-domesticated state, as discussed below. Amongst the camelids, only the *vicugna* is strongly territorial and has thus not been domesticated. Other animal species, even solitary ones, can also be subjected to domestication in the wider sense, but under quite different technical conditions such as caging or keeping in enclosures. They will never, however, attain complete docility.

The **process of domestication** of a wild species always follows a pattern in which certain stages may be defined. Initially (*Stage I*), the particular species is kept captive in small numbers with or without breeding. This may occur by herding small groups close to man, leaving them to roam freely during the day-time but restricting them by night to simple protective enclosures, or by removing young individuals or eggs from the wild population for rearing, and tending them in small enclosures. Although some breeding may occur during this stage, the captive stock requires regular replenishment from the wild and genetic material is still introduced regularly from the reproductive pool of the wild species. Such keeping of wild animals in or near human habitation may have economical reasons or may aim at having them readily available for ritual purposes. However, it may also be purely incidental. In fact the keeping of young wild animals merely as pets or for companionship is also quite common in primitive societies.

Next in the process of domestication (*Stage II*) is that the physically contained animals start breeding regularly under the care and supervision of man. As numbers of captive bred or herded individuals increase, interchange of genetic material with animals of the wild population diminishes. Such breeding might go on for many generations without any intentional selection by man. However, the captive animal stock will eventually become genetically isolated from the wild population (*Stage III*) and this constitutes an important prerequisite for full domestication. Selective breeding at this stage may either be deemed unnessary or is very limited, for example, to produce certain colorphases in furbearing animals. Even without selective breeding, such animal stock will gradually undergo morphological, physiological and behaviourial changes merely due to the different environmental circumstances.

The motives for such breeding and husbandry can be quite different: many species of *rodents* and *carnivores* are bred in captivity for fur production; certain *deer, antelopes* and *rodents* are propagated in semi-captivity for antler velvet and meat; some *bird species* have been bred for a long time only for hunting and pet purposes; various *rodent species* are being bred in large numbers for laboratory purposes; and *other wild species* have been bred for many generations in zoos and for conservation purposes.

However, prolonged breeding alone, be it in cages, pens or large enclosures, constitutes only "semi-domestication" and rarely renders the animals completely docile. Full domestication (*Stage IV*) is achieved only by long-term controlled breeding with total isolation from the wild species and the application of varying degrees of husbandry. This results in a close relationship, even interdependancy, between the particular animal species and its master. Selective breeding and husbandry aim at the promotion of distinct anatomical physiological characteristics culminating in the formation of different breeds.

Depending on the duration and severity of the selection process, one may distinguish between "domesticated" and "domestic" animals. The former are also referred to as "primitive" domestic animals, externally still resembling the progenitor species and behaving like them. Only the latter are truly "man-made" animals which associate freely with man and might bear little resemblance to the progenitor species.

The process of domestication is thus accompanied by a gradual biological change of the particular species. This is caused only to some extent by mutations in the genetic make-up of the species. Thus domestication is never a spontaneous phenomenon, but rather a long drawn out gradual change in which the entire spectrum of genetic potential and variability of the whole group of phylogenetically closely related species comes into play. Genetic characteristics are interchanged and recombined, creating varieties in external colour, size and proportions, but never a new species. This is evident from the fact that domestic animals will reproduce successfully with the progenitor species despite the most extreme external dissemblance.

The biological changes which occur in the course of domestication comprise the entire organism. They may affect all parts of the body, all organs, their functionality and performance, and also behaviour patterns.

The most typical modifications are an initial reduction in body size, variation in colour and make-up of the pelage, and deposition of fatty tissue under the skin and between muscles. Within a few generations of breeding in captivity, the skull and jaws become shorter, always associated with a relative diminution of brain size and weight. Apart from such morphological changes, domestication also affects the physiological processes, such as nutrition, growth and reproduction. Although domestic animals retain the basic behaviourial characteristics of the progenitor species as social animals, there are also important changes in the genetically determined behaviour patterns, such as sexual behaviour, aggressiveness, and activity rhythms.

**Table 1:** Wild animal species commercially utilised by captive keeping, taming, controlled breeding or domestication[1]

| | |
|---|---|
| **STAGE 1:** | **kept captive without or with occasional breeding[2]** |
| **Reptiles:** | *alligators, crocodiles* and *caimans* (1970/85; → II), *Green iguana* (1980s), *Monitor lizards* (exp. 1980s), *tegus* (exp. 1980s), *boas* and *pythons* (exp. 1980s), *div. venomous snakes* (~1910), *Green turtle* (1980; → II), *Hawksbill turtle* (exp. 1970), *Softshell turtles* (pr. 1970) |
| **Birds:** | *emu* (1976), *rheas* (1900-14; exp. 1990), *falcons* and *eagles* (pr. 3000 BC), *cormorant* (~200 BC) |
| **Mammals:** | *armadillos* (1970s), *dormouse* (200-14 BC), *Giant rat* (exp. 1976), *paka* (exp. 1984), *agoutis* (exp. 1960s), *Civet cat* (pr. 1000), *mongoose* (~1000 BC), *cheetah* (pr. 1500), *Asian elephant* (~2500 BC), *African elephant* (300 BC-400 AD), *onager* (~6000-2000 BC), *peccary* (exp. 1980s), *guanaco* (exp. 1977), *vicugna* (exp. 1958), *Tree shrews* (1970s), *tamarins* (1960s), *Squirrel monkeys* (1966), *Owl monkey* (1966), *Vervet monkeys* (1958), *chimpanzee* (1960s-80s) |
| **STAGE II:** | **kept captive with breeding, beginning genetic isolation[2]** |
| **Reptiles:** | *Siamese crocodile* (1950), *Diamond-back terrapin* (1920s-40s), *Slider turtles* (1960s) |
| **Birds:** | *ostrich* (1860; → III ), *Mute Swan* (~1100), *francolins* (exp. 1980s) |
| **Mammals:** | *European hare* (1940), *Cane rat* (1980), *capybara* (1980), *Musk deer* (1958), *moose* (200 BC; 1947), *Axis deer* (1950s), *Sambar* and *Rusa deer* (1970s), *Roe deer* (1960s), *marmosets* (1960s)[3], *macaques* (1960s)[3], *baboons* (1958)[3] |
| **STAGE III:** | **kept captive or herded, selective breeding, full genetic isolation[4], semi-domesticated[2]** |
| **Birds:** | *Chukar partridge* (pr. 1850), *Red-legged partridge* (pr. 100 BC), *Eur. quail* (pr. 1000), *Jap. quail* (~1100; → IV), *Chin. quail* (pr. 1790), *Valley quail,* (~1850), *bobwhite* (~1820), *peacock* (pr. 2000 BC), *pheasants* (pr. 1000 BC) |
| **Mammals:** | *Brown rat* (~1850; → IV), *House mouse* (pr. 1000 BC; → IV), *Golden hamster* (1930s; → IV), *Chin. hamster* (1950s), *Mong. gerbil* (1954), *nutria* (1920s), *chinchilla* (1920s; → IV), *muskrat* (1900-30s), *beaver* (1920s-30s), *Amer. mink* (~1866; → IV), *sable* (1930s), *polecat* → ferret (~400 BC; 1979), *Striped skunk* (1920s-30s), *racoon* (1920s-30s), *Red fox* (1892; → IV), *Arctic fox* (~1900; → IV), *Racoon dog* (~1900), *Fallow deer* (~2000 BC; → IV), *Sika deer* (pr. 200 BC; → IV), *Red deer* (1970s), *Père David deer* (ext.; 4000-1000 BC), *gazelles* (8000-3000 BC), *eland* (exp. 1896), *oryx* (exp. 1960s), *Gaur* → *gayal* (~3000 BC), *bison* (1960), *Musk ox* (1954) |
| **STAGE IV:** | **fully domesticated, docile, genetical changes, breeds[2]** |
| **Birds:** | *turkey* (~200 BC), *Jungle fowl* (~3500 BC), *Guinea fowl* (pr. 2500 BC), *Mallard duck* (500 BC), *Grey-lagged goose and Swan goose* (pr. 1000 BC), *Muscovy duck* (pr. 1500), *Rock pigeon* (pr. 4500 BC), *canary* (~1500), *Zebra finch* (1840), *budgerigar* (1840) |
| **Mammals:** | *European rabbit* (~300 BC), *Guinea pig* (4000/1000 BC), *Gray wolf* → *Dom. dog* (pr. 10000 BC), *Wild cat* → *Dom. cat* (pr. 2500 BC), *Wild horse* → *Dom. horse* (ext.; pr. 3000 BC), *African ass* → *Dom. ass* (ext.; 3200 BC), *Wild reindeer* → *Dom. reindeer* (pr. 8000 or 500 BC), *dromedary* (ext.; ~3000 BC), *camel* (pr. 2500 BC), *guanaco* (?) → *lama and alpaca* (5000-4000 BC), *Wild boar* → *Dom. pig* (pr. 6500 BC), *aurochs* → *cattle* (ext.; ~6000 BC), *arni* → *House and Water buffaloes* (~3,000 BC), *Wild yak* → *Dom. yak* (~1000 BC), *banteng* → *Bali cattle* (pr. 1300), *mouflon* → *Dom. sheep* (pr. 7000 BC), *Bezoar goat* → *Dom. goat* (pr. 7000 BC). |

**Legende to Table 1:**

exp. = experimental, ext. = wild progenitor extinct, pr. = prior to, (figures in brackets) = year(s) in which domestication or use commenced (- and terminated), → IV = Stage IV partly achieved.

1) not included are species bred for show purposes and the pet trade (with the exception of sliders and some pet birds); primates are bred predominantly by the research institutions for their own use and rarely commercialised; included are only species kept and bred in significant numbers; for scientific names and details see Part II; 2) full description see text; 3) although many colonies are still replenished with wild stock, in some of them breeding is on-going for several generations; 4) exceptionally, some times also intentionally, wild blood is crossed back, for example when wild males mate farmed deer hinds, or herded reindeer, oryx and gayal females.

Sources: ZEUNER (1963), HERRE and RÖHRS (1977), MASON (1984), RAETHEL (1991) (see selected literature references)

The **history of domestication** has been studied by a great number of authors and was excellently described by ZEUNER (1963), MASON (1984) and more recently CLUTTON-BROCK (1987)[1)]. There is no need to repeat this, only to relate domestication in a historical context to the subject of this book, i.e. wildlife utilisation: to domesticate an animal species was - and still is - an arbitrary choice by man, probably not always the best, out of a wide range of possibilities for exploiting wild animals to his benefit. In fact, the process of domestication itself comprises a sequence of different forms of intensified utilisation, corresponding to different traditional or socio-economic needs and technical capabilities. Tab. 1 lists all reptile, bird and mammal species which have been or are at present commercially exploited in a captive, herded, tamed or domesticated state. It is based on a worldwide survey of "game farming" conducted in 1984 by the World Conservation Monitoring Centre[2)], and supplemented by other specific sources. This survey is highly relevant because "farming" comprises all of the utilisation stages of animals from "wild" to domesticated: historically any domestication started with what is now called "farming", and this might become domestication in the future.

It is not always easy to determine what stage of intensity the utilisation of a species has reached or how far domestication has proceeded; this may vary from country to country. For example, *crocodile* farming was originally based on the collection of eggs from the wild and would thus have to be classified as Stage I. During the last decade, however, more and more farms are breeding their stock, moving their activity into Stage II. *Ostrich* farming, reached Stage III in South Africa about 50 years ago, but expanded recently into other countries and regions as Stage I or II. In other cases the progenitors of fully domesticated species, such as *wolf* and *polecat*, are now being farmed. The degree of containment of a utilised species may also vary widely in *ungulate species* which are both "ranched" and "farmed", making classification difficult. With *other species* it is difficult to decide whether the selection process has been ongoing long enough to effect genetic changes typical of Stage IV. Most of the many recently implemented farming projects are so far only experimental efforts or of short duration falling clearly still into Stage II. The intensive use of a number of species was also - even in historical times - abandoned after a while due to changing consumption needs or economics. Details of the use of species which have not yet reached Stage IV of the domestication process are discussed in Part II.

---

1) see selected literature references

2) BARZDO, J., LUXMOORE, R., BROAD; S. and D. JONES (1984): A survey of wildlife farming operations. WCMC/IUCN, Cambridge (unpubl. report)

Another important aspect of domestication to be considered in the context of this book is its **implications for the conservation of wild animal species**. Many of the more recently initiated or proposed projects of domestication claim to alleviate the pressure of indiscriminate exploitation on the particular species. The philosophy behind this is that the production of the particular animal species in captivity will satisfy the consumer demand for it and thus prevent over-exploitation of the wild population. The history of domestication shows, however, that many of the progenitor species of domestic animals have become extinct. Although the reasons for the early extermination of some of these, like *Bos primigenius, Equus przewalskii, Camelus ferus, Chinchilla laniger, Equus asinus* are not clear, the creation of their domestic forms has certainly not prevented it. Since these exterminations have occurred, wildlife management has been instituted and conservation awareness has evolved. Nevertheless, it may well be possible that farming of, for example, *crocodiles*, *Sea turtles*, *ostriches*, *Cane rats* and *others* will not curb the exploitation of the wild populations at all; their conservation depends entirely on better control and this might even be made more difficult by the marketing of the farmed animals. Therefore, in contrast to the sustainable utilisation of wild populations, farming and gradual domestication of the species concerned is not likely to contribute much to their conservation.

Occurrence of an albino in the fifth generation of Eland antelopes (*Taurotragus oryx*) under domestication at Manyoli Ranch in Rhodesia (now Zimbabwe) in 1966. (Photo: H.H. ROTH)

## 1.3 MANAGEMENT OF WILDLIFE RESOURCES

### Ecological relationships between the management of wildlife and other natural resources

Wildlife species respond directly to the quality of their environment, hence their welfare is directly dependent on the management of related natural resources.

**Forest management** operations commonly used to enhance wildlife species consist of thinning overstocked stands, mixing and managing even-aged stands in small units, maintaining food patches and open areas in coniferous and broad-leaved evergreen forests, and in some forest types controlled burning to reduce litter, stand density, and to promote forb growth. An exception to the above occurs in temperate regions where great snow depth accumulates. Coniferous growth may intercept the snow and provide the only areas free of snow in which food and cover are accessible for ground dwelling wildlife species.

Some generalities, with exceptions, pertaining to forest management and wildlife are:

- Deciduous forests produce more wildlife than coniferous forests.
- An open forest is more productive than a closed canopy forest
- A mixed age stand produces more wildlife than an even aged stand.
- A mixed species stand of trees produces more wildlife than a single species stand.
- The intermingling of small units of different age classes and different species of trees will produce more wildlife than large units of one species or even-aged stands.
- Openings in tropical forests provide more food for ground dwelling animals than continuous unaltered forests.

The **use of range forage** by domestic livestock is quite compatible with wildlife as long as neither gets overabundant or so concentrated that forage selectivity is diminished. This does not mean that the two are not in competition for available forage. Pastoral man usually grazes cattle, sheep, and goats, and in more arid country camels, in order to utilise the various grazing opportunities available to him. The mix of species of wildlife occurring in an area utilise and prefer different types of forage just as do cattle, sheep and goats. In general, all classes of herbivores prefer to use first the most succulent forage available, providing they are physically adapted to feed on it and digest it.

*Cattle* will select first certain species of succulent forbs, but as these mature they prefer grass. If forbs and grass become scarce or unavailable they turn to woody plants, commonly called browse. The normal forage preference for *sheep* consists of forbs and grasses, and they use larger quantities of browse than cattle. *Goats* use forbs and browse and small quantities of grass.

Many different species of wildlife occupy rangelands, and their welfare as well as that of domestic livestock must be considered when managing rangeland resources. Forage preferences, necessity for water, capabilities of movement from one location to another, special seasonal requirements for reproduction, potential transmission of diseases between wildlife and domestic livestock, and tolerance for other species of wildlife as well as for domestic animals must all be considered when planning for or attempting to manage rangeland wildlife species.

Normally there is more conflict between wildlife and **farm and crop production** than between wildlife and forestry or livestock production because wildlife tend to consume and damage the crops. Thus, herbivorous wildlife is almost universally considered to be undesirable by those people engaged in farming, and the implementation of control measures to reduce the threat by wildlife is just as important as measures to increase wildlife. Avoiding crop production in close proximity to wildlife ranges and cover is one possible way of decreasing conflict, but it is not infallible. Most wildlife species hesitate to wander far from protective cover, particularly during daylight hours. Consideration of the seasonal requirements of wildlife when selecting the crops to be grown can also alleviate damage. Irrigated crops during the dry part of the year are like an attractive nuisance to adjacent wildlife because the native plants available to them are dry and unappetising as well as low in nutrient content. Plants that are fertilised are more attractive to wildlife because they are more succulent and higher in nutritive content. Many of the trees used to produce fruit for man are also attractive to wildlife, because the animals like the fruit and also browse on the leaves and twigs. For example, deer and apple orchards are just not compatible forms of land use.

Game birds are less hazardous to crop production than mammals; however, they do present some problems that should be considered when contemplating the production of birds for sustained uses of land. *Ring-necked pheasants* dig up newly planted seeds, and pull sprouting grain out of the ground, necessitating replanting. They also destroy ripening tomatoes by pecking into them. Migrating flocks of *waterfowl* descend onto ripening grain fields, particularly wheat in the northern states of the United States and in Canada, and are responsible for considerable damage. Although not normally used by man as food, wildlife ecologists should be aware of the fact that non-game bird species also do considerable damage to crops around the world. *Blackbirds* in Southern Europe and North America and *queleas* in Africa are especially noted for damage to vineries and grain crops respectively because they congregate in large flocks in areas that have concentrations of such food.

The **exploitation of marine resources**, both fish and mammals, has been traditional all over the world since the earliest times. In fact the vitalities of some countries have waxed and waned with the ability of their citizens to harvest marine resources, frequently hundreds of miles from their shores. It is only in the last two decades that natural resources managers have recognised the interdependence of the different segments of the marine environment, and have been instrumental in getting laws and international agreements in place to control the harvest of marine animals, and have become concerned with the degree of use of all marine organisms.

An example of this new concept is the Convention on the Conservation of Antarctic Marine Living Resources concluded in 1980 which regulates man's harvest of *krill*, a small marine organism that influences the production of *seals, whales, penguins* and *fish*.

## Planning wildlife management

Setting definite and **well-defined goals** is as important in the management of wildlife resources as in any other business endeavour, yet is frequently overlooked or not given adequate consideration. Failure to consider and set goals, and adopt those that are realistic and reject those that are unattainable frequently leads to disappointing results for wildlife management programs. The importance of early planning, with consideration of all the biological, social, financial and legal aspects of a proposed programme, cannot be overemphasised and is essential to an evaluation of its probable success. *Primary goals* should include the major objectives of management programmes. These could include monetary returns, supply of products, recreation activities, aesthetic returns, biological improvements or modifications of wildlife species, and possibly other goals. Goals should be assigned an order of priority to facilitate management decisions as the programme progresses. *Secondary goals* would include those objectives that can be accomplished incidental to or without too much influence on the primary goals, but they nevertheless should be considered and stated in the planning stages. As the programme develops some priorities of some goals may change, perhaps as a result of new information emanating from the work, or due to societal, financial or other developments that were unanticipated in the planning stages. Hence initial secondary goals can become of primary importance.

There are some biological and other conditions that are more or less fixed and may impose **limitations on management goals**. These should be recognised and planned for accordingly. As an example, species composition of wildlife in an area is a response to many biological and physical factors over time, and is unlikely to be altered without considerable manipulation of habitat. Hence the introduction of a new species should only be undertaken with extreme caution. Usually, even with considerable efforts, introduced species fail to become established. Occasionally, however, a new species becomes too successful and competes with native wildlife or becomes detrimental to established agricultural programmes. The *dominant wildlife species* in an area will probably continue to maintain that dominance. Hence increasing a species considered to be more desirable at the expense of the current dominant species usually has little chance of success.

The financial resources available to sustain a management programme should be a major consideration in planning. Where monetary return is a major goal the period between outlay of cash and a return on the investment should be carefully evaluated. There may be better areas in which to invest money. Also the available assets may be depleted before there is sufficient return to maintain the cash flow necessary to sustain management activities, hence the programme fails. All of the above should be a part of the planning process, and the project should be undertaken only after due consideration of these important aspects.

## Implementation of management plans

Measuring progress is an important component to any undertaking and the management of wildlife is no exception. A base from which it is possible to measure changes must therefore be an initial step in the implementation of a wildlife plan. That base is established by making a wildlife habitat map of the area and making base estimates or censuses of wildlife populations.

The **mapping of wildlife habitat** can be considerably detailed or as generalised as is necessary to measure changes in vegetative composition and condition over time. It is an accepted fact that wild animals respond to their habitat; hence management of wildlife is to a great extent devoted to management of the habitat. Habitat maps for the management of small animals and those animals with small territories and limited travel capabilities need to be more detailed than maps for animals that cover great distances and those that have the capability of moving rapidly from place to place or over large areas.

All wildlife species must have food, most require some type of cover and almost all require free water. Interspersion and seasonality of these necessities are as important as abundance and hence should be noted. The requirements of the wildlife of concern must be known and the habitat map must be prepared in sufficient detail to permit the analysis of the quality and suitability of the habitat for that species.

The necessity for updating habitat maps will depend on the species of wildlife and the rapidity of change in vegetation of the area. Managing *game birds* may require an update every year, whereas for the management of *deer, antelope* or other *large ungulates* an update every two or three years may be adequate.

The **establishment of population estimates** of the animals of concern is also a necessity. The need for accuracy of the estimates will depend on the type and degree of the planned use of the animals; the potential productivity; the rate of natural and man-induced productivity of the species; and the rate of natural and man induced mortality. Other factors such as the management of rare or endangered species; catastrophic or potential changes in habitat; disease outbreaks and detection of the influence of habitat manipulation or human settlement will also influence the accuracy requirement of population estimates.

Many different types of census methods have been developed and tested for determining population levels of the various wildlife species that occur throughout the world. None are perfect and to use them effectively wildlife managers should be familiar with those census techniques that are applicable to the species of concern as well as the merits and deficiencies of each.

*Large ungulates* that occupy grassy plains or open savanna woodlands are frequently censused by direct observation from aircraft. Helicopters provide better observation platforms and travel slower than fixed-wing aircraft and thus provide better visibility; however, they are much more expensive to operate and, hence, the cost versus the results must enter into the decision as to the type of aircraft to use. The most frequently used method in open habitats is the counting of animals from a vehicle along arbitrary road-strips. In dense habitat, such as forests, the only possible way to census animals is to determine the frequency of their tracks, droppings or other signs of presence along arbitrary transects or in selected areas. The results of road strip or transect counts will furnish only relative information about the local abundance of the different species present. This may, however, with caution

Aerial census in Comoé N.P. in Côte d'Ivoire; the strip for counting the animals on the ground is delimited by flyers from the aircraft's wing struts. (Photo: B. STEINHAUER)

be extrapolated to larger areas thus giving an approximate idea of population densities. Capture, mark and recapture has also been used, particularly for *game birds* and *small mammals*. Difficulties with this techniques may result from animals that have left the area and animals that have died during the interval between the initiation and the termination of the trapping period.

The accuracy of most sampling techniques is dependent upon the assumption that the animals are randomly scattered throughout the area sampled, which any seasoned wildlife manager knows seldom, if ever, occurs. Increasing the validity of the results by increasing the degree of sampling is frequently too time consuming and costly to consider. Getting data that will withstand statistical scrutiny is therefore seldom attained. However, it is usually not necessary to determine accurately the total size of a population. It is more important to be able to determine the dynamics of populations, i.e. increases, decreases and structural changes, by repeated censuses. For this purpose, even the relative data from strip counts are quite sufficient as long as methods are strictly standardised and results thus comparable over time.

The establishment of sex and age ratios provides indices of the productivity of wildlife populations and, hence, is frequently used as a tool in management. The ratio of adult females to new born young in the population is an indication of the success of production. The ratio of adult females to sub-adults of the same population at a later date provides information on the rate of mortality that has occurred during the juvenile stages of the population, the period at which natural mortality is highest. Knowing herd composition is also important to the wildlife manager when making decisions on setting harvest quotas and hunting regulations. The ratio of adult males to adult females may have an influence on the success of breeding, the security of the herd against predation, and the degree of success of hunters, particularly when managing the population for trophy hunting.

Information on movement or migration of wildlife populations is important. The quality of habitat occupied by the animals throughout the year must be known in order for the wildlife manager to make decisions regarding habitat modification and harvest levels. It is quite common for wild animals in arid areas to use different dry and wet season grazing locations, and in temperate zones some populations use summer and winter areas that are great distances apart.

The success of wildlife management activities, particularly wildlife utilisation programmes depends to a large extent on the **programming of harvest**. There are many options open regarding the harvest of wildlife, but there are also many limitations, among which the most prominent are legal, market demand, seasonal and annual productivity.

In most parts of the world governments regulate the harvest of wild animals. In temperate regions the harvest season is limited to autumn and winter. That is the time of the year when the animals are in top condition and are not rearing young. In the tropics the production of young may occur year-round or be correlated with the rainy season, hence the harvesting can occur at any time. Governments may also place restrictions on the age, sex, or other attributes of the animals that can be taken legally, and may further restrict the number of animals that can be taken during a specified period and the methods of taking. Regulations established by government wildlife agencies are meant to limit the harvest in order to ensure an adequate carry-over of animals for the next production season; to provide recreation to hunters by preventing crowding; to distribute the available animals to be harvested among the people who are legally entitled to engage in the harvest; and to protect the public and property from possible injury and damage.

Road-strip count: two observers count animals left and right from the road and determine their distance in order to calculate the mean width of the observed strip. (Photo: H.H. ROTH)

A continuous market demand at an equitable price for the products from wildlife utilisation has at times been an obstacle to commercial wildlife enterprises. Some products, such as skins and horns can be stored or salted and dried for future processing, but the meat must be utilised fresh, or smoked, salted, dried, or canned within a relatively short period. Fresh meat is universally in demand, whereas preferences for different preservation methods and acceptance of the respective products vary from region to region.

Wildlife is a renewable natural resource; however, for it to be renewable its management must provide for sufficient brood stock to produce the subsequent years' population. It takes planning and restraint not to overharvest animal populations and this has presented problems in the past. Where unregulated, wildlife historically has been considered to be there for the taking by anyone. Hence, the general view has been that if it is not taken by me it will be taken by someone else, and that will be my loss. Thus there has been no incentive for restraint in killing it. Planned harvests have demonstrated that wildlife utilisation can be perpetual, provided that an adequate animal population is left to produce next years crop and habitat is suitable. Determination of the proper harvest must take into account the rate of natality in terms of the reproductive success, as well as the rate of mortality in terms of the natural animal loss of both the producing herd and the juvenile animals prior to reaching maturity.

Overutilisation will depress the productivity of the respective animal population below its reproductive potential. Hence the number of animals available for harvest will decrease. Setting the harvesting quota low at the initiation of a wildlife utilisation programme and then gradually increasing it, is suggested as a conservative approach. Accurate population data on numbers of animals and sex and age ratios will have to be accumulated over several seasons as a prerequisite to adjustments to the harvesting rate. Setting harvest quotas is one of the most critical exercises governing the success of any wildlife management programme in which sustainable wildlife utilisation is an integral part.

Wildlife populations, particularly large *herbivores*, have the potential of increasing in numbers to the extent that they overutilise their food supply and as a result decline in numbers. This has been abundantly documented for *antelope, deer, hippopotamus* and *elephants*: destruction of food supply resulted in lower survival, particularly of the young, followed by a decline in population. Sometimes this is a gradual process, but frequently it occurs as "population crash" induced by drought or sever winter weather or some other factor that accentuates the declining carrying capacity of the habitat.

Change in habitat frequently affects other species of wildlife associated with those animals that are altering the landscape. *Deer* and *elephants* may overutilise certain species of woody plants causing woodland or brushland to gradually change into grasslands. As a result the whole gamut of wildlife species that were dependent on woody vegetation is replaced. Grass- and forb-eating species, such as *certain antelope* and *hippopotamus* may overutilise the grass cover and thus permit the invasion of bush, causing the open grasslands to gradually decline. It is therefore important to keep large herbivore populations at a level that will preclude possible overutilisation of their major food supply.

Certain management objectives can be facilitated by harvest practices. For example, the productivity of wildlife populations can be increased by limiting the harvest of the female segment of the population. This, of course, applies only when the species of concern is polygamous and the male contributes little or nothing to the successful rearing of the young. This practice is also based on the assumption that both sexes are competing for the same food, water and cover supplies, which is not always the case. Nevertheless, this practice has been very helpful in establishing and re-establishing populations of game species throughout the world.

Notable examples are *deer, Wild turkey* and *Ring-necked pheasant* in North America, various *deer species* in Europe, Australia and New Zealand, and a*ntelope species* in Africa. Potential productivity can be doubled, and perhaps in some instances even tripled, by limiting the harvest to males only, but the total population cannot exceed habitat carrying capacity.

The management of wildlife populations to produce a high proportion of trophy-class animals is a growing industry. Trophy hunters are willing to pay high fees for the opportunity to hunt for trophy-class animals. Catering for such hunters appeals to wildlife managers because it involves fewer people to accommodate than a high harvest regime and can produce a comparable income. This practice, however, is geographically limited because trophy-class animals of a particular species tend to be limited to certain areas due to genetic characteristics and favourable habitat. The number of trophy animals produced in any one area is also limited, because an animal must be of an older age class to attain trophy size, and studies have shown that at least in some species and some localities a high proportion of potential trophy animals die of natural cause before reaching that age. Also potential trophy animals must have adequate and continuous food supplies in order to reach their maximum size, which means that competing animals must be eliminated or kept at a low number. Thus management involves the practice of harvesting many animals of less than trophy quality in order to produce a few trophy-class animals.

The **monitoring of wildlife management programmes** is essential in order to measure progress and plan for future action, like in any other business. It is also necessary to determine whether utilisation of the wildlife resources is truly sustainable. Standardised methods for censusing wildlife populations and measuring the quality of the habitat must be initiated and conducted at intervals frequent enough to provide information for decision making in setting wildlife harvest levels and desired changes in habitat improvement practices. Anticipated harvests must be projected far enough in advance to put programmes in place for marketing the products, whether it be selling meat on the open market or conducting trophy hunts. At least six months, and even up to a year is a reasonable advance planning schedule, particularly for sport hunting.

The sale or disposal of game meat also requires preliminary action. Overcoming potential objections to the harvesting of wild animals may be a necessary step to the success of the programmes, and that may require lining up support from the news media, leaders in the community who shape public opinion, government officials, and representatives from conservation organisations. Commercial meat suppliers and retailers who plan to handle the meat must also make arrangements in advance. A rather accurate estimate of the level of harvest must be made in advance so that those who plan to utilise the meat can plan accordingly. Those initiating commercial wildlife utilisation for meat should also be aware of the need for dealers to be supplied continuously, as otherwise consumer demand for game meat may decline.

Population monitoring procedures should be conducted frequently enough to detect any appreciable decline in animal numbers. Such events as disease outbreaks, food shortages, decrease in water availability, increased predation, and illegal and unplanned harvests can all lead to precipitous drops in wildlife populations.

The **manipulation of habitat** is an important aspect of the management of wildlife resources. It may serve to increase the carrying capacity for specifically desired species. However, modifications should not be undertaken without due consideration. They are frequently costly, ineffective, and may have considerable detrimental effect on non-game but otherwise valuable wildlife species. Sometimes the advantages of habitat modification are obvious; for example the provision of water in an area with an abundance of food, but no water. Even this is fraught with complexities.

In the Southwestern United States, for example, literally thousands of dollars have been spent building water facilities for the increase of *quail*. It now has become obvious through research that the species of quail that inhabit the deserts of the South-West do not require free water, and that providing this water has not resulted in increased quail populations. However, *deer* use the water and concentrate around these facilities. When the water is used up, the public demands that government wildlife agencies transport water to these animals to alleviate suffering. Although deer populations may have increased to some extent, the actual cost to the wildlife agencies for each additional deer harvested has not been determined. If it were, the cost to benefit ratio would probably be prohibitive, that is putting in the water collection device was an uneconomical exercise.

The provision of water in localities where it did not exist and where food for grazing animals was thus only available during the wet season is also very problematic. Once water is made available grazing animals will consume the forage all year long with the effect of range overutilisation and gradual degradation of the whole area. Habitat modification in this case is not only non-economical but also ecologically highly detrimental.

The benefits of manipulating habitat to produce a better food supply are even more speculative than the manipulation of water. Plants occupying an area are there for a reason. They have become established through a long period of selectivity because they have characteristics that make them better suited to that particular area than potential competitors. It certainly can be assumed that the plants that occupy any particular area are going to resist change, and replacing those plants with species deemed to be more desirable, and making them permanent or long lasting will probably be difficult and costly. Such exercises should be undertaken only after successful demonstrations on small areas, with cost accounting records so that projections can be made for cost/benefit ratios for larger areas.

The importance of **cost accounting** for a sustained wildlife utilisation programme cannot be overemphasised. Just as with any successful business, the use of resources must be offset by productivity, otherwise the business will fail. Obviously, some research and development costs are necessary and justified, but such activities can only be continued as long as there are resources available to offset those costs. Programme plans should anticipate paying off those costs with returns as the programme progresses. Cost accounting procedures should be sufficiently detailed to measure the efficiency of various programme activities thereby permitting management decisions to be made for increased efficiency in overall production at an early stage. This can mean the difference between success and failure, particularly when back-up resources are severely limited.

The future of wildlife may well rest on the success of sustainable wildlife programmes, particularly in developing countries. The success of such programmes will be judged by their contributions to making life better for local people and for all mankind. Cost accounting is a necessary element in making that judgement, and should always be included as an integral part of any wildlife management programme.

## Maintenance of biodiversity

The productivity of an ecosystem depends to a great extent on the numbers and kinds of plants and animals present, i.e. on the biological diversity. Maintenance of biodiversity is therefore an important aspect of the management of any renewable natural resource, including the management of wildlife. This is concerned with the interdependence of plant and animal populations, and the greater the biodiversity the more intricate becomes the relationship between the two. Interspecific competition may result in the decline of one species and the increase in another, or changes in habitat may bring about the reduction in some species and an increase in others. Wildlife managers must always be aware of and concerned about such ecological interactions, whether natural or man induced. Management plans to bring about maximum production of one or several large herbivores may call for an increase in their food supply and a reduction, and sometimes for economic purposes a complete elimination, of large predators. Examples are the management of selected deer species for trophy hunting and antelope for meat production; however, such management measures tend to bring about a decrease in biodiversity. On the other hand, for conservation or recreational purposes, management may be concerned with the enhancement or maintenance of optimum biodiversity. In such cases it may be necessary to reduce certain wildlife species which impinge on the habitat of others and thus threaten their existence or propagation. In such instances economics should be a secondary consideration.

An example of such a situation is shown graphically in Fig. 2 and applies to Southern African semi-arid savanna woodlands with populations of *elephant, buffalo,* and a great variety of *antelope species*. Some of these are better adapted to drought conditions and generally less habitat specific than others. Due to certain human influences, such as burning, predator control, and disturbance, these species may proliferate greatly and partly destroy the habitat of the more habitat-specific species, resulting in their decline. The more adaptable species continue to increase and concentrate locally even after the carrying capacity of the range for the total biomass of large herbivores has been exceeded and range conditions are rapidly deteriorating. If the increase of such species is not controlled early at Stage II or III in Fig. 2 biodiversity will inevitably be reduced and the population of the prolific species will ultimately also crash (Stage V in Fig. 2).

**Figure 2:** Range degradation under conditions of unmanaged game in Southern Africa

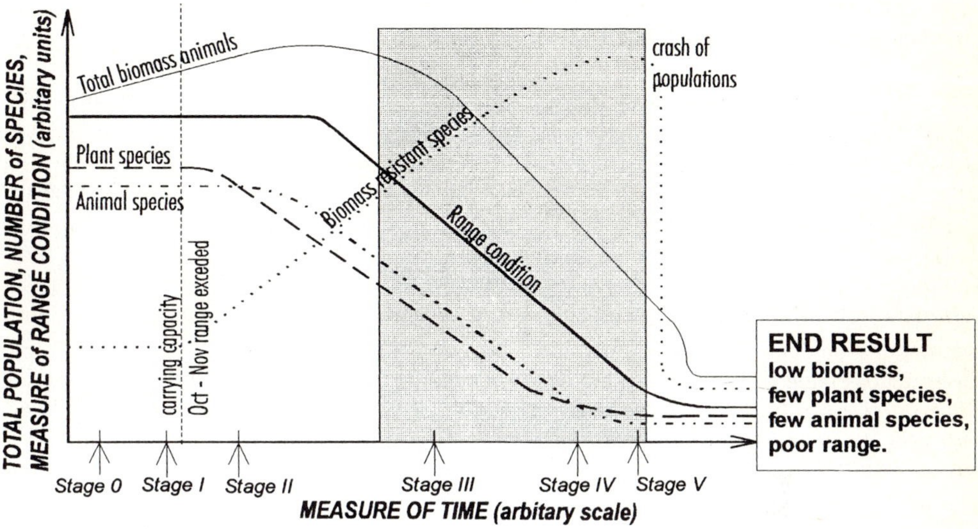


Source: SAVORY, A. (1965): Game Utilization in Southern Rhodesia. Zoologica Africana 1 (2), p. 321-337

Situations in which the wildlife range deteriorates are common in areas both where hunting occurs or which are closed to any type of animal harvesting, such as national parks. In the former, fear of overhunting or a very restrictive off-take results in overstocking and management must strive to utilise the prolific species judiciously, while maintaining at the same time optimal biodiversity. Such management occurs on many game ranches in Southern Africa. In the second situation, i.e. in non-utilisation areas, the need to control certain species such as *elephant*, *buffalo* and *hippo* must be accepted as a necessary measure to maintain biodiversity. Here the economic use of the controlled species is, however, a strictly secondary aspect of management. Some of the largest national parks in Southern Africa have been practising such management for decades thus ensuring biodiversity and at the same time producing additional financial returns to offset the cost of management.

Public objection to certain types of consumptive wildlife utilisation may be encountered in some countries, and should be considered in the planning of such an operation. Historically, *wildlife populations* have received their greatest support from the people that have hunted and utilised them. Habitat for game animals has been preserved and created primarily through the financial and political support of hunters in Europe and North America. Due to progressive and good game management, many *game species* are more abundant and occur over greater areas today than they did at the turn of the century. This, however, is often ignored by organisations and people swayed by sentiment and advocating the banishment of wildlife utilisation. It is encouraging to note that there is a growing perception that wildlife resources in developing countries can best be perpetuated by providing some benefits to those people who are forced to live in close proximity to them.

## 1.4 FORMS OF WILDLIFE UTILISATION

### Classification and definitions

The **economic use of wildlife resources** may be **a primary or secondary goal of their management.** This book is mainly concerned with the former, i.e. with the use of wild animals for the primary purpose of deriving economic benefit. Wildlife is considered to be a natural resource, the management of which must necessarily produce an economic return in one form or another if it is to be maintained. Such use is as old as mankind, but has to be better understood and adapted to socio-economic changes and ecological conditions in order to yield sustained returns. Only two forms of secondary utilisation will also be discussed: the culling of animals in conservation areas as part of the necessary regulation of wildlife populations, and the use of wildlife in these areas through commercial tourism. In both cases management of the wildlife resource is primarily aimed at conserving the diversity of the natural resource per se; economic utilisation of wildlife assumes a clearly secondary rank and must be strictly subjected to primary conservation concerns.

According to the ecological situation and economic needs, **utilisation of wildlife** may be either **extensive or intensive** in nature. The former usually implies the use of wildlife resources as a whole or over a wide and natural area, whereas the latter refers either to the use of selected species in a natural environment, or to the concentration of a few species on artificially restricted ranges, in enclosures or cages, with a view to producing meat, skins and other wildlife products from them.

According to its purpose, wildlife utilisation may be classified to be consumptive or non-consumptive. **Consumptive wildlife utilisation** refers to utilitarian purposes, i.e. wildlife is "consumed" for gaining or producing meat, oil, hides, furs, feathers and a large spectrum of other raw materials. In terms of economic importance it is globally dominant over non-consumptive uses. It may be carried out either for subsistence or for commercial purposes.This distinction is of considerable importance from a point of view of conservation.

*Subsistance use* of wild animals by indigenous rural people was formerly always extensive in nature and rarely reached intensities which would threaten the survival of animal species. It is still very important for the nutrition of, and also for limited cash returns to, rural people in many parts of the world. It is usually carried out by a few local hunters who have specialised in providing their village communities with meat from wild animals or by farmers as an occasional sideline during the agricultural offseasons. In some African countries community hunting is practised during these periods, involving large groups of beaters and hunters. As long as hunting is restricted to traditional methods and is regulated by traditional laws and customs it constitutes an ideal form of wildlife utilisation. In fact it is much more than that: the hunting and use of the prey is a most important integral part of the cultural life of the respective societies. However, with the introduction of firearms, the increase in population densities and also by rendering subsistence hunting illegal, i.e. suppressing it into an uncontrollable clandestine activity, it now often exceeds sustainable levels. With the development of cash economies, even in the remotest areas hunting has also become more and more commercially motivated. Although wildlife must be recognised as an important natural source from which local populations should rightfully benefit in one way or another, traditional hunting has to be transformed into a controlled and professionally supervised activity. Otherwise large segments of wildlife will be exterminated in the near future, particularly in wildlife-rich African countries.

*Commercial hunting* also has a long tradition all over the world, in form of fur trapping, sealing, whaling, hunting for antler velvet, musk, ivory or simply meat. What is termed poaching is also commercial, though illegal, hunting when the products of the animal killed are traded rather than consumed by the hunter. Although also extensive in nature, it usually aims at producing specific commodities for which there is high demand, leaving the rest of the animal killed to rot. In contrast to subsistence hunting it is thus often wasteful, and given the underlying profit motivation can easily become detrimental to the wildlife resources. Without control, commercially exploited species may easily be reduced below utilisable levels or even below the possibility of recovery. There are many examples in the history of wildlife utilisation which bear this out, but there are also examples to show how responsible and controlled commercial hunting can render sustainable profits to the communities and countries concerned. Commercial hunting may also be part of the control of noxious animal species and, as such, highly desirable

Another form of commercialised extensive wildlife utilisation was developed mainly in Southern Africa under the names of *"game ranching"* or *"game cropping"*. Similarly to traditional subsistence hunting, it implies the use of multi-species of game on extensive tracts of natural range, either unfenced or fenced, for meat and hides. In contrast to traditional hunting, this is carried out by the most efficient hunting methods modern and the off-take is strictly governed by annually established cropping quotas.

The success of game ranching, in economic as well as conservation terms, depends on a number of factors including the following:

1. The land on which the utilised wildlife occurs must be strictly controllable, i.e. it should preferably be uninhabited, ecologically marginal to agriculture and privately owned or leased. Although not necessarily, game ranches are increasingly being fenced to demarcate ownership and to facilitate control.
2. Wildlife legislation must recognise the right of landowners or concession holders to derive optimal benefit from the wildlife resource on their lands as an essential incentive for conserving it. Accordingly, respective governments should accept the transfer of legal responsibility for wildlife management to the owners of the land, maintaining only an advisory function. Furthermore , governments should give marketing support and render similar services to the "wildlife industry" as are usually offered to other sectors of agriculture, such as livestock production.

Although this concept of wildlife utilisation was originally developed for privately owned lands, it is also applicable to State and Communal lands if a government is prepared to lease State land to private concessionaires and/or to grant rural populations or tribal authorities the right to manage and utilise the natural resources on their communal lands. This process implies not only that the private lessee or the concerned communities assume co-responsibility with government for the management and conservation of the wildlife resources on the particular lands, but also that all returns from their sustainable utilisation will accrue directly to them.

Following agricultural concepts, the utilisation of wildlife is being intensified in various forms in most parts of the world. Selected species are propagated with artificial feed in fenced areas, enclosures or even cages. Their utilisation becomes a strictly commercial agricultural activity, aimed at the production of meat, gameskins, furs and certain other specific parts of wild animals for which there is a market. This form of utilisation is usually referred to as *"farming"* and may involve crocodiles, snakes, turtles, various birds, kangaroos, furbearers and various ungulates. Normally, farming implies the breeding of the used species for the purpose of harvesting the offspring. However, sometimes it is based on collection of eggs or young animals from the wild which are then simply reared for subsequent use.

Based on the criteria of the *Convention on International Trade in Endangered Species of Wild Fauna and Flora* (CITES) such intensified wildlife utilisation may also be referred to as "ranching". This was defined by the Parties to CITES in 1981 as the commercially oriented utilisation of a wild animal population without the genetical isolation which "farming", i.e. the breeding in captivity of wild species, normally implies. Collection of live specimens or eggs for rearing in captivity continues to link the farmed stock genetically with the wild population and is thus classified by CITES as "ranching", although the management of the stock might be quite intensive.

The wild behaviour of game animals constitutes a major constraint to their farming. Therefore all forms of wildlife farming imply to some extent the taming of the animals concerned. This in turn leads to genetic selection for docility which, of course, is the first step to domestication. The intensive utilisation of selected species may therefore involve animals in any state of transition from wild to domesticated.

In comparison with extensive "ranching", the more intensive utilisation of selected wild animal species has a number of advantages, but also some disadvantages. Cost-effective cropping of the game and hygienic processing of its meat and skin become much easier and less costly. Generally, farming activities are easier to manage and control technically and hygienically, as well as economically; they also allow for planned marketing. On the other hand, the biological productivity cannot be fully realised in economic terms and greater capital investments are required. Furthermore, major disease problems may result from the keeping of animals in high densities. Intensive use of selected species usually does not contribute to their conservation in the wild. On the contrary, it may indirectly be detrimental to it because it reduces or even removes totally the economic incentive for maintaining the wild stocks for sustainable utilisation.

The present book tends to be mainly concerned with *consumptive forms of wildlife utilisation* for the following reasons:

1. Consumptive use of wildlife resources is still of the utmost importance for the survival and well-being of rural societies in many of the less-developed countries all over the world. However, the possibilities of integration of such resource use into the develop-

ment process and its ecological implications are often inadequately understood by those involved or responsible for "development" of these societies.

2. Commercial use of wild animals is much disputed in developed countries where many people consider themselves to be the guardians of wildlife resources in the rest of the world. The topic evokes a great deal of emotion, and discussion often lacks ecological and socio-economic understanding; it is most directly related to conservation issues and ideals.

The purpose and scope of **non-consumptive wildlife utilisation** are much more complex than those of consumptive use. This is due to the fact that non-consumptive uses of wildlife originate from the special relationship between man and animals which involves much more than material aspects. Since early times man has kept animals simply for companionship or sport, has tamed and domesticated them for certain services or has hunted them for pleasure. The keeping of animals for different reasons has become increasingly popular in developed and industrialised societies. More recently, the *use of wildlife for scientific, educational and recreational purposes,* but particularly as a base for tourism, has become of major importance since its economic importance in some countries far exceeds that of consumptive forms of wildlife utilisation. Wildlife-based tourism may involve conservation areas as well as agricultural or forest lands and assumes different forms, as described in Chapt. 1.7. One of these is *sports or trophy hunting* which has become a major industry, not only in tropical regions. It involves the killing of relatively few selected animals, mainly males. If well controlled, it has, therefore, little impact on the wildlife resource and considering its primarily recreational purpose may be classified as "extensive non-consumptive".

There are also other more intensive forms of non-consumptive use of wildlife. These involve the *keeping, rearing and sometimes farming of selected species for certain services,* such as birds of prey and ferrets for hunting, cormorants for fishing, peacocks and other birds for decoration, elephants for work. Other species are intensively managed or kept in captivity for the continued production of specific raw materials such as snakes for venom, ostriches for feathers, vicuñas for wool, and Civet cats and Musk deer for musk.

## Reasons for wildlife utilisation

The concept of utilising wildlife as an integral part of natural resources is closely linked to **conservation**, but has been strongly contested by many conservationists. Although it is true that the uncontrolled and indiscriminate exploitation of wildlife has sometimes led to the destruction of wildlife, the concepts of utilisation and conservation are not necessarily opposed to each other. On the contrary, sustainable utilisation of wildlife is often the best and sometimes the only way to conserve it. Therefore, apart from the socio-economic needs of human populations living with wildlife, one of the major reasons for wildlife utilisation is conservation. The underlying philosophy is that only those resources that render man tangible benefits are being maintained and conserved by him. In other words there must be an economic incentive for conservation. This philosophy is basic to the concept of *"conservation for development"* which has been formulated and propagated since 1980 by the World

Conservation Union (IUCN), the UN Environmental Programme (UNEP) and the World Wide Fund for Nature (WWF), and which has replaced the original purely preservative approach to wildlife.

If wildlife utilisation is to serve the purposes of conservation it is, however, imperative that it is part of an ecologically based overall management activity which is legally controlled and professionally supervised. Such a management activity must address two major aspects; the *productivity* and *sustainable yield of the managed wildlife resource*. In reviewing the use of wildlife resources, particularly in developing countries, it is evident that these critical issues are frequently ignored and that wildlife is widely being commercially and indiscriminately over-exploited.

There are two main reasons for this:

1. The economic benefit derived from wildlife accrues either to the State or to individuals who do not feel responsible for the maintenance of the resource as they do not own it. Although the resource does render returns, there is no direct incentive for the owners of the land or for the communities which are working it, to conserve wildlife which, in fact, may constitute a danger or a nuisance to them.
2. In contrast to the case for forest and fish resources, wildlife is often not recognised as a natural resource which warrants special professional attention and administration. Therefore, governments in many countries still lack qualified personnel in the field of wildlife management, and ministries which have jurisdiction over wildlife and wildlands are often not given adequate funds to discharge their controlling function.

In order to successfully apply the concept of sustainable utilisation of wildlife resources to conservation, several preconditions are essential:

- Existing legislation and administrative structures are often outdated and unsuitable for wildlife utilisation concepts. They must be adapted to allow for a co-responsibility of state, land owner or tribal authorities for the management of the resource, giving at the same time adequate economic incentive for its conservation.
- Governments are not always aware of the economic importance of wildlife resources. This needs to be documented with a view to inducing them to strengthen wildlife administrations into well-functioning, professionally guided authorities capable of managing and controlling the use of wildlife resources. The direct and indirect economic benefits which may be derived from wildlife should also be emphasised in the negotiation of international technical assistance and bank loans.
- Wildlife utilisation must always be accompanied by monitoring and research, and national scientific institutions should be encouraged to include these in their research programmes.
- Economic competitiveness of wildlife utilisation with other forms of land use must be ensured by development of marketing structures and services common to other agricultural industries, and the introduction of tax benefits customary to these.

Apart from aiding conservation, there are other important **economic and ecological reasons** which mitigate for the use of wildlife resources. The greater part of the world is either unsuitable or only marginally suited for agricultural production. About two thirds of Africa are infested with Tsetse fly (*Glossina* spp.), which either precludes livestock production or renders it unproductive; other zones are too arid for crop production and allow only transhuman pastoralism; large parts of Africa, America and South-East Asia are covered with rain forest with a very limited agricultural potential, and others are too swampy or mountainous. In most of these areas, however, there is wildlife well adapted to the local ecological conditions and this often constitutes the only available local natural resource from which indigenous people can derive a livelihood. Subsistence hunting provides for food; trapping of furbearing animals gives cash returns; commercial game ranching or farming as well as wildlife-based tourism offer the only possible employment. In terms of national economies, the use of wildlife resources is of the utmost importance for engendering cash returns and work opportunities in otherwise unproductive or agriculturally marginal regions, thus preventing migration from these into urban centres. Some forms of wildlife use may also become a source of major foreign currency earnings.

In most of the agriculturally marginal zones of the world different types of livestock were introduced, either early in history or more recently during the colonial era, in order to enable man to settle in them and to derive a living from them. Livestock husbandry has thus become a major factor in the historical development of many societies around the world, dominating the socio-economic as well as cultural aspects of their life. In some areas livestock production has remained either at an extensive subsistence level or is a traditional way of displaying status and accumulating wealth. In others, however, it has been intensified for commercial purposes and is being expanded more and more into arid and ecologically fragile biomes.

Everywhere, but particularly in the semi-arid savannas of Africa, South America and Australia, livestock has come into competition with indigenous herbivorous animals. With the intensification and expansion of beef production in such areas the conflict between cattle and game has worsened to a point of active diminution and wanton destruction of wildlife. The persecution of wildlife was not only motivated by apparent competition for water and grazing, but also by fear of transmission of diseases from wildlife. During 1924-26, for example, *deer* were systematically eliminated in California, USA, in the control of Foot and Mouth Disease. In Eastern and Southern Africa millions of *herbivorous game animals* were eliminated in Tsetse fly control operations, between the 1930s and 1968, in efforts to open up semi-arid savanna lands for cattle ranching. Large parts of the abundant wildlife resources of the fragile Kalahari ecosystem in Botswana were systematically destroyed by the establishment of so-called "veterinary fences" across the country, preventing seasonal movement of game to water. As late as the 1970s no less than 4,000 African*buffalo* were eradicated in Zimbabwe in order to establish a so-called "Foot and Mouth Disease Free Zone".

As ecological knowledge of the particular ecosystems developed and the role of wildlife as a reservoir of livestock diseases was scientifically investigated, destruction of wildlife in favour of livestock development became more and more doubtful. None of the game elimination activities were ever based on objective cost-benefit considerations and the economic value of the destroyed wildlife was completely ignored. In retrospect, they were either ineffective or produced little economically in comparison with the wasted resources.

Ecologically, livestock production is based on the conversion of indigenous vegetation into animal protein. Much discussion emanated in the 1960s, particularly in Southern Africa, whether wildlife was a more efficient converter of primary into secondary biomass and should thus be utilised for meat production together with, or even in favour of, cattle. Research has shown that the semi-arid savanna biome of Africa has a particularly high secondary biological productivity. This is explained by the evolution of an exceptional diversity of large herbivorous mammals in this biome, totalling no less than 62 different species. Many are adapted to specific niches within the savanna ecosystem.

The adaptations comprise anatomical and physiological features as well as behavioral characteristics:

- Specific body features;
- Different evolution of teeth, tongue and lips, and a high degree of differentiation of the digestive tract;
- Night feeding and food selection;
- Adapted water metabolism;
- Specific metabolic capability to convert food predominantly into muscle protein and less fatty tissue;
- Intraspecific competitive behaviour regulating population densities in accordance with the carrying capacity of the ecosystem, and interspecific competition ensuring optimum use of all available food resources;
- Adaptation of reproductive behaviour to ecological conditions or specific pressures;
- Development of resistance or high degrees of tolerance against many parasites or pathogens.

Due to these diverse adaptations, the multi-species game communities of the African savannas are capable of optimum utilisation of the vegetation cover. Their total biomass may thus be much greater than that ever attained by cattle in terms of long-term carrying capacities of the same lands. However, wildlife biomass may greatly vary according to species composition and the prevailing edaphic conditions.

These facts argue strongly for the commercial utilisation of African savanna game animals for meat production, either on their own or in combination with cattle ranching, provided their great biological productivity can be realised technically and economically on a sustained yield basis. Such a possibility is attractive from a point of view of conservation as well managed game stabilises ecological conditions, contrary to cattle which have devastated large parts of Africa.

Practice during the last decades has shown, however, that the *"holistic approach"* to production from wildlife resources inherent in the "ranching" concept, as desirable as it may be, is not always possible for legal, technical and other reasons described in Part II. Everywhere, there has been a general trend towards more intensive wildlife production systems, such as intensified cropping of selected species and various forms of wildlife farming, not only in Africa, but also in Europe, North America and Australasia. Although these systems do not exploit the full potential of biological production they seem to respond best to economic needs. Some of them do utilise the special properties of wild animals, such as optimal food conversion, lean meat productivity and production of unique raw materials such as furs, antler velvet and musk. If controlled, the intensive wildlife production systems may to

Some wild herbivores, such as Elandantelope, are adapted to feeding almost exclusively on shrubs, whereas others, such as zebras, wildebeest and cattle, depend on grass. (Photo: H.H. ROTH)

some extent be beneficial also from a point of view of conservation by alleviating the pressure on wild populations of endangered species and even propagating them in captivity.

## Wildlife utilisation as a form of land use

"Land use" is taken as the utilisation of any particular tract of land for commercial or non-commercial purposes. If the land contains **natural resources** the **utilisation** of these resources may constitute either a **primary or a secondary aspect of land use**. Wildlife is a renewable natural resource and its utilisation may thus become a legitimate aspect of the use of land. In some biogeographical zones wildlife utilisation may become the primary, i.e. predominating, form of land use, with which other possible uses of that land would have to be compatible if they were to be added. In other zones or specific areas wildlife utilisation might only be a secondary pursuit, supplementing other dominating uses of land.

The use of any land is generally governed by two sets of factors: the first relates to the legal status of the land to be used, and the second concerns the ecological conditions prevailing on and around this land. The **legal status of land** is determined by land ownership and tenure, and the legal or customary provisions for the use of wildlife are often related to these. In Europe, the use of wildlife resources in the form of sport hunting was, since mediaeval times, largely the privilege of the feudal landowner. Only in more recent times have certain tracts of land been designated to the State, which has assumed all legal control over wildlife. In most of the developing tropical countries there was traditionally no private ownership of larger tracts of land. All land was communal and the tribal chiefs would allocate land and natural resources for use, retaining certain hunting privileges or demanding tributes from hunters; sometimes hunting was banned in certain sacred localities. As a result of colonisation, this system was replaced by the European concept of state or private ownership of land, and colonial administrations and later governments assumed legal control over all wildlife, including that on *Communal lands*. This change had and still has a rather negative influence on the attitude of the rural population of these countries towards the natural resources and their use. As in mediaeval Europe, they perceived the instituted control over game and hunting as the enforcement of a class privilege, alien to their needs and traditions. Consequently, conservation concepts and efforts never found broad support by the rural population. Therefore modern concepts of wildlife utilisation developed only in those countries in which large tracts of *Private land* existed, notably in Southern Africa, Argentina, Columbia and North America. There, utilisation could be based on feudal land tenureship and the strict control over the wildlife which it affords.

Private estate owners in Eastern Africa were never accorded the necessary legislative basis for the economic use of the rather abundant wildlife on their lands. Lacking the economic incentive, wildlife was either only reluctantly tolerated or eliminated, with a few notable exceptions like the Athi game ranch of HOPCRAFT. On *State land*, however, commercial consumptive wildlife utilisation programmes were attempted in various parts of Africa, either through concessionaires or parastatal agencies. Without long-term private incentive and technical efficiency most of these attempts failed economically.

It is only more recently that the economic utilisation of wildlife on communal lands inhabited by the indigenous peoples in the northern regions of North America, and to some extent in Russia, is being regulated on a sustained-yield basis to serve their traditional rights and needs. Also in Southern Africa, initiatives have been taken by some countries to apply concepts of sustainable utilisation of natural resources to communal lands involving the rural people themselves. However, this is a difficult task which implies profound changes of attitude not only of the rural people but also of wildlife administrators. It will not be achieved without education, professional guidance and tangible economic incentives to the concerned communities.

In this respect Zimbabwe has developed since 1986 the *Communal Area Management Programme for Indigenous Resources* (CAMPFIRE). In Zambia the *Luangwa Integrated Resource Development Project* (LIRDP) and the *Administrative Design for the Management* (ADMADE) of *Game Management Areas* (GMAs) have grown out of the *Lupande Development Project* which was carried out in 1985-87. Botswana has introduced the concept of *Wildlife Management Areas* (WMAs) since 1988 in which local rural authorities assume responsibility for natural resources management, and Namibia has instituted the *Community-based Natural Resources Management Programme* (CBNRM) after its independence in 1991. All of these programmes have the goal of assisting rural communities in the application of sustained utilisation of natural resources to their lands and are further described in Part II.

Non-consumptive utilisation of wildlife in the form of tourism developed rapidly after the Second World War mainly in the form of game viewing in national parks and game reserves, i.e. exclusively on *State land*, all over the world. Only more recently is this form of wildlife use also being introduced onto *Private and Communal lands*, as a supplement to other forms of land use.

More important for the economic success of wildlife utilisation than land ownership and tenure is the **ecological status of land**, which is governed by climatic and edaphic factors. The determination of the ecological suitability of land for different uses is the basis for land use planning and this should be a central concern of any economic planning, particularly in developing countries. However, land use planners very often emphasise agricultural development and neglect possibilities for the utilisation of renewable natural resources as primary or complementary forms of land use. With the development of ecological understanding and environmental concerns it should also be recognised that setting land aside for conservation and recreational purposes constitutes a perfectly legitimate and increasingly important form of land use which in fact can yield important direct and indirect economic benefits.

There are certain ecological conditions which render land particularly suitable for wildlife utilisation. The production of meat and other wildlife products as the primary goal of extensive management regimes should normally be restricted to all those lands which are ecologically either marginal or unsuited to agricultural use for climatic, topographical, hydrological or disease reasons, and which do not enjoy any particular protective status. Intensive forms of wildlife utilisation, however, such as the farming of selected species, may well be integrated into agricultural systems also on fertile or irrigated lands.

Almost all wildlands set aside for conservation and recreation lend themselves to some form of non-consumptive utilisation of wildlife, either scientific research, educational programmes or recreational tourism. These should therefore constitute an important, though secondary, aspect of the management of any such lands. In some protected areas such as forest reserves or buffer zones around national parks, the complementary use of wildlife resources may also include cropping for meat, recreational hunting and sport fishing.

## The role of hunting in wildlife utilisation

The oldest human activities for sustenance were gathering of natural produce and hunting wild animals for food and skins. Although this still applies in some parts of the world, cultural and agricultural development have generally brought about **changes in the original motivation for hunting.** As discussed, hunting in some societies became associated with social privilege and animal trophies gained importance as status symbols. The procurement of specific organs or parts of animals has assumed a great significance in ritual and medical practices which continue to play an important rôle in many developing countries. Above all, hunting in many societies is perceived as the manifestation of male courage and virility. In mediaeval Europe, hunting became the favourite sport of the male gentry, its motivation being thus far removed from the original need to kill animals for subsistence and survival. With this change ethical and other anthroposophic notions were introduced, turning hunting into a highly regulated and coded social activity reserved for a privileged minority, based either on Roman or Napoleonic Law.

With European colonisation, this refined concept of hunting was introduced into many developing countries to supersede the original objectives and traditional motives of hunting which still existed in these countries. Legal, i.e. licensed, hunting served mainly "sports" purposes; and particularly African game trophies became very valued prestige objects for the white "big game hunters"; or it served the ruthless, commercial exploitation of some species like crocodiles and elephants. The hunting traditions of the indigenous people and their need to produce food and raw material from wildlife were largely ignored and suppressed into "illegality". The ensuing conflict was most detrimental to the wildlife resources and to-date has obstructed the development of rational controlled use of these resources in many developing countries.

Only in this century with the ever-growing human identification with nature and natural life styles in the industrialised societies, the motive for hunting partly shifted back to less refined values and objectives. In particular in North America, *hunting* became an *important outdoor recreational activity* of large numbers of middle class people with an inclination to nature and wildlife conservation.

In the course of the development of the biological sciences, the ecology of game populations and their interactions with man became better understood. The need for regulation of numbers of game animals in accordance with certain environmental and socio-economic conditions was recognised and this implied the use of **hunting as a management tool**. Wildlife utilisation, being an aspect of management, depends in fact to a large extent on the skilful application of different hunting methods. Their cost efficiency in terms of hunting success has proven to be one of the most important aspects of most of the commercial forms of wildlife utilisation. The main objectives of such hunting are to kill effectively with the least damage to the product, i.e. venison, hides, fur etc., to be selective and to cause a minimum of disturbance to the game population as a whole. Although the humaneness of the kill must be ensured, there is as little room for emotionalism in game cropping as there is in an abattoir.

Apart from being the most important tool of wildlife management **hunting *per se*** has developed into a lucrative **means of deriving economic returns from wildlife resources**. Many wealthy people are ready to pay large amounts of money to get away from the stress of life in a highly industrialised society with a polluted environment, to experience individually as a hunter the thrill of wilderness, to encounter and watch large wild animals, and to relax in a natural environment. The actual hunting activity, guided by experienced hunters and culminating in the winning of a trophy, often plays only a secondary role. Such *recreational or trophy hunting* actually constitutes an increasingly important component of wildlife-based, commercialised tourism all over the world, as described in Chapt. 1.7. It is a form of economic wildlife utilisation that is restricted to countries which are prepared to lease State land to private safari operators and to auction quotas of specific trophy animals to the public. This implies that they are capable of setting up monitoring systems for the hunting activities. Returns to governments accrue from land leases and trophy fees, apart from the indirect economic benefits of tourism.

As described in relevant chapters of Part II, recreational or trophy hunting can also be successfully developed on private lands as a complement to wildlife ranching or farming. It has proven to be so lucrative that private land owners in many countries are investing much money into the re-introduction or stocking of their lands with wild animals, such as *game birds, deer, antelope* and *buffalo*.

Whereas the demand for recreational hunting has steadily risen, opportunities for it in tropical countries have diminished. This has been not only due to the depletion of wildlife, but also because of political bias against this form of wildlife utilisation, and the general unwillingness of many developing countries to involve private enterprise in natural resources management and use. However, recreational hunting in the framework of wildlife-based tourism constitutes a very real possibility for the use of available wildlife resources on communal lands which are marginal or unsuited to agriculture and livestock production, as discussed. The tribal authorities of such lands may find that direct cash returns from hunting leases are of greater benefit to the community than traditional subsistence hunting. In some countries forest authorities are leasing recreational hunting on their timber producing lands to private enterprise or individual sport hunters; this model could well be applied in many developing countries which have a large forest estate in which the wildlife resources are only subjected to illegal and uncontrolled hunting, either by a privileged elite or the commercial poachers.

### Exploitation of captive tamed animals

The **spectrum of species** which have been tamed at one stage or another for various purposes, is broad as almost any animal species can be tamed to some extent, if reared by man. The majority of species are tamed primarily as a hobby or for companionship with no economic objective. Taming for show purposes constitutes a use of wild animals which dates back to the earliest civilisations of mankind and reached its greatest popularity during the Roman empire. Although much diminished, the circus tradition has continued ever since, experiencing a new boost in the last century with the industrialisation of Western Europe. The more mankind is alienated from nature and wild animals the more fascinated it becomes by notions such as "wild", "fierce" and "dangerous" beasts. Only after the Second World War more refined forms of animal show business did evolve, such as dolphinaria or oceanaria. Other forms of commercially oriented display of animals which always entails their taming to some extent, are crocodiles, snakes and birds parks.

Another reason for taming wild animals was, and still is, to make them render services to man or to have them readily available for slaughter when their products are needed. As discussed in Chapt. 1.2 this was one of the most important reasons for the domestication of certain animal species; the keeping of the tamed animals for prolonged periods in captivity, or even breeding them there, does not necessarily constitute domestication. In fact many species have been commercially exploited in the past and are still being used by taming and keeping them simply in captivity, as Tab. 1 shows. If such a utilisation process continues for prolonged periods, either by replenishment of the tamed stock from the wild or by random breeding in captivity, it may be referred to as semi-domestication.

Exploitation of tamed animals always implies successful **capture, transport and rearing in captivity**. Capture techniques were developed with great ingenuity by primitive societies

A giraffe is captured with the aid of chemical restraint in Hwange National Park in Zimbabwe for translocation to other conservation areas. (Photo: D. BARTLETT)

since the earliest times, often linked to hunting for food. Amongst these are pitting, cage and box trapping and netting. Some species are best caught by hand, and in open country some species can be run down on horse back or with four-wheel drive vehicles until they can be restrained by a lasso. In the 1960s capture by chemical restraint was developed and became a most efficient tool for the capture of large ungulates such as *elephants, rhinoceros, hippo* and *buffalo*. Its success depends, however, on the possibility of injecting sufficient drug into the animal and being able to follow it until the drug takes its immobilising effect. The immobilisation technique is not suitable for the capture of large numbers of young animals required for stocking of game farms, for purely economical reasons. For this purpose, the method of driving whole herds either on the ground or by helicopter, along miles of brightly coloured plastic sheeting into capture pens or directly into crates was developed in South Africa in the late 1960s. Cost-effective capture of ungulates was in fact a very important factor in the development of game farming in Southern Africa.

The capture of wild animals is followed by transportation to acclimatisation or taming pens which requires just as much experience and skill as the capture operation itself. Most young animals tame down and soon start accepting captivity. However, the degree of docility an animal develops under captive or semi-captive conditions varies from species to species and even between individuals of the same species. The docility of an animal will determine to a large extent its exploitability and thus constitutes an important factor for its commercial use. Similarly, the physiological problem of adapting a species to more or less artificial feed and thus the possibility of rearing it successfully and economically in captivity differs and may exclude some species altogether from being exploited in this way. As with capture and transport, great scientific advances have been made since the 1960s in the nutrition and feeding of wild animals. Today there is extensive knowledge of the nutritional requirements of almost any wild species which can be beneficially applied to the commercial farming of wild animals.

The advent of the industrialised age in Europe and North America during the 19th century generated affluence, new aspirations and needs of man. From these came an ever increasing demand for certain wildlife products which resulted in the over-exploitation and decline of many of the wild procurer species. This fact, but also economic considerations brought about widespread efforts to produce the demanded products, for example certain furs, antler velvet and ostrich feathers, from captive-held stocks. This was the start of a **new wave of domestications of wild animals**, as discussed in Chapt. 1.2. Many of these species have by now been kept and bred successively for long periods for commercial exploitation, but none of them have become as tame as domestic animals. Although they have been englobed in man's life, subjected to planned husbandry and selective breeding, most of them may never become truly domestic animals due to their genetic make-up as solitary and non-social species. Their utilisation is described in the respective chapters of Part II.

Exploitation of captive wild animals has continued and developed further throughout the first half of the 20th century. A decline of wild stocks in conjuction with market demands and economics has prompted the breeding in captivity of ever more animals of wild species, as listed in Tab. 1, mainly for fashion and laboratory purposes. After the Second World War the breeding of game birds for hunting and a wide variety of animal species for the pet trade gained particular momentum. In the 1960s and 1970s a considerable number of additional wild species were subjected to captive exploitation, most of which is still in an experimental stage; this includes crocodiles, lizards and snakes mainly for leather production, ostrich and emu for meat and leather, capybara and Cane rats for locally preferred meat, further furbearers for specific types of fur, Musk deer for musk, deer for venison, elandantelope and oryx for production in arid zones and muskoxen in arctic regions. Most recently, the increasing demand for test animals in biomedical research and serum production has induced the captive-breeding of many primate species which were formerly procured from the wild. Thus, the production of captive-bred wild animals for one purpose or another has grown into a major and fast growing industry all over the world.

The exploitation of merely semi-tamed wild animals for commercial purposes is evidently a common and legitimate form of wildlife utilisation. The fact that wild species can well be utilised without being fully domesticated raises the question of the **desirability for further domestication of wildlife species**. In general there is a case for the utilisation of wild animals if their produce will satisfy specifically felt needs and if they are more productive than domestic animals under specific ecological conditions. It remains to be decided, however, whether domestication of any particular species is the most appropriate means to this end. Firstly, it needs to be decided whether the species lends itself to domestication in terms of its inborn social behaviour and physiological characteristics. Secondly, the special properties to be utilised need to be carefully identified and their genetics assessed. If the adaptation of a wild animal to special ecological conditions is mainly based on behavioral patterns and less on genetically founded physiological characteristics, there is every chance that the process of domestication will ultimately alter these behavioral mechanisms. If it is based on genetic characteristics, the very properties for which the species is being utilised may inadvertantly be altered or altogether lost in the course of selective breeding. Generally, this should aim at improving the usefulness of a species to man. However, in wild animals there is often no need to improve their properties as they are already optimal or they produce the specifically wanted commodity anyway.

For example, *desert antelopes* are already ideally adapted to extreme arid conditions, *elephants* have already maximum strength for work, and *fur bearers* already grow the finest and warmest possible hair coat.

There is thus no case for selective breeding other than to make the species docile for easy and economic management. This can often be achieved simply by constant contact with man from an early age and elimination of particularly nervous individuals. Therefore, with many species exploitation in a semi-domesticated state, as in the case of the *Asian elephant* or the *Eland antelope*, may be preferable to full domestication. Rendering exploited wild animals tame, and therefore more manageable, and possibly breeding them in captivity does have, of course, important positive economic implications which argue for semi-domestication where feasible. This does not necessarily signify that such a form of utilisation is also beneficial from the point of view of conserving the remaining wild population of the particular species.

Semi-domesticated eland at Manyoli Ranch in Rhodesia (now Zimbabwe) in 1965. (Photo: H.H. ROTH)

## 1.5 IMPORTANCE OF WILDLIFE UTILISATION

### Abundance and sustainability of wildlife resources

Use of wildlife, either for consumptive or non-consuptive purposes, is much more common than most people imagine because wild animals are usually inconspicuous. It is neither restricted to any particular biogeographical zones, nor to levels of cultural and economic development; it has therefore universal importance. If wildlife is to be utilised economically, the abundance and sustainability of the wildlife resources are essential factors. Where wild animals occur in large numbers it is easier and thus more economic to hunt, capture or view them. The abundance and indeed the very presence of wild animals depends to a large extent on the *density and activities of the human population* and these may thus determine the possibility and viability of wildlife utilisation.

Urban and industrial development usually eliminates wildlife habitat altogether and destroys economically viable wildlife resources. Agricultural and forestry activities often result in conflict between man and/or his livestock and wild animals. Although many species learn to adapt to human presence and habitat changes, the competition for grazing and water as well as subjective attitudes against wildlife tend to diminish its abundance and species diversity. Also over-exploitation of wildlife by indiscriminate hunting and trapping with depletion of the wildlife resources often results from the increase of the rural population. Human activities, however, may also cause an increase in a certain type of habitat which then becomes a favourable ecological niche for some animal species; or non-indigenous species are introduced into a favourable environment. There they may proliferate to such an extent that they must be reduced. Such a situation is often the ecological basis for the utilisation of certain species, as for example *hares* in Argentina. Planned conservation and management may also prove most effective for increasing abundance and species diversity for utilisation purposes.

It is important to distinguish clearly between the **exploitation and the sustainable use** of wildlife. The former may have a temporary economic importance, but the resource gradually diminishes and might be destroyed altogether. The only situation in which such use is justified is where there is a need to control wild species either for economical or ecological reasons with secondary utilisation of their products.

All other wildlife utilisation should of necessity be sustainable and its importance gauged only in terms of sustained yields. This means that out of any particular game population only that number of animals is annually cropped which is being recruited to the respective population through the natural reproductive process during the same period. Such recruitment depends on the productivity of the population which in turn depends on a number of different ecological factors such as rainfall and the carrying capacity of the habitat, inter-species competition, predation, etc. Human activities may seriously diminish productivity, but appropriate management of the wildlife resource and its supporting habitat can also enhance it considerably. Biological productivity fluctuates, therefore, and the harvest of game animals must be adjusted to it if utilisation is to be sustainable. It follows that the *"sustained yield"* of wildlife is not a static entity. What may be a sustainable level in a given year or a particular locality may well constitute overutilisation in other years or localities. In order to determine the sustainability of any consumptive use of wildlife, a good deal of biological knowledge is required coupled with monitoring at regular intervals; it should only be determined over prolonged periods which allow for temporary stages of over- or underutilisation in terms of the fluctuating productivity.

Statistics on wildlife utilisation are usually compiled on an annual basis and therefore rarely reflect sustainable yields unless data of several consecutive years are lumped together. In consequence it is most difficult to conclude the importance of consumptive wildlife utilisation in terms of sustainability merely from hunting or cropping statistics. To do this more meaningfully it is necessary to estimate the biomass of utilisable game, i.e. their total live weight per area unit, and to deduct from this the "*standing crop*". This requires inventory surveys which have to take the seasonal movements of the animals into full account; known average productivity rates are then applied to the obtained biomass figures in order to establish potential annual yields for the different species. Only if sustainability is proven is it possible to determine the actual gross economic importance of the annual crops and thus the long-term viability of the utilisation programme.

Most biomass figures as well as deductions of standing crop, however, suffer the same weakness as actual production statistics; they do not adequately reflect the dynamics of the productivity of different species under different ecological conditions and may therefore be erroneous or misleading. Ideally one has to combine both the estimation of biomass and the evaluation of actual production statistics and to compare these over periods of three and more years in order to be able to decide whether or not the level of utilisation is sustainable and to adjust it accordingly.

The importance of other non-consumptive forms of utilisation also depends on long term viability which is, however, governed by entirely different factors. The "product" to be achieved for tourist use is an optimal diversity and abundance of spectacular wildlife easily accessible for close viewing and photography, combined with good catering services; for hunting-tourists the provision of specific trophy-bearing individuals of certain species is additionally required. Long-term sustenance of the "*tourism product*" is as important for this form of wildlife utilisation as is the sustainability of cropping yields in consumptive use of wildlife. The technical aspects and problems associated with wildlife-based tourism are discussed in Chapt. 1.7.

## Range of utilised species and derived products

The economic use of wildlife in one form or another involves the majority of all mammals and also many species of birds and reptiles. The spectrum of utilised species is generally largest in the traditional use of wildlife resources as practiced in some of the tropical developing countries, in which virtually any animal may be used for food, as raw material, or for ritual and medicinal purposes. But wildlife-based tourism, also utilises the totality of wildlife although the more spectacular, large species or trophy-bearing individuals play a particularly important role.

Commercialised consumptive utilisation tends to focus on a more limited number of species of specific animal groups which render specific desired raw materials; this also applies to the non-consumptive use of some animals for certain purposes, such as the production of feathers from *ostriches*, hair from *vicuñas*, musk from *Civet cats* or services to man.

*Deer, antelope*, and *rodents* are all used preferably for commercial production of meat and generally constitute an important source of protein for human nutrition. However, in the rural areas of many developing countries almost any type of animal, including invertebrates, will be a welcome addition to the often protein-deficient daily diet of the people concerned. There are absolutely no emotional constraints against eating all and every part of wild animals; only religioius beliefs and so-called "tabus" may prevent certain species from being killed for food. *Whales, seals* and *sirenians* have been used in the past for the commercial production of oils and fats. To coastal indigenous people they are also an important source of food; only whale meat was temporarily marketed commercially and in great quantities. *Carnivores* and *some rodents* are the principal suppliers of high-priced furs and thus constitute a very important economic resource. However, the skins of *antelope, zebra* and *deer*, also assume economic importance in the context of game ranching or farming. *Ostriches, crocodiles, Monitor lizards, snakes, kangaroos* and *some suids* provide especially valued leather, but their meat is also appreciated and partly commercially marketed. *Elephants, walruses,* and *hippopotami* furnish ivory, but more recently the skins of elephants are being processed into high-priced leather products.

The utilisation of the different animal groups is discussed in Part II with tabulation of the utilised species, and the importance of the various types of products is discussed in Part III.

## Statistical evaluation of wildlife utilisation

Statistical data from the use of wildlife resources are essential for economic or financial evaluations and also for making administrative or management decisions. The statistical information may refer to "production" in the widest sense, trade or consumption of products.

*Production statistics* may originate from a rather wide variety of sources, including government authorities, professional associations, research institutions or private enterprises. They list:

- species and numbers of wild animals killed in sport hunting, game control, culling operations and other legal hunting in the form of game ranching or farming, trapping of fur-bearing animals, whaling or sealing operations etc;
- types and quantities of wildlife products legally derived from such animals; or
- numbers, circumstances and nature of confiscations of wildlife parts from poachers etc.

*Trade statistics* also originate from various government authorities, national or professional trade-monitoring institutions, and international organisations. They list:

- species and numbers of wild animals or quantities of specific wildlife products traded internally, imported or exported; or
- numbers of confiscations by veterinary, customs or CITES authorities.

*Consumption statistics* are the most difficult ones to obtain and originate only from internal market or household surveys. They list the types and quantities of wildlife consumed as food or other products utilised by individual persons or households.

The evaluation of such statistics with the objective of quantifying the importance of wildlife utilisation is much more difficult than in the case of, for example, agriculture, forestry or fisheries. This is mainly due to the fact that wildlife-derived raw materials are very diverse and, if recognised at all as a "commodity", do not fit into any one of the traditional categories of agricultural or forestry production. The administrative responsibility for wildlife varies widely and any records on wildlife production, if compiled at all, are accordingly scattered. They are rarely considered in overall national economic statistics and are difficult to obtain. Another shortcoming is the fact that in developing countries with the most important wildlife resources, their utilisation and the trade with wildlife products is a traditional and often illegal activity on which no statistics are produced at all, only "informed estimates".

Apart from these inaccuracies it is difficult to convert the numbers of hunted or otherwise produced animals into marketed or consumed weight units comparable to livestock production. Even live-weight data are very variable as they depend on sex, the involved subspecies, season and local nutritional conditions. Furthermore, they are often based on weights of killed animals and constitute therefore in reality Slaughter Weight. The relationship between the weight of live or freshly killed animals and that of their dressed-out Carcass Weight may include or exclude head, skin and feet according to local marketing customs. Any deductions of meat yields from hunting statistics are therefore bound to be only rough approximations.

In *Fallow deer*, for example, the weight of live animals relates to that of slaughtered, dressed-out animals without skin and edible meat in the ratio 100 : 74 : 58 : 42[1].

For the estimation of the total economic value of wildlife, i.e. *direct* and *indirect use values* as discussed in Chapt. 1.6, the above statistics do not suffice. Much more information is required for such economic assessments, comprising

- direct revenues from wildlife from tourist admissions, hunting licenses, leases, services or the sale of materials; and
- values of indirect socio-economic benefits accruing from wildlife-related activities, employment, commercial secondary industries.

Collection and compilation of such data is difficult and only possible through the establishment of special information systems and specialist services. They are nevertheless essential for documenting the real value of wildlife resources and for aiding management and conservation decisions.

More recently some of the international organisations have made efforts to document the value of wildlife resources. In 1963-65 the International Union for Conservation of Nature and Natural Resources (IUCN), sponsored by the Food and Agriculture Organisation of the

---

1) FREUDENREICH, P. and K. FISHER (1989): Untersuchungen zur Fleischqualität von Damtieren. Landwirtschaftliche Wildhaltung, Wildtiere in Gehegen, No. 5; p. 87

United Nations (FAO), carried out a first special survey of the economic potential of wildlife in African countries. Later it reviewed the development of game ranching in Southern Africa. It was also instrumental in establishing international monitoring of trade in wild animals and wildlife products through the TRAFFIC Network and the World Conservation Monitoring Centre (WCMC) in Cambridge. As of 1978 the FAO carried out specific regional surveys on the use of wildlife, and game meat was included in the Interlinked Computer Programme on Commodities.

## Wildlife as a source of food

The importance of wildlife as a source of food dates back to prehistoric times and despite the availability of domestic animals wildlife has continued to be a significant nutritional factor in many parts of the world. It is, however, rather difficult to quantify its present-day importance. In contrast to meat production from domestic animals, the use of wildlife for food can only be estimated approximately by computing the total dressed weight of hunted or otherwise killed wild animals, assuming that the majority of these are being used for human nutrition. The deficiencies of hunting statistics and the problems of evaluating them have been discussed above: few countries record hunting reliably and the traditional or illegal use of wild animals, which in many countries is of much greater importance, is not recorded at all. Also game birds produce, world-wide, very signficant quantities of meat which have to be considered in this context.

**Global game meat production** was estimated by FAO in 1978 at approx. 900,000 t which represented at that time only 0.7% of the total world meat production of 137 mill. t[2]. An update in 1993, discussed in greater detail in Chapt. 3.1.2, put this figure at approx. 1 mill. t against an increased world meat production of 180 mill. t (carcass weight). Game meat represented, therefore, only 0.56% of total meat production compared with 40% pork and 29.4% beef, 29.9% poultry and about 6.1% sheep and goat meat[3]. This figure is, however, certainly much too low, considering that many of the major game-meat producing countries, such as Canada, Brazil, South Africa, Zimbabwe, Namibia and the former USSR, are not included in the FAO Interlinked Computer System. *Deer* alone produce world-wide considerably more than 270,000 t of venison annually, as shown in Chapt. 2.3.12. Other animal groups, such as *marsupials, rodents, antelope* and *buffalo* produce as much or even more meat for human nutrition; *marine mammals* assume particular importance for nutrition in the arctic region. Game birds world-wide also produce very significant quantities of meat which have to be considered in this context. However, the FAO figures show that nearly 70% of the computed output of game meat was produced by developing countries which, however, use only 40% of total meat from livestock. This highlights the real importance of wildlife as a source of meat: it is available in areas which are either marginal or unsuited to conventional animal production.

---

2) KROSTITZ, W. (1979): The new international market for game meat. Unasylva, 31 (123), p. 32-36

3) ANONYMOUS (1993): World meat situation and outlook. Food and Agriculture Organisation of the United Nations, Rome; June 1993

This nutritional importance of wildlife becomes even more evident if the **domestic consumption of wild animals** by rural populations in certain regions of the world is examined in comparison with that of other meats. Relevant nutritional surveys, carried out in Northern Canada, the Amazon region and in many African countries, all show that protein from wild animals features strongly in the daily diet of rural communities in countries in which livestock production is limited.

In Northern Canada wild animals furnished more than 50% of the total meat consumed in Inuit communities[4)]. In the Ucayali region of the Amazon, game and fish together accounted for 85% of animal protein consumed[5)]. In selected West and Central African countries meat derived from wild animals constituted between 31 and 73%, and in Papua New Guinea 36% of daily meat diets, not including fish and sea food[6)]. In Rwanda the contribution of game meat to the per capita total animal protein supply was estimated at 25%, in Zambia at 13.4% and in Bostwana at 15.8%[6)]. Other studies have produced even higher figures[5)].

Surveys of locally traded *"bush meat"* from commercialised subsistence hunting in West and Central Africa also confirm the importance of game meat in the diets of urbanised populations.

In Ghana, for example, a total of 158 t of "bush meat" from 13 species were sold within 7 months in Accra on just one market[7)]. In another traditional bush meat market in Kumasi about 160 t of bush meat were marketed to domestic consumers every year[8)] and in Toumodi in Ivory Coast this figure was an estimated 30-80 t[9)].

Consumption of venison as a "*luxury commodity*" has also gained increasing importance in industrialised countries since the Second World War. In some of these, particularly in Germany, it far exceeds local production from sports hunting or game farms. These countries have thus become an important market for venison exporting countries, such as Argentina, New Zealand and South Africa.

In West Germany, for example, approximately 0.8 kg venison were consumed per capita in 1981, i.e. a total of roughly 40,000 t of which no less than 18,000 t had to be imported. Total international gross trade in game meat fluctuated in 1980-85 between 36,000 and 43,000 t[10)].

Trade and marketing of game meat are discussed in greater detail in Chapt. 3.2.1.

---

4) HAMILTON, G.D. (1981): Practical importance of moose and other wild foods to natives in a remote northern Ontario community. Alces, 17, p. 44-55

5) DE VOS, A. (1973): Game as food. Unasylva, 29 (116), p. 2-12

6) GROOMBRIDGE (ed.) (1992) (see selected literature references)

7) ASIBEY, E.O.A. (1974): The grasscutter, *Thryonomys swinderianus* Temminck, in Ghana. In: The biology of histricomorph rodents. Symposium Zoological Society; London, No. 34; p. 161-170

8) FALCONER, J. (1992): Non-timber forest products in Southern Ghana. Overseas Development Administration (ODA); Forestry Series No. 2; London

9) ROTH, H.H. and B. HOPPE-DOMINIK (1996): Le marché de viande de brousse à Toumodi en Côte d'Ivoire (publication in preparation)

10) LUXMOORE, R.A. (1989): International trade. In: Wildlife production systems, ed. HUDSON, R.J., DREW, K.R. and L.M. BASKIN; p. 28-49 (see selected literature references)

Bush meat, comprising mainly duiker and other small forest mammals, is offered in great quantities for sale on the market of Pointe-Noire in Congo. (Photo: V.J. WILSON)

## Wildlife as a source of commercial products

Apart from being a major source of food, wildlife has always been important as a producer of specific raw materials, such as hides, furskins, hair, feathers, ivory, antlers, horns, bones, and a large variety of substances used for medicinal purposes. As mankind has developed, needs and preferences for different products have changed and so has their economic importance. As discussed in Chapt. 1.4 demands for specific products - in conjunction with over-exploitation of the wild resources and economics - have resulted during the last century in a general tendency to produce wildlife products from captive or semi-domesticated stocks, in the form of "farming". This has to some extent reduced the pressure on the wild resources; but at the same time has created a steadily increasing demand for live animals for breeding that poses at times similar if not greater dangers than commercial hunting. The economic importance and trends of the production of the different wildlife commodities are discussed in Part II. Generally, the direct use value of commercial commodities derived from a wild animal is considerably greater than that of its meat.

The **total value of internationally traded wildlife and its products**, including ivory and skins, was estimated in the 1980s at about 4-5 bill. US dollars a year[11]. With skins of some rare species such as *Panda bears* worth US$ 25,000, some *live parrots* fetching US$ 50,000, some collections of birds' eggs valued at US$ 20,000[11], to mention just a few,

11) ANONYMOUS (1989): CITES - the Convention on International Trade in Endangered Species of Wild Fauna and Flora. UNEP Environment Brief No. 8; Nairobi

there are huge illegitimate profits involved. The illegal trade in wildlife products is in fact considered the world's second largest illegitimate business - only narcotics are worth more. It is vital that such trade is curbed with a view to ensuring the sustainable utilisation of wildlife resources for the lasting benefit of all mankind, as discussed below.

The importance and ramifications of the international wildlife trade were recently surveyed by WWF/TRAFFIC (US), mainly in respect of the USA[12]. Past and present trade in some of the most important wildlife products is discussed in Part III.

## Impact of wildlife utilisation on conservation

As explained there is a definite **relationship between using and conserving natural resources.** Thus wildlife utilisation may have either negative or positive importance for the conservation of wildlife resources. Indiscriminate exploitation of wild animal species has often led to their extinction or extirpation over large parts of their former range.

To mention just a few examples: The *Dodo pigeon* (*Raphus cucullatus*) of Mauritius constituted an easily accessible source of fresh meat for ships on their way to India. By the early 17th century it was extinct. The *quagga* (*Equus quagga*) in South Africa was easy to kill on the open grasslands and thus their skins were used by the early Boers as grain bags. By the mid-19th century this zebra species had vanished. The *bison* (*Bison bison*) roamed the North American prairies by the millions, until their sustained use by the Indians was turned into mass slaughter by the advancing white settlers. By the late 19th century only two small herds had survived. All three *Asian rhinoceros species* (*Rhinocerus* spp., *Dicerorhinus sumatrensis*) were brought to the verge of extinction only because their horns provided a highly priced commodity in traditional Asian medicine.

However, good management and utilisation have just as often been instrumental for the survival of game animal species, their increase in numbers or expansion of their range.

For example, the *Saiga antelope* in Southern Russia was saved from extinction through planned management and utilisation for meat and hides; near extinction, the *White rhino* in South Africa was systematically propagated and reintroduced into its former range as a valuable asset for tourism; the *Nile crocodile*, after having been extirpated in many parts of Africa, was subjected to controlled utilisation programmes and is no longer a threatened species. *Fallow* and *Père David's deer* have survived only because of management and propagation in semi-captivity; *Red deer* in Europe and *White-tailed deer* in North America are much more abundant now than in historic times; the *Wild turkey* in North America, after vanishing from 18 of 39 states has been repropagated to several million all over USA (see relevant chapters of Part II).

There is no doubt that commercial greed for some highly priced wildlife products, but also the hunger for meat, is constituting a very real danger to wildlife conservation in many parts of the world. This fact is often used to argue against consumptive utilisation of wildlife: promotion of any intensive consumptive use of wildlife is considered tantamount to its destruction. However, natural resources which do not yield tangible material benefit will, in many parts of the world, simply not be managed in one way or another, i.e. with the indiscriminate pressure of increasing human populations they will gradually disappear. Therefore, the options are "to use it or to loose it".

---

12) FITZGERALD (1989) (see selected literature references)

Economic utilisation of wildlife can certainly be beneficial for conservation in certain situations. It must be understood where and under what circumstances wildlife utilisation may contribute or may be detrimental to conservation of wildlife. The conditions under which consumptive primary wildlife utilisation may contribute to conservation have been discussed in Chapt. 1.4.

Wherever possible, *non-consumptive forms of utilisation* should be preferred, as there is less risk that they impair the resource. However, it is not sufficient to simply delimit conservation areas without assigning a specific purpose to them, such as National Parks, Recreational Game Parks, Wildlife Production Reserves, Wildlife Management Areas, Controlled Hunting or Safari Areas etc. The implied management goals can, of course, only be determined in the framework of national land use and development planning. As emphasised in Chapt. 1.7, it is unrealistic to believe that any conservation area will automatically lend itself to conventional tourist use; remote, inaccessible areas might be much more suitable for controlled hunting and safari tourism. Once the purpose for which an area has been legally set aside is defined, it is essential to actively manage it accordingly. Wildlife in unmanaged or insufficiently protected conservation areas is just as much threatened as wildlife under uncontrolled consumptive utilisation schemes.

The need to derive optimum benefit from wildlife resources applies equally to conservation areas. Only in more affluent countries will purely social and scientific values be accepted as such; in many developing countries their existence is often at stake unless they produce tangible economic returns.

Any wildlife utilisation needs to be controlled in order not to be detrimental to the long-term conservation of the resource. One of the most important mechanisms for the control of consumptive use is the **monitoring and control of trade** in the respective wildlife product, both at the national and international level. Internal statistics mainly provide an indication of the intensity of use and thus reflect the importance of wildlife utilisation in any particular area or country. Export/import statistics have become an essential tool for the monitoring of international traffic in wildlife in an attempt to prevent endangered species from being exterminated through over-exploitation.

To this end the World Conservation Union (IUCN) was instrumental in launching the Convention on International Traffic in Endangered Species of Wild Flora and Fauna (CITES) in 1975. CITES is an international treaty with the principal objective of ensuring that no species of wild fauna or flora is driven to extinction because of the pressures of international trade. It works through a trade permit system and three documents called the CITES appendices. The permit system is operated by the CITES Parties, i.e. the signatory countries, with the assistance of a Secretariat based in Lausanne, Switzerland. Producer and consumer countries share responsibility for regulating international trade in terms of the three appendices:

Appendix I: lists all species that are threatened with extinction and may be affected by trade; normally no trade is permitted.

Appendix II: includes species which may become endangered if trade were not controlled: trade is allowed if the producer country permits export.

Appendix III: comprises species for which trade is regulated within a CITES country and for which the cooperation of other Parties is sought.

CITES Parties meet every two years to update or amend the listing of species on the appendices. Apart from its regulatory function CITES has proven to be a catalyst for important background studies into the status of certain species, the rational utilisation of some species and has induced Parties to implement stricter domestic legislation regarding wildlife trade. The permit system also provides very valuable information on the international trade in wildlife which is processed and analysed by WCMC in Cambridge, mentioned above.

Since the inauguration of CITES more than 100 countries have acceeded to it and it now constitutes the most important safeguard against the possible illeffects of the commercial consumptive utilisation of wildlife. However, it does not entirely resolve the conflicts between the legitimate national interest of utilising wildlife for optimum returns and overall international consideration of endangered-species conservation. This is due to the fact that game species are often only endangered in parts of their total geographical range of distribution, whereas they might be still abundant in others. Such discrepancies are usually accompanied by striking differences in the ability of the respective countries to administer and manage wildlife utilisation in general. Such differences have evolved historically and signify that a species may well be economically utilised for certain products in some countries, but must be totally protected and not subjected to any commercial utilisation in others where the above-described positive conditions for utilisation do not exist.

An example in this respect is the *African elephant* as the main source of ivory. Throughout the 19th century this species was decimated in the course of the colonisation of Africa and a continent-wide survey in 1976-79 showed that there were only about 1.2 mill. elephants left[13)]. At the same time the ivory trade was growing rapidly from 564 t exported in 1970 to more than 1,000 t in 1980; concurrently ivory prices rocketed to unprecedented heights[14)]. Despite partial bans and attempts in 1985 to regulate the trade through a CITES quota system the price of ivory rose further and numbers of elephants declined to an estimated 600,000 in 1989[15)]. Therefore, the majority of CITES member countries decided to place the African elephant, with the Asian species, on Appendix I, thus prohibiting any trade in elephant products. Although this was a most welcome decision for countries, for example in West Africa, in which elephants are on the verge of extirpation, it was violently rejected by some Southern African countries, in which the species is still abundant and wellmanaged. These countries argue rightly that it costs them a lot of money to maintain large herds of elephants outside protected areas, that important investments were made into the technological development of product processing, and that the use of elephants for recreational hunting, viewing and producing ivory constitutes a most important economic resource for the local population; unless

---

13) CUMMING, D.H.M. and P. JACKSON (eds.) (1981): The status and conservation of Africa's elephants and rhinos. Proceedings of the Joint Meeting of IUCN/SSC African Elephant and African Rhino Specialist Groups at Hwange Safari Lodge, Zimbabwe, 30 July - 7 August 1981. IUCN Gland, Switzerland

14) RICCIUTI, E.R. (1980): The Ivory Wars. Animal Kingdom, 23, p. 4-52

15) BARBIER, E.B., BURGESS, J.C., SWANSON, T.M. and D.W. PEARCE (1990) (see selected literature references)

they are economically utilised they would be eliminated. But even within well-protected conservation areas, like the Kruger National Park in South Africa, numbers of elephants need to be controlled in order to prevent habitat destruction, and their products are paying in part for the management of such areas.

CITES attempts to cater for such discrepancies by allowing member countries to state a reservation against certain appendixed species, i.e. to become exempted from not trading them internationally. The Secretariat of CITES may also issue export quotas for appendixed species upon a well-founded and documented request by member countries wishing to utilise them economically. However, this relates only to species which are being "ranched" or "farmed" and permits other CITES member countries to legally import products from such "producer" countries. This concept has been successfully applied, for example, to the development of crocodile farming but does not resolve the elephant problem.

Whilst the international concern to conserve endangered species through curbing trade is highly commendable, there must not be any emotionalisation and victimisation as in the case of elephants. CITES must support the interests of producer countries as much as the conservation concerns of consumers in the spirit of its stated long-term objective: the rational and sustainable utilisation of the earth's living resource for human benefit.

Richard Leakey, Director of Kenya Wildlife Service, supervising the registration of 12,000 kg of ivory tusks prior to burning them in 1989 in order to support the international ivory trade ban. (Photo: E. MARTIN)

## 1.6 ECONOMICS OF WILDLIFE UTILISATION

### Utilisation value in the context of total economic value

Economics is defined as the science of allocating scarce resources among competing ends in order to maximise economic growth or social well-being. It must be distinguished from the world of financial analysis which records and measures transactions involving money. Most human decisions imply a valuation of options and an expression of preference and are therefore *economic decisions*. Money has evolved a complex role as man's primary store of wealth and is commonly used as a yardstick when resource allocation-decisions are made. The challenge to natural resource economists is to ensure that in making economic decisions all values are taken into account, including those which cannot be readily quantified or measured in money. Sound economic decisions must involve the true values to society of all possible options in terms of costs and benefits including *opportunity costs*. These are concerned with the real sacrifice involved in using resources for one option versus another. They are usually measured by considering the value which could be derived from the next-best alternative opportunity which must be foregone if the resource use in question is chosen.

Economic decisions in respect of wildlife imply determination of the economic value of wildlife as a renewable resource. This is best achieved within the concept of total economic value, which includes *direct use values* (returns from meat, hides, tourism, hunting, etc.) and *indirect use values* (benefits from retaining indigenous forests manifested through ecological system stability or watershed protection, erosion control, political stability, species diversity, water purity, etc.). It also includes *non-use values* such as option values which refer to retaining unspecified and undiscovered options for the future, as well as existence value (the pleasure derived from knowing that the resource (species) exists even if it is never to be used). *Intrinsic value*, the value the species derives from its own existence, is not strictly an economic value but becomes one if it is expressed in human preference. It is important in resource economics that the opportunity costs of resource use include the full costs and benefits to society of all these values.

Direct utilisation values are easy to measure, if manifested in production of marketed goods and services. However, direct uses which fall outside the market are often ignored in conventional financial analysis and development planning, sometimes resulting in projects which actually impoverish local communities.

This is particularly true in remote areas where, for example, *wild plants, insects, small* and *large mammals* are an important part of economic well-being, but are not necessarily part of the monetary economy. Development projects and settlements which replace or reduce wildlife habitats, may actually produce less economic value than the wilderness they are replacing. Any such development project would be economically inefficient, actually reducing economic growth and social well-being.

Indirect use values may be difficult to measure. Many development projects have failed or had negative environmental impacts because the indirect benefits and costs of the various options were overlooked. Where possible the externalities (indirect costs and benefits) should be valued and incorporated into an analysis of options, but it is essential that they be considered in as much detail as feasible.

Non-use values (for example, the value of knowing that *Polar bears* exist somewhere in the world; or the knowledge that the option to view orangutans in the wild is being preserved) are even more difficult to capture and measure, but simply listing them will help ensure that they are given some recognition in decision-making.

The *maximum sustainable yield* of wildlife occurs at population densities much lower than those of the ecological carrying capacity and is limited by environmental factors. The ecological carrying capacity may, itself, be highly unstable and fluctuate dramatically as a result of changes in weather and climate. This is especially the case in semi-arid and arid ecological systems. Renewable wildlife stocks will tend to grow towards, and stabilise at, their environmental limits, but this growth may show an exponential pattern before environmental feedback mechanisms are triggered. In such cases there may be a tendency for the population to overshoot capacity and, through successive mortality and recovery, oscillate around ecological carrying capacity. This can be accentuated if overshooting the ecological carrying capacity causes the capacity itself to be reduced through habitat damage as discussed in Chapt. 1.3. Species may be lost in this process, causing concern, but some ecologists consider these losses and changes natural. A real danger, however, is that before the habitat and species can recover naturally, pressures on the land from growing human populations and competition from expanding agriculture may prohibit their recovery.

These conflicts are particularly evident in Africa where human encroachment and elephant-induced habitat degradation is reducing the ecological carrying capacity of many animal species.

## Application of economic theory in natural resources use

Where resources are not scarce, choices for their use are not required. It is only with scarcity or the threat of potential future scarcity that the need for **sound economic choices** arises in order **to maximise social well-being.** Natural resource economics is particularly concerned with the allocation of resources over time - how much we should consume today versus how much we should conserve for tomorrow. This problem is most obvious with exhaustible resources, but also applies to renewable resources in those situations where current use can diminish future availability.

Neo-classical economics clearly shows that where the full social costs and benefits of resource use are reflected in the prices faced by individual decision-makers, *sustainable economic growth* is maximised. On the other hand central allocation systems and regulations generally result in decisions which do not maximise social well-being. One reason for this is that they apply too broadly and not to individual firms or circumstances, and thus limit the opportunity for resource users to continually adjust to changing conditions.

*Marginal analysis* is central to neo-classical economics and refers to the ratio of a marginal (very small) change in one thing to a marginal change in another. Economic growth is maximised where marginal revenue is equal to marginal cost (MR = MC); i.e. where the extra cost of using/harvesting/producing at the socially optimal level, the last unit is exactly equal to the extra benefit obtained from that last unit. When this applies individuals or companies will choose to use/harvest/produce, provided there are no price distortions. Both society and individuals consider costs and revenues with a view to maximising net benefits, i.e. to achieving the *economic optimum*. The economically optimal renewable resource use will almost always be below the maximum sustainable yield (ecological carrying capacity). This is because the maximum sustainable yield does not include the cost of using the resource (the marginal costs increase with the amount used), nor any increasing future value of that resource if it becomes more scarce. The marginal revenue (the extra benefit obtained from using each extra unit) decreases as the quantity of resources used increases, whereas the marginal costs increase as resource use increases. This is why the economically optimum yield point tends to be below the maximum sustainable yield unless all costs are zero.

If resources are being used unsustainably, the most likely reason is that the full social costs and benefits are not being paid or received by the individual user. It is important for resource managers and policy-makers to realise that in order to maximise and sustain economic growth, their best strategy is to ensure that individual users pay the full social costs and receive the full social benefits of their production/utilisation processes.

To an economist, renewable resources are regarded as *assets*. They are also inputs which, when combined with entrepreneurship, materials and labour, produce goods and services. They play the role of both capital stock and inputs for the production process. A given wildlife resource stock (and/or its habitat) is representative of the capital base which yields productive inputs to the economic system over time. It can be run down and eliminated by excessive exploitation or built up to produce more through protection and accumulation.

Natural resource economics is essentially about **optimal resource use over time.** Where society must choose between *consumption now or in the future*, this is similar to choosing between producing and consuming different goods. The interest rate and the impact of decreasing supply of the resource (good) is what affects consumption across time periods. Individuals have preferences in the timing of consumption. Capital markets increase social well-being because they allow the interchange between borrowers and savers - they permit intertemporal transfers of consumption. For society to maximise economic growth the value of the capital formed by saving must be equal to the present value of consumption sacrificed (i.e. the marginal efficiency of investment is equal to the rate of time preference which is equal to the real interest rate). If there is a market for the resources (i.e. the assets), then the *future productivity* of the asset has value and people will have incentive to maintain its viability. Depletion of the resource (and extinction of species) occurs easily where this is not commercialised and/or where access or ownership rights are not secure (see below).

This is highlighted in the case of the *African Black rhinoceros* which, despite decades of prohibition of trade in its products, is near extinction. It has declined in number from more than 13,500 to some 2,400 in less than 25 years, as discussed in greater detail in Chapt. 2.3.9. It is conceivable that there could have been many more rhino alive today if legal markets had been established which enabled individuals and countries to accrue economic benefits which compensated for the enormous costs of protecting rhino and their habitats. The apparent success achieved in the conservation of *crocodiles* in the last decade may be partly attributable to their commercial utilisation being made possible (by delisting them in CITES from Appendix I to Appendix II).

The optimum condition for the maximisation of the economic growth of natural resource use occurs where price is equal to the marginal cost of harvesting/extracting/producing from the resource *plus* the opportunity cost of depleting the stocks (of the exhaustible resources) or reducing the flow (of the renewable resources). This opportunity cost is often referred to as the *royalty or scarcity rent*. As the resource becomes more scarce, the supply decreases and the price increases. At the same time the cost of obtaining it often increases - reduced stocks of animals are more expensive to hunt, capture or view. There is therefore an incentive for individuals to hold their resource for that future time when it may become more valuable due to a rise of price as a result of its increasing scarcity. This is demonstrated in Fig. 3; however, readers are referred to the listed economics texts for theoretical analysis and proof.

**Figure 3:** The effect of incorporating opportunity costs in resource-allocation decisions

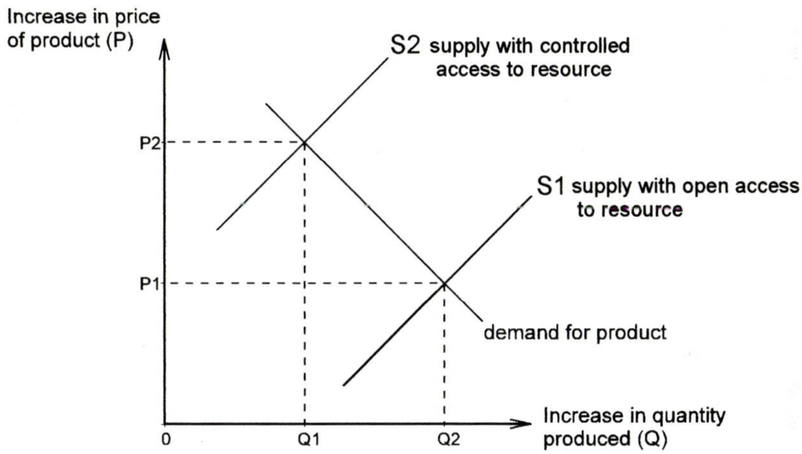


S1 = Supply where the resource user is only affected by marginal cost of use (i.e. harvesting and marketing costs)
S2 = Supply where the resource user is also affected by the future; decisions are now made on supply which include the opportunity cost of using the resource to-day instead of saving it for future use.

Another set of considerations which may significantly affect the economic value of wildlife utilisation, relates to the **price elasticities of supply and demand** or the responsiveness of supply and demand of wildlife products to changes in production costs and product prices. The nature of these price-quantity relationships among wildlife users and consumers of wildlife products can be most important. They influence the level of producer and consumer surpluses in wildlife utilisation systems. The *producer surplus* reflects those quantities which would be supplied even if the price would be lower. The *consumer surplus* reflects the difference between the costs of supply of the product and the actual price paid for it.

Any resource allocation decisions made which do not include the full social costs of using the resource, or which do not provide decision-makers with the full value of the benefits created, or which heavily discount the future will lead towards over-exploitation of renewable and non-renewable resources. This is referred to as *market failure*. This occurs where markets are imperfect because of natural monopolies or because of government policies which (usually through regulations and licenses) restrict competitive forces or use institutionally established prices which do not reflect the true social value of the resources. Many resources are under threat because of the very regulations enacted for their protection.

Resource misallocation also occurs in a perfectly competitive environment because of insecure property rights, externalities, the non-exclusive nature of many natural resources and non-existent markets for some of the values. Market failures can be addressed by policies to correct them - provided that the benefits from correction outweigh the costs of correction.

A major problem with wildlife utilisation as a land-use is **insecure property rights.** In many parts of the world it is legally and socially impossible for any individual or social group to claim ownership of wild animals. In such case individuals have no reason to make decisions which take into consideration future value and costs. It is referred to as an *open access regime* and is a major factor in the over-exploitation of resources. Open access refers to a situation where anyone may use the resource at any time with no controls. Where wild animals are state property and it is illegal for anyone to use them, they tend to be a *de facto* open access resource, particularly if it is difficult to enforce the regulations restricting use. Private property rights generally give the most control over resources, but *common property regimes* may also enable the future of natural resources to be considered by users. However, these must be based on clearly defined and widely accepted norms for the use of the resources and for the distribution of the accruing benefits and costs. They must define and effectively limit access to the resource, assuming also future access. Unless the benefits received from wildlife are greater than the costs of living with the wildlife, users will tend to over-exploit the resource and may even encourage outsiders to poach, in order to remove wild animals which cause damage.

Common property regimes are most appropriate in situations in which individual ownership of wildlife is difficult or impossible to enforce due to the movement of the game animals over large tracts of land. In such cases the owners of several smaller landholdings could decide to join in a common use of the wildlife resources; or State land could be leased out to a co-operative of individual hunters. Without owning the land, communities often have traditional rights to control natural resource use on large tracts of land for mutual benefit and this may also assume the form of a common property regime.

In a *controlled access regime*, based on either private property rights or other effective control mechanisms, the user will consider both the marginal cost of harvesting and the opportunity cost of using it today. Fig. 3 shows that this may tend to result in a lower quantity produced (S2). In an open access situation (S1) more may be produced and consumed because the producers/users have no stake in the future and, therefore, face only the costs of harvesting. Consequently there is no opportunity cost or royalty to the exploiter, although the cost may be very high for society.

For example, someone hunting *Mountain goats* in an open access regime, would only be concerned with the costs involved in hunting and would not take into consideration the future value of the goats nor the costs to society of losing this resource.

Where it is not possible for access to be controlled and future rights (tenure) to be secured, it may be necessary to establish a tax or royalty which represents the opportunity cost. The revenue from this tax should go to maintaining the resource and/or to that portion of society paying the cost. Where feasible the price of access should be bid up via, for example, auctions, to which entry should be as open as possible.

**Externalities** are those costs and benefits of resource use which do not affect the producer or consumer. Thus the decision of the user does not take account of these effects. These externalities can be either positive or negative. If individual decisions are to result in optimising sustainable development it is essential that both the positive and negative externalities are included in the prices. This is referred to as *internalising the externalities.*

A hypothetical example of this could be where a farmer has the option to use his land to produce an annual crop, such as cotton, which involves clearing a forest. Alternatively he could use the land to farm bees, to capture monkeys (for zoos) or for tourism. All these uses involve maintaining the forest. The forest has an essential role in watershed protection for a major irrigation scheme involving sugar, wheat and citrus in the valley below. For the nation, sustainable economic growth is maximised if he does not clear the forest, but the returns to the farmer are greater from cotton. If, however, those who benefit from retaining the forest (society and the farmers in the irrigation scheme) were to pay the farmer the value of retaining the forest, then he too would maximise his welfare by selecting the wildlife option. The principle is the same with respect to negative externalities and, in this case, is called the *polluter pays principle.*

In community wildlife schemes the individual participants, commonly farmers, are faced with costs created by the wildlife in terms of both crops and livestock as well as opportunity costs. These opportunity costs reflect other options for resource use which may give the farmers greater control and less risk, even if they render lower returns. If these farmers are to consider investment in the wildlife option seriously, then the benefits must outweigh the costs and the schemes developed must be sure to "internalise the externalities". This would be much more effective and fair than legislating against poaching or trying to control production practices.

**Non-exclusive, non-existent and uncompetitive markets** may inhibit economic efficiency in wildlife use. There are some goods which are termed *public goods* because they are non-exclusive. This means that it is not possible to keep people from using these goods (nor to get them to pay for them) or that one person using the resource will not affect the enjoyment of another using it (for example, air, sunsets). However, these do not have much relevance to the direct values of wildlife.

The *natural monopolies* enjoyed by some natural resources (Iguazu falls, Yellowstone hotsprings, the last Mountain gorilla population, etc.) can result in higher prices and lower use than would exist with competition, but this, again, is not a common phenomenon in wildlife economics. *Artificially created monopolies* and *barriers to entry* can, however, have a major impact on wildlife economics.

For example, a farmer may have opted to use his land for wildlife, but is only allowed to sell to the state, to a heavily controlled and regulated producer body, or to a restricted number of licensed dealers. This means that the price to him will be lower than it would otherwise have been. It may mean that he prefers to eliminate the wildlife or its habitat and produce something else instead.

Experience in the early stages of community wildlife schemes in Zimbabwe, described in Chapt. 1.4, has shown that the more open and transparent the market, the more successful the scheme. This even applies, for example, where there is a desire to restrict safari operators to those known to be senstive to communities or the environment. Generally, competitive marketing systems maximise returns and thus encourage the wildlife option. Marketing institutions should not be used to influence the market with a view to social or environmental objectives. Such other objectives can be achieved by establishing standards which all participants must adhere to, remembering that the more difficult these are, the more likely it is that they will reduce entry and prices. The other advantage of establishing more competitive systems is to reduce the potential for corruption which controlled systems foster.

## Development of wildlife economics

**Historically,** the law has considered wildlife to be an open access resource, i.e. one which is owned by noone or by the public at large. It tended to be regarded as a free good which derived from nature and could be used without any restraint. When wildlife became scarce as a result of over-exploitation or habitat eradication, it was generally the role of Royalty or the State to act as custodian of the wildlife resources. Users of these resources, either a privileged class or the public, could then gain rights of use from the custodians.

The value of wildlife to the rural populations of many countries lies still in its *use for subsistence. Recreational hunting* was historically restricted to the nobility and later to a privileged minority. Various forms of *commercial utilisation* of wildlife have now become important, as described in Chapt. 1.4, and the value of wildlife for viewing has grown significantly with the *development of world tourism.* Indirect use values and non-use values of wildlife such as option and existence values have only recently been recognised explicitly, although the creation of national parks systems throughout the world indicates a much earlier implicit recognition.

Most developments in wildlife economics have taken place in North America and these have tended to fall into two categories. The *commercial utilisation of marine fish resources* was subjected to *pioneering economic studies in the 1950s* (see listed literature). This early theoretical and empirical work on fisheries forms the basis for much of the economic theory of consumptive wildlife cropping or harvesting described above.

More recently, the predominantly *recreational use of terrestrial wildlife* has been subjected to *economic empirical analysis* where the emphasis has been on measuring the economic value of outdoor recreational pursuits such as sport hunting. Important techniques, such as contingent valuation and travel cost analysis, were developed. These were used to measure wildlife users' expenditures and consumer surpluses, and now have growing importance as techniques to measure non-use values.

Of particular interest in the context of this book is the **development of wildlife economics in Africa** which is related to historical events. *Pre-colonial wildlife utilisation* was mostly for subsistence and limited barter; access was generally open or under limited traditional control. In the *colonial period* the emphasis was on the protection from use by central national authorities. Non-utilisation values were implicitly recognised through the creation of national parks and reserves. Attempts to maximise wildlife utilisation values were generally not approved. The result was that investment in wildlife conservation was limited to that justified by non-use values, as implicitly perceived by colonial governments. No effort was made to identify or measure these.

*During the early 1960s*, the concept of using game commercially on private ranchland emerged in Rhodesia (now Zimbabwe), as described in Chapt. 1.4. However, considerably more effort was put into the study of the biological aspects of wildlife production than into the financial and economic aspects and social institutions. Therefore wildlife economics in Africa *before 1980* was virtually non-existent. Attempts to investigate the real economic value of wildlife were very rare. Financial analysis was a little more common and generally involved the development of simple financial enterprise budgets. Capital investment costs were mostly ignored and discounted measures were rarely used. There were almost no attempts to convert financial values to economic ones. *After 1980* there was increasing realisation that Africa's spectacular wildlife resource would be totally eliminated unless it could be of value to burgeoning, poverty stricken, human populations. This was accompanied by the realisation that wildlife needed economic justification to enable it to compete against land pressure. During the 1980s and early 1990s, much more work than previously, was initiated on the financial and even the economic aspects of wildlife use especially in Southern Africa. This has provided an improved, if still very incomplete, understanding of the economic characteristics of wildlife use in African environments.

*Subsistence hunting* was generally discouraged during colonial times, since wildlife resources were considered the property of the State. This illegitimacy resulted in an open access system in which all the economic costs of subsistence use (except small costs for time, materials and fines) were borne by society. Later policy has increasingly turned toward controlled *commercial use of wildlife* to provide justification for the allocation of land to wildlife resources. In East Africa emphasis has been on non-consumptive tourism, particularly in Kenya. In Southern Africa, both consumptive, and non-consumptive uses have been promoted. Wildlife use is undertaken on State, Private and Communal lands. The State or private landholder restricts access with the aim of maximising use efficiency and/or conserving the resource.

On *State land* the State or society bears some of the economic costs of use unless adequate resource rents are introduced. This applies particularly to land in and around protected areas where non-use values are high and the government aims to make them available cheaply. Wildlife use on State land results in a variety of products, ranging from wildlife-viewing tourism in and around parks to recreational hunting tourism and wildlife cropping or culling from free-ranging wildlife populations. Access is restricted. Viability varies with the type of commercial use. Large scale wildlife cropping programmes on State land to produce meat and skins have commonly failed due to poor financial profitability and technical problems. In the past, in Southern Africa, large-scale elephant culling for ivory, hides and meat was successful and found to be both financially and economically profitable. This is no longer the case since markets have collapsed. Wildlife-based tourism enterprises on State land have tended to be more financially and economically profitable. In places the economic costs of development of protected areas can be justified by the direct use values of tourism enterprises.

On *Private land,* mainly in Southern Africa, most economic costs of producing wildlife are internalised with the landowner. Production systems include wildlife ranching and farming, as described in Chapt. 1.4. Emphasis, in these systems, is on restricting access and maximising use values.

Products from Private land include meat, skins, and other products, safari hunting and wildlife viewing. Safari hunting is commonly an important contributor to financial profitability. More recently there has been a tendency, to convert financially unprofitable cattle ranching into wildlife ranching enterprises. Nevertheless, except in a few cases, mixed-species game ranching has been found to have relatively low financial profitability. This is especially the case in drier habitats with low carrying capacity. Combining wildlife, especially safari hunting and some viewing tourism with livestock production on ranches can enhance financial viability. In Zimbabwe, JANSEN and coll.[1)] have found that, while the returns to land and capital are low for both *cattle* and *wildlife*, the returns are higher for wildlife and for combined ranches than for cattle-only ranches unless the cattle farmer is able to market directly through his own retail outlets. Specialised, intensive production systems, such as those for *ostrich* and *crocodile* can be more profitable but are very capital and skills intensive and entail high risk.

*On Communal land* traditional, open access, subsistence hunting systems are increasingly commercialised, apparently in response to rises in product demand.

In West Africa subsistence use of forest and high-rainfall savanna wildlife, is tending to be commercialised in the face of high urban demand for scarce protein. In the Luangwa valley of Zambia it has been shown that there was considerable financial incentive for subsistence hunters to commercialise *elephant* and *rhino* hunting for ivory and horn by forming poaching gangs and hunting with automatic weapons. Where such conditions persist, wildlife use tends to lead to severe depletion or extinction of target wildlife species.

More recently initiated programmes to develop community management of wildlife resources are aimed at eliminating the open access problem and developing common property regimes. Rural communities are thus enabled to capture the economic benefits of commercial resource use. These include, for example, the CAMPFIRE in Zimbabwe, the LIRDP and ADMADE in Zambia and the CBNRM in Namibia, as described in Chapt. 1.4. Economic evaluations of these programmes suggest that they have been moderately successful so far, but are severely constrained by the social institutions which make it difficult to clearly link costs and benefits. Financial and economic appraisal of similar initiatives in Botswana suggests that these have the potential to be economically sound.

---

1) JANSEN, D., BOND, I. and B. CHILD (1992): Cattle, wildlife, both or neither: results of a financial and economic survey of commercial ranches in southern Zimbabwe. Project Paper No. 27, World Wide Fund for Nature, Multispecies Animal Production Systems Project; Harare, Zimbabwe

The comparative economic and financial value of different types of wildlife utilisation enterprises has received fairly detailed study in Botswana[2]. The results of some of the economic analyses of typical wildlife enterprises are shown in Tab. 2.

Financial enterprise discounted cash-flow models were developed to determine the attractiveness of wildlife use enterprises to investors. The financial models were then converted to economic ones through shadow pricing to give a measure of the economic worth of these activities for society. The economic models produced actually measure the contribution of each activity to National Income (a measure similar to Gross Domestic Product or GDP) before consideration of government expenditure and the opportunity cost of land. They also stop short of consideration of non-use values.

Generally, those enterprises which use free-ranging wildlife populations are more profitable than more capital-intensive ranching and farming of wildlife. This has a rather complex interpretation. It shows, on the one hand, that the enterprises using free-ranging wildlife on State lands are not paying the full, real rental value for the use of the resource. It also shows that ranches and farms suffer from certain diseconomies of scale - they are not able to easily stock all the higher-value wildlife species and communities. Differences between economic and financial values reflect price distortions, mostly unfavourable to the wildlife users.

**Table 2:** Comparative estimated returns to selected natural resource utilisation enterprises in Botswana, 1990

| Enterprise type | Internal rate of return (%) | | Comment |
|---|---|---|---|
| | Economic | Financial | |
| Community project, prime wildlife area | 55 | 21 | Biltong, skins, lease rentals |
| Safari hunting | 45 | 16 | Trophy off-take, 12 bed lodge |
| Tourist lodge | 35 | 18 | Exclusive 30 bed lodge |
| Community project, poor wildlife area | 28 | 21 | Biltong, skins, lease rentals |
| Ostrich farming | 14 | 19 | Skins, meat, feathers |
| Crocodile farming | 11 | 18 | Skins, meat |
| Crocodile ranching | 11 | 18 | Skins, meat |
| Game ranching | 7 | 6 | Biltong, hunting fees (9,000 ha) |

## Wildlife economics in the future

The African examples described above serve as an illustration of more **general economic principles and features of wildlife utilisation.** No resource management and conservation

---

2) ANONYMOUS (1988): The contribution of wildlife to the economy of Botswana. FGU-Kronberg/ Ministry of Commerce and Industry; Gaborone, Botswana

programme can be undertaken efficiently without incorporating economic principles. Indeed certain policies in wildlife conservation and utilisation can only succeed if decisions regarding them include an assessment of the economic costs and benefits to society and how these impact upon conservation.

Just one example of this might be the issue, referred to in Chapt. 1.5, of whether to allow legal production and trade of some wildlife products or not. In some cases promotion of legal, commercial wildlife use will displace illegal, open access poaching of a species. In other cases it will, instead, increase poaching activities for the same species, and hasten its extirpation.

Only attention to fundamental economic principles will tell us what to expect and what policy to adopt in respect of wildlife resource use. Some of the key principles we need to look at are the *differences in price elasticities of supply* (or the proportional change in the quantity supplied resulting from a proportional change in price) between commercial and poaching uses, how effective institutions are in *ensuring* current users will be able to reap *future benefits* and the *rates of growth in overall supply and demand* for the product involved. Without information and analysis of these, any decision on the effects of commercial use will be a "hit and miss" process. These basic economic considerations are dependent on a complex web of biological, social and economic factors. Research into these issues and case studies are essential if sustainable wildlife systems are to thrive. The scarcity of information on which to base sound wildlife conservation and use policies creates conflict and confusion. This is demonstrated by the controversies on trade in elephant and ivory products, rhino and rhino horn.

Non-use values (as opposed to use values) for wildlife resources are more commonly reflected in the social preferences of affluent societies in industrialised countries than in the developing world. However, it is in the developing world where a lot of wildlife diversity, the subject of these values, exists. Preliminary data suggest that in Europe and the USA, *values for maintenance of biodiversity* are high. Data from Kenya and Botswana suggest that African countries could capture significant non-use values for their biodiversity by soliciting donations from, or taxing, tourists from industrialised countries. At present there are few incentives for developing countries to invest in sustainable wildlife production systems because they do not realise the full benefits.

All this evidence points to the possibility that in developing countries where biodiversity values exist which also benefit the international community, there is significant potential for the creation of "markets" which could generate benefits to these countries in the future. So, while development of consumptive wildlife utilisation in developing countries is important for the creation of immediate incentives for investment in these resources, it should not be at the expense of biodiversity, which could also yield high, tangible benefits in the future.

Despite recent advances, wildlife economics is very much in its infancy. It is emerging as a vital component of wildlife management and conservation in the same way that wildlife ecology emerged as important earlier in the century. In an increasingly crowded world, wildlife is rapidly becoming a scarce resource, and allocation of scarce resources is the subject of economics.

## 1.7 WILDLIFE-BASED TOURISM

### Historical development

The economic utilisation of wildlife through tourism assumes an ever-increasing importance in many countries of the world. Although tourism is certainly not always, or necessarily, associated with wildlife, this has become a major factor in the development of tourism, which according to the World Travel and Tourism Council is likely to grow into the world's largest industry by the year 2000. Any wildlife, from butterflies to elephants, in any region of the world, from the Antarctic to the tropics, has a potential for economic utilisation through tourism. Generally it is rather the diversity of wildlife which fascinates man and which may thus support commercially oriented wildlife tourism. However, in some cases specific single-species wildlife has become the basis for tourism.

For example, viewing *gorillas* in the Ruwenzori Mountains, observing vast herds of migrating *wildebeest* in the Serengeti steppe, watching big herds of *elephants* at close distance at water pans in the Hwange or Kruger National Parks in Southern Africa, enjoying the sight of thousands of nesting *flamingos* in Lake Naivasha or vast swarms of *Sea birds* on the Banc d'Arguin of Mauretania, associating with thousands of *Fur seals* breeding at coastal sites in Namibia or with thousands of unsuspecting *penguins* on the Antarctic ice, are highlights of tourist travels.

The increase of wildlife-based tourism was fostered by **several concurrent historical developments**. On the one hand, the number and diversity of areas set aside for the protection of wildlife have significantly increased around the world, particularly in those countries which gained their independence after the Second World War. Although the spectacular wildlife resources in these areas were mostly not managed for any particular purpose, the general infrastructural development in the particular countries also made the wildlife resources increasingly accessible to the foreign traveller. At the same time governments in these countries became aware of the economic potential of tourism in general and - often quite unrealistically - perceived the wildlife resource as a means to promote it, thus establishing more and more "national parks" in the hope of eventual economic returns from them through foreign-exchange earning wildlife-based tourism. Such economic postulation was often politically necessary as an argument against the interests of the local population who became legally deprived of a natural resource they had previously used for subsistence.

On the other hand the perception in "western" societies that nature, and more particularly wildlife, had to be "preserved", i.e. rigorously protected against human interference including any use other than scientific research, changed gradually into the concept of nature conservation, as described in Chapt. 1.4. After the Second World War and in the wake of the economic development of the "Third World" it was realised and accepted that natural resources had to be conserved for sustainable utilisation by man. In respect of protected areas the only option for use seemed to be their development and management for tourism from which the surrounding human society was also to benefit. The above-mentioned economic interest of politicians (in newly independent countries) and the changing perception of wildlife conservationists coincided and generated growing numbers of "national parks", to be economically utilised through international tourism.

At the same time, of course, the demand for wildlife-based tourism did increase in all industrialised countries due to the estrangement of the industrialised societies from nature. Man has an innate affinity with nature, wilderness and the fierceness as well as the beauty of wild beasts. With the growing affluence, increasing spare time and the technical means of overseas travel the encounter with and observation of large herds of wild animals became one of the most desirable travel goals of tourists, in addition to the attraction of strange exotic cultures and the colourful tropical environment. Wildlife-based tourism may therefore be understood in part as a sentimental or emotional compensation for the alienation of industrialised societies from nature, enhanced by the economic needs of developing countries.

According to individual inclination, technical possibilities and financial means **wildlife tourism is pursued in different ways**:

- passively by viewing and photographing game animals from motor cars and leisurely enjoying a wild environment;
- or actively by backpacking, horse or camel trailing, canoeing or otherwise engaging in safari tours in search of specific types of wildlife species, or by hunting.

Initially it was mainly the wildlife in protected or especially managed areas which was utilised for the development of the various types of wildlife-oriented tourism, mainly national parks for game viewing and photographing and controlled hunting areas or hunting concessions for hunting tourism. However, as the considerable economic potential of wildlife tourism became apparent, more and more private landowners, and in some countries also communal authorities, started catering for both game viewing and trophy hunting in an

Canoe safaris have become particularly popular in Zimbabwe allowing to view game animals grazing on the banks of the Zambesi river. (Photo: G. THOMSON)

effort to draw direct economic benefit from the wildlife resources present. In fact, the economic incentive for many landowners who were previously only engaged in agriculture and animal husbandry, was so great that in Southern Africa alone hundreds of these started reintroducing wildlife species which had hitherto been eliminated, onto their lands. Considerable investments were and still are being made in many countries not only in the development of national parks for tourist use, but also on private land with a view to building up wildlife resources for future economic utilisation through tourism.

The most important *factors for successful development of wildlife-based tourism* have proven to be:

- a diversity of conspicuous and observable game species;
- local concentrations of game and bird species in spectacular herds or swarms;
- generally a high population density of game ensuring frequent encounters and easy observation of wild animals, implying that rigorous protection of wildlife against harassment or illegal shooting has been established for sometime;
- presence of particularly spectacular, dangerous or beautiful species, or particularly desired trophy carriers.
- development and maintenance of an infrastructure which allows easy access to a variety of different wildlife habitats and the observation of a maximum of species in the shortest possible time;
- development and maintenance of optimum catering facilities and services in keeping with the visitors varying aspirations;
- effective advertising and publicity campaigning by conservationists, wildlife photographers and writers presenting wildlife in its natural environment in conjunction with the particular conservation issues.

For the purpose of this book, three types of wildlife-based tourism which have evolved during the last decades will be described in greater detail: national parks, safari and hunting tourism.

## National parks tourism

According to the classification by the World Conservation Union (IUCN) of protected areas, national parks are defined as areas which contain exceptionally beautiful or unique landscape, fauna and flora of national or international importance. Normally national parks are of considerable size and only to a minimal degree influenced by human activity; no intensive economic use is permitted. Apart from scientific research tourism is the only form of land use compatible with the primary conservation goal. National parks are always State land and accordingly administered and managed by government agencies. Construction and subsequent use of tourist facilities may, however, be leased out to private enterprise.

*Tourist use of national parks* comprises all activities which are associated with the temporary sojourn of visitors during which catering and other offered services may be utilised by them. In those national parks which contain important rare or spectacular wildlife, the tourist activities are usually limited to game viewing in various forms and the observation and photography of specific wild animals. Apart from the recreational aspect, education in natural history and ecology is a major secondary goal of national-park management. However, tourist use of national parks must always remain secondary to the primary management goal which is to conserve the natural features and to maintain the biodiversity of the particular area. As a general rule, the total economic value of national parks is dominated by non-use values, as described in Chapt. 1.6, and use values (from tourism) are less important. Therefore, infrastructural developments and the creation of tourist facilities must remain minimal and be balanced with the conservation requirements. This is often best achieved by the zoning of national parks into areas of differing intensity of tourist use.

Recently it was realised that national parks development and tourism have also to be adjusted to and harmonised with the interests and traditions of the adjacent human population.

Observation of game animals from viewing platforms at water pans in Hwange National Park in Zimbabwe. (Photo: H.H. ROTH)

Therefore the suitability of an area for the development of wildlife-based tourism in the form of a national park has to be carefully examined in this respect, in addition to the necessary biological inventory surveys and the assessment of the ecological impact of tourism.

There has been a tendency in the past, particularly in developing countries, to wishful thinking with regard to the development of national parks tourism. Many areas qualifying from a point of view of nature conservation have been designated for development as national parks, and thus for tourist use, without considering the infrastructural and economic requirements of tourism against the present ecological constraints. In fact, many national parks around the world are unsuitable for the more profitable *mass tourism* which some countries, like Kenya, South Africa, Sri Lanka and Thailand, have successfully developed. Such mass tourism implies direct and economically feasible easy access to the national parks either by airplane or in comfortable buses, a large capacity of catering facilities and an extensive well-kept road network. These are rarely available, and even if they were realised the ensuing mass tourism would often be incompatible with the ecology of the area concerned, such as semi-arid, swamp or rain forest biomes. Mass tourism also depends to a large extent on climatic factors. In particularly hot or humid areas, accessibility to the parks, visibility of animals and comfort of the visitors may be severely constrained by seasons, rendering significant investment in tourist facilities uneconomical.

As an alternative to mass visits to national parks, some countries are developing a more individual type of national parks tourism called *eco-or adventure tourism*. This is practically safari tourism as described below and also allows the utilisation of remote and less-developed national parks. Specific wildlife and other nature attractions are made commercially accessible to small, preferably guided groups of tourists who enjoy physically more strenuous modes of transport and a minimum of comfort in a stark natural environment. However, this form of national parks tourism usually requires supplementary recreational or cultural attractions in order to become economically marketable.

Apart from the viability of the offered "product", any national parks tourism, but particularly the latter concept, requires efficient advertising and marketing services and the involvement of international tour operators if it is to be successful. The lack of these and governmental policies unfavourable to private enterprise are often the most serious constraints to the development of wildlife tourism in potentially suitable national parks.

## Safari and hunting tourism

Although safari tourism is not clearly defined, it usually refers to a more individual form of travel for the purpose of enjoying nature and wildlife while moving along through "wild" environments. The word "safari" originates from East Africa and therefore refers to mainly open savanna lands in which the main attraction is the diversity of easily observable large game animals. Hunting of some of these animals for meat and sport formed part of the classical East African safari. During the last decades, however, the *concept of safari tourism* has been extended to include any commercially organised individual group or wildlife tours in Africa, aimed at the discovery, study and enjoyment of wildlife. Depending on the natural features of the countries, guided backpack, horse or camel safaries, canoe tours and rafting may be offered, and also hunting. More recently the concept of "mini-safaris" has developed in Southern Africa during which tourists are offered the opportunity of brief stays on private wildlife ranches or farms. These may cater solely for game viewing but also for sport hunting.

In contrast to national parks tourism, safari tourism is not restricted to special wildlife reserves but utilises any land, be it state, communal or private, as long as this contains sufficient wildlife. With settlement and rapid development of agriculture, however, national parks have often become the only areas in which wildlife may be easily observed and are thus also utilised by safari tour operators. These operators provide the necessary vehicles, outfitting and local knowledge and are an essential element of safari tourism. Safari tourism is thus neither clearly distinguishable from national parks tourism nor from the more explicit hunting tourism. The latter implies the availability of specific wild animals for the purpose of sport hunting and for obtaining trophies and is usually marketed and catered for by specialist agencies. Hunting quotas are usually allocated by the wildlife or forestry authorities for special areas of State land set aside for recreational hunting, but also for Communal or Private lands on which huntable game populations occur.

*Hunting in the framework of tourism* is a purely recreational and sporting activity which large numbers of people, mainly from the western industrialised societies, like to pursue. With the affluence that has developed in these societies, some hunters are prepared to invest significant amounts of money in hunting large game species in distant countries. To cater for such hunters, a highly specialised branch of tourism has been generated.

Originally, big-game sport hunting was the privilege of high-ranking personalities or wealthy industrialists for whom the hunts would be laid on through government agencies or private landlords. It was only after the Second World War that commercialised international hunting tourism developed, usually based on license systems. In many of the former colonies, however, licensed hunting was terminated after they gained independence because it was either perceived as an unacceptable colonialist privilege or it was considered incompatible with conservation efforts. In fact, uncontrolled trophy hunting, in Africa often combined with commercial consumptive hunting, had in many regions of the world a detrimental effect on certain species. Continued selection of the biggest male trophy carriers can severely harm the socio-territorial structure of ungulate populations and diminish their productivity. However, normally the numbers of animals taken by sports hunters under controlled license systems are far too small to impair the particular game population. Controlled hunting tourism may thus be considered more or less "non-consumptive" like other wildlife-based tourism. As with any form of wildlife utilisation, it has to be integrated into an overall wildlife management plan which clearly identifies primary and secondary goals. *Trophy hunting* is usually only a secondary aspect of management which nevertheless can be particularly lucrative. Primary goals are the maintenance of diversity and, in certain cases, the production of meat and other products from the particular game population.

It is only in countries in which the management of wildlife resources became gradually better understood and in which competent wildlife authorities were instituted that hunting tourism has developed during the last 20 years into an important tourist activity; it is limited to countries where:

Safari tourists in the Serengeti grass savanna in Tanzania viewing lions at close range. (Photo: G. MERZ)

- certain big-game species valued for their trophies occur in sufficient numbers;
- commercial sport hunting is allowed and recognised as an important aspect of wildlife management and utilisation;
- private enterprise is supported and encouraged to market hunting tourism internationally and to cater for it.

In Africa only about a dozen countries offer hunting to overseas tourists, of which the most important ones are Tanzania, Zimbabwe, Botswana, South Africa, Namibia and also Cameroon and the Central African Republic. In other regions Canada and the USA, Sweden, Russia, Mongolia, Poland, Hungary, Romania, and Turkey have developed hunting tourism to varying degrees, offering usually only a few specific game species.

Although nature enjoyment, recreation and sportive physical engagement play the most important part in hunting tourism, trophies are nevertheless a central element of it. Trophies comprise mainly the horns, antlers, tusks or other "weapons" which are often mounted with the entire head. They are perceived by the hunter as a symbol of victory in his contest with nature or serve simply to remind of the exciting moments of encounter and overcoming of the wild beast. In aboriginal as well as western societies, a trophy may also have importance as a status symbol, as described in Chapt. 3.2.6. The type and quality of a trophy, i.e. its size, length and general beauty had thus already become an important aspect of hunting in early times. There are meticulous rules and extensive literature for the evaluation of a trophy. Hunting tourism therefore depends to a large degree on the types and quality of trophies which may be obtained. The processing, mounting and shipping of the trophies to the hunter's native country are an important secondary industry to hunting tourism.

The trophy cult has led to the introduction of many European game species into foreign environments where in many cases the species has proliferated and also become important for sports hunting.

*Red deer* in Argentina, for example, are considered to grow the most spectacular antlers and are thus attracting European hunters. In the USA many ranches have been stocked with *exotic ungulate species* which now extensively serve trophy hunting.

In Southern Africa game ranching and farming have also become important to the development of international hunting tourism. Although these were initially aimed at meat production, trophy hunting was soon found to be a most important economic complement. In order to enhance it and to increase the variety of marketable trophies, extensive reintroductions and restocking of private lands with different game species has taken place.

Recently the *African buffalo* has become most important in this respect. After this species had been widely extirpated in the control of Foot and Mouth Disease it is now dearly wanted on many private estates for supplementing the hunting bags of well-paying hunter tourists. In Zimbabwe, the need to offer a variety of big-game species to the hunter is also fostering the utilisation of Communal lands on which *elephants, lions* and *buffalo* still occur, through private concessionaires.

Major factors in hunting tourism are the attractiveness of the hunting area, the diversity of the offered hunting bag and the quality of the individualised catering services. Clients are met at the airport and usually taken either to adjacent hotels or specially furnished hunting lodges or base camps in the hunting areas. Usually the tour operator provides professional hunters and guides to assist the client. According to the species to be hunted, different types of hunts are offered by specialist companies or operators.

In Zimbabwe, for example, one distinguishes between Big Game, Plains Game and Ranch hunts, involving different hunting bags and different-sized hunting areas. The allocated hunting quotas are packaged by the operator into 10-21 day hunts. Ranch hunts may be only 5-10 days. Rates between US$ 250 and 1,000 are charged per day, and license or trophy fees have to be paid in addition.

## Economic importance of wildlife-based tourism

The general importance of wildlife-based tourism and its economic potential is well reflected by the **surface area of land set aside for wildlife conservation** in different countries or regions. Tab. 3 shows the total surface of protected areas in the various regions and relates them to the total size of the particular region. Tourism generally utilises any protected areas as long as these are accessible and developed for tourism. Wildlife tourism, however, is mainly based on national parks and managed reserves, such as Game Reserves or Controlled Hunting Areas. In economic terms it is difficult to distinguish between tourism in general and wildlife-based tourism. In all regions wildlife constitutes an important aspect of tourism, but only in those in which it is particularly abundant and diverse, such as East and Southern Africa, is it the dominant resource for tourism. In many countries wildlife-rich conservation areas, although designated for tourist use, have not yet been developed for tourism. In others, large areas are preserved for forestry. Where they contain wildlife or could be stocked with it, many forest reserves could well serve commercialised hunting tourism as the most lucrative form of wildlife use. Thus, there is still great potential for future development of wildlife-based tourism which is not immediately apparent from the figures in Tab. 3.

**Table 3:** Protected areas in the world's geographical regions

| Region | Total surface ($km^2$) | Surface of protected areas ($km^2$) SNR | NP/NM | MR/PL | Total protected area in % of total surface |
|---|---|---|---|---|---|
| Northern Europe | 1,256,943 | 49,004 | 100,642 | 62,663 | 16.89 |
| Central Europe | 1,643,143 | 878 | 6,825 | 63,650 | 4.34 |
| Eastern Europe | 23,348,355 | 381,018 | 71,741 | 307,333 | 3.26 |
| Mediterrania | 7,757,991 | 1,833 | 130,201 | 88,176 | 2.84 |
| Western and Central Africa | 12,117,471 | 7,029 | 242,168 | 337,765 | 4.84 |
| Eastern Africa | 6,207,482 | 350 | 203,216 | 149,947 | 5.69 |
| Southern Africa | 6,634,812 | 6,078 | 336,568 | 78,499 | 6.35 |
| Middle East | 5,369,631 | 26,086 | 22,203 | 182,582 | 4.30 |
| Eastern and Central Asia | 21,182,334 | 106,689 | 260,453 | 807,712 | 5.55 |
| Australasia | 7,680,000 | 55,282 | 323,530 | 618,413 | 12.98 |
| Northern America | 21,500,00 | 158,052 | 630,858 | 1,078,273 | 8.68 |
| Central America | 3,000,000 | 7,647 | 57,147 | 103,155 | 5.60 |
| Southern America | 17,800,000 | 61,091 | 613,139 | 274,195 | 5.33 |

SNR = Strict Nature Reserves, NP = National Parks, NM = National Monuments, MR = Managed Nature Reserves (excluding protected forests), PL = Protected Landscapes

Source: 1993 UN List of National Parks and Protected Areas, UNEP/WCMC/IUCN; Gland and Cambridge

Areas set aside for wildlife conservation and use usually comprise land in different categories such as national parks, game reserves, game production reserves, controlled hunting areas, wildlife management areas and so forth. This allows to some extent the differentiation of their use for either national parks or hunting tourism, bearing in mind that the latter is now increasingly also utilising wildlife outside designated wildlife conservation areas, i.e. on Private, Communal and Forest lands.

In Zimbabwe, for example, the total area of the Parks and Wildlife Estate comprises 12.7% of the country's surface, but hunting tourism is carried out on 17% including also agricultural and forest lands, totalling about 65,000 $km^2$[1].

The **flow of tourists to the designated wildlife areas** is rather difficult to evaluate as it is monitored reliably only in some countries. However, it is often very considerable.

1) CUMMING, D.H.M. (1989): Commercial and safari hunting in Zimbabwe. In: HUDSON, R.J., DREW, K.R. and L.M. Baskin; p. 147-169 (see selected literature references)

In the Amboseli National Park in Kenya for example, up to 95,000 motor cars are admitted annually[2)]. Green Island in the Great Barrier Reef Conservation Area of Australia registers up to 80,000 visitors per year[3)]. The number of visitors to the Hwange National Park with the adjacent Victoria Falls in Zimbabwe has risen from 46,000 in 1983 to some 158,000 in 1993[4)].

As with any tourism, wildlife-based tourism has, of course, significant **macro- and microeconomic effects**. These are, however, difficult to separate from the overall economic impact of tourism in general. The annual financial turnover of international tourism has been estimated by the UN Environmental Programme at no less than US$ 195 milliard. For Africa south of the Sahara, in which tourism is largely oriented to wildlife, an estimated 11% of export earnings are derived from tourism, in Latin America 13% and in South-East Asia 9%[5)]. It is particularly interesting to note that compared to an annual increase of about 8% in total mass tourism, the individualistic tourism orientated towards specific interests such as wildlife is believed to increase by 10 - 15% annually[6)]. Wildlife-based tourism, as a non-consumptive form of wildlife utilisation, definitely produces very considerable macroeconomic benefits. In contrast to most consumptive utilisation of wildlife for provision of food and raw materials, it earns much-wanted foreign exchange. It also creates local employment and fosters local trade and secondary industries in remote undeveloped areas. In fact, tourism is fast becoming one of the most important sources of revenue for many developing countries.

Kenya, for example, receives about 500,000 overseas visitors per year of which the majority are wildlife-interested tourists. From these it draws about US$ 400 mill. per annum[5)]. In the much smaller and impoverished Rwanda, average annual tourist spending in 1984-90 was estimated at US$ 6 - 8 mill. of which 1 mill. came only from visitors to the Valcanoes National Park[7)]. In Zimbabwe tourist arrivals have steadily increased since independence in 1980 to 879,501 in 1993, tourists spending a total of Z$ 668.5 mill. (~ US$ 90 mill.)[4)]. In Botswana in 1990 wildlife-based tourism earned about US$ 10 mill. in economic value-added which was about 0.4% of the GDP. This was an estimate based on a survey of tour operators and is likely to be an underestimate. Another survey, this time of expenditure by tourists themselves, suggested that this figure could be higher[5)].

Attempts have been made to evaluate the economic importance of the different forms of wildlife tourism. In countries which have successfully promoted safari and hunting tourism this has been found to be of greater economic importance than national parks tourism based solely on individual visits.

---

2) WESTERN, D. (1986): Tourist capacity in East African Parks. UNEP, Industry and Environment, 9 (1), p. 11-16

3) WWF-Australia (pers. comm.)

4) ANONYMOUS (1994): Zimbabwe Tourist Statistics. Zimbabwe Tourist Development Corporation (ZTDC); Harare

5) BARNES, J., BURGESS, J. and D. PEARCE (1992): Wildlife Tourism. In: SWANSON, T.M. and E.B. BARBIER; p. 136-151 (see selected literature references)

6) DIXON, J. and P. SHERMAN (1990) (see selected literature references)

7) VEDDER, A. and W. WEBER (1990): Rwanda: Mountain Gorilla Project, Valcanoes National Parks. In: KISS, A. (see selected literature references)

In Botswana, game viewing safari tourism, utilising mainly national parks, produced more than double the economic returns of hunting tourism. The latter, however, was found to be efficient in generating indirect economic benefits such as employment[5]. Tourists in Kenya spent approx. US$ 1,400 per person for locally organised safaris. This amounted to a total of about US$ 365 mill. of which about US$ 200 mill. were spent in Kenya[8]. In Zimbabwe, hunting tourism alone earned at least US$ 12.8 mill., i.e. at least 14% of the total of tourist receipts in 1993. This involved only 1,260 foreign hunters with their guests which is a very small fraction of the total number of national parks visitors, totalling 710.502 in 1993[9].

Not considering the overall economic importance of safari tourism the direct revenue of governments from hunting can in some countries be quite substantial. It depends to a large extent on the variety of species offered, rapid adjustment of trophy fees to inflation and market demands, but also on the efficiency of the private enterprise in marketing the hunting resource.

For example, in 1990 Tanzania received some US$ 4.5 mill. just from hunting licenses compared with a return of US$ 1.9 mill. from the national parks system[10]. Zimbabwe has designated nearly 40% of the State land set aside for wildlife use, totalling 18,972 km$^2$, for safari hunting. Tab. 4 shows how trophy fees were raised since 1981 in order to maintain and even increase returns from State land safari hunting which earned government Z$ 1.314 mill. (or US$ 0.8 mill.) in 1986[1] and approx. Z$ 9 mill. (or US$ 1.3 mill.) in 1993[9]. The gross economic value of safari hunting increased in the same period from US$ 7 to US$ 12.8 mill. About 75% of these returns were obtained from *elephant, buffalo, lion* and *leopard* hunts.

Micro-economic analyses are common in consumptive wildlife-utilisation enterprises all over the world, but rather rare in wildlife-based tourism. They have been attempted in just a few specific cases and have shown that protected areas may well be profitable.

For example, a study of the potential rates of return from an investment into a *game viewing lodge* in Ngamiland in Botswana revealed an internal rate of financial return of 17.5% compared with a real estate return of 12% in the private sector and only 6% in cattle ranching[5]. Tourists viewing *gorillas* in the Volcanoes National Park in Rwanda have in the past spent about US$ 200 per person, generating an income of about US$ 1 mill. annually for government coffers[7]. *Elephants* in Kenya's national parks were estimated to be worth about US$ 25 mill. in terms of tourism. Almost 50% of questioned tourists considered Kenya no longer worth visiting if elephants were significantly reduced in numbers. It was shown that elephants earn about 10times more foreign exchange through tourism than through the value of their ivory[11]. The Galapagos Islands National Park with its *unique fauna* produces sufficient revenue from tourism to support all of the other national parks of Ecuador[12]. The Monte Verde Forest Reserve in Costa Rica is capable of gaining US$ 2.4 - 2.9

---

8) BROWN, G. and W. HENRY (1989): The Economic Value of Elephants. Discussion Paper 89-12. London Environmental Economics Centre (LEEC); London

9) FERGUSSON, R.A. (1994): Sport Hunting in Zimabwe - 1993. Zimbabwe Association of Tour and Safari Operation (ZATSO/ZPHGA); Harare, Zimbabwe

10) POOLE, J.H. (1990): Elephant Conservation in Eastern Africa: A Regional Overview. World Conservation Union (IUCN/EARO); Nairobi, Kenya

11) WESTERN. D. (1989): The Ecological Value of Elephants: A Keystone Role in African Ecosystems. In: The Ivory Trade and the Future of the African Elephant; ITRG Report; 2nd Meeting of the CITES African Elephant Working Group; Gaborone

12) BOO, E. (1990): Ecotourism: The Potentials and Pitfalls. WWF-USA, Washington, DC

mill. per annum on the basis of only 3,000 visitors. If the rate of increase of foreign visitors during past years and the indirect values are considered, this conservation area is worth about US$ 12.5 mill.[13].

**Table 4:** Receipts from trophy fees for game animals hunted in Zimbabwe

| Species | Government trophy fees (in Z$) | | | Mean trophy fees (in US$) charged by safari operators | |
|---|---|---|---|---|---|
| | 1981 (1) | 1986 (1) | 1994 (2) | 1986 (1) | 1993 (3) |
| Elephant male | 2,000 | 3,500 | 15,000 | 5,806 | 7,609 |
| Buffalo male | 350 | 500 | 2,250 | 1,353 | 1,047 |
| Lion male | 1,000 | 1,500 | 5,250 | 3,311 | 2,444 |
| Leopard | 600 | 700 | 4,500 | 2,452 | 1,888 |
| Crocodile | 300 | 300 | 3,750 | 1,142 | 1,218 |
| Hippopotamus | 500 | 600 | 3,000 | 1,428 | 1,198 |
| Sable male | 500 | 800 | 2,700 | 1,545 | 1,524 |
| Nyala | - | 600 | 2,700 | n.a. | 1,142 |
| Giraffe | 400 | 400 | 2,250 | 1,372 | 689 |
| Eland | 350 | 400 | 2,025 | 915 | 689 |
| Waterbuck | 300 | 300 | 1,500 | 876 | 835 |
| Wildebeest | 130 | 200 | 900 | 638 | 450 |
| Kudu male | 200 | 250 | 900 | 760 | 544 |
| Reedbuck | 100 | 150 | 675 | 383 | 314 |
| Bushbuck | 100 | 150 | 675 | 395 | 326 |
| Klippspringer | 75 | 100 | 450 | 307 | n.a. |
| Impala male | 30 | 50 | 150 | 160 | 85 |
| Warthog | 25 | 50 | 225 | 136 | 112 |

1 US$ at fixed bank rate in 1981 was worth Z$ 0.72; in 1986, Z$ 1.70; and at floating rate in 1993, Z$ 6.70; in 1994, Z$ 8.10.

1) according to CUMMING (1989) (see footnote 1); 2) Zimbabwe Department of National Parks and Wildlife Management (pers. comm.); 3) FERGUSSON (1994) (see footnote 9)

## Impact of tourism on conservation

Like any other form of wildlife utilisation, commercially organised tourism generally serves to conserve the wildlife resources; but may also become highly detrimental to it. Whether it is beneficial or harmful depends on correct planning of the tourist development, purposeful management and the incorporation of educational programmes into tourism.

The **positive effects** of wildlife tourism result first of all from the fact that - contrary to most other wildlife utilisation - it does not imply any danger of depletion of the resource or extirpation of species. Furthermore, infrastructural development and institution of management structures, both necessary for successful development of wildlife areas for tourist use, are highly beneficial to their conservation. They contribute to the identification and administrative "existence" of designated conservation areas and, most important, are indis-

13) TOBIAS, D. and R. MENDELSOHN (1991): Valuing Ecotourism in a Tropical Rain-Forest Reserve. Ambio, 20 (2), p. 91-93

pensable for their protection against agricultural infiltration and illegal hunting. Although national parks tourism does not always produce significant direct returns, it often gives a political justification for defending conservation areas against ever increasing pressures for land, and it engenders research and education programmes. Outside conservation areas, on private, communal or forest lands, wildlife based-tourism in the form of safaris or sports hunting may also prove to be the most appropriate form of land use or an important economical complement to other uses. Any wildlife tourism also has a significant positive effect in educating overseas visitors in ecology and nature conservation.

Possible **indirect negative effects** of tourism can best be avoided at the planning stage. Apart from the aesthetic incompatibility of large conglomerates of tourist facilities, such developments may occupy areas, such as river banks, water pans, forest patches, which are crucial to certain wildlife. The road network has to be carefully laid out in accordance with ecological requirements and the volume of expected tourism. Planning must also provide for social amenities for local personnel and sufficient waste disposal facilities. The development of tourism without consideration and integration of the interests of the local rural population bears the danger of rendering the locals hostile to conservation efforts.

To prevent **direct negative effects** of tourism on the ecology of the conservation area, the numbers, activities and movements of the visiting tourists need to be carefully controlled. They have to be compatible with the ecology of the particular area, which implies good ecological knowledge on the part of management. It should always be remembered that tourism is only a secondary aspect of the management of conservation areas and must not compromise the primary goal of conserving biodiversity. Conflicts may easily arise where mass tourism is introduced. Too many visitors in ecologically sensitive areas may cause serious damage to the ecosystem by trampling, causing of fires, disturbance and littering. Some species are more sensitive than others so that tourism may cause their gradual decline in favour of less sensitive ones. Even well-guided motor traffic may cause too much dust and disturbance to be acceptable.

The possible negative ecological impact of mass tourism also has **long-term economic implications**. Although intensive mass tourism may render interesting financial returns from otherwise unproductive areas and may thus justify initial high capital investments, well protected and ecologically unimpaired conservation areas with a minimum of tourist facilities may prove to be economically much more valuable. In this way the total "package" of economic values, including both non-use and use values is more likely maximised in the long run. The world-wide increase in the demand for eco- or adventure tourism is favourable to such consideration. Instead of impairing or destroying conservation areas through heavy capital investment for mass tourism it is economically wiser to foster only ecologically acceptable forms of wildlife tourism in a greater variety of areas. Costa Rica for example, has done so and was able to double national parks tourism for five consecutive years.

Diversification of wildlife-based tourism away from national parks onto other private or communal lands, like countries in Southern Africa have been doing, seems to be another possibility for maximising ecologically acceptable economic returns from wildlife tourism.

## SELECTED LITERATURE REFERENCES FOR FURTHER READING


### History of wildlife use

ANDERSON, J.K. (1985): *Hunting in the Ancient World*. University of California Press, Berkeley, Ca., USA

BUBENIK, A.B. (1989): Sport hunting in continental Europe. In: *Wildlife production systems*, ed. HUDSON, R.J., DREW, K.R. and L.M. BASKIN; p. 115-133; Cambridge University Press, Cambridge

HARDING, R.S.O. and G. TELEKI (eds.) (1981): *Omnivorous Primates: Gathering and Hunting in Human Evolution*. Colombia University Press, New York

HIGGS, E.S. (ed.) (1975): *Palaeoeconomy*. Cambridge University Press, Cambridge

HUGHES, J.D. (1975): *Ecology in Ancient Civilisations*. University of New Mexico, Albuquerque, NM, USA

INGOLD, T. (1980): *Hunters, pastoralists and ranchers*. Cambridge University Press, Cambridge

JARMAN, M.R. (1976): Early animal husbandry. *Philosophical Transactions of the Royal Society of London*, B 276, p. 85-97

LEE, R.B. and I. de VORE (eds.) (1976): *Kalahari Hunter-Gatherers*. Harvard University Press, Cambridge, Mass., USA

MEGAW, J.V.S. (ed.) (1977): *Hunters, gatherers, and first farmers beyond Europe*. Leicester University Press, Leicester, UK

TOYNBEE, J.M.C. (1973): *Animals in Roman Life and Art*. Thames and Hudson, London

WOLFE, M.L. (1970): The history of German game administration. *Forest History*, 14, p. 6-16

### Domestication

ANONYMOUS (1991): *Microlivestock. Little-Known Small Animals with a Promising Economic Future*. BOSTID, National Research Council; Washington, DC

CLUTTON-BROCK, J. (1987): *A Natural History of Domesticated Mammals*. Cambridge University Press/British Museum (Natural History), Cambridge and London

HEMMER, H. (1990): *Domestication - the decline of environmental appreciation*. Cambridge University Press, Cambridge

HERRE, W. and M. RÖHRS (1977): Zoological considerations on the origins of farming and domestication. In: *Origins of agriculture*, ed. REED, C.A.; p. 245-280; Mouton, Hague and Paris

MASON, I.L. (ed.) (1984): *Evolution of domesticated animals*. Longman Group, London and New York

RAETHEL, H.-S. (1991): *Hühnervögel der Welt*. Natur, Augsburg, Germany

SHORT, R.V. (1976): *The introduction of new species of animals for the purpose of domestication*. Syposium of the Zoological Society London, 40, p. 321-333

ZEUNER, F.E. (1963): *A History of Domesticated Animals*. Hutchinson, London

### Wildlife Management

BEGON, M. and M. MORTIMER (1986): *Population Ecology (2nd edition)*. Blackwell Scientific, Oxford

CAUGHLEY, G. (1977): *Analysis of Vertebrate populations*. J. Wiley and Sons, New York

CONOVER, M.R. and D.J. DECKER (1991): Wildlife damage to crops: perception of agricultural and wildlife professionals in 1981 and 1987. *Wildlife Society Bulletin*, 19, p. 46-52

DELANEY, M.J. and D.C.D. HAPPOLD (1979): *Ecology of African mammals*. Longman Group, New York and London

GEIST, V. and F. WALTHER (eds.) (1974): *The behaviour of ungulates and its relation to management*. IUCN Publications, New Series No. 24; Gland, Switzerland

GETZ, W.M. and R.G. HAIGHT (1989): *Population Harvesting*. Princeton University Press, Princeton, NJ, USA

GILES, R.H. Jr, (1978): *Wildlife management*. W. H. Freeman and Co., San Francisco, Ca.

HALLS, L.K. and C. HOUSE (eds.) (1984): *White-tailed deer, management and ecology*. Stackpole Books, Harrisburg, Pa., USA

HUDSON, P.J. and M.R.W. RANDS (1988): *Ecology and management of Game Birds*. BSP Professional Books, Oxford

JEWELL, P.A., HOLT, S. and D. HART (eds.) (1981): *Problems in Management of Locally Abundant Wild Animals*. Academic Press, London

OWEN-SMITH, R.N. (1983): *Management of large mammals in African conservation areas*. Haum Educational, Pretoria, South Africa

RHINEY, T. (1982): *Study and management of large mammals*. J. Wiley and Sons, New York

SCHEMINTZ, S.D. (ed.) (1980): *Wildlife management techniques manual*. The Wildlife Society; Washington, DC

STODDARD, L.A., SMITH, A.D. and T.W. BOX (1975): *Range management*. Mc Graw Hill Book Co., New York

SWANK, W.G. (1985): Wild ungulates as competitors of livestock. In: *World animal science, parasites, pests and predators*, ed. GAAFER, S.M., HOWARD, W.E. and R.E. MARCH; p. 457-469; Elsevier Science, Amsterdam

WATSON, A. (ed.) (1969): *Animal populations in relation to their food resources*. Blackwell Scientific, Oxford

### Conservation and utilisation of wildlife resources

ANDERSEN, D. and R. GROVE (eds.) (1987): *Conservation in Africa: people, policies and practices*. Cambridge University Press, Cambridge

ANONYMOUS (1978): *Proceedings of the VIIIth World Forestry Congress: Discussion Area Agenda Item No. 8: Forestry for Food*; Oct. 1978; Jakarta, Indonesia

ANONYMOUS (1986 ): *Conservation and sustainable use of tropical rainforests: Elements of a strategy against destruction of forests in the humid tropics (in German; with English translation)*. Forschungsberichte des Bundesministeriums für wirtschaftliche Zusammenarbeit, Band 74; Weltforum, München, Köln and London

ANONYMOUS (1987): *Our Common Future (The Brundtland Report)*. World Commission on Environment and Development. Oxford University Press, Oxford

ANONYMOUS (1988): *SADCC Natural Resources and the Environment: Policies and Development Strategy*. SADCC Coordinator for Forestry, Fisheries and Wildlife; Lilongwe, Malawi

ANONYMOUS (1991): *Caring for the Earth. A Strategy for Sustainable Living*. IUCN/UNEP/WWF; Gland, Switzerland

BELL, R.H.V. and E. McSHANE-CALUZI (eds.) (1984): *Conservation and Wildlife Management in Africa*. Proceedings of a workshop organised by the US Peace Corps at Kusungu National Park, Malawi, Oct. 1984; US Peace Corps; Washington, DC

BENSON, D.E. (1991): Values and management of wildlife and recreation on private land in South Africa. *Wildlife Society Bulletin*, 19, p.497-510

BOTHMA, J. du P. (ed.) (1989): *Game Ranch Management. A practical guide on all aspects of purchasing, planning, development, management and utilisation of a modern game ranch in Southern Africa*. J.L. van Schaik, Pretoria, South Africa

BUNNING, J.L. (ed.) (1985): *Birds and Man*. Proceedings of the Birds and Man Symposium. Witwatersrand Bird Club; Johannesburg

CUTTER, S.L., RENWICK, H.L. and W.H. (1989): *Exploitation, conservation, preservation: A geographic perspective on natural resources use*. Rowan and Littlefield Publishers Inc., USA

DASSMAN, R. F. (1964): African game ranching. Pergamon Press, Oxford and London

DIAMOND, A.W. and F.L. FILION (1987): *The value of Birds*. Proceedings of the ICBP Symposium on Birds as Socio-Economic Resource. Queens University, Kingston, Canada; ICBP Technical Publications Series No. 6; Cambridge

DUFFEY, E. and A.S. WATT (eds.) (1971): *The scientific management of plant and animal communities for conservation*. 11th Symposium of the British Ecological Society. Blackwell Scientific, Oxford

EHRLICH, P.R. (1988): The Loss of Diversity: Causes and Consequences. In: *Biodiversity*, ed. WILSON, E.O. and F.M. PETER; p. 21-27; National Academy Press, Washington, DC

ELTRINGHAM, S.K. (1984): *Wildlife Resources and Economic Development*. J. Wiley and Sons, New York

FITZGERALD, S. (1989): *International Wildlife Trade: Whose Business Is It?* World Wildlife Fund (WWF-USA), Washington, DC

GEIST, V. (1985): Game ranching: threat to wildlife conservation in North America. *Wildlife Society Bulletin*, 13, p. 594-598

GROOMBRIDGE, B. (ed.) (1992): *Global Diversity: Status of the Earth's Living Resources*. World Conservation Monitoring Centre (WCMC); Chapman and Hall, London

HUDSON, R.J., DREW, K.R. and L.M. BASKIN (eds.) (1989): *Wildlife production systems: economic utilisation of wild ungulates*. Cambridge University Press, Cambridge

KISS, A. (ed.) (1990): *Living with Wildlife: Wildlife Resource Management with Local Participation in Africa*. Technical Paper No. 130; World Bank, Washington, DC

LIVINGSTON, J. (1981): *The Fallacy of Wildlife Conservation*. Mc Clelland and Stewart, Toronto, Canada

McNEELEY, J.A. and K.R. MILLER (eds.) (1984): *National Parks, Conservation, and Development. The Role of Protected Areas in Sustaining Society*. IUCN; Smithsonian Institution Press, Washington, DC

MUIR, K. and J. BOJÖ (1994): *Economic Policy, Wildlife and Land Use in Zimbabwe*. Environment Working Paper No. 68; World Bank, Washington, DC

NOVAK, M., BAKER, J.A., OBBARD, M.E. and B. MALLOCH (eds.) (1987): *Wild Furbearer Management and Conservation in North America*. Ministry of Natural Resources; Ontario, Canada

PRESCOTT-ALLEN, R. and C. (1982): *What's Wildlife Worth? Economic Contributions of Wild Plants and Animals to Developing Countries*. Earthscan, London

PRINGLE, J.A. (1982): *The conservationists and the killers*. Books of Africa, Cape Town

REID, R.L. (ed.) (1975): *Proceedings of the IIIrd World Conference on Animal Production; Theme 1: Animal Production in Relation to Conservation and Recreation*. Sydney University Press, Sydney

ROBINSON, J.G. and K.H. REDFORD (eds.) (1991): *Neotropical Widlife Use and Conservation*. Chicago University Press, Chicago and London

SCHRAMM, G. and J.J. WARFORD (eds.) (1989): *Environmental Management and Economic Development*. John Hopkins University Press, Baltimore, Md., USA

TALBOT, L.M., PAYNE, W.J.A., LEDGER, H.P., VERDCOURT, L.D. and M. TALBOT (1965): *The meat production potential of wild animals in Africa. A review of biological knowledge*. Technical Communication No. 16; Commonwealth Agricultural Bureau; Farnham Royal, UK

THORNBACK, J. and M. JENKINS (1982): *The IUCN Mammal Red Data Book*. IUCN; Gland, Switzerland

WALKER, B.H. (ed.) (1979): *Management of Semi-Arid Ecosystems*. Elsevier Scientific, Amsterdam

WHITE, R.J. (1987): *Big Game Ranching in the United States*. Wild Sheep and Goat International, Mesilla, NM, USA

WILSON, E.O. and F.M. PETER (eds.) (1988): *Biodiversity*. National Academy Press, Washington, DC

**Natural resources economics**

BARBIER, E.B., BURGESS, J.C., SWANSON, T.M. and D.W. PEARCE (1990): *Elephants, economics and ivory*. Earthscan, London

BROOKSHIRE, D.S., EUBANKS, L.S. and A. RANDALL (1993): Estimating option prices and existence values for wildlife resources. *Land economics*, 59, p. 1-15

CLARK, C.W. (1990): *Mathematical bioeconomics: the optimal management of renewable resources*. J. Wiley and Sons, New York

FISHER, A.C. (1981): *Resource and environmental economics*. Cambridge University Press, Cambridge

HARTWICK, J.M. and N.D. OLEWILER (1986): *The economics of natural resource use*. Harper and Row, New York

KRUTILLA, J.V. and A.C. FISHER (1975): *The economics of natural environments: studies in the valuation of commodity and natural resources*. Resources for the Future; Washington, DC

MITCHELL, R.C. and R.T. CARSON (1989): *Using surveys to value public goods: the contingent valuation method*. Resources for the Future; Washington, DC

NORTON, B.C. (ed.) (1986): *The Preservation of Species: The value of Biologiocal Diversity*. Princeton University Press, Princeton, NJ, USA

PEARCE, D.W. and R.K. TURNER (1990): *Economics of natural resources and the environment*. Earthscan, London

PEARCE, D.W., BARBIER, E.B. and A. MARKANDYA (1990): *Sustainable Development: Economics and Environment in the Third World*. Earthscan, London

RANDALL, A. (1991): The value of biodiversity. *Ambio*, 20, p. 64-68

SWANSON, T.M. and E.B. BARBIER (eds.) (1992): *Economics for the Wilds: Wildlife, Wildlands, Diversity and Development*. Earthscan, London

WINPENNY, J.T. (1991): *Values for the environment: guide to economic appraisal*. Overseas Development Institute; London

**Wildlife-based tourism**

ANONYMOUS (1990): *SADCC/GTZ Workshop on Wildlife-based Tourism*; held at Kafue National Park, Zambia, Aug. 1990. FGU Consulting; Königstein, Germany

ANONYMOUS (1994): 1993 *United Nations' List of National Parks and Protected Areas*. UNEP/WCMC/IUCN; Gland and Cambridge

BARNES, J., BURGESS, J. and D. PEARCE (1992): Wildlife tourism. In: *Economics for the Wilds*, ed. SWANSON, T.M. and E.B. BARBIER. Earthscan, London

CRAIG, W.A. (ed.) (1990): *International Handbook of National Parks and Nature Reserves*. Greenwood Press, New York

DIXON, J.A. and P.B. SHERMAN (1991): *Economics of protected areas: a new look at benefits and costs*. Earthscan, London

LUXMOORE, R.A. (1989): International trade. In: *Wildlife production systems*, ed. HUDSON, R.J., DREW, K.R. and L.M. BASKIN; p. 28-49; Cambridge University Press, Cambridge

MacKINNON, J. and K., CHILD, G. and J. THORSELL (compilers) (1986): *Managing Protected Areas in the Tropics*. IUCN; Gland, Switzerland

SALM, R.V. (1987): *Marine and Coastal Protected Areas. A Guide for Planners and Managers*. IUCN/UNEP; Gland, Switzerland

# PART II: UTILISATION OF SPECIFIC ANIMAL GROUPS

The principles of management and utilisation of wildlife resources, as described in Part I, are applicable to all wild animals. However, the importance of wildlife, both in a positive and negative sense, and consequently the scope of its management vary considerably between different animal orders, families or even species. This is mainly due to their zoogeographic and ecological distribution over distinct zones or areas in which climatic and socio-economic conditions differ greatly. Where animal taxa are distributed over several continents utilisation may vary accordingly between members of these. The local abundance of the wildlife resources, local traditions and utilisation, the level of cultural and technical development, and the marketability and demand for the particular products all determine which wild animals, for what purpose and in what ways they are utilised.

For the purpose of a global account of wildlife utilisation it is felt that a breakdown into zoological orders is most appropriate, although some of these are more important than others in the context of this book. Therefore, some chapters deal only with specific families or even sub-families of an order and others lump several orders together. Each "animal group" is taxonomically defined before its uses are described, and a table gives an overview of scientific names, geographical distribution and uses of the particular animal group. Taxonomic sequence and nomenclature follow the classification in GRZIMEK's Animal Life Encyclopaedia (Edition 1972-1984)[1].

In order to make reading easier each animal group is presented in a standard format which:

- firstly summarises the historical background of their use in different areas of the world;
- then describes present management and utilisation either more generally in terms of different production systems for the whole group or specifically for different species;
- assesses the economic importance and trends of the different uses;
- evaluates the implications of utilisation for the conservation of the animal group as a whole or of specific taxa;
- and lists cited literature references for further in-depth studies.

It is hoped that the reader will thus be able to consider and differentiate wildlife resources according to:

- employed types of utilisation, such as subsistence hunting, commercial hunting for meat, furs, or other products, ranching or farming, sports and trophy hunting, keeping and breeding in captivity, live sale;
- the products or services they render;
- and the socio-economic benefits derived from them by man.

The economic importance of the utilisation of the different animal groups may be expressed either for particular countries as a whole or in terms of the productivity and economics of local operations. For the reasons discussed in Chapt. 1.4 it is, however, extremely difficult to present comparative economic data even within one animal group as socio-economic, legislative and administrative conditions vary from country to country and obscure statistics. Therefore, in this respect the report can only give largely exemplary-type information.

---

1) 13 vols.: Reptiles vol. 6, Birds vols. 7-9, Mammals vols, 10-13; Van Nostrand, London

## 2.1 UTILISATION OF REPTILES

### 2.1.1 TORTOISES AND TURTLES (*TESTUDINES*)

Of the order **Testudines** 5 families are of particular interest in the context of this book: the terrestrial True tortoises (**Testudinidae**), some Freshwater turtles (**Emydidae**), Sea turtles (**Cheloniidae**), the marine Leatherback turtles (**Dermochelyidae**) and some Side-necked turtles (**Pelomedusidae**); a few taxa of the Softshell turtles (**Trionychidae**) are also used for food or traded as pets. Whereas tortoises and Freshwater turtles comprise many different species in all warmer regions there are only seven species of Sea turtles and leatherbacks inhabiting most of the tropical seas. Tab. 5 shows that all tortoises and turtles are used for food; however, the shells, particularly of the Sea turtles, also serve various utility or ornamental purposes. Some of the smaller species are traded as pet animals and attempts are made to farm some of the Freshwater and Sea turtles.

## General historical background

The only defences which tortoises and turtles have evolved against predation are their strong protective shell and their behaviour to withdraw into this, to dig into the earth or to escape and hide in the water. This was sufficient for them to survive all catastrophes and natural predators for more than 220 million years, but not the collecting man. In contrast to all other wildlife, no particular hunting or trapping effort is required by man to take these animals for food or other uses, and this fact has made this group of animals particularly vulnerable to more intensive forms of human exploitation. This applies to the tortoises on land, the terrapins and turtles in rivers and swamps, as well as to the large marine turtles: they have all constituted an easily accessible, favoured and much utilised resource since the earliest times. In most tropical countries the collection of tortoises is incidental; or in case of demand children will be sent out to find turtles or their eggs. The heavy Sea turtles can easily be waited for on the beaches when they come to lay their eggs and are completely helpless against being killed; their rookeries are taken as monopolies by nearby human populations.

There are classic examples of the early excessive utilisation of both tortoises and Sea turtles. Most of the Galapagos and Indian Ocean tropical islands were historically crowded with Giant tortoises, as we now know it only from Aldabra. The tortoises became an important source of fresh meat to the seafarers and were taken on board by the hundreds as they could be stored without water and food, staying alive for around a year. As a result most of the island tortoise populations, except those on some of the Galapagos and the Aldabra islands, had become extinct by around 1800 (1). Similarly, some of the Sea turtles, such as the Green and Hawskbill turtle, were excessively exploited since early times for food and shells as a valuable raw material until the species' survival became endangered.

**Table 5:** Overview of the utilisation of tortoises and turtles

| English name | Scientific name | Geographical distribution | Body[1] length (cm) | Utilisation |
|---|---|---|---|---|
| **FRESHWATER TURTLES and TERRAPINS - EMYDIDAE** | | | | |
| Sliders | *Pseudemys scripta* sspp. | N.America | 10-20 | farmed for pet trade |
| Diamond-back terrapin | *Malaclemys terrapin* sspp. | N.America | 20-25 | farmed for meat |
| Spotted Pond turtle | *Geoclemys hamiltoni* | W.-S.Asia | 20 | for eggs, pet trade |
| Reeves' turtle | *Chinemys reevesi* | E.Asia | 20 | for meat and |
| Box turtles | *Cuora* spp. | S.-S.E.-E.Asia | 20 | medicinal purposes |
| Asian River turtle | *Callagur borneonensis* | S.-S.E.Asia | 30-60 | for eggs, pet trade |
| Roofed turtles | *Kachuga* spp. | W.-S.Asia | 20-25 | for meat, medicine |
| Indian Black turtle | *Geoemyda trijuga* | S.Asia | 20-25 | and pet trade |
| **TRUE TORTOISES - TESTUDINIDAE** | | | | |
| Mediter. tortoises | *Testudo*[2] *hermanni, -graeca* | S.-SE.Europe | 10-20 | for pet trade |
| Steppe tortoise | *-horsefieldi* | C.-W.Asia | 20 | for meat, pet trade |
| African tortoises | *-pardalis, -sulcata, -angulata* | African south of Sahara | 25-75 | for subsistence, pet trade |
| Asian tortoises | *-elegans, -elongata, -emys* | S.-S.E.Asia | 25-40 | for meat, pet trade |
| S. American tortoises | *-carbonaria, -chilensis* | S.America | 25-50 | for meat, pet trade |
| Giant tortoises | *-gigantea, -elephantopus* | Indian Ocean and Galapagos Islands | >100 | formerly for meat, now tourism |
| Hinged tortoises | *Kinixys* spp. | W.-C.Africa | 20-30 | for subs., pet trade |
| Parrot-beaked tortoises | *Homopus* spp. | S.Africa | 10-20 | for pet trade |
| Pancake tortoise | *Malacochersus tornieri* | E.Africa | 10-15 | for pet trade |
| **SEA TURTLES - CHELONIIDAE** | | | | |
| Olive and Kemp's Ridley turtles | *Lepidochelys olivacea, -kempi* | tropical and sub-tropical oceans | 100 | for eggs, meat, leather |
| Hawksbill turtle | *Eretmochelys imbricata* | tropical oceans | 90 | for shells, meat; exp. farming |
| Loggerhead turtle | *Caretta caretta* | trop./subtr. oceans | 100 | for eggs, meat |
| Flatback turtle | *Natator depressus* | tropical oceans | | for eggs, meat |
| Green turtle | *Chelonia mydas* | tropical oceans | 100-140 | for meat, eggs, shells; farming |
| **LEATHERBACK TURTLES - DERMOCHELYIDAE** | | | | |
| Leatherback turtle | *Dermochelys coriacea* | tropical oceans | 120-180 | f. meat, leather, oi |
| **SOFTSHELL TURTLES - TRIONYCHIDAE** | | | | |
| Ind. Flapshell turtle | *Lissemys punctata* | India | 25 | for meat, medicine |
| Ind. Softshell turtles | *Aspideretes* spp. | S.Asia | 40-70 | and pet trade |
| Other Softshell turtles | *Trionyx* spp. | W.Africa, S.W.-E.Asia | 25-90 | farmed for meat, pet trade |
| **SIDE-NECKED TURTLES - PELOMEDUSIDAE** | | | | |
| Arrau and Terecay | *Pedocnemis* spp. | trop. S.America | 40-100 | for eggs and meat |
| African Side-necked turtles | *Pelomedusa* spp. *Pelusios* spp. | Africa south of Sahara, W.Africa | 20-25 | for meat, pet trade |

1) maximum length of carapace according to MLYNARSKI and WERMUTH (12); 2) the large numb of *Testudo* spp. has been divided into the following subgenera: *Testudo* (S.Europe - N.Africa), *Pseudtestudo* (N.E.Africa), *Geochelone* (N.W.-C.-S.Africa and S.Asia), *Psammobates* (S.W.-S.Afric *chersina* (s. Africa) *Asterochelys* and *Acinixys* (Madagascar), *Aldabrachelys* (Aldabra islands), *Agrionem* (W.-C.Asia), *Manouria* and *Indotestudo* (S.-S.E.Asia), *Chelonoidis* (trop. S.America, Galapagos Islands

With this general background in mind particularly careful consideration has to be given to the possibilities of sustainable and controlled forms of management and utilisation of the chelonian resources.

## TORTOISES AND FRESHWATER TURTLES

### Past and present utilisation

Most of the larger chelonians have been and are still used by collecting their eggs for food, killing them for meat, and making a variety of artefacts out of their shells; in some countries, such as India, turtle shells, blood and gall-bladders are utilised for medical purposes (5). More recently collection of live specimens for international trade and the farming of some of them have become additional important uses.

Most of the tortoises, as well as species from fresh and brackish water, have been and are still being used extensively for **subsistence purposes**. In most tropical countries any type of reptile, but particularly tortoises and turtles, is readily and even preferably consumed (7). Only where there are urban markets nearby may some of the collected turtles or their eggs also be marketed as food. Of particular importance in this respect are the *Side-necked turtles* (*Pelomedusidae*) in the waters of the Amazon rivers and the West African lagoons. In a survey of 23 restaurants in Manaus in 1976, turtle dishes (*Podocnemis* spp.) were, amongst all wild-animal food, in greatest demand and widely preferred by the general public (15). *Pelusios* and *Pelomedusa* turtles are found in numbers on every urban market along the coast of West Africa. In a survey in India it was found that of 26 turtle and tortoise species present 22 were utilised to a varying degree for food and/or medical purposes throughout their distribution range (5). Egg collection has also taken place from the nesting sites of the *River terrapin* (*Batagur baska*) and the *Asian River turtle* in South-East Asia (6, 13). *True tortoises (Testudinidae)* are taken for food anywhere in the drier parts of South America, Africa south of the Sahara, and to a lesser degree also in Asia; however, this is mostly incidental. Nevertheless, in a 12-month investigation in two localities along the Transamazon highway in 1973/74, tortoises (*Testudo (Chelonoidis)* spp.) constituted almost 2 % and 4.1 % respectively of wild-animal foods eaten (14).

Large scale **commercial exploitation** of *Podocnemis expansa* and *-unifilis*, the *Arrau* and *Terecay turtles*, formerly existed in Brazil. Perhaps more than 50 million eggs per year were harvested on the river shores of the Amazon, not only for food but mostly processed into an oil for candles (12). In Asia, where the consumption of Freshwater turtles is widely believed to have also medical effects, subsistence use has locally developed into commercialisation of turtles. Examples are *Box turtles*, and more recently mainly *Reeves' turtles*, in China, and the *Flagshell* and *Softshell turtles* in India which are systematically collected for market sale despite protective legislation. They are fished in nets, and iron pokers are used to pull out turtles burried in mud (5). In the former USSR, immense numbers of the *Steppe tortoise* have been processed into food for fox farms (3).

Apart from the immeasurable subsistence use, the most extensive utilisation of tortoises and Freshwater turtles world-wide is probably their **live capture for the pet trade**. This often concentrates on hatchlings which sustain very large mortality rates during trade. For example, in the 1960s hatchlings of *Podocnemis unifilis* were exported in large numbers to Europe and North America as pets, probably in imitation of the almost industrial farm production of *Red-eared sliders* (*Pseudemys scripta elegans*) in the USA for the pet trade (8). Hatchlings of the various subspecies of this turtle are still the most commonly traded pet turtles. They are shipped in hundreds of thousands of which less than 1% might survive captivity. Tab. 6 gives a perspective of the trade in live tortoises under the provisions of CITES in 1989 to 1993. The most heavily traded species until 1992 were *Testudo horsefieldi* and the *Kinixys* species, predominantly *-belliana*. The next in importance were still the Mediterranean *T. graeca*, *-hermanni*, *-marginata* and *-kleinmanni*, followed by the African *T. pardalis* and *-sulcata* and the Asian *T. elegans*. The formerly so important trade in Argentinean *T. chilensis* has dropped to almost nil. Exports of *T. (Indotestudo) elongata* from Malaysia and Bangladesh mainly into Hongkong, Singapore and China, was partly recorded in weight. According to the weight figures the trade in these species, most probably for food, must be substantial. Trade in *Side-necked turtles*, *Pelomedusa* and *Pelusios* spp., from Africa, has also been quite considerable.

**Table 6:** CITES registered trade in live tortoises in 1989-1993[1)]

| Species | 1989 | | 1990 | | 1991 | | 1992 | | 1993 | |
|---|---|---|---|---|---|---|---|---|---|---|
| | Imp. | Exp. | Imp. | Exp. | Imp. | Exp. | Imp. | Exp. | Imp. | Exp. |
| *T. (Testudo) graeca, -hermanni, -marginata.* | 8,455 | 1,216 | 1,821 | 303 | 2,181 | 207 | 4,436 | 887 | 878 | 171 |
| *T. (Pseudotestudo) kleinmanni* | 157 | 47 | 1,011 | 0 | 277 | 180 | 303 | 356 | 10 | 20 |
| *T. (Agrionemys) horsefieldi* | 560 | 52,328 | 10,326 | 3,698 | 1,001 | 23,184 | 7,857 | 9,216 | 0 | 1 |
| *T. (Geochelone),* mainly *-pardalis, -sulcata, -elegans* | 1,454 | 605 | 3,552 | 2,663 | 7,178 | 1,391 | 4,969 | 4,290 | 313 | 1,197 |
| *T. (Chersina) angulata* | 97 | 109 | 102 | 58 | 88 | 47 | 151 | 198 | 80 | 24 |
| *T. (Chelonoides) carbonaria, -chilensis, denticulata* | 1,809 | 1,244 | 592 | 391 | 3,188 | 1,985 | 1,691 | 1,995 | 101 | 776 |
| *T. (Indotestudo),* mainly *-elongata, -emys* | 611 | 1,284 | 933 | 1,907 | 178[2)] | 291 | 872 [3)] | 900 | 22 [4)] | 1 |
| *T. (Manouria)* spp. | 38 | 274 | 419 | 1,711 | 262 | 184 | 21 | 38 | 0 | 0 |
| *Malacochersus tornieri* | 0 | 0 | 603 | 0 | 5,572 | 301 | 740 | 172 | 0 | 0 |
| *Homopus* spp. | 17 | 82 | 36 | 38 | 24 | 14 | 55 | 79 | 15 | 2 |
| *Kinixys -belliana, -erosa, -homeana* | 15,628 | 9,422 | 9,305 | 6,376 | 7,336 | 10,720 | 4,411 | 6,700 | 0 | 3,731 |
| **Total Testudinidae** | **28,826** | **66,611** | **28,700** | **17,206** | **27,285** | **38,504** | **25,506** | **23,721** | **1,419** | **5,923** |
| **Total Pelomedusidae** | **1,785** | **4,943** | **3,287** | **5,069** | **2,905** | **5,277** | **4,725** | **7,514** | **109** | **4,323** |
| **Total Trionychidae** | **38** | **274** | **86** | **11** | **34** | **290** | **515** | **377** | **0** | **9** |

1) compiled from CITES data base extracts; 2) plus 79 t; 3) plus 280 t; 4) plus 721 t

There have been attempts to **farm turtles for commercial purposes** in addition to the above-mentioned slider farming for the pet trade and the breeding of tortoises and turtles for replenishing wild populations (8). The *Diamond-back terrapin* was farmed on the east coast of USA for the production of turtle meat for gourmets (9). Historically, of course, this species was extensively used to feed the slaves in Southwestern USA. Enclosures with water which changed at high and low tide were constructed to contain and breed this salt marsh species, sometimes attaining a body weight of more than 1 kg. The farm was profitable for 30 years after which, at the beginning of the Second World War, it was abandoned. The rearing of *Softshell turtles (Triochynidae)* has a long history in East Asian countries, particularly China and Japan. In Taiwan they have been bred and farmed for more than a century (4) for human consumption. Large-scale commercial farming of *Trionyx sinensis* in preference to the local *T. cartilagineus*, was initiated in Singapore in the 1970s. One farm cultures 30,000-40,000 turtles in ponds and sells 400-500 per month when they have attained a weight of 0.6 kg (2). They are marketed alive to local restaurants where they are made into soup which is said to have medicinal value.

Apart from the consumptive uses tortoises and turtles were **of cultural importance** to many early societies as is still reflected from mythology, art and religion in some regions. In Asian countries, such as Thailand and also Japan, large numbers of turtles (*Hieremys annandalis* and *Siebenrockiella crassicollis*) are kept in ponds around temples and castles, fed by the public and higly revered. In Thailand it is believed that the souls of people who have died in attempts to save others from drowning live in them (12).

## Economic importance

The value of tortoises and Freshwater turtles and their eggs to subsistence economies in tropical countries is probably much greater than that of the Sea turtles. This is simply due to the fact that there are some 260 species widely distributed over all continents as opposed to the 7 species of Sea turtles utilisable only on certain coasts or in some tropical waters.

In addition to this subsistence value, the international trade in live tortoises and turtles for use as pets has become most important economically since the 1960s when it became fashionable in industrialised societies to keep any type of small, convenient wild animal as a hobby (10). Initially the tortoise pet trade was largely limited to the Mediterranean species (*Testudo*). As these became rarer, large numbers of *Horsefields* or *Steppe tortoises (T. horsefieldi)* from West and Central Asia, and *Chaco tortoises (T. chilensis)* from Argentina were imported. UK imports of the Mediterranean species during 1965-1975 averaged 200,000 per year; total international trade in this period was estimated at 5-10 million tortoises (11). In 1976 some Emydidae, Trionychidae and Pelomedusidae were placed on Appendix I of CITES (no trade allowed) and all Testudinidae on Appendix II (trade only with export permits); they have thus become subject to trade restrictions. In 1984 the European Union banned the importation of the Mediterranean species altogether. Tab. 6 shows that the trade volume has since diminished greatly. The CITES statistics on which this table is based should theoretically comprise all commercial traffic in tortoises. That this is not so is clearly indicated by the great discrepancies between export and import figures and the continuance of significant internal trade in chelonians, particularly in Asian countries (6, 5). During a trade survey in India in 1991-92 more than 1,500 *Starred tortoises (Testudo elegans)* were confiscated on markets and internal trade in this species was estimated to be about 5,000 animals per year. At a price of US$ 65-300/tortoise in the United States the total export value of this trade alone was thought to be in the region of US$ 500,000 (5). Accepting that there is substantial unrecorded or illegal trade, as evidenced by occasional seizures of large falsely documented shipments, the CITES statistics nevertheless give an

Marketing of live Indian Softshell turtles, weighing up to 20 kg, for human consumption. (Photo: N.K. DAS; by courtesy of WWF-India)

indication of which tortoise species are mainly traded and what the trend of the trade has been during the period 1989-1993.

Registered exports of Testudinidae generally decreased from more than 66,000 in 1989 to about 6,000 in 1993 and the imports from about 30,000 to just 1,500. The number of specimens derived from the wild is probably lower, as the CITES export figures also include bred and re-exported tortoises. Even if one takes the figures to represent only a fraction of the total trade in live tortoises, this has dropped world-wide to economically rather insignificant levels compared with the volume of the pet trade in the 1960s and 1970s, mentioned above.

There are, however, indications that the traffic in *Freshwater turtles* in South, South-East and East Asia for consumption has been increasing. In a market survey in India in 1992 (5) this trend has been attributed to the following circumstances:

- Growth of fisheries activities in all kinds of waterbodies in which turtles are considered detrimental to the fish resources;
- With a market and pricetag, turtles provide an extra source of income to the fishermen;
- Increase of fish and meat prices rendering turtle meat a cheap and particularly healthy substitute;
- Traditional beliefs attached to the consumption of meat and other products of turtles;
- Migration of Bangladeshis into India where they constitute the bulk of buyers in Calcutta, the most important turtle market.

On 61 food markets surveyed 12 were engaged in the sale of turtle meat with prices ranging between Rp 15 and 200/kg depending on the locality. *Large Softshell turtles (Aspideretes gangeticus)*, weighing 15-20 kg, were fetching Rp 600-800/piece (approx. US$ 23-31) on the basis of the Calcutta market prices (5).

Breeding or rearing of tortoises and turtles in captivity for commercial purposes is also increasing but is still not nearly satisfying the pet market demand. Farming of turtles for human nutrition has theoretically two favourable aspects: as cold-blooded animals they consume only roughly 1/15th of the food a warm-blooded animal of the same size requires, and if herbivorous they convert plant material, otherwise non-utilised, into protein. However, even turtles of large body size, like the *Arrau*, can never compete with the productivity of other aquatic animals like fish. Therefore, it is difficult to make farming of smaller turtle species commercially profitable. One turtle farm in Singapore, for example, holds a stock worth Singapore $ 240,000-360,000 (approx. US$ 112,000-168,000) but makes only a small profit (2). Export of live turtles from this farm to Japan in 1980/81 was found to be economically not viable.

## Implications for conservation

Commercial exploitation of riverine and lacustrine turtles for food, as occurs in Asia, is a cause of worry for all those species which are either endemic or have a low reproductive capability. In India, the two brackishwater species, the Giant Softshell turtle (*Pelochelys bibroni*) and the River terrapin (*Batagur baska*) have been practically wiped out because of over-exploitation (5). Considering the longevity of tortoises and turtles even uncontrolled collection of their eggs, as has occurred for decades with the Arrau and Terecay turtles in the Amazon region for example, will certainly have a negative effect on their long-term reproductive success visible only after decades. However, the increase in human activity on tropical river courses, occupation of their banks and deforestation are probably a more important threat to the Freshwater turtles than their present exploitation. Also, there cannot be any doubt that the collection of tortoises for the international pet trade has already contributed substantially to the decline of a number of the *Testudo* species, notably *T. graeca*, *-marginata*, *-hermanni* and *-chilensis*. However, in the arid and semi-arid environments in which most of the tortoises live, fire and habitat destruction along with agricultural intensification are greatly compounding the effect of collection. In contrast to the case of the Sea turtles on which much more conservation concern is being focused, very little research has been done so far on the population dynamics of Freshwater turtles and tortoises that would allow the design of appropriate management programmes, apart from the fact that it is almost impossible to control their subsistence use. Therefore, at present it would appear to be best to curb the import of pet tortoises from the countries of origin, as the European Union has done successfully since 1984, and to promote breeding programmes as much as possible. Apart from alleviating the trade pressure breeding programmes may also eventually contribute more significantly to the replenishment of depleted species in the wild; for example, the breeding of *Testudo sulcata* in Senegal for release in protected areas of the Sahel zone.

## Literature references

1. ARNOLD, E.N. (1979): *Indian Ocean Giant Tortoises, their Systematics and Island Adaptations*. Transactions of the Royal Society, London 286 (B), p. 127-145
2. BARZDO, J., LUXMOORE, R.A., BROAD, S. and D. JONES (1994): *A survey of wildlife farming operations*. WCMC/IUCN, Cambridge (unpubl. report)
3. BRUSHKO, E. and F. KUBYKIN (1982): cited after LUXMOORE, R., GROOMBRIDGE, B. and S. BROAD (eds.) (1988): *Significant trade in Wildlife: A Review of Selected Species in CITES APPENDIX II*. Vol. 2: Reptiles and Invertebrates; IUCN/CITES, Cambridge and Lausanne, Switzerland
4. CHOU, L.M. and B.L. CHOO (1984 in press): The potential of softshell turtle culture in Singapore. *Proceedings of the International Conference on the Development and Management of Tropical Living Aquatic Resources*, Aug. 1983, Serdang, Malaysia
5. CHOUDHURY, D.C. and S. BHUPATHY (1993): Turtle Trade in India: A Study of Tortoises and Freshwater Turtles. Prepared by TRAFFIC India for WWF-India in collaboration with the Wildlife Institute of India (WII), Dehra Dun, India
6. DAS, J. (1990): The export of freshwater turtles from Bangladesh. *Oryx*, 24 (3), p. 163-166
7. DE VOS, A. (1976): Game as food. A report on its significance in Africa and Latin America. *Unasylva*, 29 (116), p. 2-12
8. HAGA, J.B. (1970): Turtle Farming. *International Turtle and Tortoise Society Journal*, 4 (2), p. 6-9
9. HILDEBRAND, S.F. (1929): Review of experiments on artifical culture of diamondback terrapins. *Bulletin US Bureau of Fisheries*, 45, p. 47-52
10. HONEGGER, R.E. (1974): The reptile trade. *International Zoo Yearbook*, 14, p. 47-52
11. INSKIPP, T. and S. WELLS (1979): *International trade in wildlife*. Earthscan, London
12. MLYNARSKI, M. and H. WERMUTH (1971): Die Schildröten. In: *GRZIMEKS Tierleben*, Vol. 6, p. 199; Kindler, Zürich
13. MOLL, E.O., MATSON, K.E. and E.B. KREHBIEL (1981): Sexual and seasonal dichromatism in the Asian river turtle *Callagur borneonensis*. *Herpetologica*, 37 (4), p. 181-194
14. SMITH, J.H. (1976): Utilisation of game along Brazil's transamazon highway. *Acta Amazonica*, 6 (4), p. 455-466
15. WETTERBERG, G.B. (1976): *Amazon fauna preferred as food*. FAO/FO:DP/BRA/ 71/545, Field Document 23; Food and Agriculture Organisation of the United Nations, Rome; 17 pp. (unpubl. report)

## SEA TURTLES

### Past and present utilisation

The Green turtle has been called "the world's most valuable reptile" (e.g. 39). As CARR (1962) put it: "Three requisites for a valuable and controllable marine food resource suggest themselves: it ought to be big; it ought to be herbivorous, or at least close to the base industry in the feeding chain; and it has got to be edible." The Green turtle meets these requirements. There are vast areas of sea grasses "growing in steady spreads over endless miles of tropical ocean bottom. Today most of these pastures are empty" (18), in part a consequence of the reduced numbers of turtles from the profusions that sustained Columbus and other explorers on their voyages (39).

All species of Sea turtles have been used in historical times by indigenous people for **subsistence purposes** and currently this continues in many parts of the world. Current utilisation of turtles varies from occasional unintentional catch in nets to systematic harvesting of turtles or their eggs. In some parts of Indonesia, a subsistence culture centres around hunting *Leatherback turtles* (22). In one village in Honduras, 72% of the households (n=93) earned some income from turtle eggs and this was estimated to account for 4.3% of the expenditure of an average household (29). The extent of local use of turtles is often difficult to quantify, but steady small scale take by individuals could add up to considerable totals. It is estimated that 10,000 turtles are killed each year by islanders in the Torres Strait (38).

In terms of the cash value of **international trade in turtle products**, the *Hawksbill* is now more important than the *Green turtle*. In 1979 the shell of a good sized Hawksbill turtle could sell for US$ 200 (20). The main exports have been from Indonesia to Japan where the shell is called "bekko" (32, 33). This could change now that Japan has withdrawn its CITES reservation on hawksbills.

Other principal uses of turtles include leather, especially from *Olive Ridley turtles* from Mexico and Ecuador, calipee of Green turtles for soup and the eggs of all species. The oil, especially of *Leatherback turtles*, is used for lamps and for medicinal purposes; turtle oil is also used in soaps and cosmetics. Stuffed turtles, usually juveniles with thinner less valuable scutes, are sold as curios. Although the Green turtle is primarily used for food, the Hawksbill for shell, and the Olive Ridley turtle for leather, these categories are not exclusive. Leather can be made from species other than the Ridley. Some Green turtle shell is sold internationally (32). Only those with no appreciation of the power of modern technology would altogether discount the possibility that some form of welding or treatment of Green turtle shell would make it an acceptable substitute for the thicker shell of hawksbills; even without special efforts, the scutes of captive reared Green turtles are thicker than those of wild turtles. All species of Sea turtles have been eaten, though there seems to be a greater chance of food poisoning from Hawksbill turtle meat.

Green Sea turtle hatchlings in an experimental farm in Suriname in 1982. (Photo: N. MROSOVSKY)

With the realisation that supplies are not limitless, a number of protected areas for the conservation of Sea turtles have been established in the last century; there have also been some **attempts to farm and ranch Sea turtles** (43, 40). At the Cayman Turtle Farm, *Green turtles* are processed at 3-4 years old when they weigh about 24 kg. The turtles are fed on modified Purina trout chows and costs of food in the early 1980s were about US$ 1.50 per kg body weight produced. Tanks are made of concrete or fibreglass; sea water is pumped in continuously. The breeding herd is kept in a large 3 m deep pond with a sand beach at one end; about 5 $m^2$ surface is allowed for each of these turtles. Once females start breeding at the Cayman Turtle Farm, their average interseasonal nesting interval is 1.6 years. Mean annual egg production is estimated at 375 (45). Second generation Green turtles have been produced in captivity (44); given time for research, large-scale production would probably be feasible. However, ranching is preferred to farming by some because it makes utilisation dependent on having a healthy population in the wild (28) and because the collection of eggs can be linked to contributions to conservation which could be greater if the considerable expense of keeping breeding herds were avoided (35).

## Economic considerations

The economic value of Sea turtles depends on which parts of the animal are used. A beautifully polished hawksbill shell might sell for several hundred dollars, making the rest of the animal relatively worthless; but in places where tourists are few, or CITES trade regulations rigorously enforced, the meat or eggs might be valued for local consumption.

An important biological point is that Sea turtles can reproduce over many seasons. A *Green turtle* has been found nesting 24 years after being first tagged (BJORNDAL, pers. comm.). Assuming an average breeding interval of 2.5 years, that gives more than 9 seasons in which eggs can be laid; at around 3 clutches a season (see (19) for elaboration), of about 100 eggs per clutch, this gives a potential lifetime production of 2,700 eggs. This is probably an underestimate because when saturation tagging of nearly all females coming ashore to lay is undertaken, nesting frequency per season is usually found to be higher than first thought (42). Of course it is not known for how many seasons an average Green turtle survives, but as an order of magnitude estimate, a *potential* of at least 3,000 eggs in a lifetime is not unreasonable. The question then arises as to whether it is best to use turtles for meat or for other purposes.

Calculations of the worth of eggs relative to other parts can be most easily done for the *Leatherback turtle* for which the value of the carapace (which lacks scutes, and rots easily) can be excluded, except perhaps for extraction of some oil. An example is given for leatherbacks on the west coast of Mexico in the late 1970s. Obviously the prices and details have to be changed for different places and times, but the example shows the kinds of considerations that those wanting to manage Sea turtles might wish to contemplate.

| Value of meat | | Value of eggs | |
|---|---|---|---|
| mean carapace length | 147 cm | clutch size | 70 eggs |
| weight | 295 kg | price per egg | 8 pesos |
| estimated weight of meat | 103 kg | value of clutch | 560 pesos |
| price of meat locally | 60 pesos/kg | | |
| total value of meat | 6,192 pesos | | |

Assuming the average leatherback lays six times per season, and assuming that one clutch is collected when the turtle is killed for meat:

total value if turtle killed = 6,192 + 560 = 6,752 pesos
total value of eggs over season = 6 x 560 = 3,360 pesos.

The basis for these estimates is given elsewhere (35), but even allowing for errors it seems likely that a Leatherback turtle only has to stay alive for two seasons for the eggs to be worth more than the meat. It comes down to the fact that leatherbacks are capable of producing thousands of eggs in their lifetimes. Killing for the meat, or to obtain the eggs without having to wait for laying, might be the best short-term strategy for an individual who may not find that turtle again or a poacher who wants to make a quick getaway. But the long-term interests of a community would be better served by harvesting eggs.

Calculations such as those above are only a guide. They would have to be adapted for particular species and locations, refined to take interest rates into account, and elaborated for species with valuable shells, or when there is a potential of the turtles attracting revenue from tourists. Much depends on over how many seasons a turtle lays. The longest recorded interval between the original date of tagging at laying and a subsequent laying is 20 years for *loggerheads* (RICHARDSON, pers. comm.), and 24 years for greens. Even though the average duration of reproductive life could be much shorter, enough is known to think that, for utilization as food, harvesting of eggs is preferable to killing turtles for meat.

## Implications for conservation

The drastic decline in numbers of Leatherback turtles nesting in Terengganu in Malaysia has been generally attributed to the overharvesting of eggs over many decades (23, 28). This may well be true but it is hard to prove. Without control areas, other factors cannot be ruled out. Many Leatherback turtles were caught in nets in Terengganu in the mid 1980s (estimated at 376 per year by (21)) and it is remarkable that almost total harvests of eggs from some other rookeries, those of Olive Ridley turtles in Honduras for example, have not wiped out the turtles there (29). Nevertheless, egg harvesting should not be undertaken lightly. Fortunately, there are some schemes that appear to be relatively safe. The most conservative is to use eggs that are naturally doomed to destruction. Tab. 7 shows that in Suriname about a third of the leatherback eggs are laid below the high-tide line and subsequently washed away. Continued monitoring of the percentage of eggs that are doomed, with attention to seasonal variations, would be a desirable component of an egg-harvesting programme. Nevertheless, the available data suggest that order-of-magnitude estimates can be obtained from limited sampling. On some beaches, difficult access or other logistic problems may make it impractical to market eggs only from nests that are actually laid below the high-tide line. Instead, as is done in Suriname, some tidally doomed eggs may be reburied higher up the beach to compensate for those taken to market (41, 35, 28). There is no evidence that particular individuals tend to lay tidally doomed eggs (27) and therefore saving such eggs should not degrade the gene pool.

**Table 7:** Percentage of Leatherback turtle eggs estimated to be laid below the high tide line (doomed) in Suriname

| Beach | Year | No. laid | % doomed | References |
|---|---|---|---|---|
| all Suriname | 1971 | 24,000 eggs[1)] | ca. 46 | SCHULZ 1975 |
| | 1972 | 36,000 eggs | ca. 44 | SCHULZ 1975 |
| | 1973 | 75,000 eggs | ca. 37 | SCHULZ 1975 |
| Krofajapasi | 1980 | 39 nests [2)] | 36 | MROSOVSKY 1983a |
| | 1982 | 196 nests | 32 | DUTTON and WHITMORE 1983 |
| Matapica | 1993 | 195 nests | 50 | GODFREY, pers. comm. |
| Baboensanti[3)] | 1982 | 9 nests | 78 | MROSOVSKY 1989 |
| | 1993 | 12 nests | 83 | MROSOVSKY, unpubl. |

1) calculations based on disposition of eggs; 2) early June; 3) this relatively inaccessible beach was visited in early May in 1982 and again in 1993

Nests laid too near the sea, though perhaps the easiest to identify, are only one source of natural mortality at an early life stage. On the island of Europa in the Indian Ocean, about 8% of Green turtle hatchlings emerging do so by day (31): few escape the sharp-eyed sharp-beaked frigate birds. By patrolling the beach in the late afternoon, a time when daylight emergences are common, hatchlings can be saved from the birds. Such hatchlings have been used to stock turtle ranches. Even if one does not know whether a particular hatchling would have escaped on its own or not, one may be confident that most of the hatchlings harvested in these circumstances would not have survived. Lest there is some biological advantage related to daylight emergences, some of the hatchlings saved from the birds should be released. To guard against changes in frequency of genes influencing particular behaviours, as management schemes become entrenched, monitoring of the percentage of hatchlings emerging by day should be continued. Also it would be desirable to learn whether particular individuals tend to produce nests that emerge by day or whether these occurrences are more related to external conditions.

Sea turtle nest that has been exposed by high seas in Suriname. Crabs and vultures are likely to eat these eggs and the rotting remains will become infested with flies. Nests classified as doomed are generally laid even lower with respect to tide lines than the present one, and are simply washed away. (Photo: N. MROSOVSKY)

Natural loss of eggs is also caused by the turtles themselves, especially by Ridley turtles whose *"arribadas"* (mass nestings) occur on short stretches of the beach. Fig. 4 shows that at Nancite, Costa Rica, ca. 17% of Olive Ridley nests are destroyed by turtles nesting later. Even if nests are not dug up, their chances of producing many viable hatchlings are small. Microbial contamination of the soil might be responsible. Whatever the cause, the huge egg wastage at arribadas offers an opportunity for utilisation without detriment and even possible benefit to populations: it is thought that reducing the density of nests will increase the survival rates of those remaining (17, 24), though definitive research is lacking.

**Figure 4:** Fate of Olive Ridley turtle nests and eggs at Nancite, Costa Rica (after CORNELIUS (23))

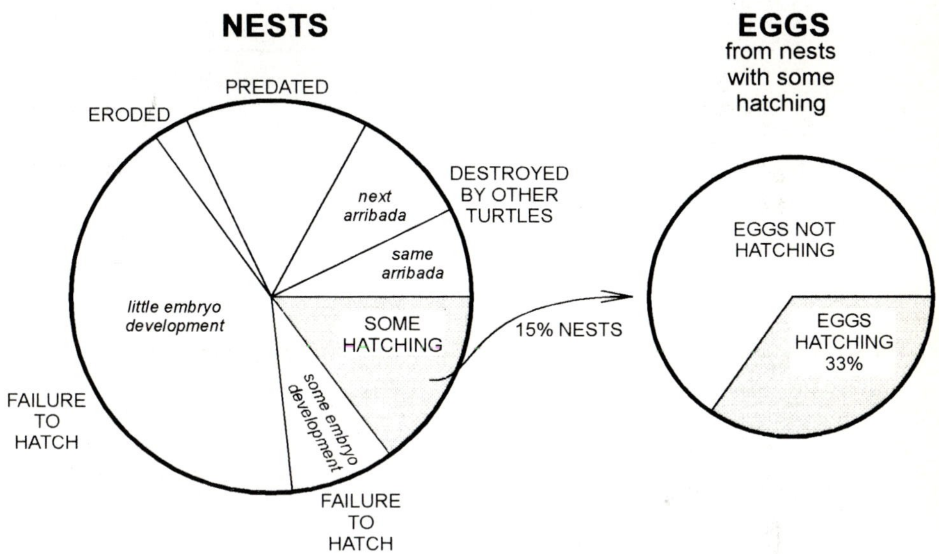


At Ostional, Costa Rica, egg taking by the villagers has been legalized during the first 36 h of an arribada nesting since eggs laid early in an arribada have a greater chance of being destroyed than those laid later. The eggs are then marketed at prices comparable to those of chicken eggs, thus undercutting egg poachers (16).

The above give some examples of naturally destroyed eggs. These particular factors are not prominent on every turtle beach. The general point is that sources of mortality at this early life stage can be studied (36). Taking steps to reduce this mortality can free eggs for human consumption without making inroads into recruitment. If some of the saved eggs are protected and the hatchlings released, recruitment may actually be increased.

Attention should be given not just to numbers in managed populations but to sexual composition. In Sea turtles, sexual differentiation is determined by temperature. This is not a problem if eggs are taken occasionally, for example to start a farm, or if only eggs that are certain to be doomed are taken. But if eggs are harvested within a fixed open season every year, with compensatory relocation of doomed eggs going on throughout the season, then there is a danger of distorting sex ratios. To avoid this some idea is needed of the seasonal sex production profiles (SSPPs), as illustrated in Fig. 5.

**Figure 5:** Estimated sex ratios of Sea turtle hatchlings in Suriname in 1982 (data from MROSOVSKY and coll. (37))

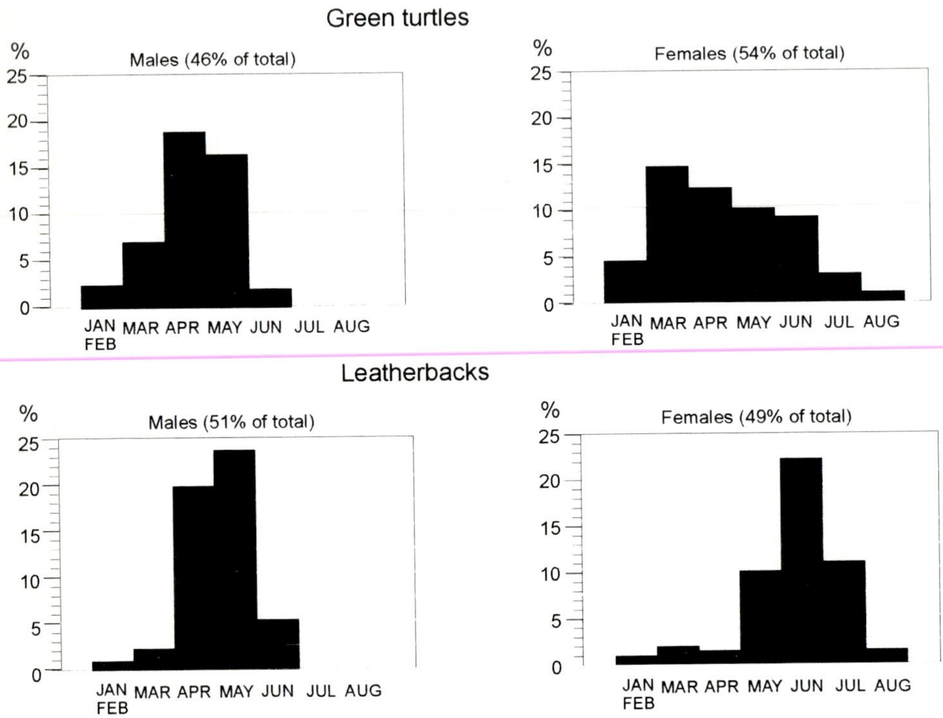


Estimates of the number of each sex produced at hatching are expressed as a percentage of the total number of hatchlings in the season and the X-axis shows laying dates. In the case above, if some harvesting were permitted for Green turtles in March and April, approximately equal numbers of male and female producing eggs would be taken, but if the same months were open for leatherback eggs, then the relative output of male leatherbacks would be decreased. SSPPs can be used in management plans, but should ideally be based on estimates from more than 1 year. This may not always be feasible because present methods of sexing hatchling Sea turtles require histology and are laborious. However, partial protection against introducing sex ratio biases can be obtained by not concentrating harvesting just at the warmest or the coolest time, combined with knowledge from laboratory work about the actual temperatures producing males and females.

Although it may eventually be possible to harvest adult turtles safely, at present it is much more risky. With egg harvesting linked to doomed eggs, it is possible to protect numbers of eggs equal to or greater than those taken. The protective measures and the take can be made in the same units: eggs. If adult turtles are taken, not enough is known about survival rates to say how many eggs should be saved or how many juvenile turtles should be head-started to compensate for those adults. But with protein-rich eggs being washed out to sea by the tides, hatchlings which emerge by day being decimated by frigate birds, and turtles digging up each other's nests at arribada sites, some utilisation is surely possible at early stages in the life cycle. Moreover, controlled utilisation can support the activities of national wildlife authorities (40). Setting up reserves without adequate enforcement is often symbolic conservation leading to unregulated unmonitored utilisation (cf. 30).

Some conservationists advocate that more attention should be paid to the most reproductively valuable stages of the life cycle, the mature turtles and the large juveniles (25). This is not disputed but it should not be taken as an argument against using doomed eggs. The question that should be asked about egg harvesting is whether it is detrimental. If it is not, then there are common interests between those harvesting eggs in a controlled way based on the high natural mortality, and those whose priorities are protection at other life stages. Those utilising Sea turtles at the early stages will want their populations to be as large as possible. They are likely to support and promote efforts directed at preserving breeding adults.

## Literature references


16. ALMENGOR, M.A., MO, C., and, E. VARGAS (1993): Preliminary evaluation of olive ridley egg commerce from Ostional wildlife refuge, Costa Rica. *Marine Turtle Newsletter*, 63, p. 10-13
17. ALVARADO, M.A. (1990): The results of more than two years of turtle egg harvests at Ostional, Costa Rica. In: *Proceedings of the 10th Annual Workshop on Sea Turtle Biology and Conservation*, comp. RICHARDSON, T.H., RICHARDSON J.I. and M. DONNELLY; p. 175-178; US National Oceanic and Atmospheric Administration (NOAA), Technical Memorandum, NMFS-SEFC-278; Miami, Fl., USA
18. CARR, A. (1962): Foreword. In: *The green turtle and man*, ed. PARSONS, J.J.; University of Florida Press, Gainesville, Fl., USA
19. CARR, A., CARR, M.H. and A.B. MEYLAN (1978): The ecolgy and migrations of sea turtles; 7. The west Caribbean green turtle colony. *Bulletin of the American Museum of Natural History*, 162, p. 1-46
20. CARR, A. and A.B. MEYLAN (1980): Extinction or rescue for the hawksbill. *Oryx*, 15, p. 449-450
21. CHAN, E.H., LIEW, H.C. and A.G. MAZLAN (1988): The incidental capture of sea turtles in fishing gear in Terengganu, Malaysia. *Biological Conservation*, 43, p. 1-7
22. COMPOST, A. (1980): *Pilot survey of exploitation of dugong and sea turtle in the Aru Islands*. Yayasan, Hijau, Indonesia; 63 pp.
23. CORNELIUS, S.E. (1986): *The sea turtles of Santa Rosa National Park*. Fundacion de Parques Nacionales, San Jose, Costa Rica
24. CORNELIUS, S.E., ALVARADO, M., CASTRO, J.C., del VALLE, M.M. and D.C. ROBINSON (1991): Management of olive ridley sea turtles (*Lepidochelys olivacea*) nesting at Playas Nancite and Ostional, Costa Rica. In: *Neotropical Wildlife Use and Conservation*, ed. ROBINSON, J.G. and K.H. REDFORD; p. 111-135; University of Chicago Press, Chicago and London
25. CROUSE, D.T., CROWDER, L.B. and H. CASWELL (1987): A stage-based population model for loggerhead sea turtles and implications for conservation. *Ecology*, 68, p. 1412-1423

26. DUTTON, P.H. and C. WHITMORE (1983): Saving doomed eggs in Suriname. *Marine Turtle Newsletter*, 24, p. 8-10
27. ECKERT, K.L. (1987): Environmental unpredictability and leatherback sea turtle (*Dermochelys coriacea*) nest loss. *Herpetologica*, 43, p. 315-323
28. GROOMBRIDGE, B. and R. LUXMOORE (1989): *The green turtle and hawksbill* (*Reptilia: Cheloniidae*): *world status, exploitation and trade*. CITES, Lausanne, Switzerland
29. LAGUEUX, C.J. (1991): Economic analysis of sea turtle eggs in a coastal community on the pacific coast of Honduras. In: *Neotropical Wildlife Use and Conservation*, ed. ROBINSON, J.G. and K.H. REDFORD; p. 136-144; University of Chicago Press, Chicago and London
30. LEADER-WILLIAMS, N. and J.D. ALBON (1988): Allocation of resources for conservation. *Nature*, 336, p. 533-535
31. LE GALL, J.-Y., LEBEAU, A. and J. KOPP (1985): Evaluation de la production de tortues vertes *Chelonia mydas* nouveaux-nées sur les sites de ponte Europa et Tromelin (océan Indien). *Oceanographie Tropicale*, 20, p. 117-133
32. LUXMOORE, R. and J. CANIN (1985): International trade in raw sea turtle shell. *TRAFFIC Bulletin*, 7, p. 30-39
33. MACK, D., DUPLAIX, N. and S. WELLS (1981): Sea turtles, animals of divisible parts: international trade in sea turtle products. In: *Biology and conservation of sea turtles*, ed. BJORNDAL, K.A.; p. 545-563; Smithsonian Insititution Press, Washington, DC
34. MROSOVSKY, N. (1983a): Ecology and nest-site selection of leatherback turtles *Dermochelys coriacea*. *Biological Conservation*, 26, p. 47-56
35. MROSOVSKY, N. (1983b): *Conserving sea turtles*. British Herpetological Society, London
36. MROSOVSKY, N. (1989): Natural mortality in sea turtles: obstacle or opportunity? In: *Proceedings of the 2nd Western Atlantic Turtle Symposium*, ed. OGREN, L.; p. 251-264; US National Oceanic and Atmospheric Administration (NOAA), Technical Memorandum, NMFS-SEFC-226; Panama City, Fl., USA
37. MROSOVSKY, N., DUTTON, P.H., and C.P. WHITMORE (1984): Sex ratios of two species of sea turtles nesting in Suriname. *Canadian Journal of Zoology*, 62, p. 2227-2239
38. PARMENTER, C.J. (1980): Environmental factors in turtle farming. In: *Management of turtle resources*; p. 23-31; Research Monograph 1, James Cook University, Queensland, Australia
39. PARSONS, J.J. (1962): *The green turtle and man*. University of Florida Press, Gainesville, Fl., USA
40. REICHART, H.A. (1981): Farming and ranching as a strategy for sea turtle conservation. In: *Biology and conservation of sea turtles*, ed. BJORNDAL, K.A.; p. 465-471; Smithsonian Institution Press, Washington, DC
41. SCHULZ, J.P. (1975): Sea turtles nesting in Surinam. *Zoologiske Verhandelingen*, 143, p. 1-143
42. TUCKER, A.D. (1989): So many turtles, so little time: underestimating fecundity and overestimating populations. In: *Proceedings of the 9th Annual Workshop on Sea Turtle Conservation and Biology*, comp. ECKERT, S.A., ECKERT, K.L. and T.H. RICHARDSON; p. 181-183; US National Oceanic and Atmospheric Administration (NOAA), Technical Memorandum, NMFS-SEFC-232; Miami, Fl., USA
43. WOOD, J.R. and F.E. WOOD (1980): Reproductive biology of captive green sea turtles *Chelonia mydas*. *American Zoology*, 20, p. 499-505
44. WOOD, F.E. and J.R. WOOD (1990): Successful production of captive F2 generation of the green sea turtle. *Marine Turtle Newsletter* 50, p. 3-4
45. WOOD, F.E. (1991): Turtle culture. In: *Production of aquatic animals*, ed. NASH, C.E.; p. 225-234; Elsevier, Amsterdam

## 2.1.2 CROCODILIANS (*CROCODYLIA*)

The order **Crocodylia** comprises 8 genera which are grouped into 3 families: the alligators and caimans (**Alligatoridae**), the True crocodiles (**Crocodylidae**) and the gharials (**Gavialidae**). They are distributed in 23 species over all tropical countries. Whereas True crocodiles occur in Africa, Asia, Australia and America, alligators are restricted to North America and China, and caimans to South and Central America; gharials occur only in India. Tab. 8 shows that species of all groups of crocodilians are utilised mainly for their skins, but also for meat. Only a few species, *Alligator mississippiensis, Caiman crocodilus, Crocodylus niloticus, -novaeguineae, -porosus* and *-siamensis* are commercially farmed in greater numbers.

**Table 8:** Overview of the utilisation of crocodilians

| English name | Scientific name | Geographical distribution | Body length (cm) | Utilisation[1)] |
|---|---|---|---|---|
| **ALLIGATORS and CAIMANS - ALLIGATORIDAE** | | | | |
| American alligator | *Alligator mississippiensis* | N.America | > 350 | for skins, meat, show, farming |
| Chinese alligator | *-sinensis* | China | 200 | for exp. farming |
| Common caiman | *Caiman crocodilus* | S.-C.America | 150 | for skins, meat, farming |
| Broad-nosed caiman | *-latirostris* | S.-C.America | 230 | for skins, farming |
| **TRUE CROCODILES - CROCODYLIDAE** | | | | |
| Nile crocodile | *Crocodylus niloticus* | Africa | > 400 | for skins, meat, |
| New Guinean crocodile | *-novaeguineae* | N.Guinea, Sulu, Philippines | > 200 | farming, curios |
| Australian crocodile | *-johnsoni* | N.Australia | > 200 | |
| Marsh crocodile or mugger | *-palustris* | S.Asia | > 300 | breeding for recovery only |
| Estuarine crocodile | *-porosus* | S.-S.E.Asia, Polynesia, N. Australia | > 450 | for meat, skins, teeth, farming |
| Siamese crocodile | *-siamensis* | S.E.Asia | > 400 | for farming, skins, meat |
| American crocodile | *-acutus* | C.-S.America | > 450 | breeding for recovery, skins |
| Cuban crocodile | *-rhombifer* | Cuba | > 400 | breeding for recovery, skins |
| False gharial | *Tomistoma schlegeli* | S.E.Asia | > 300 | for skins |
| **GHARIALS - GAVIALIDAE** | | | | |
| Gharial | *Gavialis gangeticus* | India | > 300 | breeding for recovery only |

Apart from the above some other species are kept and locally utilised in small numbers only: *Melanosuchus niger, Crocodylus intermedius, -moreleti, -mindorensis, -cataphractus, Osteolaemus tetraspis*.

1) utilisation of different parts of crocodiles varies locally

## Historical background

Crocodilians have probably been hunted for hundreds of years for their meat and skins. However, the tanning of crocodile skins and manufacture of high-quality leather goods from these requires special skills which were developed specifically in France and Italy since the 1880s, later on in Germany and after the Second World War in Japan. In fact, just a small number of specialist tanneries in these countries today handle more than 80% of the world's production of crocodile skins (15). Specialised reptile-leather industries have developed in these countries which supply most of the finished crocodile leather articles. Tanning of crocodile skins and manufacture of reptile-leather goods in the producer countries rarely attains the same sophistication and mainly serves the tourist markets.

The increasing demand from the European reptile-leather industry in the 1920s and 1930s led to ever more intensive hunting of crocodiles in Africa, America and Asia. As access to the remoter parts of these continents became easier, the efficiency of hunting increased and the shipping out of the skins was facilitated; crocodile hunting - like the procurement of ivory - became highly commercialised and was solely directed by the reptile-leather industries. As a particular species of crocodile in any particular area became depleted, intensive hunting shifted to other areas or targeted alternative species. This pattern is particularly well documented for South America where first the *American crocodile*, then the *Orinoco crocodile* and finally the *different caiman species* were exploited to near extinction (19, 20).

The international trade in crocodile skins reached its peak during the 1950s and early 1960s when an estimated 5-10 million *caiman* skins of all sizes and about half a million from the *"classic" crocodile species* were marketed each year (10, 15). Caimans provided the bulk of lower-value skins for belts, shoes and other less valuable items which can be made out of the narrow, boneless "flank" portions of the skins. The classic high-value crocodile hides originate from those species, such as most *True crocodiles* and *alligators*, which have little or no belly ossification.

Local reports highlight the tremendous drain of crocodiles during this period of indiscriminate over-exploitation. For example, according to premiums paid by the German colonial administration of modern-day Tanzania, one particular individual killed in 1910 about 1,600 crocodiles around Lake Rukwa in just two months. In 1950 no less than 12,505 skins were still being produced, mainly from that area. In the Okavango swamps in Botswana, one particular crocodile hunter earned about DM 250,000 a year (approx. US$ 62,600) in the 1960s, killing an estimated 50,000 crocodiles (9). On the lagoons of Ivory Coast one French hunter alone killed about 2,000-3,000 crocodiles during the post-war years; by 1977 the Nile crocodile was almost extirpated in these lagoons (23).

In the 1970s the trade in crocodile skins started to decrease. This was mainly due to the general decline in the wild crocodile populations around the world and the resulting difficulties of obtaining skins. But the decrease was also prompted by a positive change of attitude towards crocodiles. Whereas previously crocodiles were regarded solely as undesirable vermin, from the late 1960s onwards their ecological role and economic value started to be appreciated. This in turn led to international protection of crocodiles in 1974/76 under the provisions of CITES. At the same time, the concern of the reptile-leather industries in

Europe, the USA and Japan about the future availability of sufficient raw material for reptile-leather production was growing. It caused promotion of surveys and capital investments into farming of crocodiles in many countries.

The first initiative to rear crocodiles in ponds for commercial purposes was taken in ***South-East Asia***. Using fish, chicken and pork offal as cheap food, crocodile farms were established in Singapore in 1945, near Bangkok in 1950, and near Djakarta in 1959. All three enterprises were initially set up as tourist attractions but also produced skins for local tanning and processing. In the 1960s crocodile farming expanded further in Thailand and Malaysia, Papua New Guinea, South Africa and the USA, now predominantly for commercial production of skins and live hatchlings for other farms. The first crocodile farms in ***Africa*** were set up in Zimbabwe in 1965 and 1967, with two others following in the 1970s. Here, the concept of crocodile "ranching" was developed, i.e. the utilisation of existing wild populations through controlled collection and hatching of their eggs. Farmers were obliged to return 5% of the reared crocodiles to the wild to compensate for the removal of eggs (although this was never enforced). In countries which still have important crocodile populations unaffected by agricultural development, this type of utilisation is certainly best suited to conservation. In the ***USA*** the Louisiana Department of Wildlife and Fisheries initiated experimental farming of alligators in 1964. By 1984 13 commercial farms were operative with a total stock of 13,000 alligators (4). Farming of caimans in ***South America*** developed later, probably because of the lower value of their skins. However, since the late 1980s Colombia has the potential to become one of the most important crocodile producers, with 64 established farms holding a total of 133,000 crocodiles. In 1991 the majority of these had not yet been authorised for commercial production (17).

By the early 1980s wild crocodile populations in some countries had increased again as a result of conservation measures. This and a growing awareness of the economic potential of crocodiles led to international pressure to liberalise the trade in crocodilian products. Thus, a resolution was adopted by the parties to CITES in 1983 under which certain populations which may benefit from "ranching", could be transferred from Appendix I (no trade) to Appendix II (controlled trade allowed). In 1985 a further resolution was adopted which allowed limited quotas of skins of Appendix I species to be exported, pending transfer of the population to Appendix II in the course of the development of "ranching" in the particular country. Such quotas are set by international agreement and may include the skins of wild as well as farm-reared crocodiles. Since different species and subspecies or local populations can be and, in fact, have been transferred from Appendix I to II and vice versa, the CITES control mechanism for crocodilians has become more complex than with any other animal group. Nevertheless, as discussed below, it has furthered the economic use of crocodilians while at the same time quite effectively curbing the former large-scale indiscriminate hunting of crocodiles.

## Management and utilisation

The use of crocodilians for **subsistence purposes** is rather limited although crocodile meat is relished by many societies. In some regions live crocodiles are offered on markets for food, as for example the *West African Dwarf crocodile* (*Osteolaemus tetraspis*) in Congo

(13). Normally they are used for the commercial production of skins, and the utilisation of wild populations of crocodiles to this end is still prevalent although the establishment of captive stocks for commercial farming is rapidly gaining importance. Both types of utilisation are legitimate and may be complementary if the harvesting of the wild population is based on sound management plans.

The **cropping of wild crocodilians** may be aimed at their total removal or reduction in areas where they are considered to be a public hazard or where they conflict with agricultural development. Elsewhere, the management goal should be the sustainable harvesting of the *wild crocodile resources* for economic benefit as has been initiated in the USA in 1971 (in Louisiana) (16) and 1981 (in Florida) (11), and in Venezuela since 1983 (22). Cropping is based on a 7-8% annual off-take quota of predominantly males which does not appear to have a negative impact on the wild populations. Wild populations may also be utilised by collection of eggs or hatchlings for farming purposes. It should be noted, however, that the production of skins from wild populations has different trade implications from farming, as discussed below.

The **farming** of *crocodilians* may be based either on the artificial incubation of eggs collected from the wild or the rearing of wild-caught hatchlings. In both cases farming remains dependent on wild populations and is termed by CITES as "**ranching**". After establishment of initial stocks from the wild, farming may also aim at breeding, and in this case farming will gradually become a closed-cycle operation completely independent of wild stocks. Some farms are also being established with stock purchased from other breeding operations. The different approaches to crocodile farming depend on ecological and political conditions as well as economic considerations. Farming practices vary therefore from country to country and with species.

3-year old Nile crocodiles reared in Spencer's Crocodile Farm at Victoria Falls, Zimbabwe. (Photo: V.J. WILSON)

So far only a few farms breed crocodiles in significant numbers, although they may have started breeding while still replenishing their stock from the wild. *Alligator mississippiensis* is bred in substantial numbers in the USA, *Crocodylus niloticus* in Zimbabwe, South Africa and Botswana, *C. porosus* in Thailand and Australia, and *C. siamensis* in Thailand. Other species are bred so far mainly for conservation purposes; these include *A. sinensis* in China, *C. rhombifer* in Cuba, *C. moreleti* in Mexico, *C. palustris* in India and *C.acutus* in Honduras and Columbia (17). The breeding efforts have unfortunately resulted in an undesirable spread of crocodile species into countries far beyond their original range. In 1991, for example, there were stocks of *A. mississippiensis* held in Australia and Israel, *A. sinensis* in Cambodia, *Caiman crocodilus* in India, Thailand and Taiwan, *Crocodylus niloticus* in Brazil, Israel and France, *C. novaeguineae* in Singapore, *C. moreleti* in Thailand, and *C. rhombifer* in Thailand and Vietnam. *Feral caiman populations* from escaped animals are already causing problems in Cuba, Puerto Rico and the USA (17).

In some countries both eggs and hatchlings are collected. In the Northern Territory of Australia, for example, harvests from wild crocodiles are composed of 30% eggs and 70% hatchlings, as well as some adult nuisance animals captured alive (4). In Papua New Guinea crocodile farming was developed as a peasant activity entirely based on the collection and rearing of hatchlings (7).

If breeding is attempted, special enclosures are required in which the females can be induced to nest. Depending on the species breeding groups may comprise up to 20 females and one male, unless several males are kept with a greater number of females in a larger communal enclosure (15). Under the more favourable conditions of captivity crocodiles sometimes attain sexual maturity much earlier than in the wild; e.g. *American alligators* and *Nile crocodiles* at 6-8 years, *Estuarine crocodiles* at 8-10 years and *Common caimans* at about 3 years (15).

Skins of *Caiman crocodilus* and *Crocodylus novaeguineae* are still mainly produced from wild harvests and only small quantities come from "ranching". From the other main skin producing species more than 50% of the legally traded skins in 1989/90 originated from farming. Only for *C. porosus* and *C.siamensis* does breeding predominate over ranching (17).

Farming operations may vary greatly from region to region. In some countries, such as Papua New Guinea and Indonesia, they may entail the keeping of only a few animals in village ponds by peasants or in backyard facilities in the urban centres as a side activity. In others, such as South Africa, Zimbabwe and the USA, crocodile farming may be an industrialised and highly sophisticated enterprise with many thousands of crocodiles reared, sometimes even in temperature-controlled tiled tanks. The collection of eggs from the wild for farming is usually either concessioned, licensed or otherwise controlled. In some countries, such as Zimbabwe and Zambia, it is conditional upon the return of a certain percentage of viable juvenile crocodiles (usually 1 m long) to the wild population (5). With hole-nesting species, such as the *Nile crocodile*, locating the nesting sites requires experience, whereas the recovery of eggs from mound nests, such as *alligators* and *Estuarine crocodiles*, is much easier. In both cases eggs have to be transported very carefully in order to prevent early mortality of the embryos. Whether collected in the wild or produced in the farm, eggs are always incubated artificially under conditions of high humidity. Hatching rates may vary between less than 50 and 95% depending on the species, facilities and care. In Zimbabwe, the overall hatching rate of 22,629 eggs collected between 1967 and 1973 from the wild was 73.6% (5) but has since been improved to 80-95% (4). The Samutprakan Farm in Thailand reported only a 60% average hatching success with 5,740 eggs laid in captivity (1).

The hatchlings are placed in tanks or small open pens and sometimes have to be induced to feed. This can be achieved with live food (winged termites, tadpoles, small fish etc.) or by introduction of other small feeding animals to "teach" the newborn. Later, the young crocodiles are graded according to size and physical condition and placed in minimum-stress calculated densities into different cemented ponds for further rearing. Growth rates vary according to species, feeding regimes and climatic conditions. Efficient farmers will grow a significant percentage of their stock to 0.90-1.00 m in 12 months and 1.3-1.4 m in 18 months, but the majority will only reach 1.5 m in 24 months or longer (14). At this size, i.e. after 18-30 months of rearing, the crocodiles are usually slaughtered.

Tab. 9 shows that the majority of crocodile farms hold either between 1,000 and 3,000 animals (32%) or less than 1,000 (51%). The biggest farms maintain about 13,500 *alligators*, 31,700 *caimans* or 27,300 *classic crocodiles* (17). In order to produce 1,000 alligator skins annually, a stock of about 3,500-5,000 animals has to be maintained (4). Therefore, a large percentage of the farms are still too small to be economically viable, as discussed below.

**Table 9:** Stock held in registered commercial crocodile farms in 1990 (after LUXMOORE (17))

| Crocodile species[1)] | Number of commercial farms holding | | | | | Total stock held |
|---|---|---|---|---|---|---|
| | < 1000 | 1000-2999 | 3000-5999 | 6000-9999 | > 10000 crocs. | |
| *Alligator* spp.[2)] | 53 | 20 | 10 | 1 | 3 | **490,012** |
| *Caiman* spp. | 32 | 53 | 5 | 3 | 2 | **165,625** |
| *Crocodylus* spp. | 92 | 38 | 19 | 8 | 10 | **434,712** |
| Total number of farms | 177 (51%) | 111 (32%) | 34 (10%) | 12 (3%) | 15 (4%) | **1,090,349** |

1) excluding species which are farmed mainly for recovery programmes (*Alligator sinensis, Caiman latirostris, Crocodylus acutus, -intermedius, -rhombifer, -moreleti*); 2) not included are 224 farms in Louisiana, USA, for which no details are recorded

Slaughtered crocodiles are flayed from top to bottom keeping the belly part intact as the commercial value of a crocodile hide depends on the width of the belly portion only. Skins are then cured with salt or brine for shipment to specialised tanneries. Only exceptionally are they sold locally or even tanned at the farm, as for example, in Singapore. In some countries, such as USA, South Africa and Zimbabwe, the tail meat is processed into deep frozen fillets which are marketed to restaurants and hotels. The remainder may be used as animal feed. In Asian countries any reptile meat is popular with the indigenous population, and crocodile meat is offered either fresh or in dried form on local markets. A farm crocodile of 1.4-1.5 m length weighing with bone about 11 kg will yield about 2.3-5.5 kg fillet meat (15). Apart from skin and meat, teeth and gall-bladders are utilised in Asia for medicinal purposes and a variety of curios produced for tourists. The cloacal gland might be used for musk.

The farming success depends on three main factors: adequate water supplies, availability of cheap food, and prevention of youth mortality. All three are interlinked and the outbreak of diseases is often caused by unhygienic water conditions, wrong feeding or other management faults. However, much research has gone into the physiology and pathology of crocodilians and its results enable efficient farms to keep hatchling mortality (up to one year of age) below 5%. The nutritional factor determines not only the health and growth of the farmed crocodiles but has also an important economic implication. As carnivorous animals crocodiles compete with man in the use of animal protein. Therefore crocodile farming makes sense only where the human population does not consume all and any available animal protein, which occurs in many developing countries. Crocodile farms are economically feasible only where sufficient quantities of small or unpalatable fish, offals or waste from fisheries industries, abattoirs, poultry farms, game meat from control operations, snakes or other alternative meat sources are readily available.

## Economic importance

As with other groups of animals the economic importance of the crocodilians is well reflected by the ***trade in crocodile skins***. After this had peaked in the 1950s and 1960s it declined rapidly due to over-exploitation of the resources. By 1984 only 100,000-150,000 skins of the "classic" species were still being traded; also, the production of caiman skins, which had surged only after the American crocodile species had been depleted below exploitable levels, had probably decreased to less than a million skins per annum (10, 15). The export value of crocodile skins in the form of much needed foreign exchange was considerable. In Colombia, for example, export revenue from more than 5 million tanned and raw caiman skins between 1970 and 1978 was calculated to be US$ 33.3 mill., i.e. at an average US$ 3.7 mill. per annum (21).

Fig. 6 shows how the trade in classic skins from the four main species, *Alligator mississippiensis, Crocodylus niloticus, -novaeguineae* and *-porosus*, has gradually increased again, since implementation of CITES control to a total of about 183,000 in 1990, whereas the legal trade in caiman skins has further decreased to only 189,000. This reflects the shift from exploitation of wild populations to production in farms. Of the total of 376,323 skins entered in the legal trade statistics in 1990, 51.6% had originated from wild crocodiles and 48.4% from ranched or captive-bred animals (17). However, substantial numbers of crocodile skins from wild populations, mainly of *Caiman crocodilus* and *Crocodylus porosus*, are not recorded in the official statistics, whereas skins produced in farms are likely to be always officially registered. Therefore utilisation of wild crocodiles is still much more significant than the trade statistics suggest. However, it may be expected that the production from wild crocodiles will further decrease in favour of farming.

**Figure 6:** Estimates of world net trade in skins for the main four species of classic crocodilians (after LUXMOORE (17))

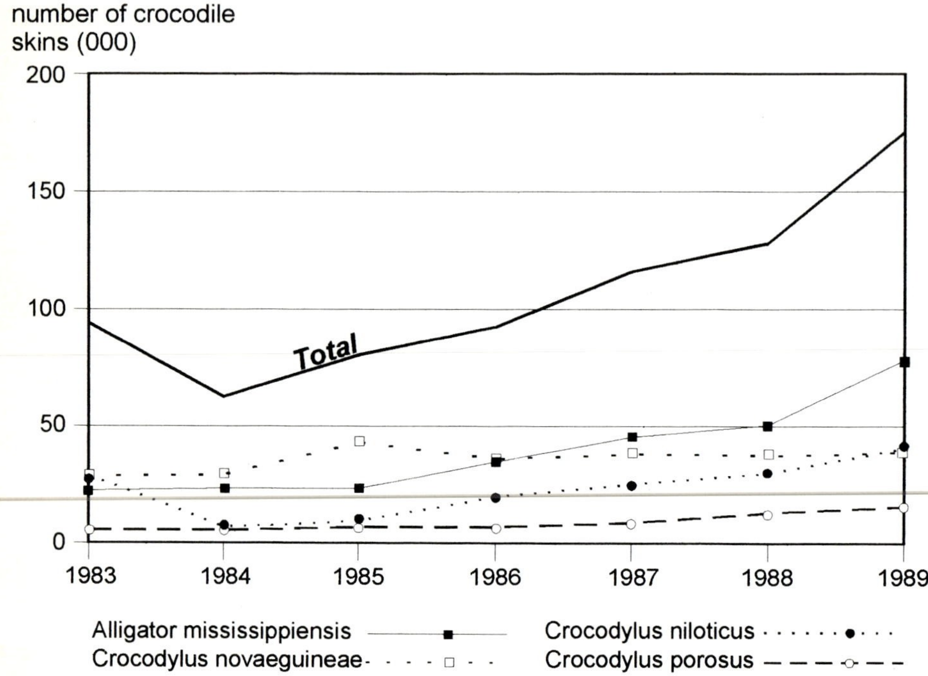


Tab. 10 shows how significantly crocodile farming has increased from 1979, when only 29 commercial farms were registered with CITES, to 1989 when there were more than 600 farms operating in 37 different countries (17). More than a million crocodiles are kept in these farms from which the above mentioned 183,000 skins were produced. Most of the stock of these farms were reared from eggs or hatchlings collected from wild populations, of which many were sold alive. The strong demand for captive-bred hatchlings indicates that the number of farms and the total crocodile stock is still growing. This implies that production, mainly of classic skins, is still increasing. While this may stimulate the fashion leather industry into increasing manufacturing, it will also have the negative effect of causing an oversupply of skins resulting in a slump in skin prices.

It is therefore important to analyse the ***economic productivity of crocodile farming*** with a view to elucidating how this might be affected by a market depression. The profitability of industrialised intensive farming of the classic crocodile species is generally governed by the amount of capital invested, the cost of feed and the returns from the sale of skins. Relatively high capital investments are required because, apart from the establishment of adequate facilities, initial running costs also have to be met for some time. Crocodile "ranching" may produce its first skins only after 2-3 years and the establishment of "closed cycle" skin production will take a minimum of 8-10 years. It is therefore desirable that crocodile enterprises at least initially seek alternative sources of revenue, such as tourism and curio sale, or are developed in association with other enterprises such as fisheries, poultry or game farms.

**Table 10:** Development of crocodile farming since 1945 (after LUXMOORE (17))

| Countries | Number of farms[1)] in | | | | Stock held in 1989/90 | | Production in 1989/90 | |
|---|---|---|---|---|---|---|---|---|
| | 1959 | 1969 | 1979 | 1990 | adult | **total** | skins | hatchlings |
| Singapore | 1 | 2 | 2 | 5 | 2,455 | **11,302** | 2,864 | 4,199 |
| Thailand | 1 | 3 | 6 | 45 | | **44,684** | 3,000 | 8,714 |
| Indonesia | 1 | 1 | 1 | 36 | 2,023 | **94,207** | 5,552 | 652 |
| Malaysia | - | 2 | 2 | 4 | 690 | **3,563** | 295 | 1,846 |
| Papua New Guinea | - | - | 3 | 6 | 192 | **27,301** | 5,526 | 1,368 |
| Australia | - | - | 1 | 13 | 1,746 | **42,421** | 3,566 | 4,854 |
| Oth. Asian-Pacific countries (6) | - | - | - | 40 | | **4,062** | 2,000 | 1,849 |
| Zimbabwe | - | 2 | 4 | 32 | | **122,854** | 17,567 | 23,348 |
| South Africa | - | - | 1 | 35 | | **33,631** | 4,747 | 7,776 |
| Oth. African countries (12) | - | 1 | 1 | 34 | | **51,062** | 8,354 | 6,235 |
| USA | - | - | 4 | 224 | | **489,844** | 45,469 | 30,089 |
| Cuba | 1 | 1 | 1 | 5 | 1,600 | **11,099** | 500 | 7,349 |
| Colombia | - | - | - | 64 | | **133,625** | 54,452 | 86,128 |
| Venezuela | - | - | - | 17 | | **21,974** | 28,136 | 5,300 |
| Oth. Lat.American countries (5) | | | | 33 | | **6,200** | 7 | 1,871 |
| Israel | - | - | - | 3 | | **3,003** | 800 | 500 |
| France | - | - | - | 1 | | **300** | - | - |
| **All countries (37)** | **4** | **12** | **29** | **597** | | **1,101,132** | **182,835** | **192,078** |

1) commercial farms only; in 1990 there were additional 75 non-commercial breeding operations registered

The Samutprakan crocodile farm in Thailand, which earned nearly US$ 1 mill. in 1982, began with a capital investment of just US$ 500 but took 10 years to become fully commercial. During initial development it was financed exclusively through tourism (8). In the USA in 1982 about US$ 150,000-300,000 had to be invested during the first three years to develop an alligator farm which would ultimately market 1,000 skins per year (2). In Botswana it was estimated in 1989 that a capital investment of between Pula 1,210,000 (approx. US$ 606,000) and Pula 1,900,000 (approx. US$ 952,000) was required for farms intending to produce 1,000-5,000 skins per annum, partly from wild eggs and breeding stock (3). A detailed study in this country determined productivity through cost-benefit analysis and some of its results are summarised in Tab. 11. Under the studied conditions it was obvious that only larger production units would be profitable. The break-even point of cost-benefit was found to be at a production level of about 2,000 skins per annum. It also shows how sensitive crocodile enterprises are to the cost of feed. If these rose above Pula 1.50/kg the return on the investment became unacceptable. For the production of 5,000 skins/annum about 234 t of feed are required. In terms of meat this would be equivalent of about 1,000 cattle, in terms of fish it would equal production from 40 ha of fish ponds, and in terms of poultry farms offals from production of about 1.6 million chicken. This highlights the economic limitations of crocodile farming in countries in which there is a general shortage of animal protein for human nutrition. Finally Tab. 11 shows that only large production units will be able to cope with market fluctuations in which the price of skins drops more than 20%. This implies that many of the established smaller crocodile farms around the world might find it difficult to survive market depressions which result from oversupply of skins.

**Table 11:** Economics of intensive crocodile farming in Botswana in 1989 (after BARNES (3))

| **a: Dependance of returns on total capital investment, i.e. scale of operation**[1] | |
|---|---|
| Scale of operation | % cash return on capital investement |
| Farm producing 1,000 skins/annum | - 19.49 |
| Farm producing 5,000 skins/annum | + 19.35 |
| Farm producting 10,000 skins/annum | + 38.96 |
| **b. Dependance of return on variation of skin prices**[2] | |
| Variation in price | % cash return on investment in farm producing 5,000 skins/annum |
| **normal price** (US$ 4.50/cm) | **19.35** (see (a) above) |
| higher prices + 10% | 25.56 |
| + 20% | 31.80 |
| lower prices - 10% | 13.13 |
| - 20% | 6.91 |
| - 30% | 0.68 |
| - 40% | - 5.54 |
| **c. Dependance of return on different food costs**[1] | |
| Cost of feed (Pula/kg) | % cash return in farm producing 5,000 skins/annum |
| 0.50 | 26.99 |
| 1.00 | 19.35 |
| 1.50 | 12.26 |
| 2.00 | 5.64 |
| 2.50 | - 0.53 |

1) based on prices in Botswana in 1989: 1 Pula = US$ 0.501; 2) normal price of first grade skin in 1989: US$ 4.50/cm

Quite different economic parameters apply to the production of the lower-value caiman skins which still constitute the bulk of the crocodile-skin trade. Rearing caimans costs less because of their smaller size, but the lower prices paid by the tanning industry have to be offset by producing greater numbers of animals and selling meat. In the long run controlled harvesting of the large caiman populations on the extensive South American wetlands, such

Local processing of crocodile leather into belts and tourist curios at Spencer's Crocodile Farm in Victoria Falls, Zimbabwe. (Photo: H.H. ROTH)

as the Llanos in Venezuela and Columbia and the Pantanal in Brazil, Bolivia and Paraguay, might be more economically viable (18). In fact, revenue to landowners in the Venezuelan Llanos from caiman harvesting in 1987 was comparable to the proceeds from cattle ranching (22).

The profitability of the more extensive village production, which has developed in Papua New Guinea and Indonesia, depends mainly on an efficient local marketing network. The widely scattered village farmers must be able to sell their young crocodiles if and when they run seasonally out of fish feed and village ponds dry up. The Papua New Guinea Government has thus set up collection points from which the young crocodiles are shifted, periodically, by light aircraft to the very large industrial crocodile farms on the coast. These hold up to tens of thousands of crocodiles and market the skins overseas. The system works only as long as the Government controlled prices for the young live animals are high enough to encourage the village hunters to spend time on capture instead of just collecting skins (6, 12).

With the possible decline of skin prices, meat assumes increasing importance as the most important byproduct of crocodile farming. Its commercialisation may significantly add to the total return from the hides. The meat of small farm-raised American alligators, for example, contributed in 1991 35-40% to their total value (15), and in South America and Africa crocodile meat is becoming a highly prized export commodity. In Asia, of course, many crocodiles are kept primarily for meat production. In Taiwan, for example, an estimated 20 farms were producing in 1991 about 20 t of meat from caimans (13). Korean importers offered US$4/kg for frozen caiman meat from Venezuela, increasing its local market value some eight fold (22).

International marketing of crocodile products always depends to a large extent on the particular CITES status which the producing country has. As described above, most crocodilian species are listed in Appendix I of CITES and international trade in their products is therefore not allowed. Only if a country has successfully applied for the transfer to Appendix II of the wild population on which ranching is based, or has been issued an interim quota pending such transfer, can skins and meat be legally traded. If they derive from breeding operations they qualify right away. Crocodile producers in non-CITES countries will find it very difficult to market their products because most of the traditional reptile-leather tanneries and manufacturers are located in CITES countries and thus not allowed to legally receive such products.

## Implications for conservation

There is no doubt that the over-exploitation of wild populations of crocodilians in the pre- and post-world war period has severely diminished these and in many areas has brought them to the verge of extirpation. However, it is equally evident that the subsequent combination of international trade control with rational management and farming of crocodiles has had a most beneficial conservation effect. This has allowed the removal of populations of several crocodile species from Appendix I to Appendix II of CITES.

Of the two farming concepts, "ranching" is clearly preferable from a point of view of conservation because it provides an economic link to the wild population. It generates a general awareness of the ecological and economic value of crocodilians and, in conjunction with the controlled cropping of the wild population, gives an economic justification for their existence. But breeding programmes have also contributed to the conservation of such species as *Crocodylus siamensis* and *Alligator sinensis*. These have survived only in captivity, where they have by now been propagated to utilisable populations. Others, such as all of the American *Crocodylus* species which had been the first to be depleted, are yet to benefit from more recently initiated captive breeding.

Wild populations of several of the classic crocodile species, such as *Crocodylus niloticus* and *Alligator mississippiensis*, as well as *Caiman* species have re-increased in parts of their range to levels at which they can again be harvested. However, the previous unregulated and often illegal cropping should be replaced by long-term sustained-yield harvesting. Good examples of such economically viable programmes which also contribute directly to the conservation of the respective species, exist in the USA and Venezuela (22).

The crocodilians thus provide an example of how controlled utilisation can foster conservation to the point at which a previously endangered species reverts to becoming an economically valuable resource.

## Literature references

1. ASHLEY, J.D. (1980): Crocodilian farming: past, present and future (*cited after* Barzdo and coll. 1984)
2. ASHLEY, J.D. (1982): An overview of Alligator farming in the United States. *Proceedings of the Symposium on Crocodile Conservation and Utilisation, Sept. 1982, Victoria Falls, Zimbabwe;* IUCN, Gland, Switzerland

3. BARNES, J.I. (1992): *Economic aspects of crocodile farming and ranching: the case of Botswana*. CSERGE Working Paper No. GEC 92-20; Centre for Social and Economic Research on the Global Environment (CSERGE), London
4. BARZDO, J., LUXMOORE, R.A., BROAD, S. and D. JONES (1984): *A survey of wildlife farming operations*. WCMC/IUCN, Cambridge (unpubl. report)
5. BLAKE, D.K. (1982): Crocodile ranching in Zimbabwe. *The Zimbabwe Science News*, 16, No. 9
6. BOLTON, M. (1980): Crocodile Management in Papua New Guinea. *World Animal Review*, 34, p. 15-22
7. BOLTON, M. and M. LAUFA (1982): The crocodile project in Papua New Guinea. *Biological Conservation*, 22, p. 169-179
8. BOONSONG, L. (1981): Wildlife resources for rural community development. *Tigerpaper*, 8 (3), p. 5-9
9. GRZIMEK, B. (1971): Das Nilkrokodil. In: *Grzimeks Tierleben*,Vol. 6, p. 135-136; Kindler, Zürich
10. HEMLEY, G. and J. CALDWELL (1986): The crocodile skin trade since 1979. In: *Crocodiles. IUCN Amphibia-Reptilia Red Data Book. Part 1: Testudines, Crocodylia, Rhynchocephalia*. IUCN, Gland, Switzerland
11. HINES, T. and C.L. ABERCROMBIE, III (1987): The management of alligators in Florida. In: *Wildlife management: Crocodiles and alligators*, ed. WEBB, G.J.W., MANOLIS, S.C., and P.J. WHITEHEAD; p. 43-47; Surrey Beatty, Chipping Norton, Australia
12. HOLLANDS, M. (1987): The management of crocodiles in Papua New Guinea. In: *Wildlife Conservation: Crocodiles and alligators*, ed. WEBB, G.J.W., MANOLIS, S.C. and P.J. WHITEHEAD; p. 73-89; Surrey Beatty, Chipping Norton, Australia
13. HUCHZERMEYER, F. (1993): *in litt.*
14. HUTTON, J.M. (1992): *in litt.*
15. HUTTON, J.M. and G.J. WEBB (eds.) (1992): *Directory of Crocodilian Farming Operations. Second Edition*, Part 1: An Introduction to the Farming of Crocodilians; p. 1-40; WCMC/IUCN, Cambridge and Gland, Switzerland
16. JOANEN, T. and L. McNEASE (1987): The management of alligators in Louisiana, USA. In: *Wildlife management: Crocodiles and alligators*, ed. WEBB, G.J.W., MANOLIS, S.C. and P.J. WHITEHEAD; p. 33-42; Surrey Beatty, Chipping Norton, Australia
17. LUXMOORE, R.A. (ed.) (1992): *Directory of Crocodilian farming operations. Second Edition*, Part 2; p. 41-336; WCMC/IUCN, Cambridge and Gland, Switzerland
18. MAGNUSSON, W.E. (1984): Economics, developing countries and the captive propagation of crocodilians. *Wildlife Society Bulletin*, 12, p. 194-197
19. MEDEM, F. (1981): *Los Crocodylia de Sur America*; Vol. 1. Universitá Nacional de Colombia, Bogota (cited after THORBJARNARSON 1991)
20. MEDEM, F. (1983): *Los Crocodylia de Sur America*; Vol. 2. Universitá Nacional de Colombia, Bogota (cited after THORBJARNARSON 1991)
21. PACHON-RIVERA, J.E. (1982): Algunos aspectos relativos a la conservación y manejo de los Crocodylia en Colombia (*unpubl. report - cited after* THORBJARNARSON 1991)
22. THORBJARNARSON, J.B. (1991): An analysis of the Spectacled Caiman (*Caiman crocodilus*) Harvest programme in Venezuela. In: *Neotropical Wildlife Use and Conservation*, ed. ROBINSON, J.G. and K.H. REDFORD; p. 217 - 235; University of Chicago Press, Chicago and London
23. WAITKUWAIT, E. (1988): Untersuchungen zur Erhaltung und Bewirtschaftung von Krokodilen in der Republik Côte d'Ivoire (Westafrika). PhD Thesis, University of Heidelberg, Germany

## 2.1.3 LIZARDS AND SNAKES (*SQUAMATA*)

Lizards (**Sauria**) and snakes (**Serpentes**) are suborders of the **Squamata**. The first one comprises numerous different families of which the large iguanas (**Iguanidae**), tegus and allies (**Teiidae**) are restricted to Latin America and the big Monitor lizards (**Varanidae**) to Africa, Asia, and Australia. Other smaller commercialised lizards belong to the **Agamidae, Gekkonidae** and **Chamaeleonidae**. The second suborder comprises 12 families with more than 2,000 species. Of these the non-venomous Sunbeam and Acrochordid snakes (**Xenopeltidae** and **Acrochordidae**) occur mainly in South-East Asia whereas the likewise non-poisonous boas and pythons (**Boidae**) are distributed across South America, Africa and Asia. The predominantly non-venomous colubrids (**Colubridae**) occur all over the world. The highly venomous cobras, kraits and their allies (**Elapidae**) are mainly distributed across Africa, Asia and Australia; the likewise highly poisonous vipers (**Viperidae**) are restricted to the Old World and the venomous Pit vipers (**Crotalidae**) to America and Asia. Seasnakes (**Hydrophiidae**) live in coastal waters of the Indian and Pacific oceans. Tab. 12 shows that iguanas, Monitor lizards and Boid snakes are widely utilised for their meat and skins. Many species of cobras, vipers and Pit vipers are used for venom extraction for medical purposes. In some regions various other snakes, including venomous species, are also eaten or used for traditional medicine. Whereas the keeping of venomous snakes for commercial purposes has been long established, more recently efforts are being made to farm iguanas and Boid snakes for leather production.

**Table 12:** Overview of the utilisation of lizards and snakes

| English name | Scientific name | Geographical distribution | Body length (m) | Utilisation[1)] |
|---|---|---|---|---|
| **SAURIA - LIZARDS** (non-poisonous) | | | | |
| Iguanas | *Iguana* spp. | S.-C.America | 1-1.5 | for meat, skins, farming |
| Tegus | *Tupinambis* spp. | S.-C.America | 0.8-1.3 | for skins, exp. farming, pet trade, |
| Caiman lizard | *Dracaena guianensis* | S.America | 0.5-0.75 | for skins |
| Monitor lizards | *Varanus* spp. | Africa, Asia, Australia | 1-2 | for skins, meat, fat, exp. farming |
| Spiny-tailed lizards | *Uromastix* spp. | N.Africa, W.-S.Asia | 0.5 | for meat, fat, medicine, skins, pets |
| Day geckos | *Phelsuma* spp. | Madagascar, Seychelles | 0.10-0.15 | for pet trade |
| Chameleons | *Chamaeleo* spp. | Africa, W.-S.Asia | 0.35-0.35? | for pet trade |
| **SERPENTES - SNAKES** | | | | |
| **SUNBEAM and ACROCHORDID SNAKES - XENOPELTIDAE - ACROCHORDIDAE** (non-poison.) | | | | |
| Sunbeam snakes | *Xenopeltis* spp. | S.-S.E.Asia | 1-1.5 | for skins |
| Acrochordid snakes | *Acrochordus* spp. | S.-S.E.Asia, Australia | 1-1.5 | for skins |

| BOID SNAKES - BOIDAE (non-poisonous) | | | | |
|---|---|---|---|---|
| Pythons | *Python-, Calabaria* spp. | Africa, Asia | 2-5 | mainly for skins, farming |
| Anaconda | *Eunectes murinus* | S.America | 2-5 | for skins, meat |
| Boas | *Boa constrictor, Epicrates-, Tropidophis-, Corallus* spp. | S.America and Asia | 1-2 | for skins, meat, exp. farming |
| Sand boas | *Eryx* spp. | N.Africa, W.-S.Asia | 0.75-1 | mainly for skins |
| **COLUBRID SNAKES - COLUBRIDAE** (non-poisonous with a few exceptions) | | | | |
| Various types of Water snakes | *Natrix-, Cerberus-, Homalopsis-, Enhydris* spp. | Europe, N.W.Africa, N.America, Asia, Australia | 0.75-1 | for skins |
| Ratsnakes | *Ptyas* spp. | S.-S.E.-E.Asia | 1.5-2 | for skins, meat |
| Tree snakes | *Boiga* spp. | S.-S.E.Asia | 1-2 | for skins |
| Trinket snakes | *Elaphe* spp. | Asia, S.Europe, N.America | 0.75-1 | for skins |
| **COBRAS, KRAITS and ALLIES - ELAPIDAE** (poisonous) | | | | |
| Cobras, kraits | *Naja-, Bungarus* spp. | Africa, Asia | 1-1.5 | for skins, meat, venom, charming |
| King cobra | *Ophiophagus hannah* | S.-S.E.-E.Asia | 2-5 | for skins, ritual use, venom |
| Death adders | *Acanthiops* spp. | Australasia | 1-2 | for venom |
| Tiger snakes | *Notechis* spp. | Australasia | 1-2 | for venom |
| Mambas | *Dendroaspis* spp. | Africa | 1.5-2.5 | for venom |
| Coral Snakes | *Micrurus* spp. | N.-C.-S.America | 0.5-1 | for venom |
| **VIPERS - VIPERIDAE** (poisonous) | | | | |
| Vipers | *Vipera-, Bitis-, Echis* spp. | S.Europe, Africa, Asia | 0.75-1 | for skins, venom |
| Puff adder | *Bitis arietans* | S.Africa | 1-1.5 | for venom, meat |
| Night adders | *Causus* spp. | Africa | 0.5-0.85 | for venom |
| **PIT VIPERS- CROTALIDAE** (poisonous) | | | | |
| Rattlesnakes | *Crotalus* spp. | N.-C.-S.America | 0.75-2 | for venom, meat, show |
| Lance-head snakes | *Bothrops* spp. | S.America | 1-2 | for venom |
| Asian Lance-heads | *Trimeresurus* spp. | S.-S.E.-E.Asia | 1-2 | for venom |
| Copperhead, moccasins | *Agkistrodon* spp. | Asia, N.America | 0.75-1 | for skins, meat, venom |
| **SEA SNAKES - HYDROPHIIDAE** (some species poisonous) | | | | |
| Various types of Sea snakes | *Hydrophis* spp.<br>*Laticauda* spp. | coasts of Indian and Pacif. Oceans | 0.75-2<br>0.75-1 | for skins, eggs, meat, oil |

1) types of utilisation vary locally

## Historical background

Palaeolithic cave paintings in Central India feature the hunting of *Monitor lizards* (13). In fact, these and many other taxa of lizards and snakes have played an important role in the subsistence economies of tribal peoples around the world since early times. In many countries they are still of considerable importance to the local communities. The larger species are relished food items, the skins are used for ornamental and ritual purposes or for making drums. Parts of some species are important as native medicine and talismans.

It was not until after the First World War that western fashion "discovered" the attraction of lizard and snake skins for leather manufacturing, although crocodile skins had been utilised much earlier in Europe for this purpose as had alligator skins in the USA (18). Lizard and snake skin was initially only utilised for upper leather in highly priced luxury fashion shoes, but later also used in mass-produced shoes, bags and other items. The rapid development of trade in squamate skins was due to the technical development of industrial tanning and the appreciation of the special properties of this commodity: it was available in a variety of tones and scalation patterns, hard wearing, and it was possible to finish it in a variety of colours. Thus large numbers of squamate skins started to enter commerce from 1926 onwards and in 1932 an estimated 12 million squamate reptiles were killed for the sake of their skins (24). The international trade in reptile skins reached a peak after the Second World War: it is reported that between 1949 and 1957 Britain alone imported an average of 12 million reptile skins per year and the USA 4-8 million. At least half of these were lizard and snake skins (23). The ever increasing demand for lizard and snake skins generated cash returns to the tribal hunters in the countries of origin of the skins, changing their lives drastically. This in turn escalated the exploitation of reptiles into the most remote areas of these countries, the main targeted species being *Monitor lizards, iguanas, pythons, boas, cobras* and *ratsnakes*.

Apart from the international skin trade, the consumption of reptiles as food has also developed into a commercial activity in some regions of the world. In Central and some South American countries *iguanas* have been eaten almost out of existence. In Asia there has always been a market for reptiles as food and specific *parts of snakes as traditional medicines*, and the commerce in these items is flourishing today in the big cities of China, Taiwan, Thailand and in Hong Kong. The fat and meat of *Monitor* and *Spiny-tailed lizards* are considered strength givers and aphrodisiacs in India. Snake blood is drunk and gall bladders are eaten throughout Asia for the same reasons and other purported remedies.

More recently the zoo and pet trade has become another significant consumer of a wide array of live lizards and snakes. It has often exploited the larger, spectacular and rare species that cannot take much pressure. Scientific research has also made particularly heavy use of lizards and snakes. One particular study in Borneo, for example, sacrificed over 3,000 lizards; another one utilised 14,000 Sea snakes of one species alone.

This rather indiscriminate use of lizards and snakes in combination with the continuous loss or change of habitat, has led to a drastic reduction of these reptiles in many parts of the world. As a result, many countries, for example, India in 1976, have banned lizard and snake skin exports altogether. Others, such as Panama, Costa Rica and Columbia have initiated iguana rehabilitation and farming projects. Unfortunately, unlike the crocodile skin trade, the international lizard and snake skin trade is not yet sufficiently monitored nor have efforts been made towards sustainable utilisation of this resource. In general, conservation awareness and legislative measures are usually first afforded to the larger more popular mammals and birds, whereas lizards and snakes are ignored or even considered undesirable vermin.

## Management and utilization

The great majority of traded lizard and snake skins are still produced through **trapping and collecting** of the animals in the wild, either by professional hunters or by villagers as a side-line. According to the intended use they will either take them alive or kill them for skinning and for food. In the remoter areas of many developing countries collecting snakes provides (like other wildlife use) one of the few opportunities for earning cash. The ban on exports of reptile skins which was imposed in many countries during the 1980s as a counter measure against over-exploitation, has thus affected a large number of village hunters whose primary source of income had been snake catching for skin exports. A project to off-set this negative impact on the Irula tribe near Madras in India, stands out as a model for the sustained yield utilisation of snakes in the context of community managment of wildlife resources (30). The Irulas (among the poorest tribal people) were encouraged to form a Snake Catchers Co-operative licensed by Government to catch snakes, not for killing them but for extraction of venom and subsequent release. Since 1978 about 60,000 snakes of the four most common venomous species, *cobra, krait, Russell's viper* and *Saw-scaled viper*, have been caught, had their venom extracted and then been released back into the wild. The Coop may hold around 600 snakes at peak times and houses the snakes temporarily for venom extraction. While the snakes are held in captivity for three weeks they are fed just glucose water. According to the records of the Coop, over 1,500 g of dried venom has thus been produced and sold to antivenom-serum manufacturers. In Argentina the harvesting of *tegu lizards* for their skins is a widely established tradition. The hunters, some professional but mostly occasional, track the tegus to their burrows with the aid of trained dogs where they are dug out to be taken alive. About half of the families involved eat the tegu meat, and tegu fat is valued for medicinal purposes. The skins are sold or bartered to roving tradesmen, so-called "acopiadores" who resell them either to other middlemen or direct to the tanneries. Thus millions of these lizards are annually exploited and provide significant monetary returns to the rural populations (16).

The trapping or hunting of wild lizards and snakes primarily for food is still common. This involves in Mexico, the Carribean and some South American countries mainly *Iguana iguana,* in Asian countries mainly *Varanus* and *Uromastix* spp. and in the Far East most snakes including the venomous species. It usually involves trapping and marketing them alive. Some particularly cruel practices are employed such as boiling Spiny-tailed lizards alive and extracting fat from living Monitor lizards. *Sea snakes (Hydrophiidae)*, whose meat has a protein content of over 68% and is considered to have aphrodisiac qualities, are also used. Their gall bladders are supposedly helpful for stomach troubles and the oil derived from the fat is prized. In the Philippines, for example, *Laticauda* and *Hydrophis* spp. are caught in great numbers on or near isolated islets and reefs where they congregate to breed (15).

Efforts at **farming lizards and snakes** for meat and skin production on a sustained-yield basis are still in their infancy. In Colombia, the keeping of *Iguana lizards (Iguana iguana)* and *boas (Boa constrictor)* has been widely integrated with crocodile farming. According to a survey by the World Conservation Monitoring Centre in 1989/91 both are reared from wild-collected eggs in 31 farms under the supervision of the government's Instituto de Recursos Naturales y del Ambiente (INDERENA) (19). Of these, however, only eight had started commercial production by 1991. In Panama and Costa Rica the Foundation for the Green Iguana, after five years of research, set up an Iguana Management Project (IMP) in the 1980s which is aiming to help farmers and landowners to raise Green iguanas for harvesting and marketing as food (25). The iguanas are hatched from wild-collected eggs and reared in captivity to the age of 7-9 months. After branding they are then released on Private land and can be harvested at 2-3 years of age for the local consumption of meat and for skin export. Over 10,000 young iguanas have been released so far resulting in a 45-fold survivorship of the naturally laid eggs and hatchlings. It is estimated that current management techniques will produce a more than five-fold increase in the adult iguana population (28). Considering the extensive use of *Monitor lizard* skins, efforts are also being made to farm *Varanus* species in Thailand, Philippines, Pakistan and India. However, their eggs require eight months or more incubation. This fact and their aggressive and combative behaviour have so far prevented Monitor lizard farming from becoming an economic venture (1). In Argentina, efforts are being made at several places to rear *tegus* (*Tupinambis teguixin* and - *rufescens*) in captivity. The aim is to replace eventually the widespread traditional hunting of tegus by the indigenous communities and colonists in the northern part of the country with small scale family-run farming, or simply to produce tegu skins more economically on a large scale for export. Reproduction in captivity was achieved in 1987/88 (16). *Pythons* are bred in greater numbers for commercial purposes at the Samutprakan Crocodile Farm in Thailand (10, 19). With a view to the special preference for snake meat and organs in China, farming of a *variety of snakes*, including *Python* spp., and *Naja* spp., has been inititated in that country. One particular farm in Wuzhou, Guangxi Province, is allegedly producing 500,000 snakes per year to supply gastronomic and medicinal demands in Hong Kong and Macau. Other smaller farms exist in the Fujian Province, one holding about 1,000 snakes and supplying within two years of operation 500 kg snake meat and 2 kg gall bladders for local markets and 3,000 skins for

Lizard catchers in Bangladesh offering Yellow and Common Monitor lizards for sale. (Photo: R. WHITAKER)

export (10). Snakes are said to be reared from wild-collected eggs and kept either in open fields or pits, however, it is likely that most snakes are wild caught, as rearing baby snakes in large numbers is costly and difficult.

**Table 13:** Venom yields for anti-serum production in the Butantan Institute in Brazil during 1948-1951 (after BÜCHERL (11))

| Snake species | Venom extractions | | | | |
|---|---|---|---|---|---|
| | 1st | 2nd | 3rd | 4th | 5th |
| ***Crotalus terrificus*** | | | | | |
| No. of snakes used | 4,775 | 4,499 | 2,298 | 600 | 28 |
| Liquid venom/snake (in ml) | 0.165 | 0.079 | 0.032 | 0.058 | 0.096 |
| (dry venom/snake (in g)) | (0.033) | (0.016) | (0.007) | (0.014) | (0.023) |
| Greatest individual yield (dried) | 0.185 g | 0.088 g | 0.049 g | 0.035 g | 0.025 g |
| ***Bothrops jararaca*** | | | | | |
| No. of snakes used | 11,184 | 7,157 | 926 | 183 | |
| Liquid venom/snake (in ml) | 0.136 | 0.084 | 0.122 | 0.135 | |
| (dry venom/snake (in g)) | (0.036) | (0.021) | (0.028) | (0.030) | |
| Greatest individual yield (dried) | 0.185 g | 0.100 | 0.060 | 0.100 | |
| ***Bothrops alternatus*** | | | | | |
| No. of snakes used | 712 | 628 | 142 | 54 | |
| Liquid venom/snake (in ml) | 0.36 | 0.26 | 0.298 | 0.25 | |
| (dry venom/snake (in g)) | (0.070) | (0.048) | (0.063) | (0.051) | |
| Greatest individual yield (dried) | 0.435 g | | | | |
| ***Bothrops cotiara*** | | | | | |
| No. of snakes used | 969 | 577 | 346 | 221 | |
| Liquid venom/snake (in ml) | 0.200 | 0.123 | 0.164 | 0.175 | |
| (dry venom/snake (in g)) | (0.041) | (0.035) | (0.039) | (0.037) | |
| Greatest individual yield (dried) | 0.185 g | | | | |
| ***Micrurus corallinus*** | | | | | |
| No. of snakes used | 116 | | | | |
| Liquid venom/snake (in ml) | 0.082 | | | | |
| (dry venom/snake (in g) | (0.015) | | | | |
| Greatest individual yield (dried) | 0.065 g | | | | |

Apart from the above-mentioned Irula project in India, the commercial **production of snake venom** is well established in many countries either as a sideline of private herpetologists or in combination with the commercialised display of snakes in so-called "snake parks". Some of the major pharmaceutical and scientific institutions such as the various local branches of the Pasteur Institute, the South African Institute for Medical Research in Africa, and the Butantan Institute in Brazil used to maintain their own serpentaria for venom production as a primary activity. None of them, however, bred their stock and all depended entirely on supplies from the wild. The Butantan Institute in Brazil, for example, organised delivery of live snakes through an agreement with the Federal Railways. About 16,000 persons all over Brazil voluntarily collected snakes for venom production and the Railways transported them free of charge to the Butantan Institute in Saõ Paulo. The Institute provided the volunteers with catching equipment, transport boxes and detailed instructions. The intake of snakes increased gradually from 2,100 in 1910 to about 22,000 in 1935. From that time it started decreasing to about 8,000-9,000 per year in 1951 (11).

About 83% of these snakes were either *Bothrops jararaca* or *Crotalus terrificus*. Snakes in the Butantan Institute are kept in large walled enclosures covered with lawn, and equipped with water basins and shade. Here they are displayed to the many thousands of visitors every year. Others are confined to large wooden terraria for venom production only. Venom is extracted either mechanically or electrically from their glands by handling the snakes with special hooks or forceps. Tab. 13 shows the yields of mechanical extraction for some species at the Butantan Institute. The snakes would not normally be kept alive longer than 15-16 weeks without feeding. During this time up to five venom extractions were possible without a marked decrease in quantity. Electrical extraction, however, may yield up to five times more venom per snake due to the fact that they may then be kept, with feeding, for more than a year, venom being extracted about every three weeks. The electrically produced venom is also clearer and more uniform. Even with such improvements large scale venom production in these institutions is extremely wasteful and certainly not sustainable as the snakes die soon and must be replaced.

Manual extraction of venom from a Lance-head viper at the Butantan Institute in Brazil; the venom is queezed out into a glass beaker for anti-serum production. (Photo: W. BÜCHERL)

Venom is either dried into powder form or lyophilised for immunisation of horses from which the anti-snake-bite serum is then produced. For the immunisation of one horse, up to 3.12 g of mixed venom may be required producing about 300 ampoules of 10 ml ea of marketable anti-serum. Depending on the species of snake and the extraction technique used, between 14 and 68 snakes are thus consumed to immunise one horse, i.e. one snake will produce between 4 and 27 ampoules of serum (11).

## Economic importance

Lizards and snakes have major economic importance as producers of skins for leather goods. This is reflected by **trade statistics**, summarised in Tab. 14. Until 1976 when exports were banned, ***India*** dominated the lizard and snake skin trade. In the peak years of the

1950s and 1960s, exports were measured in the millions and comprised mainly the *Monitor lizard*, the harmless and most effective rodent-destroying *ratsnake*, the *Spectacled cobra, Russell's viper, Checkered Keelback watersnake, Common Sand boa*, and of course the large-size *Indian python*. The proportions of skins of these different species held in 1984 by dealers in South India were 67%, 22.5%, 3% and 1% respectively (2). A special compilation by the WCMC in 1991 (21) shows that ratsnake skins are still one of the most heavily traded items. Between 1984 and 1989 CITES countries, mainly the USA, Belgium, Italy, Spain, Taiwan and Germany, imported no less than 0.7-2.65 million *Coluber mucosus* skins per year. In 1990 on the reptile-skin market in Germany a ratsnake skin fetched a price between DM 17 and 23 (US$ 11-14), whereas the tribal catcher in India would earn only the equivalent of DM 1.00 (US$ 0.62) per fresh skin, risking going to gaol for up to a year if caught. Tanneries in Madras claim to have tanned up to 5,000 snake skins a day before exports were banned in 1976. In their place small clandestine "cottage industry" tanneries established themselves. Between 1979 and 1987 an estimated 5 million lizard and snake skins, mainly ratsnake and cobra, were seized by Customs and Wildlife authorities. Most of these were turned over to the Bharat Leather Corporation (BLC), until 1990 a Government licensed company, to be processed for export. The obvious loophole which this procedure created in India's effort to control the snake-skin trade resulted in the non-renewal of the BLC license. In 1990, 70,000 confiscated lizard and 650,000 confiscated snake skins were demonstratively burnt by the Wildlife authorities in Madras.

In ***other Asian countries***, such as Bangladesh, the skin trade, legally or illegally, is still thriving. It has been the main supplier of *Monitor lizards* for the past 20 years, particularly after India banned exports in 1976 (29). Tab. 14 (a) shows that exports reached a peak of almost 2 million skins of *Varanus flavescens*, *-bengalensis* and *-salvator* in 1978/79, despite a Monitor lizard skin export ban imposed since 1973. In Indonesia efforts are made to control the skin trade through a quota system. Hong Kong, as a major conduit for the skin trade, has implemented controls for *Rat snakes, cobra* and *King cobras*. This has in turn resulted in a 40% drop in supplies and a price rise of 20% (7). A ban in 1981 on wildlife trade in Pakistan, another important supplier of *Monitor lizard* skins, seems to have been more effective in reducing exports from that country. There, approximate population densities, as an essential base for any type of utilisation, were established for *Varanus bengalensis* at 9 animals per $km^2$ in desert scrub, *V. griseus* at 32 animals per $km^2$ in wetland/borderland and *V. flavescens* at 46 animals per $km^2$ in the same habitat (9).

In ***Africa*** only limited statistical data on lizard and snake exploitation has become available since the early 1980's. Two *Monitor lizards, V. niloticus* and *exanthematicus,* have been prominent in the skin trade over the years. From 1975 to 1986 an estimated 3 million Nile and 680,000 Scrub Monitor lizard skins were shipped to Europe, Japan and USA (18). Tab. 14 (b) shows the annual average number of skins exported from Africa since 1983 to have fluctuated around 500,000. The most commonly traded snake skin is that of the *African rock python*. The smaller *Royal python* is popular in the pet trade and an average of 20,000 per year were exported from Africa between 1983 and 1988. Some dealers in West Africa have started to breed them successfully in captivity.

**Table 14:** Trade in lizard and snake skins (in thousands)
(a) Export from selected Asian countries

| Genera | Source | Countries | 1975/1976 | 1977/1978 | 1979/1980 | 1981/1982 | 1983/1984 | 1985/1986 | 1987/1988 |
|---|---|---|---|---|---|---|---|---|---|
| *Varanus* | 1) | CITES | 450.6 | 421.4 | 581.8 | 1,659.0 | 1,751.0 | 2,658.4 | n.a. |
| | 2) | Bangl. | | 1,890.6 | | | | | |
| *Uromastix* | 1) | CITES | | | | | | 40.0 | n.a. |
| *Python* | 1) | Thail. | | | | | ~170.3 4) | 370.1 | 125.9 4) |
| | 1) | Indon. | | | | | | 391.3 | 964.4 |
| | 1) | Malay. | | 5.5 4) | 7.4 | 4.2 | 5.9 | 13.4 | |
| *Ptyas, Cerberus, Homalopsis, Enhydris, Elaphe, Boiga, Natrix* | 1) | Thail. | | | | | 2.3 4) | 1,074.7 | 954.0 |
| | 1) | Indon. | | | | | | 4,142.4 | 2,927.0 |
| *Ophiophagus, Naja* | 1) | Thail. | | | | | | 266.3 | 46.3 4) |
| | 1) | Indon. | | | | | | 250.2 | 114.6 |
| *Vipera* | 1) | Thail. | | | | | | 144.4 | 8.7 4) |
| | 1) | Indon. | | | | | | 60.0 5) | |
| *Agkistrodon* | 1) | Indon. | | | | | | 6.0 | |
| *Laticauda, Hydrophis* | 3) | Philippines | 1971/74 740.0 | | | | | | 185.0 |
| *Acrochordus* | 1) | Indon. | | | | | | 828.3 | 457.1 |
| *Xenopeltis* | 1) | Indon. | | | | | | 130.8 | 125.0 |

Bangl. = Bangladesh, Thail. = Thailand, Indon. = Indonesia, Malay. = peninsular Malaysia

1) minimum net trade (in whole skins only) reported in CITES Annual Reports (quoted from LUXMOORE and GROOMBRIDGE (18) and GROOMBRIDGE and LUXMOORE (17); 2) quoted from WHITAKER and HIKIDA (29); 3) quoted from DUNSON (15); 4) figure for one year only

(b) Export from African countries

| Genera |
|---|
| *Varanus* |
| *Chamaeleo* |
| *Phelsuma* |
| *Uromastix* |

1) see Table 14 (a), 5) data from IUCN Trade Specialist Group; figures are only approximative because some of the original statistics were given in meters and kilograms of skins from which extrapolations were made into numbers of animals/skins

(c) Export from some Central and South American countries

| Genera | Source | 1977/ 1978 | 1979/ 1980 | 1981/ 1982 | 1983/ 1984 | 1985/ 1986 | 1987/ 1988 |
|---|---|---|---|---|---|---|---|
| *Tupinambis* | 1) | 1,703.0 | 2,877.9 | 5,392.0 | 3,548.8 | 2,678.1 | |
| | 5) | | | | 4,288.3 | 3,029.7 | 3,691.7 |
| *Iguana* | 1) | 0.01 | 61.3 | 39.9 | 1.1 | 0.8 | |
| | 5) | | | | 100.7 | 145.8 | 178.8 |
| *Dracaena* | 1) | 3.9 | 11.9 | 157.8 | 86.1 | 51.7 | |
| | | | | | 100.3 | 54.9 | 3.5 |
| *Boa* | 5) | | | | 433.2 | 98.1 | 22.6 |
| *Eunectes* | 5) | | | | 90.0 | 65.3 | 61.4 |
| *Corallus* | 5) | | | | 3.8 | 5.2 | 3.9 |
| *Epicrates* | 5) | | | | 3.8 | 1.3 | 2.6 |

1) see Table 14 (a), 5) see Table 14 (b)

Most ***South American countries*** are exporters of large quantities of mainly smaller species of lizards and snakes for the American and European pet trade. This has put pressure on some species, but it is the skin trade that has the maximum impact on the resource. The most heavily exploited lizards are the *tegus* in Argentina. Between 1975 and 1986 over 16 million skins of *Tupinambis rufescens* and *teguixin* were legally exported (18). Tab. 14 (c) shows that from 1983 to 1987 an average of no less than 1.83 million tegu skins were traded every year. This trade earns the Argentine national economy millions of dollars, but has also considerable socio-economic importance for the rural population of Northern Argentina where tegu hunting is an important source of income (16). Another important skin in the trade is the *Caiman lizard* from Brazil, Peru and Ecuador. Exports from these countries have been prohibited, but nevertheless between 1983 and 1988 an average of 26,000 skins have reached Japan and the USA every year (20). Paraguay, another important lizard-skin exporter, banned most wildlife exports in 1987 (4). Similarly, Colombia prohibited all trade in wildlife in 1989 with the exception of the *Tegu lizards* and pre-CITES registerd *boa* skins. Suriname contributes significantly to the trade in *iguana* skins which fluctuates around 70,000 per year. Export quotas were raised in 1989 from 20,000 to 30,000 per annum on the grounds of complaints of damage to crops (6).

The ***USA*** has been one of the most important importers of lizard and snake skins over the years. Between 1977 and 1983 about 250,000 live snakes and 618,625 skins of Boid snakes, plus another 1.7 million lizard and snake-skin articles, such as handbags and shoes, were imported (14). However, the USA also produces a considerable quantity of *rattlesnake* skins for commercial purposes. An estimated 300,000-500,000 rattlesnakes are taken every year in the so-called "rattlesnake round-ups", mainly in Oklahoma and Texas, for making into belts, wallets and other artefacts. The meat is also eaten locally (27).

Compared with the skin trade **other uses** of lizards and snakes, for food and venom production, are economically rather insignificant. Of the more than 2,000 snake species only about 400 are venomous. WHO lists 38 species as medically important, i.e. they commonly cause death or serious disability by bites (3) and the respective genera are included in Tab. 12 as venom producers. The total annual number of fatal snake-bite cases world-wide was estimated in the 1950s at about 30,000-40,000 of which the majority occurred in South and South-East Asia and in South America (26). Since then mortality rates have declined due to more efficient production, distribution and application of anti-venoms. However, the average number of deaths per annum in India alone was nearly 9,000 in 1975 and still more than thousand in 1981 (3). There are now anti-sera available against almost all snake bites and these are commercially produced by some 50 different laboratories all over the world (12). Although the prices paid for powdered venoms may be as high as US$ 150/g locally and US$ 500/g in Europe, the market demand is rather limited and has been decreasing during the last decade (22).

Unfortunately, none of the efforts to convert the exploitation of lizards and snakes described above into sustainable forms of utilisation have as yet reached an important economic level. Snake catching for venom exctraction with subsequent release, as practised by the Irulas in India, is marginally self-financing. About 250 tribal families subsist on this "tribal technology", and the project is still dependent on grants for development (30). The Iguana Management Projects in Panama and Costa Rica are designed primarily to promote biodiversity and soil conservation. The snake farms in China are reported to be self-financing or even profitable, but an independent assessment is needed to prove this. In Guangzhou, it is reported that at least 10 t of snake meat are consumed a day and new snake markets open all the time to keep up with the demand. Poisonous snakes, prized by the Cantonese for their "warming" properties in winter, sell for up to US$ 21/kg, while non-poisonous species are cheaper (8).

## Implications for conservation

The staggering numbers of lizards and snakes killed for one reason or another year after year all over the world certainly suggests that this resource is being over-exploited, although few if any species have been exterminated as yet as a direct result of being collected for commercial purposes. In fact, the biggest threat to most of these same reptiles is the rapid loss of habitat they are experiencing. The greatest danger of the present exploitation of lizards and snakes is the fact that it is completely indiscriminate. This is to a large extent due to the great diversity of this reptile order which makes scientific identification of the traded species in the countries of origin difficult. Many of the commonly traded species, for example the Caiman lizard, are scientifically hardly known. Others are caught and traded together with widely abundant species, although they may be extremely rare and endangered. For example, in 1987 18 specimens of one of the world's rarest snakes, Boelen's python (*Python boeleni*) from New Guinea, were found in a shipment of other more common reptile species at the Los Angeles International Airport (5). CITES regulations can be easily circumvented by wrongly declaring species and origin, the monitoring of which usually requires specialist taxonomic knowledge.

Due to disparate population status and population dynamics of the various species concerned and the different types of use, it is not possible to determine the effects of the skin exploitation in general terms for lizards and snakes as a whole. A special study of Monitor lizards concluded "that although some local declines around areas of high human populations are reported, there is no evidence that any species as a whole is threatened by the (skin) trade" (18). However, it is apparent that the Water Monitor lizard has drastically declined due to the loss of its mangrove habitat in large areas of South-East Asia. Another detailed study was carried out in 1990 on South-East Asian pythons listed in Appendix I or II of CITES. Whereas *Python curtus* may have been affected only locally by the intensification of the snake-skin trade, there is good evidence that populations of *P.molurus molurus* in India and Bangladesh have decreased significantly and are also depleted to some extent in other Asian countries; *P.m. bivittatus* appears to have declined only in parts of its range. *P.reticulatus* provides the great majority of Asian python skins in the international trade and there is reasonable evidence for a significant decline of this species in Bangladesh and depletion in much of the remainder of its range. It is concluded that "none of the South-East Asian python species appear to be threatened with extinction in the near future" but "appropriate management is required to ensure that populations of target species are maintained and that harvesting is carried out in a sustainable manner" (17).

Many of the commercially valuable lizards and snakes could certainly be cropped in the wild on a sustained-yield basis if and when their biology and ecology is better understood. Scientific knowledge is an absolute prerequisite for any type of sustainable resource utilisation, and as long as this is not sufficiently available lizards and snakes should not be taken indiscriminately from the wild.

Other serious obstacles for the rational management of lizards and snakes as a natural resource are, of course, the superstitions associated with them and the widespread emotional hostility towards them. They would have to be given the same appreciation for their 'place' in nature, legislative status and conservation education as afforded to the more charismatic mega-fauna. The great and growing number of people interested in keeping of lizards and snakes as pets or study animals gives hope in this respect. These people need to be educated, however, about the species which are unlikely to survive in captivity. Generally, the necessary knowledge and technology for keeping and breeding reptiles is developing, and this could be applied to more intensive forms of lizard and snake utilisation, as in the case of crocodilians.

## Literature references


1. ANDREWS, H.V. and M. GAULKE (1990): Observations on the reproductive biology and growth of the Water Monitor (*Varanus salvator*) at the Madras Crocodile Bank. *Hamadryad*, 15(1), p. 1-5
2. ANDREWS, J. and C. BIRKINSHAW (1988): India's snake skin trade. *TRAFFIC Bulletin*, 9 (4), p. 66-71
3. ANONYMOUS (1981): *Progress in the characterisation of Venoms and Standardisation of Anti-venoms*. WHO Offset Publication No. 58; World Health Organisation, Geneva
4. ANONYMOUS (1987a): Paraguay suspends export ban on two snake species. *TRAFFIC Bulletin*, 9(2/3), p. 31
5. ANONYMOUS (1987b): Rare pythons in U.S. *TRAFFIC Bulletin*, 9(2/3), p. 33
6. ANONYMOUS (1990): Suriname raises Iguana quota. *TRAFFIC Bulletin*, 11(2/3), p. 32
7. ANONYMOUS (1991): Hongkong-snakes. *TRAFFIC Bulletin*, 12(1/2), p. 22

8. ANONYMOUS (1993): Chinese snap up snake snacks. *Reuter News Agency release*, March 1993, Beijing
9. AUFFENBERG, W. (1989) : Utilisation of monitor lizards in Pakistan. *TRAFFIC Bulletin*, 2 (1), p. 8-12
10. BARZDO, J., LUXMOORE, R., BROAD, S. and D. JONES (1984): *A survey of wildlife farming operations*. WCMC/IUCN, Cambridge (unpubl. report)
11. BÜCHERL, W. (1963): Über die Ermittlung von Durchschnitts- und Höchstgiftmengen bei den häufigsten Giftschlangen Südamerikas, ein Beitrag zur Serumtherapie von Giftschlangenbissen. In: *Die Giftschlangen der Erde*; Behringwerk-Mitteilungen; Sonderband; p. 67-120; Elwert, Marburg/Lahn, Germany
12. CHIPPAUX, J.P. and M. GOYFFON (1983): Producers of Anti-venomous Sera. *Toxicon*, 21 (6), p. 739-752
13. DAS, I. (1989): Indian monitor lizards: A review of human utilisation patterns. *Hamadryad*, 14(1), p.16-19
14. DODD, C. (1986): Importation of live snakes and snake products into the U.S. 1977-1983. *Herpetological Review*, 17(4), p.76-78
15. DUNSON, W.A. (ed.) (1975): *Biology of the sea snakes*. University Park Press, Maryland, USA
16. FITZGERALD, L.A., CHANI, M.J. and O.E. DONADIO (1991): *Tupinambis* Lizards in Argentina: Implementing Management of a Traditionally Exploited Resource. In: *Neotropical Wildlife Use and Conservation*, ed. ROBINSON, J.A. and K.H. REDFORD; p. 303-315; Univesity of Chicago Press, Chicago and London
17. GROOMBRIDGE, B. and R. LUXMOORE (1991): *Pythons in South-East Asia*. Report to the CITES Secretariat, Aug. 1990. WCMC/IUCN, Cambridge
18. LUXMOORE, R. and B. GROOMBRIDGE (1989): *Asian monitor lizards: a review of distributions, status, exploitation and trade in four selected species*. WCMC/IUCN Cambridge (draft report)
19. LUXMOORE, R. (ed.) (1992): *Directory of Crocodilian Farming Operations. Second edition*. WCMC/IUCN, Gland and Cambridge
20. LUXMOORE, R. (1988): The caiman lizard. *TRAFFIC Bulletin*, 9 (4), p. 80-81
21. LUXMOORE, R. (1991): *in litt.*
22. MITTAINE, Y. (1953): Director of LATOXAN (furnishers of toxins and alkaloids), Rosans, France; *in litt.*
23. POPE, C.H. (1965): *The Giant Snakes*. A. Knopf, New York
24. SMITH, M.A. (1935): *Fauna of British India*; Vol. 2, p. 16-17; Taylor and Francis, London
25. STONEY, C. (1987): The day of the iguana. *VITA News*; Oct. 1987; Food and Agriculture Organisation of the United Nations, Rome; p. 4-7
26. SWAROOP, S. and E. GRAB (1956): The Snake bite mortality problem in the world. In: *Venoms*, ed. BUCKLEY, E.E. and N. PORGES; p. 439; American Association for the Advancement of Science, Washington, DC
27. WARWICK, C., STEEDMAN, C. and T. HOLDFORD (1991): Rattlesnake collection drives - their implications for species and environmental conservation. *Oryx*, 25 (1), p. 39-44
28. WERNER, D. I. (1991): The Rational Use of Green Iguanas. In: *Neotropical Wildlife Use and Conservation*, ed. ROBINSON, J.G. and K.H. REDFORD; p. 181-201; University of Chicago Press, Chicago and London
29. WHITAKER, R. and T. HIKIDA (1989): *Monitor lizards and prevention of food losses in Bangladesh*. Report of Project Formulation Mission to Bangladesh (20 June - 22 July 1989); FAO Action Program for the prevention of food losses (unpubl. report)
30. WHITAKER, R. and H.V. ANDREWS (1995): The Irula Co-operative Venom Centre, India. *Oryx*, 29 (2), p. 129-135

## 2.2 UTILISATION OF BIRDS

### 2.2.1 OSTRICH-LIKE BIRDS (*STRUTHIONIFORMES*)

Ostrich-like birds or **Ratites** comprise 6 quite distinct genera with 9 species: the rheas (*Rhea americana* and *Pterocnemia pennata*), the ostrich (*Struthio camelus*), the emu (*Dromaius novaehollandiae*), the cassowaries (*Casuarius* spp.), and kiwis (*Apteryx* spp.). The Greater rhea inhabits the low grasslands of South America from Brazil to Argentina whereas the Lesser Darwin's rhea occurs only in the Southern Andes and in Patagonia. The ostrich was historically distributed over all of Africa, its range extending into the deserts of Western Asia as far as Persia. The emu was found all over arid and semi-arid Australia and Tasmania but cassowaries are restricted to the tropical humid forests of New Guinea and Northeastern Australia; the small nocturnal kiwis are part of the forest fauna of New Zealand. Tab. 15 shows that all ratites are utilised for their meat and feathers, but only ostriches and emus are commercially farmed.

**Table 15:** Overview of the utilisation of ratites

| English name | Scientific name | Geographical distribution | Body weight (kg) | Utilisation |
|---|---|---|---|---|
| Greater rhea | *Rhea americana* | South. S. America | 35 | for meat, hides, feathers |
| Darwin's rhea | *Pterocnemia pennata* | South. S. America (Patagonia, Andes) | 25 | for meat, hides, feathers |
| African ostrich | *Struthio camelus* | South. and E. Africa | > 150 | for meat, hides, feathers; farmed |
| Emu | *Dromaius novaehollandiae* | Australia | 55 | for meat, oil, hides; farmed |
| Cassowaries | *Casuarius* spp. | N. Australia, New Guinea | 85 | for meat, feathers |
| Kiwis | *Apteryx* spp. | New Zealand | 1-4 | only tradition. use |

## Historical background

The use of the *ostrich* for its meat, eggs and particularly as a source of feathers, dates back to the early civilisations of the Assyrians, Babylonians and Egyptians. In fact, the Egyptian hieroglyphs frequently refer to ostrich feathers, which were perceived as a symbol of justice because the shaft of the ostrich feather runs exactly through the centre of the plume and leaves even flues on either side. Ostriches were probably already tamed and kept semi-domesticated at that time. Ostrich eggs were traded all over the Mediterranean, being used as vessels or converted into beads (30). During the Roman era, captive birds were fattened for the table and noble ladies rode ostriches on ceremonial occasions (23). In the 13th century the ostrich plume was introduced to Western Europe by the returning crusaders as an article of adornment of the nobles, and its beauty has been appreciated ever since in the European civilisation. Until the middle of the 19th century, ostrich feathers were mainly supplied from the Near East and North Africa from birds which were hunted specifically for their feathers. This has certainly contributed to the early decline of ostriches in Syria and Northern Africa where they were all but extirpated by 1930 (29). Today they are still fairly abundant only in East and Southern Africa.

In South Africa the Dutch settlers started to tame ostriches by rearing their chicks as early as 1775 (10) and in 1856-1864 farming of ostriches for the commercial purpose of "reaping" feathers from live birds commenced both in Algeria and South Africa (23). During the subsequent decades ostriches were systematically bred in captivity as an agricultural activity, although farmers continued to utilise eggs and chicks from wild birds to increase their stocks until the beginning of the 20th century. In contrast to the situation in North Africa, the farming of ostriches in South Africa quickly became popular and proved to be a profitable enterprise as it required little capital and labour. With the introduction of artificial incubation in 1869, the ostrich stock increased from 80 tame birds in 1865 to 253,463 in 1895, giving rise to the development of a veritable ostrich industry despite several slumps caused by droughts and epidemics (25). As early as 1884 the production from ostrich farms more or less replaced the trade in feathers from wild birds. In 1911 a special Ostrich Section was established at the Grootfontein Agricultural College which became an important first centre for research in support of ostrich farming (25). The early South African ostrich industry reached its peak in 1913 when more than a million pounds of feathers were exported at a total value of almost £st 3 mill. (approx. US$ 8 mill.) (25). At that time ostrich feathers exported from South Africa ranked fourth in value after gold, diamonds and wool. Although live ostriches were exported to Asia, Australia, New Zealand, Europe, North and South America where by 1910 altogether more than 20,000 tame ostriches had been established (29), only South Africa succeeded in large-scale economic production of ostrich feathers and virtually established a world monopoly in this commodity.

During the First World War the ostrich industry crashed and declined steadily until 1945, mainly due to the diminishing demand for ostrich feathers. Stock in South Africa decreased from about a million birds in 1914 to 400,000 in 1916 and a mere 23,528 in 1930. By that time ostrich breeding was only maintained in the Karoo region of South Africa where ostrich farming continued throughout to be a relatively important agricultural activity. After the Second World War the market for ostrich feathers improved again and demand for ostrich hides for leather production and for meat consumption in the form of the popular sun-dried and salted "biltong" arose. The total value of ostrich produce handled by the Klein Karoo Agricultural Cooperative in 1962, the only legally established marketing agency for ostrich farming, amounted to R 1.3 mill. (approx. US$ 1.8 mill.), but had risen to R 11.6 mill. (approx. US$ 11.6 mill.)by 1980 mainly due to the increase in the value of hides (9).

During the 1970s ostrich farming spread again to other parts of South Africa and, as of 1980, was also taken up vigorously by the neighbouring Botswana, Zimbabwe and Namibia, and by Israel. However, the USA and Australia also recognised the market chances for producing lean ostrich meat and luxury ostrich leather. In both countries Ostrich Farming Associations were established and breeding stock or fertilised eggs have been systematically imported from Botswana, Zimbabwe and Tanzania since the early 1980s.

The history of ostrich farming is particularly interesting as it represents one of the few efforts at truly domesticating an African wild animal species. During the peak period of South African ostrich farming between 1890 and 1914, breeders generally applied animal husbandry methods which are characteristic of domestication. They rigidly selected for improvement of feather quality and particular feather features. Various distinct strains of domesticated ostriches emanated which were entered into a special Ostrich Section of the South African Studbook and which attained a high degree of hereditary purity through inbreeding. Most of these recorded strains could be traced back to cross-breeding with wild birds which had successively been imported in considerable numbers both from Northern Africa (the "Barbary states") and Syria in 1886, 1888 and 1903 (26). The crossbreeding was well controlled and resulted in a marked improvement in the feather quality of the South African ostriches. There were many well-known and prized cocks which were used for decades and sold at prices of £st 1,000 at that time. Unfortunately, the research into the husbandry and cross-breeding of ostriches was discontinued when the ostrich industry collapsed in 1914 and only a few of the domesticated blood lines were maintained through the recession to the revival of the industry in the 1960s. The more recent development of ostrich farming in other countries, however, is not based at all on the genetic material of these early domesticated ostriches. On the one hand, stock in these countries is derived from eggs or chicks collected from wild populations or by breeding with imported birds of unknown genetic origin. On the other, feather quality is no longer a prime consideration and the process of domestication which had been initiated more than a hundred years ago in South Africa is being restarted with a view to achieving maximum meat and hide productivity.

The *other ratite species* were also historically an important resource. They were utilised extensively for food and utility by the respective indigenous peoples; the emus by the Australian Aboriginals, the rheas by the Patagonian Indians, and the cassowaries by the New Guinean indigenous tribes. Only when Australia and South America were settled by Europeans were the ratites persecuted into decline because they were perceived as competitors of the introduced sheep. *Emus* were extirpated in Tasmania and in large parts of Southern and Eastern Australia in the course of the 19th century. In the 1930s, premiums for emu heads were introduced in Western Australia and still in 1964 the elimination of 14,476 birds was encouraged through payment of public funds (21): now sheep pastures and grain fields are being protected by fencing. It was only in the late 1960s that interest arose in the farming of emus, probably from observation of the South African ostrich industry. Considering the special relationship between the Australian Aboriginal population and the Australian native fauna, emu farming was seen as a possible way of involving the Aboriginals in commercially oriented production (28). The government in Western Australia in 1976 sponsored investigations which resulted in the development of a first Aboriginal-controlled emu farm at Wiluna which was the start of emu farming as discussed below. However, commercially motivated breeding of emus was initiated in the 1980s in the USA, New Zealand and France and has also resulted there in the development of emu farming (17). *Rheas* in Patagonia were indiscriminately hunted by the landlords and "gauchos" from horseback with "boleros" for sport and meat. Before the First World War they were farmed both in Argentinia and Brazil together with imported African ostriches, but their feathers never fetched the same prices as those from the ostrich. With the collapse of the feather market, farming was abandoned. Continued hunting pressure eventually rendered the rheas so rare that by 1981 Darwin's rhea had to be placed on Appendix I and the Greater rhea on Appendix II of CITES.

In New Guinea, *cassowaries* are the only "big game" and remain an important resource for the indigenous peoples today. Along with the feral or semi-domestic pigs, cassowaries are still an important source of meat and their feathers are used for adornment and their leg bones and claws to make weapons. The highland species, *Casuarius benetti*, has been depleted in some areas since the 1950s (18) because of mounting hunting pressure in conjunction with extensive capture of chicks and sub-adults for raising and captive keeping in the villages. Only after contact with the Australians had developed, after the Second World War, did live cassowaries become a most precious possession and the symbol for social status and wealth for most highlanders. Extensive trade in live birds developed through the 1960s. These were accumulated for prestige or used for compensation payments, in barter trade and as a bride price. In 1974/75 the total number of cassowaries kept by villagers in the Mendi Sub-Province, with a total population of about 36,000, was estimated at 950-1,000 birds of which 80% were *C. benetti* (18). They were killed only at special occasions such as ceremonial functions and festivities to provide meat for eating and feathers for use in traditional head-dresses, so called "bilas" (19). The greatest number of kills, however, were made in settlement of inter-clan disputes, so-called "races", as a substitute for tribal warfare. During 1970-74, in only 61 out of 90 villages in the Mendi Sub-Province a total of 1,600 birds were killed, i.e. about 320 each year, of which 61.9% were used for dispute "races", 28.7% for ceremonies and 9.4% for other reasons (18). Since the late 1970s the use of cassowaries in dispute races, which has drawn heavily on the cassowary resource, seems to have ceased and the socio-economic importance of cassowaries has generally diminished in the wake of cattle development and the arrival of a cash economy (19).

## Management and utilisation

The **utilisation of wild ratite populations** has become rather limited. Only *cassowaries* are still quite extensively, though opportunistically, hunted by the indigenous people in Papua New Guinea for subsistence purposes. This includes all cassowary species in the Highlands as well as the Lowlands where villagers were still killing them extensively in 1979 with bows, spears, dogs and pit traps (19). The use of wild *ostriches* and *emus* is limited to the collection of eggs and hatchlings for farming purposes in some areas only. The latter are also occasionally hunted by the Aboriginals of Australia for subsistence. Considerable numbers of emus are still killed in pest-control operations or die along vermin fences in Western Australia, but these cannot at present be utilised commercially unless and until a "management programme" under the Wildlife Protection (Regulation of Exports and Imports) Act is approved (28).

The **farming** of *ostriches* and *emus* involves the keeping of tamed or semi-domesticated flocks in enclosures of various sizes for agricultural purposes. Similarly to crocodile farming it may be initially based on the collection of eggs or hatchlings from wild populations for rearing; but where farming is already well established, stock is exclusively bred from farm birds. Only during the last three decades did ostrich farming, originally aimed at feather production, become diversified in response to an increasing demand for ostrich

Ostrich eggs in a Buckeye incubator at an Ostrich farm in Zimbabwe; one chick is "pipping" the egg shell. (Photo: K. WORSLEY)

leather and game meat. Today production of hides and meat combined with the sale of live chicks for stocking are economically much more important than feathers. In emu farming, production of oil from the subcutaneous fat gives additional returns. This is reputed to have a remarkable ability to penetrate the human skin and also to be hypoallergenic. Accordingly it has potential in both cosmetics and medicines.

The management of ostriches on a classic South African ostrich farm is governed by the reproductive cycle and behaviour of ostriches, the growth of their plumage and annual seasons. Cocks and hens are run in separate fenced camps until the mating season starts in March/April when they are "camped-off", i.e. selected breeders are turned into small breeding paddocks to nest. A cock is paired with only 1-2 mature hens. Whereas wild hens don't start breeding until the age of 3-4 years, semi-domesticated birds may attain sexual maturity as early as 20-30 months (25, 14). The hens will lay 12-16 eggs on alternate days, but will continue laying right through the breeding season if eggs are removed and the hens well fed. Thus, it is possible to obtain up to a hundred eggs per hen per season and its reproductive capacity may be maintained for more than 40 years (11). In most farms, eggs are collected, stored under carefully controlled conditions and artificially incubated in batches. Chicks are placed in small rearing pens where they receive special dietary and hygienic attention for a period of 3-4 months when they are ready to fend for themselves. Their plumage is fully developed at 6 months with quills being ripe 2 months later. Feathers can thus be cropped from 8 months at intervals of 8-12 months, but cropping has to be synchronised with the annual moulting cycle. It may be done by plucking the whole feathers or by clipping only the plumes in prime condition, in which case the bird has to be "quilled" two months later to stimulate regrowth of the plumage. Ostriches are physically fully grown at the age of 16-18 months, but full sexual distinction may not be reached until 2 years. At the second plucking, i.e. at 16-20 months of age, farmers can already judge which birds would be worthwhile keeping in their flocks from a point of view of docility, physical properties and feather production. Those which do not conform will be slaughtered for hide and meat production.

Hand-reared ratites will become quite tame and ostriches may even be ridden. However, the majority of the ratite stock remains difficult to manage and some cocks can be extremely dangerous. Therefore special facilities for handling, plucking and crating are essential. Birds are killed humanely by a quick breaking of their necks or electrocution, and processing is carried out in especially built abattoirs. These are a prerequisite for profitable marketing of meat and hides. The first ostrich abattoir was established in 1965 at Oudtshoorn, the centre of the Karoo region of South Africa, but special slaughter facilities are now being built in all countries in which farming of ratites has reached a production stage.

Hides are flayed and salted and must contain the wings up to the elbow joint, the neck and the whole legskin down to the toe joint without damaging the tarsal scales as exclusive articles are made from this part of the skin. The meat of ratites is dark and fibrous resembling beef rather than poultry. Most of it is still marketed locally through restaurants; only South Africa (9) and to a smaller extent Zimbabwe and Israel have been able to capture export markets.

The intensity and the production goals of management and husbandry regimes may vary greatly from country to country and even within them depending on the availability of land, water, feed and markets for different products. For example, if ostriches are predominantly farmed for meat they may be slaughtered at an earlier age than in feather production where at least two pluckings are required. In emu farming, production of oil calls for delayed slaughtering as fat is only deposited from 6 months onwards attaining a maximum at 12-15 months. As with deer farming an ample literature on all aspects of ostrich farming has been produced (14, 16), and the Universities of Cape Town and Pretoria have both compiled comprehensive bibliographies (22, 27); handbooks and practical guidance are available from the special farmers associations which now exist in all countries where ratite farming is developing (4, 5, 12, 15). Although adult ostriches are extremely hardy, their chicks are, however, most vulnerable to species-specific parasite infestations as well as bacterial and fungal lung infections (13). Unless very well attended, mortality rates of up to 40% may well occur.

*Rheas* were farmed quite extensively in Argentina before the First World War and rhea feathers were traded in considerable quantities as described in Chapt. 3.2.5. Renewed interest in commercial farming of both rhea species has arisen in the 1970s, but this has not passed an experimental stage as yet. Attempts to farm *Rhea americana* were made in Brazil under the supervision of the Universidade de Paraiba (7) and are in progress in the Buenos Aires Province in Argentina mainly with a view to producing feathers by plucking. CITES import data show that an average of 465 kg per annum of feathers were traded legally in 1983-1989. The possibilities of farming *Pterocnemia pennata* in Uruguay were investigated by an FAO Consultant in the 1980s (8), and the Instituto Nacional de Tecnologia Agropecuaria (INTA) has to this end been running a research station in Chubut in Patagonia since 1977 (7).

Having considered the particular value of *cassowaries* to the natives of Papua New Guinea, as described above, the government tried to promote commercially oriented farming of cassowaries by setting up experimental farms in the 1970s in Mendi, Southern Highlands, and in the East Sepik Province (7, 18). Cassowary chicks or sub-adults had been traditionally caught and raised in captivity until they were traded, sold or killed. Birds were kept by villagers singly in backyards, hutches or pens, or were sometimes left to wander around the gardens without any breeding attempts. The aim of the Government farms was to teach and demonstrate basic husbandry techniques with a view to converting this wasteful and uneconomic practice into more economical "farming". However, cassowaries differ from other ratites in that they are strictly territorial, solitary and highly aggressive rain-forest dwellers. The cocks join females only when these are receptive to mating and hens are determined egg layers, i.e. they will stop laying after 3-4 eggs. Nutritionally they are fruit dependent which makes artificial feeding more difficult and costly. Although the farming attempts have produced interesting growth data, it was found in conclusion that cassowaries are quite unsuitable for commercial large-scale captive breeding and farming (20).

## Economic importance

Based entirely on feather production ***ostrich farming*** had a major economic importance at the beginning of this century, as related above. In present day ostrich farming, however, the use of feathers, as described in Chapt. 3.2.5, has become an economically marginal factor. Nevertheless, there is no doubt that it is developing again into a major, innovative poultry industry that will supply markets of the industrialised countries with much wanted and preferred lean and biologically valuable meat and produce more and more specialty leather for luxury purposes. Whereas in the early 1960s the ratio of monetary returns from ostrich products in South Africa was still 77% from feathers, 17% from hides and only 4.8% from meat turned into "biltong" (25), it had changed by 1980 to 50% from hides, 40% from feathers and 10% from meat (9). In Australia it is now rated at 47% from meat marketed in restaurants, 39% from hides, and only 14% from feathers (5). Ostrich farming is developing fast in Southern Africa, the USA, Australia, and also in European countries, Israel, Canada and New Zealand. Tab. 16 shows the status of ostrich farming in 1989-1993. South Africa is still by far the most important ostrich producer. The marketing monopoly for ostrich products of the Klein Karoo Agriculture Cooperative, established in 1959, was lifted in 1993 and this will facilitate further growth of the industry (2). In other countries the trend is also upward: in Australia for example, an increase of ostriches to 70,000 in 1996/97 and possibly to 100,000 in 1998/99, with about 40,000 slaughterings per annum, is envisioned (3, 5). However, except for the Southern African countries, ostrich farming is still in an establishment phase. Monetary returns are at present largely derived from the production and sale of live chicks. The prices of 4-month-old chicks have steadily increased since the 1980s and varied in 1993 between US$ 700-1,000 in Zimbabwe and Austr.$ 2,500-3,000 (approx. US$ 1,680-2,020) in Australia. In the USA a fully adult breeding pair may cost up to US$ 30,000-50,000 (3). The rapid growth of ostrich farming during the last decade was therefore based mainly on the strong demand for live chicks. This bears the danger of overproduction in the future with a crash of the market for chicks and also ostrich products. Then only those farms which have comparative advantages and high profitability will survive.

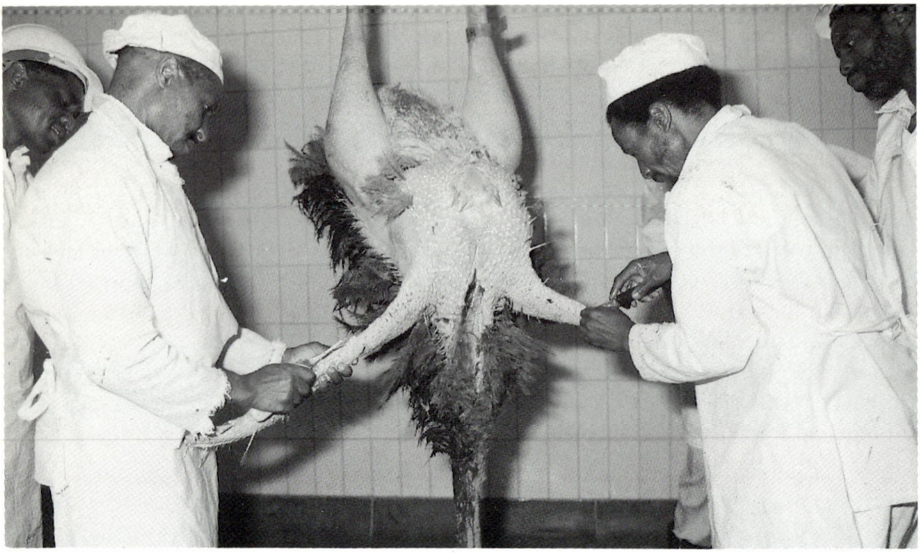

Ostrich slaughtered at an abattoir in Zimbabwe being processed for export. (Photo: K WILSON)

Apart from market demand the *economics of ostrich farming* depend largely on the intensity and cost efficiency of the operation (15). As with the utilisation of many wild animals the trend in ostrich farming has been away from extensive regimes as they used to exist in the Karoo region of South Africa (25) to semi-intensive or intensive production systems depending on irrigation possibilities. Stocking rates of 10-15 birds/ha may be achieved on established lucerne, a flock of 50 birds requiring a paddock of 3.4-5 ha. Under irrigation rotational grazing may be employed which reduces paddock sizes to only 1.5-2.5 ha (25). Thus an average farm in South Africa of a few hundred hectares may run up to 2,000-2,500 birds which will give sufficient turnover rates. In Zimbabwe, in woodland savanna, ostrich farming has proven to be competitive with cattle ranching in that it remained unaffected by drought. When compared with crocodile farming, the food constraint in ostrich farming is much less severe. In general, financial break-even may be expected to be somewhere between 1,000-2,000 eggs throughput with a comfortable level of between 1,500 and 2,500 eggs per annum (6). Tab. 17 compares the 1993 productivity of ostrich farming in Australia with other livestock production including deer and emu farming.

***Emu farming*** was initially slow to develop mainly because of legislative reasons. Only in exceptional cases did the Australian Government permit the capture and raising of native fauna within Australia. In Western Australia, the breeding of emu chicks at the first Government sponsored Aboriginal-controlled farming enterprise at Wiluna, mentioned above, led to the more general establishment of *emu* farming as of 1987. Participation was open to both Aboriginal and non-Aboriginal farmers (24). In Queensland, permission to capture wild stock was granted in 1989 to the Cherbourg Aboriginal Community near Murgon on the condition that the same number of progeny would eventually be returned to the wild. There, emu farming, including production, processing and marketing, has been developed to the stage where by 1993 2,000 birds were kept. Following the authorisation of the sale of birds in 1994, farming is now spreading in all Australian states so that by early 1995 there were over 1,200 farms registered all over Australia with more in the establishment stage. Initially centered in Western Australia, rapid expansion has occurred in Victoria which in 1995 had even 700 farms with more than 150,000 birds. In 1995 there were over 87,000 breeders and over 240,000 chicks kept on Australian emu farms.

As in the case for the ostrich industry in Australia, emu farming is still in an establishment stage. Production of live chicks, at prices in 1993 of Austr.$ 250-300 (US$ 170-200), constitutes an important economic factor. Under intensive management emus become as tame as ostriches and can be transported to efficient slaughtering facilities currently being developed. Tab. 17 shows that farmed emus are generally smaller than ostriches and accordingly produce less meat which, however, is as much appreciated as that of ostriches; as is emu

**Table 16:** Status of ostrich farming in 1989 - 1993

| Country | Year | Number of farms | Ostrich flock | | | Production |
|---|---|---|---|---|---|---|
| | | | Total | Adults | Juv. | |
| South Africa | 1993 | ca. 400[1)] | 169,895[2)] | 31,620 | | 5,290 birds slaughtered |
| Zimbabwe | 1993 | ca. 200 | 12,800 | 3,300 | 9,500 | 2 abattoirs in operation |
| Zambia | 1993 | 9 | 500 | 150 | 350 | no slaughter yet |
| Botswana | 1993 | ca. 18 | ca. 4,500 | n.a. | n.a. | no slaughter yet, export of eggs and chicks |
| Namibia | 1993 | 160-180 | 8,000-10,000 | n.a. | n.a. | no slaughter yet, export of eggs |
| Tanzania | 1989 | few | insignificant | | | export of chicks |
| Kenya | 1989 | few | insignificant | | | |
| Mexico | 1993 | n.a. | 200 | n.a. | n.a. | |
| USA | 1993 | >1,000[3)] | ca. 30,000 | n.a. | n.a. | slaughter for |
| Canada | 1993 | n.a. | 500 | n.a. | n.a. | restaurants only |
| Australia | 1993 | ca. 1,000[3)] | ca. 15,000 | n.a. | ca. 6,000 | expected to begin 1995 |
| New Zealand | 1989 | few | insignificant | | | |
| EU[4)] | 1993 | >1,600 | 4,860-5,880 | >1,800 | >2,000 | only France and Portugal prod. >50 t meat |
| Israel | 1990 | n.a. | ca. 8,600 | 1,500 | n.a. | 1,000 birds slaughtered in 1990/91 and 2,500 in 1991/92 |
| **Total** | | **ca. 4,300** | **>254,255** | | | |

1) in 1989 79% were located in the Karoo and Cape Province, 43% of farms keeping >200 birds; 2) in the Karoo region the number of breeding birds has increased since 1987 by about 45%; 3) members of the Breeders' Association, not necessarily farmers; 4) most important countries are France, Netherlands, Italy

Sources: BARZDO and coll. (7), ANONYMOUS (3, 4, 5) and New Zealand Press Agency (in litt)

leather. Emu farmers usually budget on returns net of slaughtering costs of about Austr.$ 300 (US$ 200) per bird based on approximately equal returns from the meat, leather and oil. The latter is considered by many farmers as having the potential to keep emu farming profitable when prices for chicks drop. In Aboriginal communities returns from carving the egg shells, for which they have developed an extraordinary skill, may add up to Austr.$ 500 (US$ 340) per egg.

**Table 17:** Comparative potential productivity of ratite farming in Australia in 1993

| Characteristic | Ostrich | Emu | Fallow deer | Cattle | Lamb | Pig |
|---|---|---|---|---|---|---|
| Liveweight (in kg) | 110-120 | 35-40 | 52 | 350 | 44 | 120-140 |
| Dressed weight (in kg) | 71 | 18-20 | 28-30 | 200 | 16-20 | 60-70 |
| Saleable meat yield (in kg) | 30 | 11 | 20 | 120 | 13 | 44 |
| Wholesale meat value (per kg) | $20 | $13 | $3 | $2.70 | $2.50 | $2.50 |
| Hide value[1)] | $350-500 | $0-107 | $3-5 | $53 | $1-3 | n.a. |
| Estimated ex-farm gate return[2)] | $500-900 | $150[3)] | $60-90 | $370 | $27 | $150 |
| Current breeder value | $50,000[4)] | $160-250 | $120 (does) | $370 (cow) | $27 | $150 |
| Area needed | 1 acre | 1 acre | 1 acre | 6 acres | 1 acre | 1 acre |
| Estimated ex-farm gate return per breeder within 18 months on current product values (if all prog. slaught.) | $7,500 (at 15 chicks per breeding pair) | $1,200 (at 8 chicks per breeding pair) | $60 (progeny for slaughter at 3 years[5)]) | $400 (bull/cows x 0.9) | $59 (ram/ewe x 1.4) | $1,800 |

The weights/prices for species other than ostrich are indicative only since these are dependent upon a number of factors such as quality, age and sex; all values in Austr.$ (1 Austr.$ = 0.674 US$)
1) dependent on size, condition and quality; 2) dependent on size and quality; 3) includes $30-65 oil value (most sales around $30); 4) average market price as of 1/1/94 - greatly inflated by present high demand for breeding stock; 5) this return increases if progeny slaughtered under 3 years

Source: Australian Ostrich Association (ANONYMOUS (5))

In the past the ***use of semi-domesticated cassowaries*** by villagers in the highlands of Papua New Guinea was relatively of great economic importance, as described above. In the 1970s *live cassowaries* were exchanged for money, and/or traded for items of value at an average price of Austr.$ 36 (approx. US$ 30) for young chicks and Austr.$ 218 (approx. US$ 175) for an adult bird; some fetched prices of up to Austr.$ 500 (approx. US$ 400). Thus the cash equivalent of captive kept cassowaries killed annually for different purposes was close to Austr.$ 80,000 (approx. US$ 64,000) per annum (18). However, a survey in 1979 showed how this importance had waned; only in 39% of the questioned 425 villages were cassowaries still considered to be of importance to the local economy (19). Since commercial farming of cassowaries - in contrast to that of other ratites - does not seem to be practicable for biological reasons, the present and future resource value of the cassowaries depends on the control and regulation of subsistence hunting and trapping.

## Implications for conservation

The trade in ostrich feathers during the 18th and 19th centuries has probably contributed significantly to the decline of the North African and Asian subspecies of *Struthio camelus*. The most perfect and thus the most sought after plumes came from Alleppo (10); i.e. they originated from ostriches in the Syrian desert which in turn had become very rare as early as 1840 and extinct in about 1910 (26). By 1875 there were hardly any ostriches left in Northern Africa, the alternative source of supply. By that time, however, ostrich farming in South Africa had begun to supply sufficient feathers for the international trade and this has definitely prevented a further decline of the remaining wild populations. The more recent upsurge of ostrich farming could have constituted a new threat to ostriches as it was largely based on the collection of eggs and hatchlings from the wild. However, as was the case a hundred years ago in South Africa, this was only a transitional phase and today sufficient semi-domesticated stock has been established outside South Africa to satisfy demand for farming stock around the world through captive breeding. Also, those countries which still have significant wild populations of ostriches are regulating their use as a wildlife resource (1).

The farming of emus, and, if successfully further developed, that of rheas, will also certainly have only positive implications for the conservation of their wild populations by reducing hunting pressure on these. Semi-domesticated emus have already been returned to Tasmania where the species was extirpated 160 years ago. The situation with cassowaries, however, is somewhat different as these still constitute a fairly important resource for subsistence hunting. Past traditional use of captive cassowaries was not based on breeding and was therefore very wasteful of the wild cassowary populations. As it seems unlikely for biological reasons that this use can be converted into a commercially oriented farming activity based on captive breeding, the conservation of cassowaries can only be aided by resource management, i.e. regulation and control of the hunting and trapping activities of the local population and the industry which has developed around it.

## Literature references


1. ANONYMOUS (1991): *Management Plan for Ostriches in Zimbabwe*. Department of National Parks and Wildlife Management, Harare, Zimbabwe
2. ANONYMOUS (1993a): Scheme for ostriches and ostrich products discontinued. *Agricultural News*, Nov. 1993, No 43; Pretoria, South Africa
3. ANONYMOUS (1993b): *Verslag: Bedryfsondersoek oor die Volstruisbedryf in die RSA*. Department van Lanbou, November 1993, Pretoria, South Africa
4. ANONYMOUS (1994a): *American Ostrich*. Official Publication of the American Ostrich Association, Fort Worth, Ta., USA

5. ANONYMOUS (1994b): *Ostrich Farming in Australia:* Information Leaflets. Australian Ostrich Association, Jamison Centre, ACT 2614, Australia
6. BARNES, J. (1989): *Report to the Department of Wildlife and National Parks, Gaberone, Botswana* (unpublished)
7. BARZDO, J., LUXMOORE, R., BROAD, S. and D. JONES (1984): *A survey of wildlife farming operations.* WCMC/IUCN, Cambridge (unpubl. report)
8. CHILD, G. (1994): in litt. (*cited after* BARZDO and coll. 1984)
9. CONROY, A.M. and I.G. GAIGHER (1982): Venison, aquaculture and ostrich meat production: Action 2003. *South African Journal of Animal Science*, 12, p. 219-233
10. DE MOSENTHAL, J. and J.E. HARTING (1877): *Ostriches and Ostrich Farming.* Trubner, London
11. DUERDEN, J.E. (1905-1925): results of extensive research into ostriches and ostrich breeding at the Grootfontein Agricultural College published in series of Reports on Ostrich Experiments and in various periodicals (*cited from* SMIT 1963)
12. HALLAM, M.G. (1992): *The TOPAZ Introduction to Practical Ostrich Farming.* The Ostrich Producers Assocation of Zimbabwe, Harare, Zimbabwe
13. HUCHZERMEYER, F. W. (1994): *Ostrich diseases.* Agricultural Research Council; Onderstepoort Veterinary Institute, Pretoria, South Africa
14. OSTERHOFF, D.R. (1979): Ostrich farming in South Africa. *World Review of Animal Production*, 15 (2), p. 19-30
15. PISTORIUS, J.J. (1991): Economic Analysis of Ostrich Production in Zimbabwe. In: *Handbook of TOPAZ*; The Ostrich Producers Association of Zimbabwe, Harare, Zimbabwe
16. MELLET, F.D. (1993): Ostrich production and products. In: *Livestock Production Systems - Principles and Practice*, ed. MAREE, C. and N.H. CASEY; p. 187-194; Agri Development Foundation, Pretoria, South Africa
17. RAMSAY, B.J. (1994): *Commercial use of wild animals in Australia.* Bureau for Resource Sciences, Australian Government Publishing Service, Canberra, Australia
18. REID, B. (1978/79): History of domestication of the cassowary in Mendi Valley, Southern Highlands, Papua New Guinea. *Ethnomedicine*, V (3/4), p. 407-432
19. REID, B. (1981/82): The cassowary and the highlanders. Present day contribution and value to village life of a traditionally important wildlife resource in Papua New Guinea. *Ethnomedicine*, VII (1/4), p. 149-240
20. REID, B. (1994): *in litt.*
21. SANFT, K. and B. GRZIMEK (1968): Emus. In: *Grzimeks Tierleben*, Vol. 7, p. 100; Kindler, Zürich
22. SIEGFRIED, W.R. and R.K. BROOKE (1983): *The Ostrich and Ostrich Farming: a bibliography.* Fitzpatrick Institute, University of Cape Town, Cape Town, South Africa
23. SIEGFRIED, W.R. (1984): 55. Ostrich. In: *Evolution of domesticated animals*, ed. MASON, I.L.; p. 365-366; Longman Group, New York and London
24. SMETANA, P. (1992): Emu farming: Background information. *Miscellaneous Publication No. 38/92*; Department of Agriculture, Western Australia, Perth, Australia
25. SMIT, D.J.v.Z. (1963): Ostrich farming in the Little Karoo. *Department of Agricultural Technical Services, Bull. No. 358*; Government Printer, Pretoria, South Africa
26. THORNTON, R.W., DUERDEN, J.E. and F.S. SMITH (1912): Report on the North African ostrich imports into South Africa (*cited from* SMIT 1963)
27. VAN DER WESTHUIZEN, E. and A. EARLE (comp.) (1993): *Ostrich Bibliography.* University of Pretoria, Academic Information Service, Pretoria, South Africa
28. WILSON, G.R., McNEE, A. and P. PLATTS (1982): *Wild animal resources: their use by Aboriginal communities.* Australian Government Publishing Service, Canberra, Australia
29. WORMSER, M.F. (1930): The Ostrich Industry in South Africa (*cited after* SMIT 1963)
30. ZEUNER, F.E. (1963): *A History of Domesticated Animals.* Hutchinson, London

## 2.2.2 BIRDS OF PREY (*FALCONIFORMES*)

Birds of prey or raptors include the nocturnal owls (**Strigiformes**) and the diurnal birds of prey (**Falconiformes**). Only the latter are considered in this book. Traditionally the order has included 5 families with a total of 292 species: the New World condors (**Carthartidae**), the osprey or Fish hawk (**Pandionidae**), the Secretary bird (**Sagitariidae**), the eagles, vultures, kites, buzzards, hawks, and harriers (**Accipitridae**), and the falcons and caracaras (**Falconidae**). All of the most commonly used raptors are wide-ranging species with continental or intercontinental distributions, mainly in the Northern Hemisphere. One, the peregrine, is virtually cosmopolitan. Tab. 18 shows that only some of the accipitrid and falconid raptors are commonly used by man, mainly for falconry and related activities.

**Table 18:** Overview of the utilisation of birds of prey

| English name | Scientific name | Geographical distribution | Utilisation[1)] |
|---|---|---|---|
| **EAGLES, BUZZARDS, and HAWKS - ACCIPITRIDAE** | | | |
| Golden eagle | *Aquila chrysaetos* | Holarctic | falconry, occ. breed. |
| Tawny eagle | *-rapax* | Africa | falconry, occ. breed. |
| Red-tailed hawk | *Buteo jamaicensis* | N.America | falconry, comm. breed. |
| Ferruginous hawk | *-regalis* | N.America | falconry, occ. breed. |
| Harris' hawk | *Parabuteo unicinctus* | C.-S.America | falconry, comm. breed. |
| Goshawk | *Accipiter gentilis* | Holarctic | falconry, occ. breed. |
| Sparrowhawk | *-nisus* | Eurasia | falconry, occ. breed. |
| Cooper's hawk | *-cooperii* | N.America | falconry, occ. breed. |
| Sharp-shinned hawk | *-striatus* | N.America | falconry, occ. breed.; formerly for food |
| **FALCONS - FALCONIDAE** | | | |
| Gyrfalcon | *Falco rusticolus* | Holarctic | falconry, comm. breed. |
| Saker | *-cherrug* | Eurasia | falconry, comm. breed. |
| Lanner | *-biarmicus* | Europe and Africa | falconry, comm. breed. |
| Prairie falcon | *-mexicanus* | N.America | falconry, comm. breed. |
| Peregrine | *-peregrinus* | Cosmopolitan | falconry, comm. breed. |
| Merlin | *-columbarius* | Holarctic | falconry, occ. breed. |
| Eleonora's falcon | *-eleonorae* | Europe and Africa | falconry, occ. breed. |

1) breeding solely for conservation purposes not included; occ. = occasional, comm. = commercial breeding

## Historical background

Probably the first and most basic human use of raptors was for food. Ethnographers have reported that a number of aboriginal peoples from such disparate regions as Australia, the Americas, Africa, and Eurasia ate several kinds of raptors, although never as common items of their diets. Some of this consumption was probably ritualistic while in other cultures there were taboos against killing certain birds of prey. Native tribes also sometimes kept live birds of prey, especially big ones, as totems or as objects for ritual sacrifice, and as sources of moulted feathers for ceremonial attire (6) (see also Chapt. 3.2.5).

Even civilised men occasionally kill raptors for food. Before raptors enjoyed legal protection, in Eastern North America at such famous "hawk shoots" as Cape May Point, New Jersey, hunters sometimes took small species such as the *Sharp-shinned hawk* home for dinner. However, most of the hawks at these migratory locations were shot "for sport" or as "vermin". A similar situation still exists on Malta, Sicily, in Lebanon, and other places. Perhaps the most persistent and large-scale use of a raptor for human food has been the taking for many centuries of fat, succulent nestlings of *Eleonora's falcon* from its Mediterranean breeding colonies by hungry fishermen, a practice which lingers to the present day. The Inuit hunters of the Arctic also relish *baby falcons*.

Man's principal use of diurnal birds of prey has, however, been as instruments to obtain food - as trained hunters for the sport of falconry - and, indeed, that is still one of the main uses today. Historically the raptors most commonly employed in falconry were several *species of eagles*, mainly the circumboreal *Golden eagle, True hawks* of the genus *Accipiter*, mainly the *goshawk* and *sparrowhawk*, and several kinds of *falcons*, notably the *gyrfalcon, saker, lanner, peregrine*, and *merlin*. These are all active pursuers - species capable of killing quarry that humans consider to be suitable game animals such as rabbits and hares among mammals, and waterfowl, shorebirds, and upland game birds among feathered quarry. In mediaeval times trained falcons were also pitted against stronger quarry such as herons, cranes, and even other birds of prey (11).

The origin of falconry no doubt predates the first historical records of its occurrence, circa 700 BC, and it probably arose as a part of the culture of the steppe nomads of Central Asia. It may have been coincident with the development of horse-riding about 5,000 years ago, as these people traditionally hawked from horseback. Certainly the Mongols were ardent falconers, and it seems likely that Atilla the Hun or one of his predecessors introduced the practice into Europe in the 4th or 5th century AD, where it had been unknown to the Greeks and Romans.

Falconry was widespread and common throughout Eurasia in the Middle Ages. MARCO POLO'S description of the falconry establishment of the Great Khan of Cathay in the 13th century vividly portrays the importance which the later Mongol rulers attached to the sport in China, Mongolia, India, and throughout Central Asia. The Japanese had a separate, rich tradition of falconry, as did the Bedouin Arabs of Arabia and North Africa. From 500 AD to the 18th century falconry flourished in Europe as the chief pastime of the aristocracy. It was peculiarly suited to the feudal society of the Middle Ages, and its practice reached its zenith during the reign of the Holy Roman Emperor, Frederick II of Hohenstaufen (1194-1250), who with his son, King Manfred of Sicily, wrote an outstanding treatise on the sport, *De Arte Venandi cum Avibus* (11). This book describes the birds of prey, the quarry, and methods of training used in that age, and was written in the spirit of modern science long before the Italian Renaissance.

The birds of prey used in falconry were taken from the wild, either trapped as fully grown "passage" and "haggard" birds, or taken as "eyasses" from the nest. Until very recently no effort was made to breed any of these feathered hunters in captivity or to domesticate them because wild birds were plentiful. Consequently, a considerable trade and industry developed in ancient and mediaeval times to supply birds of prey for falconry. One of the duties of a feudal vassal was to supply a certain number of hawks and falcons to his lord each year, and rare falcons were often demanded as ransom for the release of important political prisoners. Famous hawk markets such as the one in Damascus existed in various parts of the world. The Russian Czars maintained an elaborate falconry establishment in Moscow and patronised an extensive system of trappers and collectors who were sent into Siberia and the far reaches of the empire to obtain gyrfalcons, peregrines, and rare White goshawks from Kamchatka (4). Many of these birds were traded or sent as gifts to various courts in Europe and Asia.

Similarly, during more than five centuries of Danish rule over Iceland beginning in 1280, landholders on the island annually harvested gyrfalcons which were sent to Copenhagen and from there distributed to all parts of Europe (4). Unusually detailed export records exist for the latter part of this period (1731-1739) when interest in falconry was already on the wane, and show that 4,848 falcons were exported in a period of 61 years, an average of nearly 80 gyrfalcons per year (NIELSEN, unpubl. ms.). It is probable that trade of this magnitude occurred for about 500 years.

Migration corridors where peregrines concentrate, such as at Valkenswaard in Holland, provided a "cottage industry" for families such as the Mollens who for many generations supplied passage peregrines to European falconers. This one family was active from the 18th well into the 20th century.

It is difficult to estimate the economic importance of falconry in former times, but judging by its pervasive practice throughout most of the civilised world, by the international commerce in trapped birds, and by the fact that rank of falconer was accorded high social status at court, it must have been considerable. Hawking also provided fresh meat for the table for both commoner and aristocrat, so that it had more than a recreational significance for society.

## Recent development, practice, and importance of present uses

After the middle of the 18th century with the advent of gunning and major societal changes, **falconry** went into severe decline in Europe although its practice continued unabated among the steppe nomads of Asia, the maharajahs of India, and the Bedouin tribes of Arabia and North Africa. Interest in the sport never completely died out in Europe; it slowly re-emerged there in the late 1800s and spread to the Americas in the 1900s. Especially since the Second World War the number of practising falconers has increased dramatically.

Today there are 4,000 or more falconers in North America (8) and a similar number in Europe (but many more "keepers" of live raptors there). There are active groups of falconers spread all over the globe in such places as Mexico, Argentina, Peru, Australia, New Zealand, South Africa, Zimbabwe, and Kenya, in addition to the old traditional centres still active in Russia, Central Asia, China, Japan, and Arabia (where there are probably 5,000 or more falconers including the sheikhs and all their trainers). After nearly dying out in India and Pakistan following independence from British rule and the consequent loss of power and wealth of the maharajahs, the sport is slowly reviving there as well. It is hard to estimate how many falconers there are in the decade of the 1990s, but the number would fall somewhere between 15,000 and 25,000, far more than there were at any time in the past 300 years.

While the number of falconers has dramatically increased during the last 50 years, legal commerce in wild birds of prey for falconry has been greatly curtailed since about 1970. Most nations have enacted laws to protect birds of prey from being taken except under limited permits, and in some countries it is illegal under any circumstances to buy and sell wild raptors. As an example, PARRY-JONES (7) provides a thorough description of the legal situation in Great Britain. International treaties such as the Convention on International Trade in Endangered Species of Wild Fauna and Flora (CITES) further restrict traffic in wild birds of prey.

The largest traffic in wild birds of prey for falconry in recent decades has been to the oil-rich sheikhs and princes of the Middle East, many of whom now make up annual hawking expeditions that rival MARCO POLO'S description of the hunting excursions of Kublai Khan. The number of birds involved (mostly sakers, some peregrines, and a few gyrfalcons) has been estimated at 3,000 in use each year with 2,000 annual replacements (PLATT in (3)). The high rate of replacement results from the loss of birds or their deliberate release at the end of the hunting season. A more recent estimate increased these figures to 8,600 held and 3,000 annual replacements (9). It should be noted, however, that two special falcon hospitals in the United Arab Emirates see between 3,500 and 4,000 falcons each year, and these figures reveal nothing about numbers in Qatar, Bahrain, Kuwait, and Saudi Arabia which could have twice as many as the Emirates. There may well be 10,000 falcons held in all of Arabia.

Red grouse hawkers with a cadge of peregrines on a Scottish moor. (Photo: S. KENT; by courtesy of the Carnie Archives of North American Falconry)

Prices paid for these falcons range from a few thousand to a few tens of thousands of US dollars. If one takes US$ 5,000-10,000 as the range for an average price, then this Middle Eastern trade amounts to US$ 15 mill.-30 mill. per year.

One of the main sources of these birds is Pakistan, a region which has been a traditional supplier of falcons (mostly sakers) to the Arabs for a very long time. Some birds come literally from all over the world, sometimes surreptitiously in diplomatic pouches from countries such as Mexico, Peru, Chile, and Argentina, but also in various ways from North America and Europe (5). Most of this traffic in wild raptors has been illegal for the past 30 years, but it continues, especially from Eastern Europe and countries of the former USSR following the demise of communist authority.

From time to time there have been limited attempts by contemporary governments to allow a legal, commercial harvest of raptors for falconry. Two Canadian jurisdictions have experimented with commercial harvesting of gyrfalcons but they met stiff opposition from protectionists. Many state governments in the USA allow individual falconers under permit to take wild raptors for their personal use, but trade in these birds is strictly prohibited under the U.S. Migratory Bird Treaty Act. In Europe, a limited number of permits are still issued by some governments for falconers to take traditionally used hunting hawks such as the goshawk, sparrowhawk, peregrine and merlin.

Another largely illegal use of birds of prey involves traffic in taxidermic mounts, as discussed in Chapt. 3.2.6. This activity is said to be larger than the trade in live birds of prey and seems to be centred in places such as Damascus, Hong Kong, and Singapore with principal markets in Japan and Europe. There is also still some illegal trade in raptors' eggs.

The **breeding of birds of prey in captivity** has assumed increasing economic importance in the last 20 years (1, 7). With the catastrophic reduction in numbers of Peregrine falcons and several other species of raptors resulting from excessive use of pesticides such as DDT and dieldrin in the 1950s and 1960s, biologists and falconers alike began to look to artificial propagation as a technique for preserving endangered species. The goal for conservation was to produce suitable captive stocks that could be released back into natural habitats to restore lost populations or to reinforce greatly diminished wild ones. The goal for falconry was to provide a supply of domestically propagated birds that falconers could use in place of the traditionally taken wild birds.

A number of organisational and private individual initiatives led by The Peregrine Fund, Inc. in the United States achieved both goals in the period between 1970 and 1990 (3). More than 90 species of Falconiformes have now been bred successfully in captivity (1) and a number of species, particularly those of importance for falconry, now have self-sustaining captive populations that have produced hundreds to thousands of birds since 1970. Notable among these species are the *peregrine, gyrfalcon, saker, lanner, Prairie falcon, Harris' hawk, Red-tailed hawk*, several other *buzzard species*, and, to a lesser extent, *merlin goshawk, sparrowhawk*, and *Golden eagle* (see Tab. 18).

Several species of raptors have benefitted from restoration efforts resulting in the establishment of new populations derived either from captive-produced birds or translocated wild ones. Notable examples are the *Peregrine falcon* (several populations in Europe and North America) and the *Mauritius kestrel (Falco punctatus)*. More than 6,000 peregrines have been released in North America alone and the equivalent of several tens of millions of US dollars has been expended in the execution of these conservation projects.

Since about 1984 commerce in captive-produced birds of prey has become legal in several countries, and a number of commercial breeders in North America and Europe, and a few elsewhere, are now annually producing several thousand birds - mostly large falcons and Harris' hawks - for the falconry market. For example, a recent trading catalogue mainly serving the USA and Canada (Northwoods Limited, Catalogue No. 7, 1994-5, Rainer, Washington) listed offerings from 32 breeding establishments. The following kinds of birds were advertised: Peregrine falcon (several subspecies including *Barbary falcon*); *gyrfalcon* (several colour varieties); *Prairie falcon; saker; lanner; merlin*; several kinds of hybrid falcons (especially the popular *gyrfalcon x peregrine cross*); *Harris' hawk; Ferruginous hawk; Red-tailed hawk; goshawk; Cooper's hawk; Harris' x Cooper's hawk hybrids; Harris' x Ferruginous hawk hybrids; Tawny Eagle;* and *Golden Eagle*. In December 1994 the North American Falconers' Association published a registry of 102 raptor breeders in North America (*Hawk Chalk* XXXIII(3) p. 43-46). Similar advertisements can be found in European journals such as the *Newsletter of the British Falconers' Club*.

Although the Arab falconers have traditionally preferred to hunt only with wild-caught passage or haggard falcons, a significant number of these "domesticated" raptors are now being sold to Middle Eastern clients as well, and there are also several private falcon breeding facilities in Arabia. Top prices paid in Arabia are around US$ 20,000 to US$ 40,000, but otherwise prices generally range from a few hundred to a few thousand US dollars. Total annual sales world-wide probably exceed 4,000 birds and 10 mill. dollars. It seems likely that the market for captivity-produced raptors will soon eclipse the legal and illegal trade in wild-caught birds for falconry, and this change alone may be the most important contribution that captive breeding will make to the conservation of birds of prey.

Several **other activities involving birds of prey** have economic importance. They include falconry schools, public exhibitions of live birds of prey, research and conservation, rehabilitation centres, use of falconry-trained raptors to rid airport runways and other places of nuisance birds, and ecotourism.

Schools for training novices to become falconers are especially popular in Great Britain and to a lesser extent in other European countries. They are just starting to make an appearance in North America where the old, traditional European system of apprenticeship has been the main way that aspiring falconers learn their art. One of the better known courses is offered at the British School of Falconry at the Gleneagles Hotel in Perthshire, Scotland.

On the European continent, especially in Germany and Austria, exhibitions of live raptors and demonstrations of free-flying, trained birds are more popular than schools. These "adlerhofs" and "falkenhofs" are often associated with old castles or country estates and take on a mediaeval or renaissance atmosphere. In Britain one of the best known of these centres is the National Birds of Prey Centre (formerly the Falconry Centre) near Newent in

Gloucestershire, where some 65 species (including owls) are displayed (7). Zoos also often have a bird of prey demonstration involving trained birds.

A growing number of research and conservation organisations specialise in work on birds of prey. The Raptor Research Foundation, Inc. produced a world directory of more than 80 such organisations in 1988. Although generally in the category of "non-profit" public service organisations, a considerable economic component attaches to their operations (2). Annual operating budgets for the larger of these organisations range from a few hundreds of thousands to a few million US dollars. In a somewhat different but related category are the numerous "rehabilitation centres" that attempt to take care of sick or injured birds of prey with the goal of returning fit individuals back to nature.

Collision of birds with aircraft taking off and landing has been a serious problem at airports where certain species congregate in large numbers. Damage to a single jet engine can run into the hundreds of thousands of dollars, not to mention the hazard to human life. Falconry-trained hawks or falcons have been used with varying degrees of success to frighten troublesome birds away.

Perhaps the most successful of these operations was the one carried out for a number of years at the US Air Force base at Torrejon, Spain, by the late Felix Rodriguez de la Fuente and his associates. The problem species there was the Little bustard (*Tetrax tetrax*) and the key to success was to have the falcons well-enough trained so that they actually caught some of the bustards (10). One famous peregrine, Dona Aldonza, killed more than 900 bustards over several years. When such a falcon flies and waits on overhead, all the bustards recognise that it is in a serious "hunting mode" and they depart.

Similarly, trained falcons have been used to rid buildings of pest species such as pigeons, starlings, and crows. The crows that infest the Kremlin walls have been dispersed in this way. In parts of China rice farmers have traditionally used trained sparrowhawks to chase off marauding flocks of small birds, and there have been recent attempts in New Zealand and Australia to do the same thing in vineyards and orchards.

In India and Tibet wild vultures are sometimes used for disposal of human corpses (6).

Birds of prey are also increasingly the focus of the "ecotourism" business. Special areas where raptors occur in unusual numbers and variety have organisations which cater to viewers' interests, e.g. Hawk Mountain Sanctuary Association in Pennsylvania, Cape May Bird Observatory in New Jersey, Snake River Birds of Prey National Conservation Area in Idaho, Matopo National Park in Zimbabwe, and the International Bird Watching Centre, Eilat, Israel. Special feeding programmes such as those for the Mauritius kestrel in the Black River Gorge and for vultures in the Massif Central of France also attract visitors.

## Implications for conservation

Birds of prey are one of the most popular groups of birds. They vie with parrots, cranes, waterfowl, and upland gamebirds for public interest, sympathy, and support. It seems likely that human involvement with raptors in various economic and utilitarian ways will continue to increase. For example, far from being an arcane sport practised by a few romantics steeped in mediaeval lore, falconry is becoming more popular all the time and enjoys many modern innovations of the past four decades. Falconry is an inefficient form of hunting that makes minimal demands on the wildlife resources and is, therefore, a sustainable use. Long after social pressures and environmental constraints combine to make gun-hunting unacceptable some time in the 21st century, falconers will still be able to go afield with their domestically reared hawks and falcons to participate in the pursuit of quarry -- to join with their dogs and their hawks in the primal contest between hunter and hunted and to enjoy their special communion with animals and nature.

These human uses of raptors are actually the driving force behind the conservation of these birds. Because there are falconers and breeders and scientists and rehabilitators and exhibitors and bird-watchers, birds of prey will continue to enjoy a secure place in our increasingly managed world simply because so many human beings with their varied interests in raptors want to see these great birds maintained as a part of their environment and heritage.

## Literature References

1. CADE, T.J. (1986): Propagating diurnal raptors in captivity: a review. *International Zoo Yearbook*, 24/25, p. 1-20
2. CADE, T.J. (1988): Using science and technology to re-establish species lost in nature. In: *Biodiversity*, ed. WILSON, E.O. and F.M. PETER; p.279-288; National Academy Press, Washington, DC
3. CADE, T.J., ENDERSON, J.H., THELANDER, C.G., and C.M. WHITE (eds.) (1988): *Peregrine falcon populations, their management and recovery*. The Peregrine Fund, Inc., Boise, Idaho
4. DEMENTIEV, G.P. (1960): *Der Gerfalke (Falco gyrfalco = Falco rusticolus)*. Neue Brehm-Bücherei No. 264. Ziemsen, Wittenberg -Lutherstadt, Germany (Can. Wildl. Serv. Translation TR-Ger-28)
5. McKAY, P. (1989): *The pilgrim and the cowboy*. McGraw Hill, New York
6. NEWTON, I., and P. OLSEN (eds.) (1990): *Birds of Prey*. Merehurst, London
7. PARRY-JONES, J. (1988): *Falconry, care, captive breeding and conservation*. David and Charles, London
8. PEYTON, R.B., VORRO, J., GRISE, L., TOBIN, R. and R. EBERHARDT (1995): A profile of falconers in the United States: Falconry practices, attitudes and conservation behaviours. *Transactions of the 60th North American Wildlife and Natural Resources Conference, 24-29 March, 1995, Minneapolis, Minn.*; Wildlife Management Institute, Washington, DC
9. RIDDLE, K.E., and J.D. REMPLE (1994): Use of the Saker and other large falcons in Middle East falconry. In: *Raptor Conservation Today*, ed. MEYBURG, B.-U. and R.D. CHANCELLOR; p. 415-420; World Working Group on Birds of Prey and Owls; Helm Information (Pica Press), Mountfield, East Sussex, UK
10. RODRIGUES DE LA FUENTE, F. (1970): *El Arte de Catreria. Segunda Edicion*. Nauta, Barcelona, Spain
11. WOOD, C.A. and F.M. FYFE. (1943): *The art of falconry, being "De Arte Venandi cum Avibus of Frederick II" of Hohenstaufen*. Stanford University Press, Stanford, Ca., USA

## 2.2.3 FOWL-LIKE BIRDS (*GALLIFORMES*)

The order Galliformes includes birds known traditionally as "gamebirds". It comprises 256 species of which the majority belong to the pheasant family (**Phasianidae**). Other less diverse families are the megapodes (**Megapodiidae**), curassows and allies (**Cracidae**), grouse (**Tetraonidae**), guineafowl (**Numididae**), and turkeys (**Meleagridae**). Tab. 19 shows that there original distributions of several galliform species have been vastly expanded by man. World-wide, 74 of them have been introduced to locations outside their normal range, more than in any other group of wild animals (21, 29). Four species, the Indian jungle fowl, the Japanese quail, the American wild turkey and the African guineafowl, have been domesticated for centuries, but many others are bred in captivity or have been semi-domesticated either for release as gamebirds, for recreational purposes or for production of eggs, meat and feathers. Almost all wild galliforms have been or still are being extensively hunted for subsistence, sport or commercial reasons.

### Historical background

Galliform birds are predominantly terrestrial and are thus relatively easy to hunt and trap. Fossil evidence shows that they were consumed by prehistoric man, and they have been associated with hunting and other human activities as far back as historical records exist (34).

The centre of domestication of the *Red jungle fowl*, the ancestor of the domestic hen, is thought to be South-East Asia. Primitive slash-and-burn agriculture brought humans into close contact with the bird, and its domestication provided eggs, meat and entertainment in the form of cock-fighting. The bird's spread east and west probably began before the second millennium BC and it now occurs throughout the world.

*Peacocks* were kept and bred in ancient India for aesthetic reasons and traders brought them to the Eastern Mediterranean probably as early as 2000 BC. The Romans prized them as food and by the end of the 1st century AD were rearing them in large quantities. From Italy they reached France, Germany and England, remaining scarce until the late Middle Ages. As park fowl, peacocks have been semi-domesticated and are now found world-wide.

The *Ring-necked pheasant* was originally exploited mainly for food. Greek mythology states that it was brought to Greece by the Argonauts from the river Phasis in Colchis, now part of Georgia. It was first mentioned by the Greek poet Aristophanes around 400 BC. During the reign of Ptolemy Euergetes II (145-116 BC), pheasants were bred for the royal table, and from the 1st century onwards they were eaten by Roman noblemen. Charles the Great (742-814) ordered the pheasant to be bred in his palaces along with other fowl. Thereafter it became a regular inhabitant of the pheasantries of mediaeval courts eventually becoming feral in many parts of Europe. It is now one of the most popular sporting gamebirds. Since 1882, successful introductions have led to its establishment across much of temperate North

America. Pheasant numbers peaked in the 1950s, then declined with the intensification and mechanisation of farming.

**Table 19:** Overview of the utilisation of Fowl-like birds

| English name | | | Body weight (kg )1) | Distribution and utilisation by man |
|---|---|---|---|---|
| **MEGAPODES - MEGAPODIIDAE** | | Australasia | | exploited for eggs |
| **CURASSOWS and GUANS - CRACIDAE** | | C. and S.America | | hunted for food |
| **GROUSE - TETRAONIDAE** | | | | |
| Capercaillie | *Tetrao urogallus* | N.Eurasia, Alps | 1.8/4.2 | reintroduced to Scotland (1837/38) for hunting |
| Black grouse | *-tetrix* | N.Europe, Alps | 0.9/1.3 | traditional use of feathers |
| Hazel grouse | *Tetrastes bonasia,* | N.Eurasia, Alps | 0.4 | and hunted for sport |
| Ruffed grouse | *-umbellus* | N.America | 0.6 | hunted for sport |
| Willow/Red grouse | *Lagopus lagopus* sspp. | N.Eurasia, N.America | 0.6 | *L. l. scoticus* introduced to England, Belgium (1890s) for sport hunting |
| Ptarmigan | *-mutus* | Arctic, Alps | 0.5 | hunted for sport |
| Blue and | *Dendragapus ob-* | N.America | 0.8/1.3 | hunted for sport |
| Spruce grouses | *scurus, -canadensis* | N.America | 0.5/0.6 | hunted for sport |
| Sage grouse | *Centrocercus urophasianus* | N.America | 1.4/2.4 | hunted for sport |
| Sharp-tailed grouse | *Tympanuchus* spp. | N.America | 0.8/0.9 | hunted for sport |
| and Prairie chicken | | N.America | 0.8/1.0 | hunted for sport |
| Jungle fowl | *Gallus gallus* | C. and S.E.Asia | 0.8/1.0 | domest. BC into dom. fowl |
| Blue peacock | *Pavo cristatus* | India | 3.5/5.0 | introduced BC to Europe; semi-domest. as park fowl |
| Ring-necked pheasant | *Phasianus colchicus* | W.,C. and E.Asia | 0.9/1.2 | introduced BC to Europe; ext. bred for release for hunting; and for meat |
| Grey partridge | *Perdix perdix* | Eurasia | 0.4 | hunted for sport; introduced to USA and Chile for hunting |
| Chukar partridge | *Alectoris chukar* | C. and S.Europe to C.Asia | 0.5 | semi-domesticated; introduced for hunting |
| Red-legged partridge | *-rufa* | W.Europe | 0.5 | introd. to England (1770), Atlantic Islands for hunting |
| Francolins | *Francolinus* spp. | Africa, S.Asia | 0.2-1.2 | hunted for meat, exp. farm. |
| Common quail | *Coturnix coturnix* | S.Europe, C. Asia to Far East, N., E. and S.Africa | 0.1 | intensively hunted for sport; *C.c. japonicus* semi-domesticated for meat and eggs |
| Bobwhite | *Colinus virginianus* | East. USA | 0.17 | intensively bred in captivity for release; semi-domestic. |
| California quail | *Callipepla californica* | West. USA | 0.17 | introd. to Canada, Hawaii, New Zealand, Chile; pet bird |
| **GUINEAFOWL - NUMIDIDAE** | | | | |
| **TURKEYS - MELEAGRIDAE** | | | | |

1) weights are expressed as species means, or as a female/male pair of means for dimorphic species

The *Grey partridge*, which originally inhabited the steppes of Central Europe and Asia, is another important gamebird in Europe. During the Middle Ages, deforestation and agricultural development in Europe increased the amount of habitat suitable for this species leading to an enormous expansion of its range. Its spread was accelerated by numerous introductions as a gamebird into almost all countries of the northern temperate zone, and some of the southern hemisphere. The most successful attempts were in North America where it is established in the northern half of the USA and in Southwestern Canada. Across most of its global range, however, it has declined by over 80% since the Second World War because of agricultural intensification (33). As bags declined, keepering became uneconomic and ceased on many estates so that the situation was aggravated by an increase in predation pressure. The overall result is that in Eastern, Central and Western Europe, the Grey partridge has largely been replaced as a quarry species by the pheasant.

The *Alectoris partridges* are the principal sporting gamebirds in Mediterranean countries. They occupy arid and mountainous habitats from the Alps to Manchuria. Since ancient times, cocks of the *Chukar partridge* have been tamed and used for cock fights in Asia. Through introductions after the First World War, chukars have become established on rocky grassland slopes in the sagebrush belt of the Western USA (a habitat not used by any native gamebird), in Hawaii and in New Zealand. The *Red-legged partridge* is also a very popular gamebird but is restricted to less mountainous and less arid areas in Western Europe only. French redlegs were successfully introduced to England around 1770.

The *European* and the *Japanese quails* are the only truly migratory species among the galliforms. Since ancient times, the vast numbers migrating south from Europe and Western Asia in the autumn have been hunted and trapped for food along the Southern and Eastern Mediterranean coast. The Old Testament (Exodus 16:13) tells how the fleeing Israelites caught quails for food on the Sinai Peninsula. Starting in 1885, Egypt exported netted quails to Europe; exports peaked at over 3 million in 1920 but collapsed to nothing by 1939 as such harvest levels were unsustainable. Nowadays, the Japanese quail has replaced the European quail as a commercially traded food item, having undergone domestication in Japan or the East Asian continent since the 11th century. In 1910, Kotaro Oda began selection for egg-laying and is considered the founder of Japan's modern quail industry. In the 1950s, quail farming was introduced into Europe and has gained increasing economic importance in most countries of the world.

The *Helmeted guineafowl* has been known to the Greeks since the 4th century BC, having been imported from the Sudan (*Numida m. meleagris*) and Morocco (*N. m. sabyi*). The birds were kept primarily for ritual purposes. After 110 BC, guineafowl were also kept in Rome and highly esteemed as a delicacy. They were well known in Byzantium, as shown by numerous depictions and mosaics from the 4th century AD. After the fall of the Roman and Byzantine empires, guineafowl were apparently forgotten until Portuguese seafarers brought *N. m galeata* from the Guinea coast to Europe in the 15th century. It was intensively bred for meat and first domesticated in France in the 16th century. During that time it was exported by the Portuguese to the Cape Verde Islands and by the Spaniards to the Antilles where they are still the most popular gamebirds for hunting. In Africa, guineafowl were never truly domesticated although they are extensively used for food. They were introduced to the Cape Province of South Africa, mainly for the control of pest insects and weeds in orchards and vineyards.

Indians in the Mexican state of Michoacán had domesticated the local subspecies of the *Wild turkey* long before the Spanish conquistadors arrived in Mexico (34). When Columbus discovered the Antilles in 1492, the domestic turkey was already well established with the local Karaib Indians. The so-called "Indian chicken" was brought to Europe in 1520 by the Spaniards and first appeared on an English Christmas menu in 1585. European settlers in North America relentlessly hunted Wild turkeys, the US population was tallied of which in 1930 at 30,000 birds and falling. In this critical situation hunters lobbied for the enactment of the Pittman-Robertson Act in 1937 which levied a special tax on hunting equipment in order to assist wildlife conservation. Over the last 50 years this Act has generated more than US$ 1,700 mill. for turkey restoration projects, and the development of trap-and-transfer methods, regrowth of eastern forests, protection from excessive hunting, and habitat improvements have resulted in an enormous come-back of this species. Wild turkeys have also been introduced into Hawaii and New Zealand for hunting.

## Management and utilisation

Management techniques vary from country to country depending on the quarry species, method of exploitation, local hunting rights and land ownership. These determine to a large extent the scope for habitat and predation management. In Portugal, Italy, Greece and Turkey, hunting is open on all land except national parks, reserves and some agricultural land (although regional authorities may restrict hunting in some areas); there is correspondingly little incentive for management. The same situation pertains on public land in the USA. In other countries, especially Germany and in Eastern Europe, hunting rights are delegated to hunters' associations: only members may hunt and practical wildlife management is a requirement of membership. In countries with private hunting rights, the landowner may exercise those rights himself or leave them to other individuals or hunting syndicates. Because the landowner is able to benefit financially from shooting, there is a vested interest in managing game and game habitat in a way that will produce a high sustained income.

The recent declines of partridges and many other gamebirds are mostly the result of habitat loss or alteration and increased predation, which can be reversed by appropriate management. For instance, field margins can be made more attractive to *Grey partridges* by enhancing the field edge (hedges, grassy bank) as nesting cover and the edge of the cereal crop itself as brood-rearing habitat (33). In the same way, the densities of *pheasants* can be increased by planting areas of suitable shrubby cover and increasing the amount of woodland edge (17). In Scotland, management of *Red grouse* traditionally includes heather burning, and methods that combat parasites have been developed (18); slight changes in forestry practices can benefit *capercaillie* and other *woodland grouse* (43). Often small adjoining estates are treated as one unit for the purpose of management.

Predators have been shown to reduce populations of harvestable game birds (11, 26) and their control has been the traditional role of the gamekeeper. However, excessive control in the last century and the early 1900s drastically reduced numbers of many predator species in

Day-old Ring-necked pheasant chicks reared artificially. Advances in poultry science have enabled several galliform species to be bred intensively in captivity, in some cases on an industrial scale, for eggs and meat. (Photo: T. BLANK; by courtesy of the Game Conservancy, Fordingbridge, UK)

Europe. Nowadays, all raptors and owls are protected in most countries. Legal control of mammalian predators and corvids is still an important component of partridge, pheasant and grouse management in many countries.

The releasing of reared gamebirds into the wild was developed in an attempt to restock declining galliform populations and thereby maintain sporting revenues. However, research has shown that restocking is of no benefit to wild populations of partridges and pheasants (7). The survival rate of reared birds in the wild is so poor that few are able to breed. Moreover, releasing reared birds may introduce disease into the wild stock and encourage an artificially high harvest rate that further reduces it. Finally, released birds of dubious origin can influence the genetic composition, behaviour and reproductive ability of the wild birds if they interbreed. Nevertheless, releasing has gathered its own momentum because of the underlying economics, and in many countries *partridges* and *pheasants* especially are released specifically for shooting.

The commercial rearing of gamebirds for food and eggs is also increasing throughout the world. By 1970 more than 2 million *quail hens (Coturnix c. japonicus)* were kept in industrial farms in Japan, and the total quail egg production in Japan is of a size comparable to hen egg production (34). India and Pakistan are also large producers of quail. In Europe, large-scale production of quail for eggs and meat takes place in France, Spain and Italy, with France consuming 260 million quail a year. Consumption is much lower elsewhere in Europe, being for instance 0.5 million quail a year in the UK (15). *Chukar-redleg hybrids* have been found to have a higher egg-laying capacity than their parent species and this has stimulated the farming of chukar crosses for consumptive purposes. In competition with the Japanese quail, the farming of the chukar partridge is fast gaining economic importance particularly in France, Italy and Spain. Other gamebirds reared for meat are *pheasants, francolins* and *guineafowl*, though none of these approache the scale of operations of quail farming. There are small outlets for the commercial sale of non-meat products, for example feathers from *pheasants* and *Jungle fowl*, which are used for fishing flies and decorative purposes, as described in Chapt. 3.2.5.

## Economic importance

In many countries the shooting of gamebirds is a popular sport. There are currently over 9 million shooters in Europe and 18 million in North America (44, 28, 11). Of these, the proportions that hunt gamebirds range from approximately 50% in Scandinavian countries to over 90% in the Mediterranean ones. The regions with the highest proportions of shooters in the population are North America and Fennoscandia followed by the Mediterranean. Hunting pressure, as measured by the number of hunters per $km^2$, is highest in Malta (13), Cyprus (24) and Italy (28) (over 4 shooters per $km^2$), whereas the hunting pressure tends to be lowest in Nordic and Alpine countries (2, 44). The USA ranks 9th, and Canada last, for numbers of shooters per unit area.

The ***sporting harvests*** of gamebirds in ***Europe*** are shown in Tab. 20. They reflect the north-south transition from boreal forests to increasingly dry agricultural or mountainous landscapes. In northern countries the bag comprises predominantly grouse species. Approximately 250,000 *Black grouse*, 96,000 *capercaillie*, 500,000 *Willow grouse* and 120,000 *Hazel grouse* are killed annually, mainly in Norway, Sweden and Finland; a further 450,000 *Red grouse* are shot in Northern England and Scotland.

In lowland Britain and Central Europe from France eastwards, the *Ring-necked pheasant* and *Grey partridge* have become the most popular game species with total bags of around 22.4 million and 2.8 million respectively. Over three quarters of the Grey partridges are shot in France, while the greatest number of pheasants (about 12 million annually) are shot in Britain. In Southwestern Europe the *Red-legged partridge* dominates with an overall bag that may be as high as 10 million. The *European quail* is the other favourite gamebird of Southern Europe, but the estimated total annual bag of 2.8 million is conservative as no bag statistics are available from Italy or Portugal. The overall annual gamebird bag for Europe lies in the region of 40 million birds of which over half are pheasants.

In ***North America*** about 24 million gamebirds are shot annually (12, 33, 41). Half of these are introduced species, mostly the *Ring-necked pheasant* (9.5 million killed) which is the foremost gamebird in the mid-west and Central Basin regions. The *Grey partridge* bag has fallen below 1 million and has been overtaken by the *chukar* which has become the principal quarry species in Nevada (over 1 million shot annually). Most of the indigenous game-

**Table 20:** Numbers (in thousands) of gamebirds harvested annually in Europe

| Country | *Tetrao urogallus* | *Tetrao tetrix* | *Tetrastes bonasia* | *Lagopus* spp. | *Phasianus colchicus* | *Alectoris rufa* | *Perdix perdix* | *Coturnix coturnix* |
|---|---|---|---|---|---|---|---|---|
| Austria | <1 | 2 | 0 | 0 | 213 | - | 8 | - |
| Belgium | - | - | - | - | 300<br>*(500)* | - | <40<br>*(25)* | - |
| Czechoslovakia | ? | ? | ? | - | 652<br>*(>1,000)* | - | 0 | 0 |
| Denmark | - | - | - | - | 772<br>*(450)* | - | 77<br>*(35)* | 0 |
| Finland | 40 | 132 | 86 | 76 | ? | - | 3 | - |
| France | ? | 15 | 2 | ? | 6,153<br>*(8,000)* | 1,166<br>*(800)* | 2,181<br>*(1,200)* | 644<br>*(1,000)* |
| Germany | ? | ? | 0 | - | 390 | - | 22 | 0 |
| Hungary | 0 | 0 | 0 | - | 654<br>*(1,300)* | - | 0 | 0 |
| Ireland | - | - | - | - | >180<br>*(>110)* | - | 3<br>*(5)* | - |
| Italy | ? | ? | 0 | ? | ?<br>*(6,000)* | ? | 100<br>*(500)* | ? |
| Luxembourg | - | - | 0 | - | 1 | - | - | - |
| Malta | - | - | - | - | - | - | - | 24 |
| Netherlands | - | 0 | - | - | 200<br>*(54)* | - | 12 | 0 |
| Norway | 19 | 51 | 8 | 479 | ? | - | - | 0 |
| Poland | ? | ? | ? | - | 141 | - | 100 | 0 |
| Portugal | - | - | - | - | ? | 6,200<br>*(?)* | - | ? |
| Romania | ? | 0 | ? | - | 100 | - | <36 | ? |
| Spain | ? | - | - | - | ? | 2,808<br>*(>100)* | ? | 2,103 |
| Sweden | 36 | 48 | 25 | 19 | 19 | - | 6 | 0 |
| Switzerland | 0 | <1 | 0 | <1 | <1<br>*(5)* | - | 0 | 0 |
| United Kingdom | <1 | 2 | - | 450 | 12,000<br>*(20,000)* | 450<br>*(2,000)* | 187<br>*(100)* | 0 |
| former Yugoslavia | ? | ? | ? | - | 620<br>*(>300)* | - | 45 | ? |
| **Total (in thousands)** | **96** | **251** | **121** | **1025** | **22,396**<br>***(37,719)*** | **10,624**<br>***(2,900)*** | **2,820**<br>***(1,865)*** | **2,771**<br>***(1,000)*** |

Figures in parentheses indicate numbers (in thousands) released for shooting

Sources: References (2), (5), (6), (7), (13), (16), (22), (28), (30), (31), (32), (37), (40), (44), and pers. comm. by BEER, J. V., KAVANAGH, B., POTTS, G.R., ROBERTSON, P.A. and T. SZEP

bird species killed are New World quails. The one most frequently hunted in the USA is the *Northern bobwhite*, with more than 4 million birds taken each autumn in the eastern half of the country. In the Southcentral and Southwestern USA it is replaced by the western quails, mainly *Scaled (Callipepla squamata), Gambel's (C.gambellii)* and *California quail (C. californica)*, which together amount to another 5.5 million birds shot per annum. In the forests of Canada and the Northern USA, about a million *Ruffed grouse* and a third of a million *Blue* and *Spruce grouses* are taken annually. *Sharp-tailed grouse (Tympanuchus phasianellus)* are the most popular gamebirds of the grasslands and brushlands of the northern half of North America, though some *Prairie chickens (T.cupido)* and *Sage grouse* are also taken making a total of 1.2 million. The annual harvest of *Wild turkeys* is about 340,000.

The main reason for the larger total harvest in Europe than in North America despite fewer hunters is the scale of the releasing of reared birds in Europe. The three main species involved are *Ring-necked pheasant, Red-legged partridge* and *Grey partridge* (see Tab. 20). Releases in North America amount to approximately 1 million pheasants; in Europe the total exceeds 37 million pheasants and approaches 5 million partridges. The bulk of the European releases occur in the UK, France and Italy, which together account for 90% of released pheasants and over 95% of partridges. Small numbers of other *Alectoris partridges (A. chukar, -graeca, -barbara)* have been released for hunting purposes in many European countries both within and outside their normal range.

In 1970 the ***overall carcass value*** for game of 15 European countries was estimated at about US$ 100 mill. of which gamebirds contributed US$ 11 mill. (30). The same calculation applied to the recent data in Tab. 20 gives a total carcass value for gamebirds of US$ 23 mill., a two-fold increase in real terms over 20 years. Including annual hunting expenses, the total value of hunting in Europe was estimated to be US$ 3,500 mill. of which about half is attributable to gamebird hunters (28). In Canada, a survey of the importance of wildlife to Canadians carried out in 1987 indicated that hunter expenditure totalled US$ 796 mill. of which gamebird hunters probably contributed half (14). This had impacts at the local, provincial and national levels of the Canadian economy by generating an estimated US$ 826 mill. of business production, contributing US$ 488 mill. to Canada's GDP and supporting some 16,000 jobs providing US$ 278 mill. in personal income. The governmental tax receipts were US$ 188 mill.

In the United Kingdom, a detailed examination of the economic significance of countryside sports allows a closer look at the ***economics of gamebird hunting*** standardised to 1990 values (10). Gamebird shooters were estimated to spend US$ 437 mill. per annum and organisers of the sport US$ 82 mill. Direct employment comprised 5,000 gamekeepers and 8,000 casual jobs, while trades and services sustained 14,500 jobs. The sale of the produce (87% pheasants, 8% grouse, 4% partridges) amounted to US$ 21 mill. Revenue to local and central Government from direct and indirect taxation was US$ 179 mill. The indirect expenditure and employment generated by the sport was US$ 428 mill. and 21,550 jobs giving a total expenditure of US$ 948 mill. and total gamebird-related employment of about 50,000 jobs. For landowners, the leasing of shooting rights brings in US$ 9-14/ha for good-quality rough shooting and US$ 14-37/ha for good-quality driven shooting. The sales of 140 game farmers in the UK in 1981 were valued at US$ 18 mill. for an expenditure of US$ 13 mill. (in 1990 terms). The sale of pheasant eggs, chicks and poults formed an important part of their revenue (60% for poults alone) so considering that 8.3 million pheasant eggs, chicks, poults and adults were sold in 1981 and that releases are today estimated at 20 million, the figures for sales and expenditure should be at least doubled to give a true picture for 1990.

In other countries gamebird shooting provides vital ***extra revenue on marginal land*** that is agriculturally poor. In Spain, the rent for *Red-legged partridge* shooting may represent a quarter of a small landowners's earnings per hectare on his farm and rent can be as high as US$ 40/ha (GULLICK *in litt.*). In the Southwestern USA the leasing of land for *quail* hunting (US$ 10/ha) is an important economic factor for many ranches. The provision of guides, dogs and accommodation for quail hunters has built up into a thriving industry, particularly in the Southeastern states (7). In South Africa, commercial shooting of *Grey-wing francolin (Francolinus africanus)* brings revenue into rural areas by providing farmers in the East Cape Province with a sustainable product worth US$ 40 per bird compared with US$ 45 per sheep on the hoof (20).

More generally, francolins, guineafowl, and quail are exploited for ***subsistence*** throughout the African continent. The birds are often trapped and sold live at roadsides or local markets (7, 27). The eggs are also taken for food or reared under domestic fowl giving rise to *semi-feral flocks of guineafowl* kept for meat (3). In Egypt, around 200,000 *European quails* are captured annually for local consumption on the north coast during autumn migration (4). In Asia and Latin America, most of the *phasianid* and *cracid species* are shot or trapped primarily for subsistence, but also for sale at local markets or for the bird trade. The harvest is extremely difficult to evaluate, although some 0.5 million birds are killed for food annually in the Amazonas state of Brazil (35). In Australasia, the eggs of *megapodes* which lay large clutches in mounds heated by sun, thermal activity or rotting vegetation have long been collected for human consumption. In Papua New Guinea, this traditional activity has become an enormous industry involving almost 5 million eggs of the *scrubfowl (Megapodius freycinet)* (8).

## Implications for conservation

World-wide, 27% of the 256 galliform species are globally threatened, proportionately more than twice as many as for birds generally. Out of the 68 species in question, 42 are pheasants, 16 cracids and 8 megapodes, while 57 of them live in Asia or the Neotropics. Many of these globally threatened species are single-country endemics often confined to small areas of relict natural habitat, and deforestation is the single most important threat affecting them. Where they inhabit farmed landscapes (predominantly partridges and francolins), intensification and development are major threats (7). Overall, in most cases the primary reason for the birds' threatened status is habitat loss or degradation. Nevertheless, the impact of exploitation must not be underestimated, as birds perfectly able to withstand a high hunting pressure in good-quality habitat are unable to tolerate even a low level of exploitation in degraded or fragmented habitat. In much of Asia, Africa and the Neotropics, subsistence hunting can aggravate a situation already rendered unfavourable by deforestation, overgrazing or agricultural development.

In Asia, where all of the threatened pheasants are concentrated, overhunting is frequently cited as a cause of population declines (19, 37, 38, 36, 9, 7) and over-exploitation of eggs as another (23, 7). In the *USA*, hunting pressure on public lands is often unsustainable and populations are maintained through immigration from nearby private lands where densities are relatively high. Growing concern about the impact of hunting on wildlife, not least the

Wild turkey, led to the creation in 1937 of a co-operative programme for wildlife management as described above. This accounts to a large extent for the current abundance of wildlife. In Europe, overhunting has been the main cause of gamebird declines in Mediterranean countries such as Italy, Greece, Turkey and Cyprus. Steps have, however, been taken to remedy the situation, with the establishment of protected areas and regional or national protection of selected species (e.g. chukar and Grey partridges in Greece and Turkey (25)). For species where habitat loss or degradation is the main problem, shooting can actually be beneficial as the sporting interest leads to investments in habitat management that would otherwise be absent. This can lead to the apparent paradox of sustained higher densities when the birds are shot than when they are not (1). It is particularly relevant for species that have become rare or endangered. In Scotland, the sporting interest in capercaillie is the main factor preventing the conversion of its primary habitat, mature Caledonian pinewoods, to commercial forestry. Moreover, habitat protection is not enough to ensure the survival of this species: it is highly susceptible to predation during the breeding season and practically total breeding failure has been observed in nature reserves containing prime habitat but lacking predation control.

The benefits to non-game wildlife of gamebird management is an area whose potential has scarcely been realised. The provision of nesting, brood-rearing and overwintering cover for pheasants and partridges provides food and shelter for songbirds, small mammals and invertebrates that also suffer from the effects of agricultural intensification (39). Accordingly, major habitat development programmes for farmland wildlife, primarily pheasants and partridges, have been initiated in the USA (12). Within the European Union, recent regulations allow management of set-aside land for small game and other wildlife. Gamebirds may also benefit from the Union's promotion of low-intensity agriculture under its agri-environment programme and from agricultural extensification schemes financed by individual countries or regions, often combined with private initiatives at the local level (44).

The role of sustainable utilisation of gamebirds as a means of promoting their conservation in developing countries has recently gathered momentum. The key to success is the participation of local communities who, by sharing in financial rewards of sustainable utilisation, become defenders as well as users of the resource. Programmes to this end have been initiated in Zimbabwe in the framework of CAMPFIRE (7), in South Africa by the African Gamebird Research, Education and Development Trust (42) and in Nepal by the World Pheasant Association (19). For galliform species at risk because of the demand for their meat, farming has been proposed as a means of relieving the pressure on wild populations in India (9) and in China (7). All of these programmes have in common the fact that they recognise gamebirds as a resource which, if valued and managed correctly, can motivate local people to husband and conserve it. This notion of wise use offers the best long-term chances of survival for many of the world's Fowl-like birds.

## Literature references


1. AEBISCHER, N.J. (1991): Sustainable yields: gamebirds as a harvestable resource. *Gibier Faune Sauvage*, 8, p. 335-351
2. ANONONYMOUS (1993): *Eidgenössische Jagdstatistik 1992*. Bundesamt für Umwelt, Wald und Landschaft; Bern, Switzerland
3. AYENI, J.S.O. (1981): Aspects of the biology of the Helmeted Guineafowl (*Numida meleagris galeata* Pallas) in Nigeria. *World Pheasant Association Journal*, 6, p. 31-39
4. BAHA EL DIN, S.M. and W. SALAMA (1991): The Catching of Birds in North Sinai, Autumn 1990. *International Council for Bird Preservation; Study Report. No. 45*; Cambridge
5. BERGSTRÖM, R., HULDT, H. and U. NILSSON (1992): *Swedish Game: Biology and Management*. Svenska Jägareförbundet; Spånga, Sweden
6. BERTELSEN, J. and N.H. SIMONSEN (1986): *Documentation on Bird Hunting and the Conservation Status of the Species Involved: Situation in 1986*. Landbrugsministeriets Vildtforvaltning; Rønde, Denmark
7. BIRKAN, M., POTTS, G.R., AEBISCHER, N.J. and S.D. DOWELL (eds.) (1992): Proceedings of Perdix VI, Ist International Symposium on Partridges, Quails and Francolins, Fordingbridge, UK. *Gibier Faune Sauvage*, 9, p. 281-918
8. BISHOP, K.D. (1980): Birds of the volcanoes - the scrubfowl of West New Britain. *World Pheasant Association Journal*, 5, p. 80-90
9. CHANDOLA-SAKLANI, A, BHADAURIA, R.S., JAIN, M.S. and P. LAKHERA (1990): Game management and captive breeding as a conservation measure in Himalayan hills. In: *Transactions of the XIXth Congress of the International Union of Game Biologists*, ed. MYRBERGET, S.; p. 677-685; Norwegian Institute for Nature Research, Trondheim, Norway
10. COBHAM RESOURCE CONSULTANTS (1992): *Countryside Sports: Their Economic and Conservation Significance*. The Standing Conference on Countryside Sports, Reading, UK
11. CSÁNYI, S. and J. ERNHAFT (eds.) (1992): *Transactions of the XXth Congress of the International Union of Game Biologists*. University of Agricultural Sciences; Gödöllö, Hungary
12. DUMKE, R.T., STIEHL, R.B. and R.B. KAHL (eds.) (1984): *Proceedings of Perdix III: Gray Partridge and Ring-necked Pheasant Workshop, Campbellsport, USA*; Wisconsin Department of Natural Resources, Madison, USA
13. FENECH, N. (1992): *Fatal Flight: The Maltese Obsession with Killing Birds*. Quiller, London
14. FILION, F.L., JACQUEMOT, A., BOXALL, P.C., REID, R., BOUCHARD, P., DUWORS, E. and P.A. GRAY (1990): *The Importance of Wildlife to Canadians in 1987: The Economic Significance of Wildlife-related Recreational Activities*. Canadian Wildlife Service; Ottawa, Canada
15. HAYWOOD, M.P.S. (1981): *Quail Production*. Agricultural Development and Advisory Service; Ministry of Agriculture, Fisheries and Food; London
16. HELL, P. (ed.) (1983): *Transactions of the XVIth Congress of the International Union of Game Biologists*. Vysoké Tatry, Strbské Pleso, Czechoslovakia
17. HILL, D.A. and P.A. ROBERTSON (1988): *The Pheasant: Ecology, Management and Conservation*. Blackwell Scientific, Oxford
18. HUDSON, P.J. (1992): *Grouse in Space and Time: The Population Biology of a Managed Gamebird*. The Game Conservancy Ltd., Fordingbridge, UK
19. JENKINS, D. (ed.) (1986): Pheasants in Asia 1986. *Proceedings of the IIIrd International Symposium on Pheasants in Asia, Chiang Mai, Thailand*. World Pheasant Association, Reading, UK
20. LITTLE, R.M. and T.M. CROWE (1993): In the bag. *Our Living World, September 1993*: p. 12-13
21. LONG, J.L. (1981): *Introduced Birds of the World*. David and Charles, Newton Abbot and London

22. LONGVA, S. (1993): *Jaktstatistikk 1992*. Norges Offisielle Statistikk Nos C92, Statistisk Sentralbyrå, Oslo-Kongsvinger, Norway
23. MACKINNON, J. (1978): Sulawesi megapodes. *World Pheasant Association Journal*, 3, p. 96-103
24. MAGNIN, G. (1987): An Account of the Illegal Catching and Shooting of Birds in Cyprus during 1986. *International Council for Bird Preservation; Study Report No. 21*; Cambridge
25. MAGNIN, G. (1989). Falconry and Hunting in Turkey during 1987. *International Council for Bird Preservation; Study Report No. 34*; Cambridge
26. MARCSTRÖM, V., KENWARD, R.E. and E. ENGREN (1988): The impact of predation on boreal tetraonids during vole cycles: an experimental study. *Journal of Animal Ecology*, 57, p. 859-872
27. MARTIN, G.H.G. (1993): Bush Meat in Nigeria as a natural resource with environmental implications. *Environmental Conservation*, 10, p. 125-132
28. MYRBERGET, S. (1990): *Wildlife Management in Europe Outside the Soviet Union*. Norway Institute for Nature Research, Trondheim, Norway
29. NIETHAMMER, G. (1963): *Die Einbürgerung von Säugetieren und Vögeln in Europa*. Parey, Hamburg and Berlin
30. NÜSSLEIN, F. (1974): Der ökonomische Wert der Wildproduktion in Europa (ohne Sowjetunion). *Zeitschrift für Jagdwissenschaft*, 20, p. 85-95
31. OFFICE NATIONAL DE LA CHASSE (1986); Enquête Nationale sur les Tableaux de Chasse à Tir, Saison 1983-1984. *Bulletin Mensuel de l'Office National de la Chasse*, 108, p. 1-88.
32. ORTONI, F. (1992): Un albo nazionale per tutelare il settore. *Rivista di Avicoltura*, 4, p. 24-26
33. POTTS, G.R. (1986): *The Partridge: Pesticides, Predation and Conservation*. Collins, London
34. RAETHEL, H.-S. (1988): *Hühnervögel der Welt*. Neumann-Neudamm, Melsungen, Germany
35. REDFORD, K.H. and J.G. ROBINSON (1991): Subsistence and commercial uses of wildlife in Latin America. In: *Neotropical Wildlife Use and Conservation*, ed. ROBINSON, J.G. and K.H. REDFORD; p. 6-23; University of Chicago Press, Chicago and London
36. ROUND, P.D. (1988): *Resident Forest Birds in Thailand: Their Status and Conservation*. International Council for the Preservation of Birds; Monograph No. 2; Cambridge
37. SAVAGE, C.D.W. (ed.) (1980): Pheasants in Asia 1979: *Proceedings of the 1st International Symposium on Pheasants in Asia, Kathmandu, Nepal*. World Pheasant Assocication, Reading, UK
38. SAVAGE, C.D.W and M.W. RIDLEY (eds.) (1983): Pheasants in Asia 1982. *Proceedings of the IInd International Symposium on Pheasants in Asia, Srinagar, Kashmir*. World Pheasant Association, Reading, UK
39. SOTHERTON, N.W. (1991): Conservation Headlands: a practical combination of intensive cereal farming and conservation. In: *The Ecology of Temperate Cereal Fields*, ed. FIRBANK, L.G., CARTER, N., DERBYSHIRE, J.F. and G.R. POTTS; p. 373-397; Blackwell Scientific, Oxford
40. TAPPER, S.C. (1992): *Game Heritage: An Ecological Review from Shooting and Gamekeeping Records*. Game Conservancy Ltd., Fordingbridge, UK
41. US FISH AND WILDLIFE SERVICE (1983): *A summary of selected fish and wildlife characteristics of the 50 states*. Washington, DC
42. WANNENBURGH, A. (1992): *African Gamebird: Field Report No. 1*. The African Gamebird Research, Education and Development Trust, Johannesburg, South Africa
43. WEGGE, P. (1984): Skogburket og storfuglbiotopene. *Viltrapport*, 36, p. 84-90
44. WIESE, M. (ed.) (1994): *DJV-Handbuch 1994*. Deutscher Jagdschutz-Verband, Bonn

## 2.2.4 OTHER BIRDS

There are approximately 9,000 species of bird in the world, divided between 28 orders. Birds are perhaps the best known taxa of any form of animal and plant life. They are widely distributed, very visible (the majority being diurnal), popular amongst millions of people, in developed countries for aesthetic reasons, in developing countries for socio-economic and cultural reasons, and have for centuries had a close association with man. Tab. 21 shows that birds and their products are used for a highly diverse set of purposes by man. This chapter briefly reviews this diversity. It focuses primarily on the more direct uses and avoids any detail on those particular groups of birds highlighted in previous chapters, namely ostriches, birds of prey and galliforms.

Any review of the use of birds by man leads us to conclude that people have been fascinated by birds for centuries. They have been observed, used, painted and written and sung about. People have kept them, protected them and revered them since the earliest times from which we have any record. Thus our concern for birds is, as DIAMOND (3) put it, an unbroken link with our earliest ancestors. It is a major theme of mankind's cultural, aesthetic and spiritual development, as well as a source of vital insight into present-day environments which we share with birds and all biodiversity.

### Historical background of bird utilisation

Historical evidence suggests that man has used birds since time immemorial. Cave paintings by Stone Age man have been found in Spain and France that are 12,000 to 15,000 years old and depict birds, including the use of *birds of prey, doves* and some *galliforms*. The first person to use birds, as what we would today call, monitors of the environment was the Greek philosopher Aristotle (384-322 BC). From the Christian religion we find perhaps one of the most widely known uses of birds, namely the dove that brought the olive branch to tell Noah that land had emerged from the flood. Similar reference is made to the value of birds in the Koran and many other religious texts.

The process of domestication of birds began some 5,000 to 8,000 years ago. From ancient written and pictorial material and the discovery of bones at archaeological sites, evidence of the earliest bird domestication comes from the oriental region centred on India from the period 3,000 BC. The first species to be domesticated here were the *Grey-lagged goose (Anser anser), Rock dove (Columba livia)* and *Red jungle-fowl* (*Gallus* spp.). Several centuries later the *mallard (Anas platyrhynchos)* was domesticated in the Near East, the *Swan goose (Anser cygnoides)* in Eastern Asia and the *Moscovy duck (Cairina moschata)* and *turkey (Meleagris gallopavo)* in Central America (1).

The use of bird products has an equally long history. It includes such diverse uses as the harvesting of guano for fertiliser to the use of feathers for ornamentation and down to the use of *cormorants*, for example, to catch fish in China, a practice used since the Song Dynasty (960-1,298 AD) (23).

**Table 21:** Overview of the utilisation of other birds[1)]

| Zoological Order | Prominently utilised species and genera[2)] | Utilisation[3)] |
|---|---|---|
| TINAMIFORMES Tinamous | Tinamous (*Tinamus*- and *Rhynchotus* spp.) | hunted for meat, attempted introduction to Europe for sport |
| SPENISCIFORMES Penguins | Various species | formerly exploited for oil |
| PROCELLARII-FORMES Albatrosses and shearwaters | Albatrosses (*Diomeda* spp.), Arctic fulmar (*Fulmarus glacialis*), shearwaters (*Puffinus* spp.) | chicks for meat, eggs and down, oil |
| PELICANIFORMES Pelicans, cormorants and boobies | Pelicans (*Pelicanus* spp.), cormorants (*Phalacrocorax* spp.), boobies (*Sula* spp.) | for guano, fishing and down |
| CICONIIFORMES Storks, egrets and herons | Egrets (*Egretta* spp.), herons (*Mesophoxy intermedia, Caserodius alba*), ibises (*Ibis* spp.), jabiru (*Ephippiorhynchus mycteria*), marabou (*Leptoptilos crumaniferus*) | for feathers |
| PHOENICOPTERI-FORMES Flamingos | Red flamingo (*Phoenicopterus ruber*) | for feathers |
| ANSERIFORMES Waterfowl | Wild geese (*Anser-, Branta* spp.), Common eider (*Somateria mollissima*), Common shelduck (*Tadorna tadorna*), ducks (*Anas* spp.), goosander (*Mergus merganser*) | for down, eggs, hunted for meat and sport |
| CHARADRIIFORMES Plovers, gulls and auks | Lapwings and plovers (*Vanellus-, Charadrius* spp.), sniper and woodcocks (*Gallinago-, Scolopax* spp.), gulls and terns (*Larus* spp.), auks (*Uria-, Cepphus-, Alca-, Alle-, Fratercola* spp.) | for sports, hunting, eggs and chicks, down, trapping adults for meat |
| COLUMBIFORMES Pigeon-like birds | Various dove species (*Columba-, Streptotelia* spp.), Crowned pigeons (*Goura* spp.) | hunting for meat, feathers |
| PSITTACIFORMES Parrot-like birds | Cockatoos (C*akatoe* spp.), lories (*Trichoglossus* spp.), Love birds (*Agapornis* spp.), parrots (*Psittacus-, Poicephalus-, Amazona* spp.), macaws (*Ara* spp.) | for feathers and pet trade, breeding |
| APODIFORMES and TROCHILIFORMES | Swiftlets (*Collacalia* spp.), hummingbirds (*Trochilidae*) | nests for food, guano, for feathers |
| TROGONIFORMES Trogons | Resplendant quetzal (*Pharomachrus mocino*), trogons (*Trogon* spp.) | for feathers |
| PICIFORMES Woodpeckers | Honeyguides (*Indicator* spp.) | for guidance to bee nests |
| PASSERIFORMES Finch-like birds | Blackbirds (*Turdus* spp.), buntings, tanagers (*Emberizidae*), finches (*Fringillidae*), sparrows, weavers, whydas (*Ploceidae*), wax-bills (*Estrildidae*), starlings and mynahs (*Sturnidae*), birds of paradise (*Paradisaeidae*), | for feathers and pet trade, breeding and keeping for recreation, netting and liming for food |

1) other than STRUTHIOFORMES, FALCONIFORMES and GALLIFORMES; 2) utilisation of some of the listed species has ceased and is mentioned in an historical context only; 3) use of feathers is detailed in Chapt. 3.2.5.

Not surprisingly the greatest utilisation of birds and their products recorded in history is for their food value. Wild birds provide food in a host of different societies throughout the world.

As society has "developed" the diversity of uses of birds has increased with little reduction in their utilisation for man's basic needs, such as food, but a greater awareness of ways in which birds can bring pleasure to millions of people and be used as early warning systems of environmental change. The science of ornithology now has a key role to play in the whole area of environmental conservation and in some parts of the world it is recognised as being at the forefront of scientific research. Similarly, much of the debate surrounding decisions made on subjects such as sustainable development and long-term integrated land-use planning makes use of our knowledge of birds as measures of sustainability (11).

## Present utilisation practices and their importance

As with other animal classes the utilisation of birds may be consumptive, i.e. birds and/or their eggs or hatchlings are removed from the wild populations, or non-consumptive, i.e. indirect without reduction of numbers of birds but nevertheless sometimes with repercussions on their population dynamics and ecology. Serving man's material needs, consumptive use is the more original use and is presently mainly limited to developing countries. Most of the non-consumptive uses have developed only during recent decades.

### Consumptive uses

All over the world birds are an important source of food providing either meat or eggs. In a host of different societies wild birds are **hunted or trapped for their meat**. To the forest dwelling peoples of Amazonia, for example, this is estimated to contribute 2.5 to 3% of all their food (3). *Seabird* colonies have proved particularly valuable to human societies, perhaps because they are often tame, abundant and easy to harvest. Around Tasmania, for example, over half a million chicks of the *Short-tailed shearwater (Puffinus tenuirostris)*, locally known as the *muttonbird*, are taken annually by a well-organised industry (19). Investigation of this has shown it is a sustainable harvest in the sense that it does not affect the subsequent shearwater population size, although harvesting is beginning to destroy the habitat and thereby reduce the quantity and quality of breeding sites (20). In Europe many *wildfowl* as well as *rooks* and *pigeons* have traditionally been harvested for food. Similarly, *smaller passerines* on migration are trapped and shot for food in very large numbers in the Southern Mediterranean and some Middle Eastern countries. Shooting, trapping, liming and netting are widely practised and it is estimated that hundreds of millions of birds, representing some 15% of those migrating through the Mediterranean, are killed annually. *Turtle doves* (*Streptotelia* spp.) and *thrushes* (*Turdus* spp.) are the birds most favoured for eating, but significant quantities of all migrants are killed in this way (5).

Harvesting for food is not confined to adults. The chicks of many species are prized as food and this is, for example, believed to be one of the primary reasons for the domestication of the *Rock dove (Columba livia)*. From many of the *procellariid seabird* colonies, but particularly from colonies of the *auks and guillemots (Alcidae)*, chicks, which become very fat, have been and still are collected in great numbers for food. These regular harvests have certainly contributed to the decline of these seabirds although efforts were made early to regulate their harvests (14). Before the grain-eating swarms of *Quelea birds (Quelea quelea)* in Africa were controlled by helicopter spraying villagers harvested hatchlings from the quelea breeding colonies by the thousands (16).

The other major use of birds for food which is potentially more sustainable is the **harvesting of their eggs**. Again *seabirds*, because of the accessibility of their nests and the density of colonies, are the most heavily exploited. For example, throughout the tropics *Sooty terns (Sterna fuscata)* yield over a million eggs per year and in some areas harvesting the eggs has become a significant factor in small-island-nation economies. Where, as is the case in many islands in the South Pacific and others in the North Atlantic, communities become dependent on such resources, cultural traditions to control ownership of nesting sites and to regulate cropping have evolved. One of the best documented examples is on the island of St. Kilda, off Scotland, where the harvesting of a variety of *seabird species*, most notably the *fulmar (Fulmar glacialis)*, played a key role in the entire human community's life. Harvesting of *swiftlet* (*Collacalia* spp.) nests is another, rather unusual, example of bird products being used by people. However, given the considerable economic value of such food, which is considered a great delicacy by the Chinese, some rural communities in South-East Asia have generated very valuable incomes from harvesting this resource.

Harvesting that was initially based on a need to provide food has in some cultures, developed into a major **use of birds for sporting purposes**. This is best known in richer, more developed nations where hunting of *wildfowl* or the shooting of *gamebirds* is of major economic and social importance and often has a well-developed culture associated with it. Benefits of sport hunting, both for people and wider environmental issues, are best exemplified by the gamebirds and are considered in Chapt. 2.2.3. Much of the bird killing that goes on at bottle-neck sites for migration, for example of *raptors* through the Mediterranean, the Middle East and parts of Central America, may have started for food but is now seen by those who practice it as much more a form of entertainment. Similarly, the harvesting of seabirds in certain parts of the world, for example the enormous destruction of *auks* (especially *Brünnich's guillemots (Uria lomvia)*) in Greenland, has in recent years become a sport, where once it was primarily for subsistence purposes. Other sporting uses of birds include falconry, another specialised form of hunting discussed in greater detail in Chapt. 2.2.2 and cockfighting, the gruesome practice of getting two cockerels to fight to the death.

As well as their food value, birds have always been important as a **source of feathers**. These were and still are used for a great variety of purposes as described in Chapt. 3.2.5. They are mostly obtained from birds killed in conjunction with sports hunting or harvesting of birds for meat, but in some cases, such as the *quetzals (Pharaomachrus mocino)* and *Paradise birds (Paradisaeidae)*, they are obtained as a primary commercial product. Only the feathers of *ratites* are plucked from live farm birds, as discussed in Chapt. 2.2.1.

One of the most important consumptive uses of wild birds is their **live capture for the international pet bird trade**. Live exotic birds have been traded in large numbers since at least the mid-19th century and this trade was in response to the recreational needs of the industrialising societies in Europe and North America. Only since the 1960s has the bird trade developed into a high-profile industry, producing major commercial gains but also giving rise to serious animal welfare and conservation concerns. It was thus subjected to an in-depth study by TRAFFIC INTERNATIONAL in 1990-91 (22). Over 2,600 bird species have been recorded in the international trade during the past 20 years (9) and in 1975 over 7 million birds were transported between countries: 80% of these were *passerines* and the next most important group were *psittacines*. The major markets are the USA, UK, Netherlands, Germany and Japan, but the international trade only accounts for a small fraction of the total trade since the majority takes place within countries. It has been estimated that the total worldwide wild-bird trade probably involves at least 2 million birds per year and involves at least one in five of all the world's bird species.

**Non-consumptive uses**

Apart from being consumed, wild birds have been and are still extensively used for the collection of certain commercial commodities which they produce, for rendering various services to man and increasingly to satisfy recreational needs in the industrialised countries. **Feathers** plucked from live birds or collected from their nests, and the sundried droppings or guano of certain seabirds are the most important commercial avian commodities. **Guano deposits** have been exploited for centuries for use as fertilizer. Such use is known from many parts of the world, ranging from cave-nesting swiftlets in South-East Asia to non-breeding roosts of seabirds, for example the *Socotra cormorant (Phalacrocorax nigrogularis)* off the Arabian coast. The main guano producing region, however, is the off-shore islands of Peru. Colonies dominated by *Guanay cormorants (Phalacrocorax bougainvillei), Peruvian boobies (Sula variegata)* and *Brown pelicans (Pelecanus occidentalis)* yield up to 300,000 tons of guano annually and are harvested in a renewable way by visiting colonies every two to three years. Recently, however, overfishing has depleted anchovy stocks and this had a negative impact on seabird numbers greatly reducing guano production. Elsewhere in the world guano harvesting has been unsustainable and has had a significant negative impact on seabird populations and their habitats.

Guano exploitation has an interesting economic and legal history. For example, on Nauru in the Western Pacific, employment of the 100,000 to 800,000 inhabitants who harvest and export 2 million tons of phosphate annually has enabled local communities to make considerable investments abroad in order to provide for their future. By contrast, the inhabitants of Ocean Island were forced to evacuate their homes to allow further guano extraction in 1942. They bought an island in the Fiji archipelago from which they brought an action against the British Government and the Phosphate Commission for the rehabilitation of Ocean Island and for higher royalties. This action, lasting 225 days and costing about £750,000, was at the time the longest and most expensive law suit in British judicial history (1).

In both China and Japan *cormorants* have been semi-domesticated and **trained to catch fish** and return to their owners. These birds have a collar around their neck to ensure they do not swallow the fish (23). The behaviour of the *Indicator bird (Indicator indicator)* in Africa **as a guide to honey** has been similarly utilised by man (12). This bird has a symbiotic relationship with the Honey badger. The birds, through their behaviour, guide the mammals to the bees' nests so that the Honey badger can break open the nest and eat most of the contents. The Honeyguide then eats the remaining larvae and the wax which would have remained inaccessible without the Honey badger. It has been shown that there is real nutritional and economic value to the people who have adopted the role of the Honey badger in this relationship. A most unusual use of birds has been **to locate leaks in oil pipelines** in the Americas. When a leak is suspected a foul smelling chemical is pumped into the pipeline which seeps out through leaks and attracts the *Turkey vulture* (*Cathartes aura*), which can then be located circling around the location of the leak (1).

The **use of birds in pest control** is in itself the subject of many books. The most successful documented insect pest control has arisen from introducing non-native species to areas where they will be effective in controlling pests. The best example is the introduction of the *House sparrow (Passer domesticus)* to North America to control the dropworm and other insect pests. Most examples, including this one, have not been effective in controlling the pest, although in more recent years integrated pest management involving the management of native species has become more popular, and probably more effective.

Birds have been at the forefront of some areas of **scientific exploration and discovery**. Studies of the process of evolution, from the pioneering work of DARWIN in his Origin of Species through to today's behaviourial ecologists, taxonomists and evolutionary biologists (to name just three scientific disciplines), have made particular use of birds in their analysis and thinking. Similarly, other research areas such as biomedical studies or the development of vaccines have owed much to the use of birds, particularly domesti cated varieties.

Perhaps the most important non-consumptive use of birds is their **role in recreation**. Billions of people in the world derive simple pleasure from observing birds. This is by no means confined to the developed world although there the use of garden bird tables, nest-boxes, bird baths and so forth has become a multi-million dollar industry, but the marvel of such phenomena as migration, especially where it is concentrated, to produce some of the most spectacular natural sights in the world, has captivated people throughout all continents for thousands of years. Bird tourism is a strong and economically valuable component of eco-tourism, itself a growth industry upon which many countries are coming to depend as a vital source of income. The value, for example, of the Galapagos Islands to Ecuador is internationally recognised and brings significant revenue to the Ecuadorian economy.

Watching birds from a hidden observation site in Italy. Bird watching has become one of the most pupular recreational activities in industrialised countries. (Photo: G. MERZ)

If we look at Canada, for example, more than 84% of the population participate in wildlife-related recreation activities during the course of the year. Wild birds, and particularly *wildfowl* are a major source of such benefits. A study by JACQUEMOT and FILION (13) concludes that millions of participants in bird-related recreational activities receive direct benefits estimated to be worth US$ 347 mill. a year. The overall expenditures that Canadians make on bird-related recreational activities, they conclude, amounts to almost US$ 2 bill. and these provide a significant annual stimulus to Canada's economy in terms of Gross National Product. Whilst Canada may be one of those nations in the world where bird-related recreational activities are higher than most, it is far from unique. Many European nations as well as Japan, the USA and Australia, have similarly buoyant recreational markets and other nations are developing these. Furthermore, with the improvement in global transport systems, much of the benefits of bird-related recreational activities are being felt by developing nations that house the richest and most diverse avifauna. Many island nations, for example the Seychelles in the Indian Ocean, are beginning to derive significant financial assets from such international bird tourism.

Of all life forms, birds are used perhaps more than any other as **indicators of environmental conditions** (15, 6). For example, they have been used extensively both as indicators and predictors of the intensity of use of agricultural pesticides in developed and some developing nations. In the United Kingdom the development of pesticides for use on farmland and the assessment of hazards to wildlife from such agro-chemical use has made great use of *captive birds* in laboratory toxicity tests, *wild bird populations* for monitoring during field experimentation and bird deaths on farmland, which are investigated as part of a government-wide surveillance of pesticide use (8).

The use of birds as **monitors of pollutants** has received enormous study. One of the clearest examples over the last three decades is the study of organo-chlorines, which accumulate in top predators. *Birds of prey*, being at the top of the food chain, have been particularly effectively used to act as what have been described as "sentinel" species for the harmful environmental effects of organo-chlorines (17). This has relied on the existence of base-line population survey data for these birds of prey prior to the major environmental impact. Studies have not been confined to birds of prey, however; *seabirds*, for example *Common terns (Sterna hirundo)* and *Herring gulls (Larus argentatus)*, have successfully been used to monitor organo-chlorine contamination in the North Sea. Pollution by heavy metals, particularly mercury, cadmium and lead which are highly toxic, plastics and radionuclides have all been the subject of useful studies and investigations, using birds as indicators of environmental damage (see (7) for a detailed review). One specific form of pollution for which birds have made an invaluable contribution is that of assessing water quality. People the world over are dependent on fresh water and it has long been recognised that, although chemical in nature, biologically damaging pollutants can be detected using biological rather than chemical criteria. Such biological indicators have the advantage of integrating exposure to total pollution over a span of time, whereas intermittent chemical sampling may miss short-lived and episodic bouts of pollution. Studies of *dippers (Cinclidae)* in the UK have shown the species to be a very sensitive indicator of acidity, both within catchment areas and between them (18).

A rather different use of birds as indicators comes primarily from marine eco-systems. For millennia humans have followed birds at sea **to locate fish and mammals**; as highly visible, wide-ranging consumers *seabirds* are good indicators of marine productivity. Interestingly, seabird behaviourial ecology can be monitored to enhance fish-stock assessments by providing estimates of predator-induced mortality and technology independent indices of fish abundance. Seabird energetics models have been used to estimate mortality due to avian predation and, when incorporated into multi-species analysis, to develop conservative harvesting strategies (15).

A rather different form of monitoring is **weather forecasting**, and bird behaviour has traditionally been used in this way. For example, the Montagnais North American native people, who inhabited Newfoundland in the early 17th century, believed that a particular call of the *Great Northern diver (Gavia immer)* forecast a windy day. *Swifts* and *swallows* flying high to catch their insect prey are a sign of good settled weather, whereas low-flying is supposed to indicate a change to poor weather. Whilst there are many anecdotal observations of this nature there has been little scientific study of the ability of bird behaviour to predict weather conditions.

Finally, birds have been very successfully used for evaluating the **overall conservation value of habitats**, for example, the richness and diversity of forests in North America. In a pioneering study of the distribution of restricted-range species BirdLife International was able to identify centres of avian endemism around the world (10). These centres are not only important for birds; comparisons of bird data with other life forms show that they are of significance for wider biodiversity. Thus, in this case, birds can be used as indicators of overall biodiversity and so be used to identify areas of outstanding richness, worthy of appropriate conservation measures if the diversity of unique species on earth is to be maintained.

**Keeping and breeding birds in captivity for non-productive reasons** has a long history and surprisingly diverse goals, such as enjoyment of the aesthetic sight of *peacocks* in a park, the pleasure derived from the singing of *canaries* or the companionship of a "speaking" *parrot*, the satisfaction of breeding colourful hybrids or new varieties of a finch species, the use of the homing ability of *pigeons*, and so on. However, most of the keeping of birds, like that of other pet animals, may be seen as a compensatory activity for the alienation from nature of industrialised man. "Aviculture" has now become a major industry and is the basis of the above-mentioned consumptive bird trade as breeding does not nearly satisfy demand for exotic live birds. Some captive breeding has been justified on conservation grounds and has in a very small number of instances been integral to successful reintroduction programmes for endangered species (4).

## Implications for conservation

The latest global review of the trends and factors in endangerment of birds shows that 11% of the world's avifauna is at risk of global extinction, with a further 1% having insufficient information to be able to determine their status and another 9% considered near threatened (2). The primary threats to these species have been analysed and reveal that human uses have a rather small but significant impact. For example, hunting, persecution, accidental trapping and subsistence egg collecting is believed to account for 7.6% of the species threatened, whilst trade and commercial egg collecting account for a further 2.6%. Often these uses become harmful only in conjunction with other causes, such as habitat degradation, insecticide abuse or oil pests.

Very few species of birds have been brought to extinction by human use alone. The Great auk *(Pinguinus impennis)* is one example of a species hunted to extinction for its food, which of course was following in the footsteps of the dodo *(Raphus cucullatus)* from Mauritius. The only species believed to have been brought to extinction by sports hunting is the Pink-headed duck *(Rhodonessa caryophyllacea)*, although a number of others such as the Eskimo curlew *(Numenius borealis)* have been brought to the brink of extinction by this means. The wild-bird trade, which claims millions of birds' lives every year, has not, in comparison with a threat such as habitat destruction, been very important in terms of its impact upon global bird conservation. However, a number of groups of birds, most notably parrots, have suffered significantly as a result of the wild-bird trade and several parrot species are now considered globally threatened because of the trade in them. In the past egg and skin collecting, and possibly some other uses such as falconry, have had negative impacts upon bird populations and similarly the harvesting for food where it sometimes changes to a sporting recreation can and in some cases has become unsustainable.

It must be accepted, however, that birds, like other wildlife, constitute a very important natural resource which needs to be utilised for the benefit of man, be it for consumptive or non-consumptive purposes. It is simply vital that the various uses of birds - still largely haphazard and indiscriminate - be channelled into planned management providing the basis for programmes of sustainable utilisation. Such change has historically occurred already with several groups of birds such as galliforms and waterfowl, and many attempts are made around the world to render bird utilisation sustainable; for example, controlled programmes of parrot trapping for trade (21), the regulation of egg collection from seabirds (14) and the harvesting of downs from eiderduck breeding sites.

Overall, the benefits of human use in terms of providing important resources for local communities and of creating awareness and pleasure probably outweigh the costs borne by birds and their populations. However, man and birds can continue to survive together to each others mutual benefit only if sufficient concern is maintained for the sustainability of human utilisation.

## Literature references

1. BROOKE, M. de L. and T.R. BIRKHEAD (1991): *The Cambridge Encyclopedia of Ornithology*. Cambridge University Press, Cambridge
2. COLLAR, N.J., CROSBY, M.J. and A.J. STATTERSFIELD (1994): *Birds to Watch: The World Checklist of Threatened Birds*. BirdLife International, Cambridge
3. DIAMOND, A.J. (1987): A global view of cultural and economic uses of birds. In: *The Value of Birds*, ed. DIAMOND, A.W. and F. FILION; p. 99-112; International Council for Bird Preservation, Technical Publication 6, Cambridge
4. DIAMOND, A.W., SCHREIBER, R.L., ATTENBOROUGH, D. and I. PRESTT (1987): *Save the Birds*. Cambridge University Press, Cambridge
5. FENECH, N. (1992): *Fatal flight: the Maltese obsession with killing birds*. Quiller, London
6. FURNESS, R.W. and J.J.D. GREENWOOD (1993): *Birds as monitors of environmental change*. Chapman and Hall, London

7. FURNESS, R.W. (1993): Birds as monitors of pollutants. In: *Birds as monitors of environmental change*, ed. FURNESS, R.W. and J.J.D. GREENWOOD; p. 42-85; Chapman and Hall, London
8. HARDY, A.R., STANLEY, P.I. and P.W. GREIG-SMITH (1987): Birds as indicators of the intensity of use of agricultural pesticides in the U.K. In: *The Value of Birds*, ed. DIAMOND, A.W. and F. FILION; p. 119-131; International Council for Bird Preservation, Technical Publication 6, Cambridge
9. INSKIPP, T.P., BROAD, S.R. and R. LUXMOORE (eds.) (1988): *Significant trade in wildlife: A Review of selected species in CITES APPENDIX II*. Vol. 3, Birds. IUCN/CITES, Gland and Lausanne, Switzerland
10. INTERNATIONAL COUNCIL FOR BIRD PRESERVATION (1992): *Putting biodiversity on the map: priority areas for global conservation*. International Council for Bird Preservation, Cambridge
11. IMBODEN, C. (1994): Birds as indicators of un-sustainability. In: *Conserving Europe's natural heritage: towards a European ecological network*, ed. BENNETT, G.; p. 61-68; Graham and Trotman, London
12. ISACK, H.A. (1987): The cultural and economic importance of birds among the Boran people of Northern Kenya. In: *The Value of Birds*, ed. DIAMOND, A.W. and F. FILION; p. 99-112; International Council for Bird Preservation, Technical Publication 6, Cambridge
13. JACQUEMOT, A. and F.L. FILION (1987): The economic use of birds in Canada. In: *The Value of Birds*, ed. DIAMOND, A.W. and F. FILION; p. 15-24; International Council for Bird Preservation, Technical Publication 6, Cambridge
14. KARTASCHEW, N.N. (1960): *Die Alkenvögel des Nordatlantik*. Neue Brehm-Bücherei No. 257; Ziemsen, Wittenberg-Lutherstadt, Germany
15. MORRISON, M.L. (1986): Bird populations as indicators of environmental change. *Current Ornithology*, 3, p. 429-451
16. MUNDY, P.J. and M.J.F. JARVIS (eds.) (1989): *Africa's feathered locust*. Baobab Books, Harare, Zimbabwe
17. NEWTON, I., WYLLIE, I. and A. ASHER (1993): Long-term trends in organochlorines and mercury residues in some predatory birds in Britain. *Environmental Pollution*, 79, p. 143-152
18. ORMEROD, S.J. and S.J. TYLER (1993): Birds as indicators of changes in water quality. In: *Birds as monitors of environmental change*, ed. FURNESS, R.W. and J.J.D. GREENWOOD; p. 179-216; Chapman and Hall, London
19. RAMSAY, B.J. (1994): *Commercial use of wild animals in Australia*. Bureau of Resource Sciences, Australian Government Publishing Service, Canberra, Australia
20. SKIRA, I.J., WAPSTRA, J.E., TAWNEY, G.N. and J.A. NAARDING (1986): Conservation of the Short-tailed Shearwater *Puffinus tenuirostris* in Tasmania, Australia. *Biological Conservation*, 37, p. 225-236.
21. THOMSON, J.B. and A. BRAUTIGAM (1991): Sustainable use of Neotropic parrots. In: *Neotropical Wildlife Use and Conservation*, ed. ROBINSON, J.G. and K.H. REDFORD; p. 359-379; University of Chicago Press, Chicago and London
22. THOMSON, J.B., EDWARDS, R.S. and T.A. MULLIKEN (1992): *Perception, conservation and management of wild birds in trade*. TRAFFIC International, Cambridge
23. ZHANG ZHONG-GE (1984): 56. Cormorant. In: *The evolution of domesticated animals*, ed. MASON, I.L. (1984); p. 367-368; Longhurst Group, New York and London

## 2.3 UTILISATION OF MAMMALS

### 2.3.1 MARSUPIALS (*MARSUPIALIA*)

The marsupials are primitive mammals and comprise numerous different families. The most important in the context of this book are the opossums (**Didelphidae**), the bandicoots (**Peramelidae** and **Peroryctidae**), phalangers and possums (**Phalangeridae** and **Pseudocheiridae**), and kangaroos and wallabies (**Macropodidae**). The opossums occur only in the Americas from where they were also introduced to Madagascar. All other marsupials are restricted to Australasia. In Australia the herbivorous macropodids, comprising about 42 different species, utilise a similar diversity of habitats as the antelopes in Africa. Tab. 22 shows, however, that only a small number of them are commercially utilised for meat and skins. The possibility of ranching and/or farming some of the kangaroo species is under investigation.

**Table 22:** Overview of the utilisation of marsupials

| English name | Scientific name | Geographical distribution | Body length (cm) | Utilisation[1)] |
|---|---|---|---|---|
| **OPOSSUMS - DIDELPHIDAE** | | | | |
| Virginia opossum | *Didelphis virginiana* | N.America | 40-45 | for fur, food |
| Other opossums | -spp. | C.and S.America | 40-45 | for food |
| **BANDICOOTS - PERAMELIDAE and PERORYCTIDAE** | | | | |
| Long-nosed bandicoots | *Peramelis* spp. | Australia, Tasmania | 21-43 | for skins, food |
| N. Guinea bandicoots | *Peroryctes* spp. | New Guinea | 17-50 | for skins, food |
| **POSSUMS and PHALANGERS - PHALANGERIDAE and PSEUDOCHEIRIDAE** | | | | |
| Brushtail possums | *Trichosurus* spp. | Australia, Tasmania[2)] | 32-58 | for fur, food |
| Phalangers or cuscuses | *Phalanger* spp. | Australia, Tasmania, N.Guinea, Irian Jaya | 27-65 | for skins, food |
| Ring-tailed possums | *Pseudocheirus* spp. | Australia, N.Guinea | 19-45 | for skins, food |
| **WALLABIES and KANGAROOS - MACROPODIDAE** | | | | |
| Tree kangaroos | *Dendrolagus* spp. | N.Guinea, N.Australia | 52-81 | for skins, food |
| Forest wallabies | *Dorcopsis* spp. | N.Guinea | 49-80 | for skins, food |
| Pademelons | *Thylogale* spp. | N.Guinea, Australia, Tasmania | 53-77 | for skins, food |
| Wallabies, wallaroos and kangaroos | *Macropus* spp. | Tasmania[3)]<br>N.Guinea, Australia | 45-105<br>100-160 | for skins, sinews, food and meat; farming |

1) after BREEDEN and BREEDEN (4) and FLANNERY (10); 2) *T.vulpecula* introduced to New Zealand (HOWARD (18)); 3) several attempts were made in the 1890s to introduce Bennett's wallabies into Germany and England; although biologically successful the established free-ranging populations did not long withstand poaching

## Historical background

Although marsupials have reached their greatest development in Australia and New Guinea in terms of adaptations, one family, the *opossums (Didelphidae)*, also occurs in significant numbers in South and Central America; however, only *Didelphis virginiana* ranges as far north as Canada. Opossums are easily trapped and have always constituted an important food resource for the indigenous peoples in the Americas. Mainly in North America, they also became early on an important fur animal which is still being commercially exploited today. In Australasia marsupials have evolved in similar ecological niches to those occupied by ungulates in Africa and partly by rodents in South America. They have thus acquired similar physiological capabilities by convergence. Consequently, marsupials have had and still have a similar importance for subsistence to the indigenous peoples in Australasia as ungulates and rodents have in Africa and in South America. They are hunted extensively for food and their skins are used as cloaks and water bags, sinews for binding, bones for barbs and spear tips, and teeth for decoration (3).

The early European settlers in Australia also used *kangaroos* and *wallabies* as a source of food. However, as agriculture and the raising of livestock became well established they were increasingly considered to be competitors to livestock and ruthlessly destroyed. For example, no less than 90,000 kangaroos were poisoned at watering points on just one large estate in the Pilbara District of Northwestern Australia during a five year period. On another estate in Western Australia, totalling 84,000 ha and carrying only one sheep per 36 ha, 12,834 wallaroos were destroyed by poisoning in less than 10 weeks (22). Only when the superior quality of kangaroo skins began to be appreciated was commercial utilisation of kangaroos started. An industry based on the taking of hides developed in the 1920s well before kangaroos were harvested for meat, and during subsequent decades processed annual harvests of 300,000-400,000 skins (19). From Tasmania alone about 2 million wallaby skins were exported between 1923 and 1955 (22).

It was not until 1958 that the kangaroo industry also started to utilise meat, both for home markets and export. Fauna-protection legislation then in force permitted liberal shooting of kangaroos in most States. Between July 1958 and June 1962, 7,500 t of kangaroo meat were exported. As only hind quarters were exported and more meat was marketed locally, this represented a kill of about 1-2 million kangaroos, mainly in just one State, Western Australia. By 1964 50 t of kangaroo meat were being exported from Australia weekly with another 10 t processed into pet food for which about 8,000 kangaroos had to be harvested (22).

Around the turn of the century hunters, encouraged by international demand for their furs, also began killing large numbers of *koalas (Phascolarctos cinereum)* and *possums* (*Trichosurus* and *Pseudocheirus* spp.). In 1924, more than 2 million koala pelts were exported from the eastern states of Australia (26). In 1927, when the Queensland government declared an open season on both koalas and possums, 10,000 trappers disposed of 600,000 koalas and over a million *Brushtail possums (Trichosurus vulpecula)* within months. The pelts of the latter had become fashionable and were traded as "Adelaide chinchilla" or "Australian beaver" (12). Koalas, threatened by extinction by 1930, are now fully protected, and Brushtail possums are commercially hunted only in Tasmania. There, and also in New Zealand where they were introduced in 1858, they came into conflict with expanding agriculture and with silviculture, so that control measures had to be initiated (18). Based on these the regular harvesting of Brushtail possums has developed into a source of considerable income (6).

## Management and utilisation

The various indigenous peoples still widely use marsupials for **subsistence purposes**, mainly *opossums* in South America, *wallabies* in New Guinea and *kangaroos* in parts of Australia. As elsewhere, subsistence hunting and trapping of marsupials has intensified as firearms have become more readily available and cash economies developed. Thus, hunting pressure in some areas, for example on wallabies in Papua New Guinea, has been building up to non-sustainable levels (9).

Only opossums in North America, kangaroos and wallaroos in Australia and possums in Tasmania and New Zealand are **trapped and hunted for commercial purposes** on a large scale for meat and skins. *Virginia opossums* are regularly harvested in the USA and Canada mainly by live trapping for fur production (21). Commercial cropping of *kangaroos* involves mainly four species: the *Red kangaroo (Macropus rufus)*, the *Eastern Grey kangaroo (M. giganteus)*, the *Western Grey kangaroo (M. fuliginosus)* and the *Common wallaroo (M. robustus)*. These species are locally more abundant now than they were before European settlement took place and Fig. 7 shows their increase since 1981. This increase in numbers and range has come about by changes wrought by pastoralists, creation of more permanent watering points, changes in pasture composition, the control of dingos and a reduction of Aboriginal hunting pressure. In Queensland the *Whiptail wallaby (M. parryi)* and in Tasmania *Bennett's wallaby (M. rufogriseus)* and the *Tasmanian pademelon* or *Rufous wallaby (Thylogale billardieri)* are harvested in limited numbers (20). Although *possums* remain locally abundant on the mainland, commercial harvesting of *Brushtail possums (Trichosurus vulpecula)* is limited to Tasmania, partly for pest control. There it has a long tradition and has been regulated since 1918 for procuring furskins and pet meat (6). Introduced possums in New Zealand have become a serious pest and are suspected to be a reservoir for bovine tuberculosis (8). Despite eradication efforts including extensive poisoning campaigns, populations have flourished and substantial trade in raw and finished possum skins has developed over the past century (20).

The main concentrations of kangaroos are on the sheep rangelands of inland Australia. Their numbers fluctuate widely in response to the rainfall of preceding years: a succession of good rainfall years leads to large increases and drought leads to large decreases. Thus, considerable short term fluctuations in numbers may take place. Low densities in settled areas are attributed to intensive agricultural practices, small landholdings, lack of tree cover, the absence of permanent watering points and the lack of palatable grasses (25).

The management of kangaroos is the responsibility of the States and Territories, while the Commonwealth Government controls the export of their meat and skins. In 1990, kangaroos were commerically harvested for their meat and skins in every Australian State (except Victoria, the Northern and Australian Capital Territories, where kangaroos can only be

**Figure 7:** Development of kangaroo resources in Australia in 1981-1990 (after RAMSAY (20))

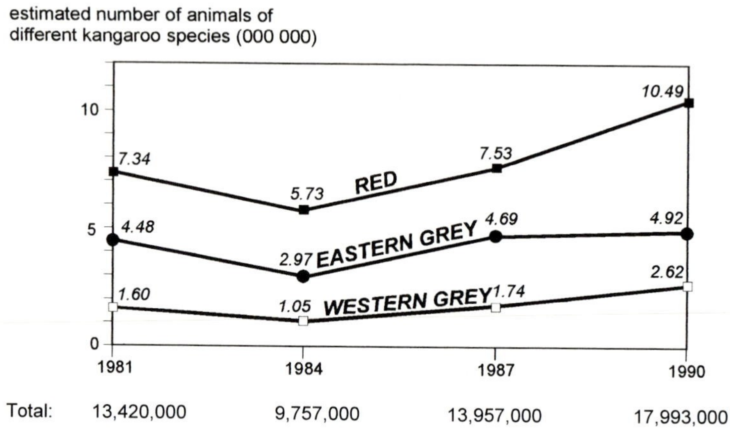


taken for pest control). Each State government declares open seasons and annual quotas for each species that can be harvested, a responsibility that demands a yearly assessment of the capacity of populations to withstand harvesting. Open seasons are declared for each species for such periods as are considered suitable. Quotas are established each year and the numbers taken are governed by tags applied to killed kangaroos. The objectives for kangaroo management as stated by the Australian Council for Nature Conservation Ministers in 1985 are:

- To maintain populations of the designated species of Macropodidae over their natural ranges;
- To contain the deleterious effects of kangaroos on pastoral and agricultural production;
- To make the best possible use of the kangaroos harvested.

The primary objective of administrative control is to avoid over-exploitation, and the secondary is to nurse the industry through bad times (24). Todays legislation relating to kangaroo harvesting, industrial practices and community attitudes seeks to conserve the harvested species of kangaroos. However, all legislation recognises that the large kangaroo species are highly successful and abundant, and represent a pest problem to rural enterprise.

There is substantial variation between States in the way they control the killing of kangaroos and in the justification they advance for permitting it. Management programmes for the different States are set out in Part B of a National Plan of Management for Kangaroos' (1). Two States rely heavily on mitigation of damage to crops and fences, and competition with sheep for pasture, to justify killing. Queensland's plan rests on harvesting kangaroos as a renewable resource; conflict with agriculture is rendered merely as a supporting reason. As the primary objective of administrative control is to avoid over-exploitation, there are limits to the number of kangaroos that may be taken.

All kangaroo shooters and dealers are required to be registered annually, pay license fees and name the properties or areas in which they intend to operate. They have to declare monthly the number of each species taken or handled. Non-reusable, numbered, self-locking tags must be attached to each harvested animal. Royalties on the four main commercial species are payable to most States. Despite the introduction of quotas, it is desirable to ensure an even distribution of the harvest in both space and time. Shootering licenses help to ensure a reasonable distribution in terms of space. In addition to aerial and ground surveys, using line-transect methodology to determine population levels, species distribution records obtained from shooters are of importance to the understanding of the status of exploited species. Graziers are generally satisfied to have shooters concentrate their greatest efforts on those paddocks that are being spelled from stock grazing. Fig. 8 shows the relevant statistical data that are collected. The sex and age distribution of the take is a measure of shooter selectivity and as such is essentially a form of catch-effort data (7). At present, the legal kill of kangaroos is about 12% of the total population (15). The greatest long-term yield for *Red kangaroos* is between 10 and 15% of the population each year. This is a little lower for *Grey kangaroos* (24).

Kangaroos are shot at night from 4-wheel drive vehicles, manned with 1-4 shooters, a spotlighter and a driver. A night's shooting normally begins a few hours after dusk and continues through the night until shortly after daylight. Skins are salted or pegged out. Kangaroo hides are mostly tanned to produce leather which has outstanding strength to weight. It is

**Figure 8:** Sample of kangaroo shooters' returns prescribed in Queensland

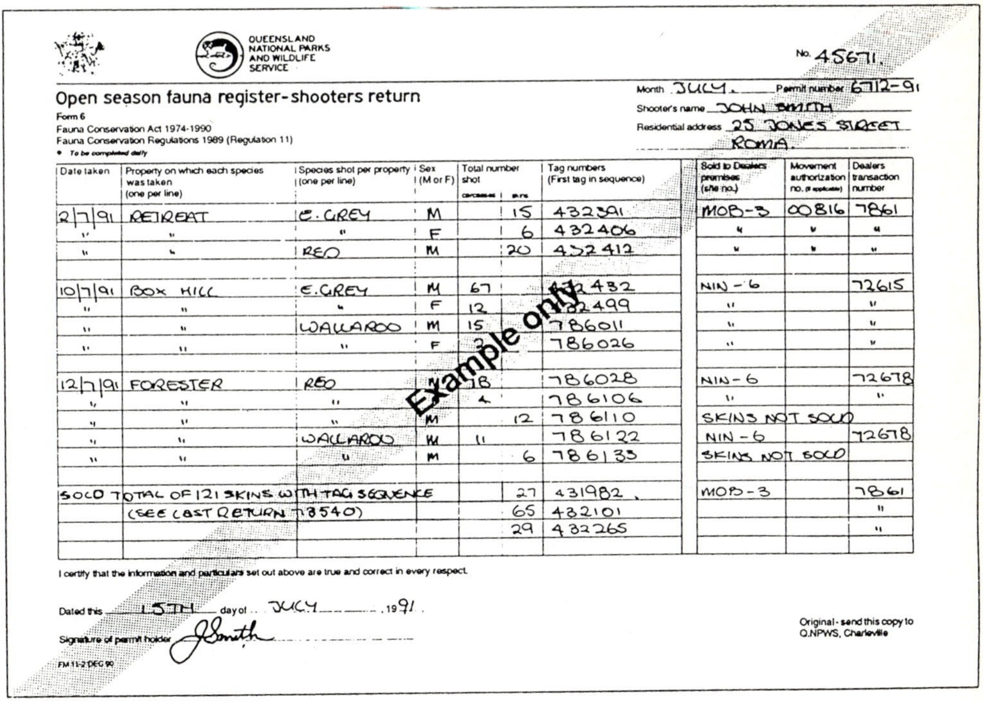

QUEENSLAND NATIONAL PARKS AND WILDLIFE SERVICE

No. 45671.

Open season fauna register-shooters return

Form 6

Fauna Conservation Act 1974-1990

Fauna Conservation Regulations 1989 (Regulation 11)

• To be completed daily

Month JULY. Permit number 6712-91

Shooter's name JOHN SMITH

Residential address 25 JONES STREET ROMA.

| Date taken | Property on which each species was taken (one per line) | Species shot per property (one per line) | Sex (M or F) | Total number shot: carcasses | Total number shot: skins | Tag numbers (First tag in sequence) | Sold to Dealers premises (site no.) | Movement authorization no. (if applicable) | Dealers transaction number |
|---|---|---|---|---|---|---|---|---|---|
| 2/7/91 | RETREAT | E. GREY | M | | 15 | 432391 | MOB-3 | 00816 | 7861 |
| " | " | " | F | | 6 | 432406 | " | " | " |
| " | " | RED | M | | 20 | 432412 | " | " | " |
| | | | | | | | | | |
| 10/7/91 | BOX HILL | E. GREY | M | 67 | | 432432 | NIN-6 | | 72615 |
| " | " | " | F | 12 | | 432499 | " | | " |
| " | " | WALLAROO | M | 15 | | 786011 | " | | " |
| " | " | " | F | 2 | | 786026 | " | | " |
| | | | | | | | | | |
| 12/7/91 | FORESTER | RED | M | 18 | | 786028 | NIN-6 | | 72678 |
| " | " | " | | 4 | | 786106 | " | | " |
| " | " | " | M | | 12 | 786110 | SKINS NOT SOLD | | |
| " | " | WALLAROO | M | 11 | | 786122 | NIN-6 | | 72678 |
| " | " | " | M | | 6 | 786133 | SKINS NOT SOLD | | |
| | | | | | | | | | |
| SOLD TOTAL OF 121 SKINS WITH TAG SEQUENCE | | | | | 27 | 431982. | MOB-3 | | 7861 |
| | (SEE LAST RETURN 73540) | | | | 65 | 432101 | | | " |
| | | | | | 29 | 432265 | | | " |
| | | | | | | | | | |

I certify that the information and particulars set out above are true and correct in every respect.

Dated this 15TH day of JULY 1991.

Signature of permit holder J Smith

FM 11-2 DEC 90

Original - send this copy to Q.NPWS, Charleville

Example only

Source: by courtesey of Queensland National Parks and Wildlife Service preferably used for ski clothing and shoes. Furred hides may be used as floor mats and rugs, and for the manufacture of souvenirs.

Field-dressed carcasses are placed in mobile refrigerated compartments and taken to an abattoir for further processing. Preparation of a carcass for pet food involves the removal of the head, tail, feet, forearms and viscera; carcasses taken for game meat export are prepared by removal of the alimentary tract only. Meat for export must be frozen within a few hours of slaughter and must pass veterinary inspection before processing. Bones and unusable meat may be reduced to blood-and-bone fertilizer, and sometimes also to meat meal. The body of kangaroos consists of a large proportion of carcass muscle and only negligible fat. The average carcass weight is about 15 kg.

The kangaroo industry is now largely in the hands of relatively few large firms and the quality of the industry and its products is gradually improving. Trade in whole carcasses as against skins alone is now more common and is actually required by the regulations of some States. Stringent veterinary requirements are being applied to the control of kangaroo meat taken for human consumption (16). These prescribe *post-mortem* inspection, field dressing procedures, chilling and hanging of carcasses, and detail the procedures and facilities for carcass boning, meat packing, refrigeration and transport to processing plants.

As is the case for some antelope species in Southern Africa, the possibilities of **ranching and/or farming of kangaroos** on large private estates has been under serious consideration (24, 25). Aspects limiting the possibility of ranching kangaroos include the mobility of the animals concerned, handling difficulties, fencing problems, relatively slow rates of growth and reproduction, and the absence of reliable techniques for censusing populations on small

Harvested kangaroos are weighed and placed into a refrigerated truck for transport to centralised processing plants. (Photo: BUREAU OF RESOURCE SCIENCES, Australia)

areas. Administrative problems are also considerable. Landowners have no proprietary rights over kangaroos on their land. In order to successfully ranch kangaroos, a pastoralist should have an interest sufficient to ensure that his investment in raising these animals cannot be lost to a third party without adequate compensation. Kangaroo-proof fencing is not economical on rangelands and mustering kangaroos for branding is not feasible. A partial solution to the mobility problem could be to enlarge the area to boundary ratio, either by enlarging a single property or by an agreement between adjoining landowners to run kangaroos as a joint enterprise.

The possibility should be considered that the productivity of rangelands may be better utilised by grazing a combination of sheep and macropods. Although competition for food between kangaroos and sheep does occur in certain situations, pastures may be maintained in a more ecologically desirable equilibrium under a mixed grazing regime than when grazed by sheep or macropods alone. The extent of interactions between macropods and sheep appears to be a function of both grazing intensity and pasture availability and conditions. Kangaroos are generally better adapted to utilise very poor quality roughage than sheep (11). However, the *quokka (Setonix brachyurus)* was found to be considerably less efficient than ruminants in digesting crude fibre (5). Further investigations will be necessary before kangaroo ranching can develop into a practical form of land use.

## Economic importance

The utilisation of *opossums* for fur production in North America has considerable economic importance. The annual harvests since 1970 had peaks of more than 1.5 million pelts and the number of skins marketed in 1982/83 was estimated to be 990,330. With a value of a little over Can.$ 3 mill. (approx. US$ 2.43 mill.) opossums ranked twelfth in total value among the furbearers of the USA and Canada (21).

The commercial utilisation of *kangaroos* is largely governed by public concerns about conservation, and by cultural, biological and land management considerations; economics are therefore only a secondary aspect. When assessing the economic importance of the utilisation of large kangaroos, various aspects have thus to be considered. In the first place there is the financial return to shooters and dealers in kangaroo products. Few hunters, however, are wholly dependent for their income on cropping kangaroos. Secondly, there is a benefit to private landholders because competition with livestock for fodder is reduced. Thirdly, there is the benefit of preserving a gene pool, since kangaroo populations are harvested on a sustained-yield basis. If kangaroos were to be ranched in the future in the absence of hoofed domestic stock there would be an added advantage of reduced soil erosion, since they have padded feet. The current utilisation of kangaroos is therefore definitely and significantly contributing to the national economy, providing employment and generating income in domestic and export markets. However, most landowners view kangaroos as a threat to the rural economy and competition with livestock is especially evident during drought. A recent survey put the annual damages and production losses perceived by farmers in Australia through kangaroos as high as US$ 90 mill. (13). But land-owners do accept that if populations of kangaroos are left unchecked this will ultimately adversely affect agricultural, forestry and pastoral operations.

**Figure 9:** Commercial harvest quotas and harvests of kangaroos in 1984-1993 (after RAMSAY (20))

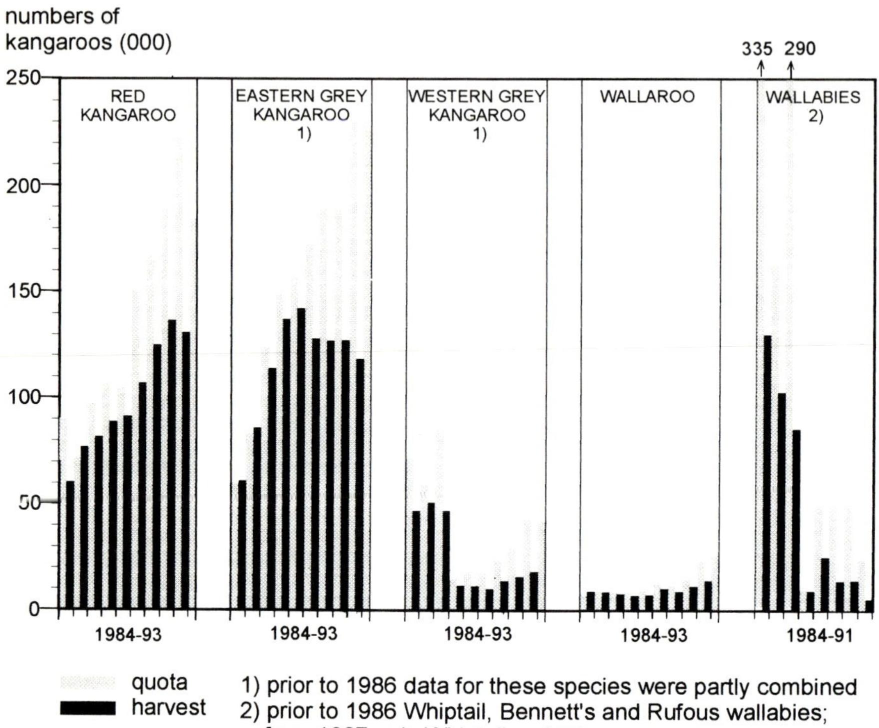


1) prior to 1986 data for these species were partly combined
2) prior to 1986 Whiptail, Bennett's and Rufous wallabies; from 1987 only Whiptail wallabies

Fig. 9 shows the annual harvests of kangaroos against the allowed quotas. Generally the kangaroo industry cannot take enough animals to prevent a build-up of numbers occasioned by a run of good seasons. This often leaves a gap between the required and the actual off-take. In comparison with the commerical harvests, however, the number of kangaroos killed for damage mitigation were low, fluctuating between 118,194 in 1986 and 18,592 in 1989 (20). Market forces operate independently of damage mitigation needs and are therefore often out of phase with them. No matter how kangaroos are removed, the costs of removal, monitoring and enforcement should be deducted from any profits made. The net figure must be balanced against the cost of pastoral production if kangaroos are to remain. Also, the income derived from kangaroo harvesting should be made to offset income lost from reducing sheep numbers which in turn will help to halt land degradation (14, 15). Dealers in kangaroo products will buy only if they can make a profit from selling these products. The number of kangaroos cropped therefore depends first of all on the availability of a market for skins and meat.

Fig. 10 shows the production of kangaroo skins and meat since 1981/82 in terms of exports. The skins are sold on local and international markets, but prices paid often fluctuate sharply and vast stock piling of unsold skins may occur. Close substitutes (calf and kid) for kangaroo skins exist and when these are readily available prices for kangaroo skins fall.

**Figure 10:** Export of kangaroo skins and meat in 1981/82-1991/92 (after RAMSAY (20))

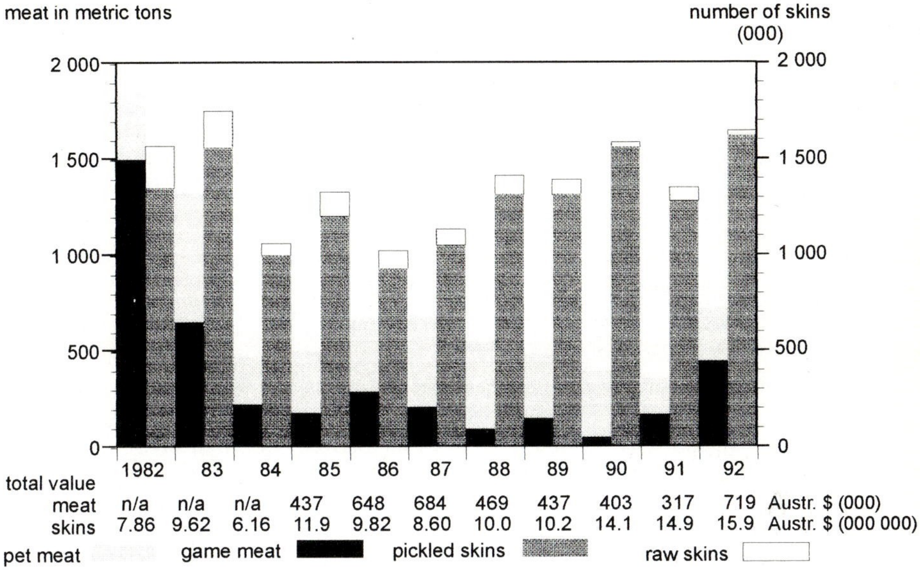


Nevertheless, the total export value of kangaroo skins has increased from Austr.$ 4.7 mill. (approx. US$ 4.1 mill.) in 1980/81 to Austr.$ 15.9 mill. (approx. US$ 11.9 mill.) in 1991/92, mainly due to increases in prices for pickled skins (20). Since the late 1980s substantial quantities of finished kangaroo leather have been produced in Australia mainly for export for footwear manufacture. This export has doubled the total value of kangaroo products exported from Australia to almost Austr.$ 30 mill. (approx. US$ 22.5 mill.) per year in 1991/92 (20).

Compared with this total value the production of kangaroo meat for exports has remained rather insignificant, although this has a considerable potential for sale at high prices: it is highly nutritional, palatable and lean, over 50% of the empty body weight is muscle and the muscle mass is concentrated in the region of the loin (17). Fig. 10 shows that this potential has, however, not been realised as yet because of human prejudices and market forces. After the initial export of kangaroo meat for human consumption collapsed in the 1960s due to a lack of regulation of the hygiene standards, it peaked again in the early 1980s to more than 1,000 t/annum. Thereafter however, it steadily decreased to the values of kangaroo meat exports for pet food. The main use of kangaroo meat remains therefore for pet food on domestic and international markets. The meat market is highly competitive and kangaroo meat has to be sold against low-value meats. Thus meat exports in 1991/92 for human consumption as well as for pet food, mainly to Japan and other Asian countries, have realised only some Austr.$ 700,000 (approx. US$ 525,000) (20). To realise the full value of kangaroo meat a marketing drive is needed to lift it from the pet food to the gourmet category. This change is now gradually taking place.

Whereas the formerly rather significant export of *wallaby* skins (*M. rufogriseus* and *T. billardieri*) from Tasmania was terminated in 1986, the utilisation of *Brushtail possums* for fur export has continued to-date. Export of possum skins fluctuated strongly according to prices, reaching peaks of more than 200,000 skins in the 1970s and 1980s, but dropping to only 14,900 in 1990 (6). The average annual income to the local hunters dropped accordingly from Austr.$ 1,109 (approx. US$ 920) per hunter in 1976 to Austr.$ 277 (approx. US$ 220) in 1989 and the trade seems to be barely profitable to the dealers. An export market for possum meat has so far not developed although it is locally marketed as pet food (2). In comparison the New Zealand possum industry is much larger, having exported up to 3 million skins during peak years. It also is involved in dressing and manufacturing possum furs for local and export markets (20).

## Implications for conservation

Considerable controversy has been raging during recent decades about whether or not commercial exploitation of large kangaroos should be permitted. This is still in progress, although a sentiment in favour of utilisation is gradually prevailing as public opinion is becoming more aware of the issues at stake. Views vary between being in favour of total protection to being opposed to unnecessary killing yet being prepared to support scientifically based management (19). The question is: should kangaroos be destroyed as pests, protected from destruction because they are uniquely Australian animals, or be managed as a renewable resource? The aim of present legislation is to attempt to perpetuate self-sustaining populations of each species throughout its preferred range. Commercial utilisation is regarded only as a tool for controlling numbers of kangaroos on private lands and is aimed at keeping their numbers within limits that are tolerable. Research has indicated that it is weather rather than shooting which determines short-term fluctuations in numbers and that it is habitat alienation which determines long-term changes and spells doom for regional populations (15). Kangaroos are a potentially very valuable resource, which is presently undervalued. Ever increasing pressure from introduced domestic herbivores threatens to cut their habitat away from underneath them. Kangaroos should be used on a sustained yield basis in order to save the land from further degradation. It can be expected that under improved management and utilisation procedures more economic benefits will be realised from kangaroos; particularly if kangaroo meat were to be legalised for sale for human consumption throughout Australia.

## Literature references

1. ANONYMOUS (1985): *National Plan of Management for Kangaroos*. Council of Nature Conservation Ministers (CONCOM), Canberra, Australia
2. ANONYMOUS (1985): Possum Meat for Sale. *TRAFFIC Bulletin*, 7(3/4), p.62
3. ARCHER, M., FLANNERY, T.F. and G.C. GRIGG (1985): *The Kangaroo*. Kevin Weldon, Sydney; 263 pp.
4. BREEDEN, S. and K. (1970): Living Marsupials. Collins, Sydney
5. CALABY, J.H. (1958): Studies on marsupial nutrition II. *Australian Journal of Biological Science*, 11, p. 571-580
6. CALLISTER, D.J. (1991): A review of the Tasmanian Brushtail Possum Industry. *TRAFFIC Bulletin*, 12(3), p. 49-58

7. CAUGHLEY, G., SHEPHERD, N. and J. SHORT (eds.) (1987): *Kangaroos - their ecology and management in the sheep rangelands of Australia.* Cambridge University Press, Cambridge
8. COLEMAN, J.D. (1988): Distribution, prevalence and epidemiology of bovine tuberculosis in brushtail possums, *Trichosurus vulpecula*, in the Hohonu range, New Zealand. *Australian Wildlife Research*, 15, p. 651-663
9. DAWNES, M.C. (1969): *Report of the Wildlife Ecology Section 1968/69.* Department of Agriculture, Stock and Fisheries, Papua New Guinea; 12 pp. (unpubl. report)
10. FLANNERY, T. (1990): *Mammals of New Guinea.* R. Brown and Associates, Carina, Queensland, Australia
11. FOOT, J.Z. and B. ROMBERG (1965): The utilisation of roughage by sheep and the red kangaroo (*Macropus rufus* Desmarest). *Australian Journal of Agricultural Research*, 16, p. 429-435
12. GEWALT, W. (1967): Die Kusus. In: *Grzimeks Tierleben*, Vol. 10, p. 97; Kindler, Zürich
13. GIBSON, L.M. and M.D. YOUNG, (1988): *Kangaroos: counting the cost.* Commonwealth Scientific and Industrial Research Organisation (CSIRO), Canberra, Australia
14. GRIGG, G.C. (1988): Kangaroo harvesting and the conservation of the sheep rangelands. *Australian Zoology*, 24, (3), p. 124-128
15. GRIGG, G.C. (1989): *Kangaroo harvesting and the conservation of arid and semi-arid rangelands.* Conservation Biology, 3, p. 194-197
16. HOPWOOD, P.R. (undated): *Recommendations to the Minister of Health, New South Wales Government, Commonwealth of Australia*
17. HOPWOOD, P.R., HILMI, M., and R.M. BUTTERFIELD (1976): A comparative study of the carcass composition of kangaroos and sheep. *Australian Journal of Zoology*, 24, p. 1-6
18. HOWARD, W.E. (1965): *Control of introduced mammals in New Zealand.* New Zealand Department of Scientific and Industrial Research; Information Series, No. 45, Wellington, New Zealand
19. POOLE, W.E. (1984): *Management of Kangaroo harvesting in Australia.* Occasional Paper No. 9; Australian National Parks and Wildlife Service, Canberra, Australia
20. RAMSAY, B.J. (1994): *Commercial use of wild animals in Australia.* Bureau of Resource Sciences; Australian Government Publishing Service, Canberra, Australia
21. SEIDENSTICKER, J., O'CONNELL, M.A. and A.J. JOHNSINGH (1987): 23. Virginia Opossum. In: *Wild Furbearer Management and Conservation in North America*, ed. NOVAK, M., BAKER, J.A., OBBARD, M.E. and B. MALLOCH; p. 246-263; Ontario Ministry of Natural Resources, Canada
22. SHARMAN, G. and H.J. FRITH (1967): *cited from Grzimeks Tierleben*, Vol. 10; p. 150-152; Kindler, Zürich
23. SHEPHERD, N.C. (1983): The feasibility of farming kangaroos. *Australian Rangeland Journal*, 5, p. 35-44
24. SHEPHERD, N.C. and G. CAUGHLEY (1987): Options for management of kangaroos . In: *Kangaroos - their ecology and management in the sheep rangelands of Australia*, ed. CAUGHLEY, G., SHEPHERD, N. and J. SHORT; Cambridge University Press, Cambridge
25. SHORT, J. and G.C. GRIGG (1982): The abundance of kangaroos in suboptimal habitats : wheat, intensive pastoral, mallee. *Australian Wildlife Research*, 9, p. 221-227

## 2.3.2 MONKEYS AND APES (*PRIMATES*)

The Primates are divided into the more primitive prosimians (**Prosimiae**) and monkeys and apes (**Simiae**). The former include the lemurs, makis, loris, galagos, and tarsiers. The latter are further subdivided into the **Platyrrhini** (New World monkeys) and the **Catarrhini** (Old World monkeys, apes, and man). The platyrrhines are restricted to South and Central America and include three families: Capuchin-like monkeys (**Cebidae**), the Goeldi monkey (**Callimiconidae**), and marmosets and tamarins (**Callitrichidae**). The catarrhines are distributed over Africa and Asia and include four families: Old-world monkeys (**Cercopithecidae**) including the Guenon-like (**Cercopithecinae**) and the Leaf-eating monkeys (**Colobinae**), gibbons (**Hylobatidae**), apes (**Pongidae**), and man (**Hominidae**). Most monkeys and apes are used for food by indigenous peoples who in many areas relish their meat. However, Tab. 23 shows that the trade in live primates for laboratory purposes or pets has also gained considerable economic importance during the last decades.

### Historical background

As man's closest biological relatives, primates have always received much attention. While appearing as gods in Ancient Egypt, India or Mesoamerica, they were considered a kind of devil in Mediaeval Europe. In habitat countries primates probably always have been an important source of protein for human populations, especially in tropical rainforest areas. Trade of primates dates back as far as 5,000 years (19), and they were brought to Europe and later to North America for exhibition in menageries, zoos and circuses, and as pets. In the early 20th century the intensive use of primates as experimental animals in biomedical research began, a use that culminated between the 1950s and 1970s when large numbers of primates were required for the production and testing of vaccines (12). Primates are now of much concern to conservationists as populations of most species are declining and many species are threatened by extinction. Whether they can be used economically as a wildlife resource and whether this use can be sustainable requires careful examination.

### Management and utilisation

Like most wildlife resources primates are used economically in two fundamentally different ways: either by extracting them from their habitats or by using them *in situ*. Extraction from the habitat is by live trapping or by hunting, depending on the purpose for which primates are used.

**Live trapping** is generally done to provide primates for local, national, and international markets. On the local and national markets primates are mainly traded as pets; on the international market the primary destination is their exploitation as experimental animals in biomedical research. Other purposes of the ***trade of live primates*** include the supply of animals for zoos, entertainment (shows, circuses, etc.) and private collections. This listing is not necessarily complete, but comprises those reasons that probably account for the majority of primates removed from their natural habitat by live trapping. The international primate trade has recently been reviewed by MACK and MITTERMEIER (23) and KAVANAGH and coll. (19).

**Table 23:** Overview of the utilisation of primates

| English name | Scientific name | Geographical distribution | Utilisation |
|---|---|---|---|
| **PROSIMIANS - PROSIMIAE** | | | |
| Lemurs | *Lemur* spp. | Madagascar | hunted for meat |
| Sifaka | *Propithecus diadema* | Madagascar | hunted for meat |
| Slender lori | *Loris tardigradus* | Asia | hunted for folk medicine |
| **MONKEYS and APES - SIMIAE** | | | |
| **New World Monkeys - Platyrrhini** | | | |
| Night monkeys | *Aotus* spp. | S. and C.America | live for biomedicine |
| Squirrel monkeys | *Saimiri* spp. | S. and C.America | live for biomedicine and as pets |
| Saki monkeys | *Pithecia* spp. | S.America | hunted for meat and fur |
| Bearded monkeys | *Chiropotes* spp. | S.America | hunted for meat |
| Capuchin monkeys | *Cebus* spp. | S. and C.America | hunted for meat and as pets |
| Spider monkeys | *Ateles* spp. | S. and C.America | hunted for meat and as pets |
| Woolly monkeys | *Lagothrix* spp. | S.America | hunted for meat and as pets |
| Tamarins | *Saguinus* spp. | S. and C.America | live for biomedicine and as pets |
| Lion tamarins | *Leontopithecus* spp. | S.America | live as pets |
| **Old World Monkeys and Apes - Catarrhini** | | | |
| Macaques | *Macaca* spp. | Asia | hunted for meat; live for biomedicine and coconut harvesting |
| Guenons, vervets | *Cercopithecus* spp. | Africa | hunted for meat and fur; live for biomedicine |
| Mangabeys | *Cercocebus* spp. | Africa | hunted for meat |
| Baboons | *Papio* spp. | Africa | live for biomedicine |
| Mandrill, drill | *Mandrillus* spp. | Africa | hunted for meat |
| Langurs | *Presbytis* spp. | Asia | hunted for meat and folk medicine |
| Colobus monkeys | *Colobus* spp. | Africa | hunted for fur and meat |
| Gibbons | *Hylobates* spp. | Asia | hunted for meat; live as pets |
| Orang-utan | *Pongo pygmaeus* | Asia | hunted for meat; live as pets |
| Gorilla | *Gorilla gorilla* | Africa | hunted for meat and trophies; live for tourism |
| Chimpanzee | *Pan troglodytes* | Africa | hunted for meat; live for biomedicine, as pets and for tourism |

**Hunting** of primates has several reasons, too. Primates are hunted for meat; to obtain body parts, e.g. fur for ornamentation and clothing or for medical purposes (folk medicine); and to obtain infant primates as pets. Reasons for hunting are not necessarily exclusive. Adult primates are hunted for meat, but infants that stay alive when their mother is shot are sold as pets as is the case, for example, with *Woolly monkeys* and *chimpanzees* (28, 35). *Saki monkeys* are hunted for meat, too, and their bushy tails are used as dusters or sold as ornaments in tourist shops (28). Hunting of primates has been extensively reviewed by MITTERMEIER (28).

In contrast to the number of purposes for the extractive or consumptive use of wild primates, only one type can be identified where primates are used economically *in situ*, i.e. while remaining in their natural habitat. This is the **use of primates as tourist attraction**. The best-known example is provided by the Mountain Gorilla Project. In the Parc National des Volcans in Rwanda, groups of *Mountain gorillas (Gorilla gorilla beringei)* have been habituated to the presence of humans and since 1984, small groups of paying visitors are allowed to watch gorillas at close range. To avoid disturbing the gorillas, each group is visited only once a day by not more than six tourists. Similar projects have been initiated with *Mountain gorillas* and *chimpanzees* in Eastern Zaïre (3). Other, less spectacular primate species like *baboons* and *Vervet monkeys* can also serve as tourist attractions, but if not carefully managed, they may become serious pests.

## Economic importance

**Live trapping and trade** of wild primates can be divided into two sectors, the legal sector that complies with the regulations established by the CITES and the illegal sector that cir-

Live Rhesus monkeys for sale at a market in South-East Asia. (Photo: J. ADLER)

cumvents these regulations. The volume of the legal trade and its economic importance can be estimated from the annual trade statistics that are maintained by the parties to CITES as shown in Tab. 24. However, these figures are likely to underestimate the real volume since not all parties do report annual statistics (18). The principal exporting countries are Indonesia and the Philippines, while the USA is the major importer (12), followed by Japan, Great Britain, France, the Netherlands and former Yugoslavia. Trade statistics for the USA are likely to reflect the major trends in the worldwide primate trade. The *Long-tailed macaque (Macaca fascicularis)* was the most intensively traded species. From 1988 to 1991, 43,217 Long-tailed macaques were imported to the USA (1, 2, 4, 5), and this species has largely replaced the *Rhesus monkey (M. mulatta)* as an experimental animal in biomedical research. However, in 1994 Indonesia has banned the export of wild-caught *Long-tailed and Pig-tailed macaques (M. fascicularis* and *-nemestrina)* and this is likely to cause substantial trade changes (6). Other species imported to the USA in larger quantities between 1988 and 1991 were *Squirrel monkeys* (*Saimiri sciureus*: 4,682), *Rhesus monkeys* (*Macaca mulatta:* 3,540), *Vervet monkeys* (*Cercopithecus aethiops*: 2,746) and *baboons* (*Papio* spp.: 2,372) (1, 2, 4, 5). Primates represent a major proportion of the trade of live mammals that is regulated by CITES; neotropical primates made up 99% of all live mammals (total N = 7,011) imported to the USA from Latin America between 1982 and 1984 (17). The legal sector of the international primate trade is governed by the demand for primates as experimental animals in biomedical research. The pet trade and the trade for zoo animals has drastically declined after the implementation of CITES; e.g. 79% of primates imported to Japan in 1973 were intended for pet use, but only 7% between 1981 and 1983 (27).

**Table 24:** Trade in live primates in 1989[1)]

| Main exporters[2)] | Main importers[2)] | Total exports | Total imports |
|---|---|---|---|
| AFRICA:<br>Mauritius, Tanzania, Kenya, Senegal, Ethiopia | | 9,613 | 17 |
| ASIA:<br>Indonesia, Philippines, China, | Japan, former USSR | 26,918 | 6,445 |
| EUROPE: | Great Britain, France, Netherlands, former Yugoslavia, Italy, Belgium, Sweden | 3 | 14,626 |
| NORTH AND CENTRAL AMERICA:<br>Barbados, Honduras | USA, Canada | 1,515 | 20,914 |
| SOUTH AMERICA:<br>Guyana, Peru | | 3,501 | 110 |
| OCEANIA AND OTHER:<br>Australia | | 699 | 137 |
| **TOTAL WORLD TRADE:** | | **approx. 42,249 exports/imports** | |

1) as reported by parties to CITES only; 2) accounting for >90% of exports/imports in sequence of importance

Source: World Conservation Monitoring Centre (37)

It is difficult to estimate the economic revenue to exporting countries of the international primate trade because there is generally a long chain from the trapper through intermediates to the final receiver, but it is likely that much if not most of the profit remains with animal-dealing companies in the importing countries rather than in the source countries (cf. 24).

Despite international and national regulations of wildlife trade, illegal trapping and trade of primates is ongoing. Live primates are found for sale on many markets in the habitat countries and are offered by internationally operating animal dealers. Illegally imported young *chimpanzees* are used in some countries by beach photographers to pose with tourists (19). The illegal trade includes many of the primate species that are listed on Appendix 1 of CITES and which therefore must not be traded for commercial purposes; these include *orang-utans* and *chimpanzees* (35). The volume of the illegal trade cannot be quantified but it is likely that for many species it far exceeds the volume of the legal trade. It is estimated that each year about 1,000 chimpanzees are still exported illegally from Africa to the USA, Europe and Japan (30).

As with the illegal live trade of primates, no statistics are available to quantify the amount of **hunting**. The hunting for meat varies between areas and within areas from species to species (26). It also strongly depends on the availability of alternative sources of protein from other wildlife (ungulates, large birds, fish) or from domestic animals. Particularly in rainforest areas, primates can be important prey items for human populations as Tab. 25 shows. On many markets, particularly in the Neotropics and in Africa, primate meat is sold as a much appreciated wildlife product (9). In tropical rain forests, primates often represent a large proportion of the mammalian biomass, and due to their arboreal, group living habits are easier to locate and to hunt than other rain-forest mammals that live solitary or in small groups.

**Table 25:** Examples for the relative importance of primates in subsistence hunting of indigenous people in some rainforest areas

| Indigenous tribes | Kills 1) | Primates 2) | Source |
|---|---|---|---|
| Siona Secoya (Ecuador) | 1,300 | 27.1% | VICKERS (1991) |
| Waorani (Ecuador) | 3,165 | 36.0% | YOST and KELLEY (1983) |
| Trio Indians (Suriname) | 426 | 33.6% | LENSELINK (1972) |
| Semaq Beri (Malayisa) | 185 | 96.8% | KUCHIKURA (1988) |

1) total number of kills observed; 2) percentage of kills representing primates

The **use of primates as tourist attraction** can render considerable economic benefits as exemplified by the *Mountain gorilla* projects running in Rwanda and Eastern Zaïre. Since the initiation of these projects, the number of tourists attracted to the national parks (Parc National des Volcans in Rwanda, Virunga National Park in Zaïre) has strongly increased and so has the revenue from tourists. In the mid 1980s tourism to the Parc National des Volcans rendered US$ 250,000 per year and has made it the third largest source of foreign currency for Rwanda (8). In Zaïre, the number of tourists attracted by the Mountain gorillas trebled from 1986 to 1990 and revenues amounted to more than US$ 400,000 in 1990 (3).

Another use of *Pig-tailed macaques (Macaca nemestrina)* in South-East Asia, is for **harvesting coconuts**. Young monkeys, probably obtained by killing the mother, are trained to pick ripe coconuts from the palms. Well-trained monkeys can harvest 500 nuts/day on average and are said to work for about 15 years.

In several areas of Africa and Asia, *baboons, Vervet monkeys,* and *some species of macaques* have become **agricultural pests** by raiding crops (11). Instead of killing them in extensive "monkey drives" these primates could be live-trapped and used to replace animals from other populations that don't need to be controlled (19). However, their being pests should not be considered *per se* as a justification for use as experimental animals.

Recent field observations have shown that several primate species are capable of self-medication by the **consumption of plant material with pharmacological properties** (29). Intensification of eco-ethological field studies on primates together with the screening of the plant material could increase the knowledge of plants with such properties, with subsequent economic benefits.

**Wildlife films** could be another form of *in-situ* use of primates that renders substantial economic benefit (10).

Although by definition captive bred primates are not a wildlife resource, **captive breeding** is included here since it represents an important sector in the economic use of primates *per se*. Export bans imposed by habitat countries in response to declining populations of wild primates have forced the biomedical research community to search for alternative sources of primates for use as laboratory animals. Primate centres and commercial breeding operations have been established in the USA, Europe and Japan where primates are used in biomedical research, and - on smaller scales - also in several habitat countries. Generally, the number of primates bred in captivity does not cover the need of the biomedical research community. For example, on average 55,000 primates per year were used in experimentations in the USA between 1979 and 1984, but the number of bred primates was only 8,151 in 1984 (13). Captive-bred primates are more expensive than primates captured from the wild. Therefore, wild primates are frequently used by biomedical researchers even when captive-bred animals are available. Occasionally, primates are offered and sold as captive bred, although numbers obviously exceed breeding capacities of the operation and therefore at least some of the animals must stem from the wild. An alternative to both captive-breeding and capture from the wild is the breeding of primates under semi-free conditions on islands, as implemented in the last years in seveal habitat countries, for example Peru and Indonesia (e.g. (15), (21)).

## Implications for conservation

At least one third of the more than 200 extant primate species are considered as being endangered and several are actually threatened by extinction. Although habitat destruction is the most important reason for the decline of primate populations, hunting and live trapping of primates contribute to the precarious situation of many primate species. One of the reasons for this is that most primates have lower recruitment rates compared to other mammals (e.g. 33), due to long interbirth intervals, small litter size, and slow maturation. Therefore, the potential for recovery after a population has been encroached is limited. The intensity of the economic use strongly influences the degree of sustainability and thus the impact of that use on populations of wild primates and their future survival.

If current hunting practices of traditional-living rain-forest people are equal or similar to former practices, then primates have always been a source of protein for human populations. There is no means to evaluate the impact hunting had in historical times, but the fact that most rain-forest primates thrived until rain-forest destruction became a serious threat indicates that traditional hunting of primates was sustainable. However, even traditional hunting may lead to a local decline of populations of large primates as has been shown for Siona-Secoya hunters in Ecuador (36). With the introduction of firearms and the growth of human populations, hunting pressure on primates has increased and is now a serious threat to many primate populations, along with the increasing destruction of natural habitats. Larger primates are particularly susceptible to this pressure. A comparison of population densities at hunted and non-hunted sites in Amazonia revealed considerably lower densities of larger primates (< 3 kg) at hunted sites or even the local extinction of species such as the Woolly monkey (31, 32).

Hunting is particularly detrimental to primate populations if the purposes of obtaining meat and pets are combined: by shooting down a mother with her infant, two animals are removed simultaneously, one that is of actual reproductive value (mother) and the other of future reproductive value (infant). Higher primates often behave protectively if an infant is threatened. As a consequence, in the process of hunting to obtain an infant, up to five other individuals are killed as in the case of chimpanzees (35).

Gorillas in Zaire are not only illegally killed for meat but their skulls may be used for ritual purposes. (Photo: G. MERZ)

For most species listed in the IUCN Red Data Book on Threatened Primates of Africa (16), hunting is a major threat to survival being second only to habitat destruction. For populations that are already reduced or isolated by habitat destruction and fragmentation, hunting can give the final blow. Overall, given that many primates are threatened through increasing habitat destruction, hunting can no longer be considered a type of sustainable use of wild primates.

Like hunting, live trapping may also have severe consequences for wild primate populations. Examples are provided by the Cotton-top tamarin (*Saguinus oedipus*) in Colombia and the Rhesus monkey (*Macaca mulatta*) in India, both of which were trapped to provide animals for biomedical research. Along with habitat destruction, excessive trapping has brought down populations of Cotton-top tamarins to small numbers and the species is now threatened by extinction. Although not endangered, excessive trapping has reduced populations of Rhesus monkeys by 50% and more (25). Sustainability of live trapping has been demonstrated only for two small primate species, the Moustached tamarin (*Saguinus mystax*) and the Night monkey (*Aotus nancymae*) (7, 14). These two species have been trapped only in small numbers, unlike the Long-tailed macaque which is the currently the most heavily used species. A decline of populations of Long-tailed macaques on the Philippine Islands has been attributed to excessive trapping (34). Given the experiences with the Cotton-top tamarins and Rhesus monkeys, it may be antisipated that the ongoing removal of large numbers along with shrinking of the habitats will make this species vulnerable in the future. This case may be even more likely since trade figures generally underestimate the total number of animals removed from the wild. It has been estimated that for every Long-tailed macaque that reaches the USA approximately one more dies during capture and holding in the exporting country (DARSONO 1979 quoted in (19)). Trapping can be particularly detrimental if it concerns species that are already depleted by habitat destruction, as in the case of Golden-headed lion-tamarins (*Leontopithecus chrysomelas*). The trade of some 50 individuals of this species in 1983 removed a substantial portion of the remaining population from the wild (26). The trade of live primates seems to be sustainable only if done on a very small scale and in species whose populations are not encroached by other threats (habitat destruction, hunting).

The use of primates as tourist attraction is a relatively recent form of use and little experience is available. Compared to the potential risks of this form of use (e.g. transmission of infectious disease from visitors to the primates), the conservation benefits are substantial: the daily monitoring of gorilla groups protects them from poaching, and the economic benefits from tourism help to improve conservation efforts (8). Since the beginnings of the projects, the sizes of the respective gorilla populations and the number of immature individuals have increased. The Mountain Gorilla Project thus demonstrates that the use of the primates as tourist attraction is a sustainable form of economic exploitation of primates. The future has to show whether similar projects can be established successfully with other primate species.

## Literature References


1. ANOYMOUS (1990a): *Convention on International Trade in Endangered Species of Wild Fauna and Flora; Annual Report - 1988*. US Department of the Interior, Fish and Wildlife Service, Washington, DC
2. ANOYMOUS (1990b): *Convention on International Trade in Endangered Species of Wild Fauna and Flora; Annual Report - 1989*. US Department of the Interior, Fish and Wildlife Service, Washington, DC
3. ANONYMOUS (1991a): Schutz von Gorillas und Schimpansen im östlichen Zaire. *Mitteilungen der Zoologischen Gesellschaft Frankfurt*, 3(91), p. 8-10
4. ANONYMOUS (1991b): Primate imports to the United States, 1990. *International Primate Protection League Newsletter*, 18(3), p. 24
5. ANONYMOUS (1992): Primate imports to the United States, 1991. *International Primate Protection League Newsletter*, 19(3), p. 19
6. ANONYMOUS (1994): Indonesia export bans. *Asian Primates*, 3(3/4), p. 7
7. AQUINO, R., PUERTAS, P. and F. ENCARNACIÓN (1993): Effects of cropping on the *Aotus nancymae* population in a forest of Peruvian Amazonia. *Primate Report*, 37, p. 31-40
8. AVELING, R. and C. (1987): Report from the Zaire gorilla conservation project. *Primate Conservation*, 8, p. 162-164
9. CASTRO, N., REVILLA, J. and M NEVILLE (1990): Carne de monte como una fuente de proteinas en Iquitos, con referencia especial a monos. In: *La Primatología en el Perú: Investigaciones Primatológicas (1973-1985)*, ed. PROYECTO DE PRIMATOLOGIA; p. 17-35; Imprenta Propaceb, Lima
10. DOBSON, A and R. ABSHER (1991): How to pay for tropical rain forests. *Trends in Ecology and Evolution*, 6, p. 348-351
11. ELSE, J.G. (1991): Nonhuman primates as pests. In: *Primate Responses to Environmental Change*, ed. BOX, H.O.; p. 155-165; Chapman and Hill, London
12. EUDEY, A. and D. MACK, (1984): Use of primates and captive breeding programs in the United States. In: *The International Primate Trade*, ed. MACK, D. and R.A. MITTERMEIER, Vol. 1: Legislation, Trade and Captive Breeding; p. 153-180; TRAFFIC (USA), Washington, DC
13. FULLER, D. (1985): Trends in primate imports into the United States, 1984. *ILAR News*, 28(4), p. 4-7
14. GLANDER, K.E., TAPIA, R.J. and T.A. FACHIN (1984): The impact of cropping on wild populations of *Saguinus mystax* and *Saguinus fuscicollis* in Peru. *American Journal of Primatology*, 7, p. 89-97
15. GOZALO, A., MOYA, L., IQUE, C., MORO, J. and F. ENCARNACIÓN (1991): Breeding the moustached tamarin (*Saguinus mystax*) on islands. In: *Primatology Today*, ed. EHARA, A., KIMURA, T., TAKENAKA, O. and M. IWAMOTO; p. 401-402; Elsevier, Amsterdam
16. IUCN (1988): *Threatened Primates of Africa. The IUCN Red Data Book*. IUCN, Gland. Switzerland
17. JORGENSON, J.P. and A.B. (1991): Imports of CITES-regulated mammals into the United States from Latin America: 1982-1984. In: *Latin American Mammalogy*, ed. MARES, M.A. and D.J. SCHMIDLY; p. 322-335; University of Oklahoma Press, Norman, Okla., USA
18. KAVANAGH, M. (1984): A review of the international primate trade. In: *The International Primate Trade*, ed. MACK, D. and R.A. MITTERMEIER, Vol. 1: Legislation, Trade and Captive Breeding; p. 49-90; TRAFFIC (USA), Washington, DC
19. KAVANAGH, M., EUDEY, A.A. and D. MACK (1987): The effects of live trapping and trade on primate populations. In: *Primate Conservation in the Tropical Rain Forest*, ed. MARSH, C.W. and R. A. MITTERMEIER; p. 147-177; Alan R. Liss, New York

20. KUCHIKURA, Y. (1988): Efficiency and focus of blowpipe hunting among Semaq Beri hunter-gatherers of Peninsular Malaysia. *Human Ecology*, 16, p. 271-305
21. KYES, R.C. (1993): Survey of the long-tailed macaques introduced onto Tinjil Island, Indonesia. *American Journal of Primatology*, 31, p. 77-83
22. LENSELINK, J. (1972): De jacht op brengst in een Surinaams Trio-dorp. *De Surinaamse Landbouw*, 20, p. 37-41
23. MACK, D. and R. A. MITTERMEIER (eds.) (1984a): *The International Primate Trade*, Vol. 1: Legislation, Trade and Captive Breeding. TRAFFIC (USA), Washington, DC
24. MACK, D. and R. A. MITTERMEIER (1984b): Introduction. In: *The International Primate Trade*, ed. MACK, D. and R.A. MITTERMEIER, Vol. 1: Legislation, Trade and Captive Breeding; p. 15-17; TRAFFIC (USA), Washington, DC
25. MALIK, I. (1992): Consequences of export and trapping of monkeys. *Primate Report*, 24, p.5-11
26. MALLINSON, J.J.C. (1984): Golden-headed lion tamarin contraband a major conservation problem. *IUCN/SSC Primate Specialist Group Newsletter*, 4, p. 23-25
27. MATSUBAYASHI, K., GOTOH, S. and J. SUZUKI (1986): Changes in import of non-human primates after ratification of CITES (Washington Convention) in Japan. *Primates*, 27, p. 125-135
28. MITTERMEIER, R.A. (1987): Effects of hunting on rain forest primates. In: *Primate Conservation in the Tropical Rain Forest*, ed. MARSH, C.W. and R.A. MITTERMEIER; , p. 109-146; Alan R. Liss, New York
29. NEWTON, P. (1991): The use of medicinal plants by primates: a missing link? *Trends in Ecology and Evolution*, 6, p. 297-299
30. PASCALI, V.L., DESTRO BISOL, G., DOBOSZ, M., D'ALOJA, E., PAONESSA, G. and U. MEREU (1994): Chimpanzee DNA profiles on trial. *Nature*, 367 (6465), p. 692-693
31. PERES, C.A., (1990): Effects of hunting on western Amazonian primate communities. *Biological Conservation*, 54, p. 47-59
32. PERES, C.A. (1991): Humboldt's woolly monkeys decimated by hunting in Amazonia. *Oryx*, 25, p. 89-95
33. ROBINSON, J.G. and K.H. REDFORD (1988): Intrinsic rate of natural increase in Neotropical forest mammals: relationships to phylogeny and diet. *Oecologia*, 68, p. 516-520
34. RUBIO, R.P., ALVIOLA, III, P.L. and M. R. FELIZARDO (1990): *Biological and economic implications in the trade and commercial exploitation of the Philippine long-tailed macaques (Completion Report)*; World Wildlife Fund (USA); New York - The Haribon Foundation, Department of Environment and Natural Resources; 135 pp.
35. TELEKI, G. (1989): Population status of wild chimpanzees (*Pan troglodytes*) and threats to survival. In: *Understanding Chimpanzees*, ed. HELTNE, P.G. and L.A. MARQUARDT; p. 312-353; Harvard University Press, Cambridge, USA
36. VICKERS, W.T. (1991): Hunting yields and game composition over ten years in an Amazon Indian territory. In: *Neotropical Wildlife Use and Conservation*, ed. ROBINSON, J.G. and K.H. REDFORD; p. 53-81; University of Chicago Press, Chicago and London
37. WORLD CONSERVATION MONITORING CENTRE. (1992): *Global Biodiversity: Status of the Earth's Living Resources*; p. 382-383; Chapman and Hall, London
38. YOST, J.A. and P.M. KELLEY (1983): Shotguns, blowguns, and spears: the analysis of technological efficiency. In: *Adaptive Responses of Native Amazonians*, ed. HAMES, R.B. and W.T. VICKERS; p. 189-224; Academic Press, New York

## 2.3.3 HARES, RABBITS AND RODENTS (*LAGOMORPHA AND RODENTIA*)

Hares and rabbits were formerly regarded as rodents, but constitute a distinct mammal order (**Lagomorpha**) which resembles rodents (**Rodentia**) only superficially. Lagomorphs inhabit more open landscapes and are exclusively herbivorous. Both groups have large rootless incisors which enable them to "gnaw". Their flesh is particularly tender and tasty and the skin of some species furnishes valuable furs. Because of the similarities of the two groups they are lumped together here. True hares (*Lepus* spp.) occur in various species on all continents. Rabbits comprise the European rabbit (*Oryctolagus cuniculus*), which was restricted to Europe and North-West Africa before it was spread by man, and the Cottontail rabbits (*Sylvilagus* spp.) on the American continent. With almost 3,000 species rodents are the largest and most diverse group of mammals distributed all over the world. They are subdivided into Squirrel-like rodents (**Sciumorpha**), beavers (**Castorimorpha**), Scaly-tailed squirrels and allies (**Anomaluromorpha**), Mouse-like (**Myomorpha**), Dormice-like (**Glirimorpha**) and Porcupine-like rodents (**Hystricognathi**) including the Cavy-like rodents (**Cavioidea**). Tab. 26 shows that species of all suborders are used either for food or for their pelts. The rabbit and cavy have been domesticated; other species are being farmed or bred for laboratory purposes.

### Historical background

Archaeological sites throughout Europe and also in North America have revealed that prehistoric man used lagomorphs for food (34, 57), although large game animals, like deer, moose and bison predominated in his diet. In Germanic mythology *hares* had a particular importance which is still reflected by many German fairy tales, idioms and figures of speech. Hare hunting and roast hare for special festive occasions have remained popular in Europe to-date although since the late Middle Ages they were partly replaced by the *rabbit*. In contrast to the European hare this had only survived the glacial period in Europe on the Iberian Peninsula and in North-West Africa where it was first discovered by the Phoenician seafarers around 1100 BC. Later the Romans introduced the rabbit to Italy as a special delicacy. They started to keep and to breed rabbits together with hares in large pens, so-called "leporaria". This was the start of the domestication of the rabbit which was later successfully pursued in French cloisters (40). Concurrently the Wild rabbit was spread throughout Europe for hunting purposes. Portuguese ships took them to the Atlantic Islands and the Normans introduced them to England in the 12th century. In Germany Wild rabbits seem to have become known only after the domestic form had already arrived from France in 1149 (7). During the 19th and early 20th centuries both species became widely distributed overseas either by deliberate introduction for hunting purposes or by escape from game parks: hares and rabbits to Australia, New Zealand, Chile and some areas in the USA, hares only to Argentina, parts of Canada, Southern Siberia and the Far East (36). In some of these countries they have subsequently multiplied into nuisance populations which are now being commercially exploited, as described below.

**Table 26:** Overview of the utilisation of hares, rabbits and rodents

| English name | Scientific name | Original geographical distribution | Weight (kg) | Utilisation |
|---|---|---|---|---|
| Hares | *Lepus* spp. | Europe, Asia, Africa, N.America | 2.5-6 | for sport hunting, meat, fur |
| European rabbit | *Oryctolagus cuniculus* | S.Europe, N.W.Africa | 1.5-2.5 | for sport hunting, meat, fur; domestic., pet use |
| Cottontail rabbits | *Sylvilagus* spp. | N. and S.America | 1-2 | for sport hunting, meat |
| **SQUIRREL-LIKE RODENTS - SCIUMORPHA** | | | | |
| Bobaks | *Marmota bobak* sspp. | C.Asia | | for meat, fat, fur |
| Ground squirrels | *Citellus* spp. | C.Asia, N.America | | for fur, meat |
| Red squirrel | *Tamiasciurus hudsonicus* | N.America | | for fur, meat |
| Eurasian squirrel | *Sciurus vulgaris* | North. Eurasia | | for fur |
| **SCALY-TAILED SQUIRRELS and allies - ANOMALUROMORPHA** | | | | |
| **MOUSE-LIKE RODENTS - MYOMORPHA** | | | | |
| House mouse | *Mus domesticus* | S.Europe - E.Asia | <0.1 | semi-domesticated |
| Brown rat | *Rattus norvegicus* | E.Asia | 0.2-0.4 | for laboratory and |
| Golden hamster | *Mesocricetus auratus* | W.Asia | ~0.1 | pet use |
| Chinese hamster | *Cricetulus griseus* | E.Asia | <0.1 | for pet and laborat. use |
| Common hamster | *Cricetus cricetus* | Eurasia | 0.3-0.5 | for meat, fat and fur |
| Giant rat | *Cricetomys gambianus* | W.-E.Africa | 0.8-1.2 | for meat; semi-dom.[1)] |
| Muskrat | *Ondatra zibethicus* | S.America | 0.8-1.5 | for fur and meat |
| Mongolian gerbil | *Meriones unguiculatus* | E.Asia | ~0.1 | for pet and laborat. use |
| Fat dormouse | *Glis glis* | Europe, Asia min. | <0.1-0.1 | for meat, fat |
| **PORCUPINE-LIKE RODENTS - HYSTRICOGNATHI** | | | | |
| Porcupines | | | | |
| Brush-tailed porcup. | | | | |
| Cane rat | | | | |
| Amer. porcupine | | | | |
| Viscacha | | | | |
| Mountain viscacha | | | | |
| Chinchilla | | | | |
| **Superfamily Cavy-like rodents - Cavioidea** | | | | |
| Wild cavy | | | | |
| Mara | | | | |
| Capybara | | | | |
| Paka, agoutis | | | | |
| Acouchis | | | | |
| Nutria | | | | |

1) only in experimental stage of domestication

Strangely only one of the many Cavy-like rodents utilised for food in South America, the *Wild cavy* or *aperea*, was domesticated and became the domestic "guinea pig". It was common in the Inca Empire long before the Spaniards conquered it in the 16th century. In the Andean villages guinea pigs are often kept free and become feral. These semi-domesticated as well as the Wild cavies are extensively utilised by the rural communities, one killed animal conveniently providing a day's meat ration for the whole family.

Amongst the larger rodents the *beaver* and to a lesser extent also the *muskrat* had great importance to prehistoric man in North America, providing meat and fat for food, incisors for tool making, and fur for protective clothing (57). Other smaller rodents, such as *Ground squirrels, marmots* and *chipmunks*, were eaten where they occurred in numbers, but not used much for furs (38). The arrival of the Europeans in North America in the 16th century had a profound influence on the utilisation of the faunal resources by the native peoples. Although they clearly appreciated fur quality, they valued any wildlife first of all in terms of their subsistence needs and the spiritual constraints to its use. The Europeans induced them to trapping fur animals for cash according to European market demands enabling them to buy goods offered by the European traders. This change in values is clearly evident, for example, from the marked increase in all historic strata of archaeological deposits of beaver bones, but particularly from the abundance of remains of muskrat, a species predominantly trapped for the European fur trade; prehistorically, beaver had predominated (57). Thus the subsistence economy of the Indians changed into a cash economy and this as in other parts of the world had very negative effects on the resource.

The historical development of the use of furbearing animals and the associated trade are discussed in Chapt. 2.3.4 as most of them belong to the order of Carnivora. However, the trapping of *beaver* played a key role in the early exploration and colonisation of North America and in the fur trade from the 16th to the 19th century which needs to be mentioned here. Around 1550 felt hats became fashionable in Europe and the principal raw material for their manufacture was the underwool of beaver pelts. This was preferred because beaver hair could be bonded together into extremely high-grade felt. Thus the felt hat industry became the driving force behind the fur trade and beaver pelts a most valuable staple commodity (47). Other peltries were also traded but did not rival beaver in total value. Until the beginning of the 19th century the beaver-based trade expanded all over North America, but by 1821 when the major fur-trading companies ended competition by merger, beaver had been seriously overtrapped. *Nutria* pelts furnished from Argentina, started to substitute beaver wool in felt making. However, in the 1840s fashion changed to silk and this now became the key fashion article causing the trade in beaver pelts to collapse. In the USA muskrat trapping became most important in numbers although raccoon pelts took the first position in terms of value. In subarctic Canada, however, where the possibilities of agricultural development were limited, the beaver regained its importance in total value by 1876. At the end of the 19th century captive breeding of furbearers commenced, and by the 1920s, ranching and farming of the large furbearing rodent species was started. However, only nutria farming proved to be competitive in the long run with the trapping of wild populations in South America and feral stock in the Southern USA and the former USSR where the species had been introduced in the 1930s (4).

The particularly fine fur of the *chinchilla* was used by the Incas by clipping the wool of captive-held animals. In the 19th century it was, like other furbearers, ruthlessly over-exploited until 1910 when all hunting and trapping was prohibited by the governments of the Andes countries. A mine employee in Bolivia, Mathias Chapman, started to breed them 1918 in captivity and eventually transported eleven animals to California. This gave rise to modern chinchilla farming which spread from the 1930s onwards all over the world (27).

Many other rodent species have been used historically both in a wild or captive state for meat, fat and fur. Just as they did to hares and rabbits the Romans also cultivated *dormice* as a culinary delicacy and these are still eaten in France and the Balkan countries (44). *Alpine marmots* were extensively used until the early 20th century for their meat and fat to which medicinal properties were ascribed. MARCO POLO described in the 13th century how *Bobak marmots* were hunted and trapped by the Mongols, and the use of this species is still of great economic importance in Mongolia (22). Furskins from the *Eurasian squirrel* were, next to beaver pelts, the most important raw material derived from rodents during the 15th and 16th centuries when furskins were an important staple commodity in the trade of the Hanseatic League along the Baltic coast (16).

Last but not least, the development of medicine and microbiology since the 1880s and experimental genetics during this century brought about an enormous, ever increasing demand for laboratory animals. This resulted in the intensification of captive breeding of semi-domesticated strains of the *Brown rat* and the *House mouse*. After the Second World War other wild rodent species, the *Golden hamster* and the *Mongolian gerbil*, were added to the spectrum of laboratory animals and also becoming popular pet species. The former had been known scientifically for almost 100 years but was first caught alive and bred in captivity from the 1930s onwards (43). The latter was first bred in Japan from 20 pairs collected in 1935 in Mongolia and was imported to the USA in 1954 where a breeding colony was established. From this colony gerbils became distributed throughout the United States and Continental Europe as an attractive children's pet before being introduced into biomedical research (55).

## Management and utilisation

Hares, rabbits and large rodents are all extensively utilised for their meat or fur, or both, in the form of either subsistence or commercial hunting and trapping. The lagomorphs are also hunted for sport and some furbearers are ranched or farmed. The meat of lagomorphs and rodents is widely considered to be particularly tasty and tender, and is eaten in preference to domestic animals. They produce fat, are easy to hunt or trap and are therefore of particular **importance in subsistence societies** in almost all regions of the world. In North America Indians have formerly relied heavily on *beavers* and *muskrats* (38); in South America the many different species of *Cavy-like rodents* (18, 8), in West Africa *Giant* and *Cane rats* as well as the different types of *porcupines* (11, 3, 48) and in South Africa *springhares* (17) constitute major sources of food. Also, in Central Asia millions of *Ground squirrels* and *bobaks* are eaten every year by the rural people (22).

Many rodent species are competitors with man and livestock for cereals and vegetable food or are otherwise noxious. Therefore they often become a nuisance or even a pest and their numbers need to be controlled. This necessity is often the basis for **large-scale commercial exploitation** of the particular species. Although explosive multiplication of rodents has been known since time immemorial, some of the situations in which lagomorphs or larger rodents have become an agricultural problem are caused by man. Apart from deliberate introductions of non-indigenous species for sports hunting and trapping, or the incidental establishment of feral populations from escapees, it is often the intensification of agriculture which creates ecologically favourable conditions even for indigenous species, causing them to multiply into nuisance populations. This has occurred, for example, in the case of the European rabbit and hare, the muskrat, beaver and Cane rat, all of which are in conflict with man and exploited commercially in this context, as described below.

After introduction of the *European rabbit* into Australia in 1859 it rapidly colonised 4 million km$^2$ of Southern and Central Australia. By the turn of the century rabbits had become a major pest, but concerted efforts to eliminate them by poisoning, trapping and shooting had only limited local effect. Control was combined with commercial exploitation of the rodent resource and this developed into a major domestic and export industry. It reached a peak in 1948/49 when about 50 million rabbits were killed and exported (46). After the introduction of the *Myxoma* virus in 1950 as a biological control agent, rabbit populations diminished drastically and their commercial use was greatly reduced. Today only about 2-3 million rabbits are commercially harvested every year concurrently with on-going pest control operations. Previously the majority of them were marketed in Australia for domestic con-

European rabbits were introduced to Australia during the 19th century and are now a valuable export commodity. (Photo: BUREAU OF RESOURCE SCIENCES, Australia)

sumption, but since 1989 increasing quantities of rabbit meat are also exported (1,081 t in 1991/92). Skins are used for felt production, mainly in Australia (46).

*European hares*, first introduced into Argentina in 1888, rapidly colonised most of the pampas region of this country causing extensive damage and the gradual displacement of the indigenous "*pampas hare*" or *mara*. Later, the hares expanded their range northwards into Uruguay and Brazil and to the south into Patagonia and on to the Falkland Islands. Gradually hares became the resource base of a major industry which annually utilises 5-10 million hares (6). During the 1940s and 1950s this was focused on the export of pelts, but from the early 1960s the sale of meat to European countries became more important. A network of collection points has been established where the carcasses are received from the rural hunters *in toto* and refrigerated. From there they are shipped daily to a few large processing plants around Buenos Aires, where they are skinned, degutted, deep frozen and packaged in different cuts for export. Skins are only a by-product and are used mainly for felt production (23, 32, 29).

Also in Australia, the introduced hares, having become widely established by the 1870s, initially became a pest and were controlled and exploited accordingly. However, during the latter part of the first decade of the 20th century, populations declined for ecological reasons and have remained low ever since (46). Today commercial exploitation is only opportunistic, together with rabbit and fox shooting, and hares are considered more as a recreational resource.

*Muskrats* are aquatic animals indigenous to North America where they were always hunted and trapped as valuable furbearing animals (38). Only when agriculture is intensified and water systems are regulated does the muskrat become a harmful animal; it may cause very considerable damage by burrowing into and destroying river banks and dams. Deliberate liberation for hunting purposes as far back as 1905 and escapes from fur farms in most European countries during the 1920s have resulted in a rapid spread of feral muskrat populations all over Europe. Most West European countries perceived muskrats only as a pest and have prohibited imports and farming. However, except for Great Britain, attempts to eliminate the muskrats have remained unsuccessful despite continued and costly control efforts (30). In Eastern European countries control is limited and muskrats are commercially cropped on a limited scale. Only in Finland and the former USSR, with their vast largely unoccupied river and swamp systems, were muskrats deliberately and repeatedly liberated in large numbers with a view to their utilisation for fur production (5). They are now well established from Eastern Europe into Mongolia and China, and have become a major resource for fur trapping. In Argentina muskrats were liberated from 1940 onwards and have spread to Chile, being harvested for fur together with nutria in both countries (56). The meat of muskrats is traditionally used in North America for human consumption; elsewhere it is used mainly for feeding fur animals in farms (5). Also the musk from the male preputial glands is used in limited quantities by the North American perfume industry (30).

Similarly to muskrats *beavers* interfere with the water systems they inhabit and come into conflict with man's regulation and use of these. Originally distributed all over Eurasia and North America, they were drastically reduced in numbers and in some parts of their range totally eliminated during the 19th century. Their decline was mainly due to overtrapping, probably by the Indians, in response to the strong demand for beaver pelts. However, the decrease of habitat has certainly also been a major reason (41). With the beginning of the 20th century protective legislation was introduced in North America and the species has since made a dramatic recovery, becoming one of the best studied furbearers and an example of successful management. Once near extinction, it has spread and multiplied under human control into stable populations, over much of North America, which are being sustainably utilised. Harvesting is strictly controlled by the wildlife authorities and regulated on the basis of quota and license systems (39).

The *nutria* or *coypu*, the second most important furbearing rodent, is indigenous only to the southern parts of South America where it occurs in swampy habitats. Nutria are commercially hunted either by shooting or spearing the swimming animals, mainly for their particularly fine fur. Until 1910 approximately 10 million pelts were exported annually which resulted in a serious decline of nutria in South America (50). This and the continuing fashion demand gave rise to the production of nutria pelts from captive-held animals as of 1922 in the USA and Europe (10). Escapes from the fur farms, but also intentional liberations, led to the establishment of numerous small feral populations in Western Europe, the USA and Canada and also in the Near East, Africa and Japan (25). In the former USSR nutria, like other non-indigenous furbearers, were deliberately introduced from 1930 onwards and became well established in the extensive reed swamps between the Caspian Sea and Turkey (4). Most of the feral populations in Western Europe succumbed or were eliminated as undesirable, but in Louisiana and the former USSR they have become a valuable, sustainably managed resource for fur harvesting. In Louisiana this is done by the landowners or individual trappers with guidance by the wildlife authorities (35). In 1992/93 129,545 pelts were harvested and some 200 t of meat produced at a total value of about US$ 340,000.

Although *Cane rats* or *grasscutters* occur all over Africa South of the Sahara, their commercial utilisation is limited to West and parts of Central Africa. There, this use has evolved from the traditional subsistence hunting of the species because of a strong local preference for Cane rat meat but also due to ecological factors. Being originally limited to moist river vegetation in the dry savanna zone, the opening up and modification of the moist forest zone for agriculture has created unlimited suitable habitat in which Cane rats have multiplied into particularly abundant and productive populations. These are exploited traditionally by village hunters either by shotgun hunting or trapping. Without being skinned the animals are also marketed fresh or smoked to urban centres where they command much higher prices than beef (11, 19, 48).

Another rodent species which is hunted extensively for its meat is the *capybara*, *chiguïre* or *carpincho* of South America which weighs up to 60 kg. It is the largest rodent and inhabits river and swamp systems from which it forages into the adjacent grasslands. As was the case for kangaroos in Australia and antelopes in Africa, capybara were perceived by the European settlers as competitors to the introduced livestock and were therefore mercilessly pursued, only their hides being utilised for leather production. Only since the 1960s has the productivity and commercial value of capybara been recognised and this has led to controlled utilisation in the form of ranching of the species on the Llanos, the seasonally inundated grasslands in Colombia and Venezuela (42).

After deer and game birds lagomorphs are probably the most popular animals for **sports hunting** in Europe and North America. The desire to have these readily available for hunting has greatly influenced their distribution and local abundance since Roman and Mediaeval times. In Europe *hares* were not only introduced into countries in which they did not occur, such as Ireland and Scandinavia, but were also extensively transplanted from eastern countries in which they were abundant to countries where they were rare, such as France and Italy. As mentioned above, the *European rabbit* was also initially spread and propagated all over Europe exclusively for hunting purposes and, despite the negative experiences in Australia since the 19th century, further introductions for hunting were made elsewhere. Ever since the Middle Ages the Wild rabbit has remained one of the most favourite prey animals for sport hunters in Europe. After the Second World War, before the introduction of myxomatosis, no less than 1.86 million licenses were annually sold to rabbit hunters in France alone (28). In the USA the hunting of *Cottontail rabbits* is as popular as rabbit hunting in Europe. It is estimated that about 25 million are killed every year (7). Nevertheless an attempt was made in 1951 to introduce the European rabbit into the USA, but this was less "successful" than previous liberations in Australia, New Zealand and Chile.

Commercially oriented **captive breeding and propagation** of lagomorphs and rodents, usually referred to as **farming**, is carried out for different purposes. Since the 1960s several European countries have started to produce live *hares* for release on hunting grounds. Previously these were captured and exported by countries with an abundance of hares. After overcoming heavy neonatal mortality the farming of hares in cages or open runs has developed into a marginally profitable enterprise mainly in France. In 1980 some 800 farms have produced 100,000-150,000 hares (14). In the Ukraine two large areas were fenced in 1959 in which hares are raised and netted for release in hunting areas (15). In Sweden *Blue hares* (*Lepus timidus*) have been bred since 1940 for release on hunting estates and islands (13).

Large scale fur farming commenced in the 1920s in North America and Europe in response to the demand for especially fine furs in different colour variations. It was attempted with all of the important rodent furbearers but only *nutria* and *chinchilla* farming developed into an economically viable industry. Nutria were initially kept in large natural pens, but after the Second World War farming was greatly intensified, mainly in Poland and the former USSR, focusing on selective breeding of different colour mutants (25). Chinchilla pelts are now exclusively produced from semi-domesticated stock as wild chinchilla are very rare and strictly protected. They are easily kept and bred in small cages, but have a relatively low reproductive rate. This explains why it took 30 years for the chinchilla farming industry to develop. Initially most extensive in the USA, it is now also established in Canada, the UK, Scandinavia and China (13). *Muskrat* farming in the 1920s, ranging from extensive management to pen or cage breeding, was either prohibited because of the danger of animals escaping and forming feral populations, or given up in the 1930s because it could not compete with pelt production from the wild. With the recovery of wild beaver populations in North America, *beaver* farming, extensive or intensive, also became unprofitable. Whereas in 1948 there were still some 440 farms in the USA and Canada, only a few very large ranching enterprises, holding several thousands of beaver, are still operative in the USA. There selective breeding has been practised for more than 20 years and several colour mutants, from black through gold to white have been developed. Target production in 1982 was 15,000 pelts/annum, by-products being tail leather, castorium for perfume, and meat for gourmet restaurants (13).

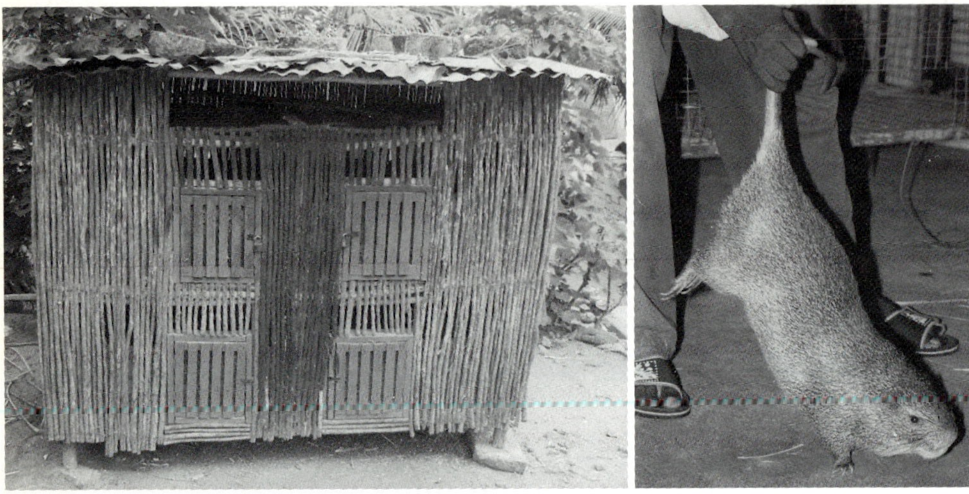

Rural facility for breeding Cane rats, made out of local materials and placed in the court yard. Adult male Cane rats can be as heavy as 7-8 kg. (Photo: by courtesy of the German Office for Technical Co-operation (GTZ), Eschborn)

More recently various attempts have been made to "domesticate" some large rodents, other than the domestic guinea pigs and rabbits, for meat production. These efforts are based on the advantage of using locally well-adapted and small mammals for meat production and the marked preference of the consumers for these local rodents. In South America captive breeding of *capybara* in conjunction with crocodile farming has developed in Colombia into a near commercial stage (24, 37), whereas *paka* and *agouti* farming is so far only sporadic and experimental (13, 52). In West Africa farming of the *Giant rat* was experimented with in Nigeria (1, 2) and domestication of the *Cane rat* by controlled and selective cage breeding has been seriously pursued since 1982 in Benin (12).

The breeding of **laboratory animals** has developed into an activity of major importance to research in many different scientific fields. In the Bethseda Medical Centre in the USA alone no less than 300,000 *rats* and 800,000 *mice* in addition to 200,000 *guinea pigs* were kept in 1965 for research and diagnostic purposes (21). The propagation of many hundreds of specific strains and sub-strains, their identification, and the breeding of new strains with known genetic make-up under sterile conditions for specific scientific purposes has become a special academic discipline which produces annually a wealth of scientific literature. Since 1977 a special periodical, the Rat Newsletter, informs on genetics, new mutants, development and location of new or formerly established strains.

## Economic importance

Except for the furbearing species it is difficult to assess the economic importance of lagomorphs and rodents. Their utilisation is often a spontaneous response either to overabundance or to special market demands, or it is a traditional subsistence activity. It is therefore rarely planned and harvests of lagomorphs and rodents may fluctuate greatly, exceeding temporarily sustainable levels. However, almost all over-exploited rodent populations recover very quickly if protected. Furthermore, the benefit derived from commercial utilisation of wild rodent populations has to be compared with the damage which they often cause. Tab. 27 gives examples of the volume of utilisation of some species in various countries at different times as a very rough indication of the economic importance of their use.

**Table 27:** Selected examples for the volume of utilisation of wild and feral populations of lagomorphs and rodents

| Species | Region or country | Period | Harvest per year (nos.) | Main use | Literature reference |
|---|---|---|---|---|---|
| European hare | Europe[1)] | 1987/90 | ~2.3 mill. | sport | (9) |
| | Argentina | 1970s | >6 mill. | meat | (32) |
| | Australia | 1990s | 20,000+ | meat | (46) |
| European rabbit | Germany | 1987/90 | 658,655 | sport | (9) |
| | Australia | 1989 | 3 mill. | meat | (46) |
| Cottontail rabbit | USA | 1960s | ~25 mill. | sport | (7) |
| Bobak marmot | Mongolia | 1960s | ~2.0 mill. | meat | (22) |
| | former USSR[2)] | 1980 | ~100,000 | meat | (26) |
| Ground squirrels | former USSR | 1980/90 | ~2 mill. | fur | (*) |
| Red squirrel | Canada | 1982/83 | 348,678 | fur | (53) |
| Euras. squirrel | former USSR | 1985 | 4.3 mill. | fur | (45) |
| Beaver | N. America | 1982/83 | 641,088 | fur | (53) |
| | former USSR | 1985 | 21,000 | fur | (45) |
| Springhare | Botswana | 1970s | ~2.5 mill. | meat | (17) |
| Muskrat | N. America | 1982/83 | 7.4 mill. | fur | (53) |
| | Germany | 1946/57 | 62,000[3)] | control | (30) |
| | former USSR | 1985[4)] | 1.3 mill. | fur | (45) |
| Porcupines | Ivory Coast[5)] | 1989/92 | ~4,000 | meat | (48) |
| Cane rat | Ivory Coast[5)] | 1989/92 | ~11,000 | meat | (48) |
| | Ghana[6)] | 1978/89 | ~48,000 | meat | (11), (19) |
| | Benin[7)] | 1980 | 191,829 | meat | (31) |
| Capybara | Venezuela[8)] | 1977/84 | ~57,000 | meat | (42) |
| Nutria | USA | 1977/84 | ~1.3 mill. | fur | (35) |
| | Argentina | 1977/84 | ~2.8 mill. | fur | (35) |
| Guinea pig | Peru | unspecified | ~50 mill. | meat | (8) |

*) BEKENOV (1994): pers. comm.

1) includes only Germany, Denmark, Sweden, Finland, Holland, Hungary, Austria, Slovenia, Switzerland; 2) *Marmota bobak, -sibirica* and *-menzbieri* are used; 3) official control figures, not all utilised; 4) harvests decreased since 1970s due to major irrigation developments in Central Asian habitats (51); 5) only animals sold on Toumodi bush meat market; 6) only animals sold on Accra and Kumasi bush meat markets; 7) estimate for whole country; harvest decreased sharply as of 1985; 8) only animals marketed throughout the country

If one converts the animal numbers listed in Tab. 27 into mass weight using the minimum average live weights of Tab. 26 just the selected examples of some ***hare, rabbit and rodent harvests*** amount to more than 110,000 t per year. This does not even include 16,000-17,000 t of meat derived annually from domesticated or semi-domesticated *guinea pigs* in Peru alone (8). Although a large proportion of the listed animals are killed primarily for their fur or for sport, their meat is also used locally in one way or another. Considering that the cited examples comprise only a very small fraction of all hares, rabbits and rodents utilised worldwide it becomes evident that these small mammals, due to their great fertility and adaptability, are certainly as important for human nutrition as deer and antelopes.

Apart from the food aspect, exploitation of lagomorphs and rodents has in some countries major national economic importance, comparable with that of branches of livestock production. For example in Argentina the *hare* industry annually exports 10,000-14,000 t of meat worth millions of dollars, and provides a livelihood for thousands of people. At the same time Argentina is also the largest producer of *nutria* pelts. The value of all wildlife exports together equalled in 1977-79 that of canned beef (23). The former USSR produced more than 5 million *muskrat* pelts in 1960 and expected annual harvests to increase to 8.5 million by 1970 (5). Only 3.3 million pelts, however, were exported (49). In North America the total number of *muskrat* pelts taken in 1982/83 was 7.4 million with a value of Can.$ 32 mill. (approx. US$ 25.9 mill.), second only to that of raccoons. *Beaver* relised about Can.$ 13.7 mill. (approx. US$ 11.1 mill.) (53). However, in all cases the economic benefit has to be gauged in terms of the damage caused by the respective species and that which it might cause if it remained unexploited.

The above evaluation does not include the ***farming*** of the various species of rodents which is developing in South America and West Africa. Although this may in future become as important for human nutrition as the use of the domesticated guinea pigs (mainly in South America), it is still economically insignificant compared with the utilisation of wild rodent populations. The only rodent species which produce major economic returns from farming are *nutria* and *chinchilla*. Of the total number of marketed nutria pelts, about 38%, i.e. 2.5 million, originated in the 1980s from farming (35), and the trade is polarising into relatively cheaper furskins from wild or feral production and the more costly high-quality mutant skins from intensive selective breeding (25). Chinchilla production, from farms only, is much lower but the pelts are very valuable. In 1994 about 250,000 pelts were marketed, originating mainly from the USA and realising about US$ 5.5 mill..

The market which has developed since the 1960s for *live hares* due to the apparent decline of hares in Western Europe is economically quite interesting (54). Just as for game birds, hares are released by sport hunters to improve their sport. Considering the willingness of sport hunters to pay well for their hobby this market is likely to become as lucrative as breeding gamebirds or supplying live deer. As far back as 1967/68 Poland exported 171,000 live hares to France and Italy (20). Argentina started supplying live hares to Europe in 1981 as a lucrative sideline of the haremeat industry (32), and Australia is contemplating the same (46).

The breeding of semi-domesticated *laboratory rodents* has grown into a relatively important activity. In Germany, for example, about 700,000 rats and 1.3 million mice were utilised for research and scientific purposes in 1989, not including those killed for harvesting of organs or used for vaccine testing. These figures have been decreasing to about 500,000 and 973,000 respectively in 1993 (33). Much of this breeding, however, is carried out directly by the consuming scientific institutions and has thus limited commercial importance. Commercial breeders furnish special strains or varieties upon order, being strictly supervised by Government.

## Implications for conservation

Hares, rabbits and rodents are amongst the most productive mammals and most of them are adaptable to changing ecological conditions. Despite intensive exploitation over long periods only few species have been seriously depleted by man and apart from wild chinchilla, none of the commercially utilised species are subject to CITES trade restrictions.

Because of their fertility and adaptability hare, rabbit and rodent populations can multiply and expand so rapidly that they sometimes become a serious nuisance to agriculture, forestry or other human endeavours. Thus the need to control such populations often dominates and makes it difficult to manage them for sustainable yields like ungulates. In fact, very little of the present utilisation of lagomorphs or rodents is based on scientific knowledge and planned management, most of it is haphazard and empirical. This implies certain dangers for conservation as without knowledge of population dynamics, increases in market demands combined with ecological factors may result in over-exploitation and sudden depletion of a species. This occurred in the 19th century when beaver in North America and nutria and chinchilla in South America were depleted because of overtrapping for the European fur trade. On the Caribbean islands the hutias and zagoutis (Capromyidae), cavy-like rodents, have been almost exterminated by intensive use for food, compounded by the massive ecological changes in the course of agricultural development and the introduction of mongooses. In fact, as with most other animal groups, habitat change and in some regions climatic constraints are usually the most important population regulatory factors. They may cause decline, but also spectacular increases of rodent populations and their utilisation has to be always adjusted to them.

If population decline is predominantly caused by over-exploitation, protective measures are necessary and will result in rapid recovery of the population due to the high rate of reproduction of lagomorphs and rodents. This has been demonstrated with beaver in North America after they were near extinction at the end of the 19th century. Under human control they made a relatively rapid recovery and based on research are now successfully managed for sustained yield. This should serve as an example for the utilisation of other potentially harmful but economically valuable species. In many cases planned utilisation should replace empirical control and the financial returns from this will compensate or even exceed the cost of control.

Farming of rodents is not a valid alternative to the need for planned management of wild populations, although it may alleviate locally the hunting pressure on the particular species. It may nevertheless be socio-economically desirable and prevent the extirpation of a species that has become rare, like the chinchilla in South America.

## Literature references


1. AJAYI, S. (1974): *Domestication of the African Giant rat* (*Cricetomys gambianus* Waterhouse). University of Ibadan, Department of Forest Resources Management; p. 1-44
2. AJAYI, S., TEWE, O.O. and E.O. FATUROTI (1978): Behavourial changes in the African Giant rat under domestication. *East African Wildlife Journal*, 16, p. 137-143
3. AJAYI, S.S. (1979): *Utilisation of Forest Wildlife in West Africa*. FAO FO:MISC /79/26; Food and Agriculture Organisation of the United Nations, Rome; 75 pp.
4. ALIEV, F.F. (1965): Dispersal of nutria in the USSR. *Journal of Mammalogy*, 46, p. 106-102
5. ALIEV, F. (1969): Population structure of the muskrat, *Ondatra zibethicus* (Linné, 1766), and its numerical fluctuations in the USSR. *Säugetierkundliche Mitteilungen*, 17(2), p. 152-155
6. AMAYA, J.W. (1979): The European hare in Argentina. In: *Proceedings of the World Lagomorph Conference, Aug. 1979, Guelph, Ontario*, ed. MYERS, K.; p. 493-494; University of Guelph, Canada
7. ANGERMANN, R. (1972): Ordnung Hasentiere. In: *Grzimeks Tierleben*, Vol. 12, p. 440-443 and 453; Kindler, Zürich
8. ANONYMOUS (1991): *Microlivestock. Little-known Small Animals with a Promising Economic Future*. BOSTICK, National Research Council; Washington, DC; 447 pp.
9. ANONYMOUS (1993): *DJV-Handbuch 1993*: Jagd aktuell. Deutscher Jagdschutz-Verband e.V. Bonn, Germany
10 ASHBROOK, F.G. (1948): Nutria grow in the United States. *Journal of Wildlife Managment*, 12, p. 87-95
11. ASIBEY, E.O.A. (1974): Wildlife as a source of protein in Africa south of the Sahara. *Biological Conservation*, 6(1), p. 32-39
12. BAPTIST, R. and G.A. MENSAH (1986): Benin and West Africa: The Cane rat: farm animal for the future? *World Animal Review*, 60, p. 2-6
13. BARZDO, J., LUXMOORE, R., BROAD, S. and D. JONES (1984): *A survey of wildlife farming operations*. WCMC/IUCN, Cambridge (unpubl. report)
14. BIADI, F. (undated): L'elevage du gibier (*cited after* BARZDO and coll. 1984)
15. BOLDENKOV, S.V., KRAINEV, E.D. and V.A. GALAKA (1973): Common hare (*Lepus europaeus* Pall.) rearing in the forest of the Ukrainia agricultural economy. *Transactions of the Xth Congress of the International Union of Game Biologists*; Office National de la Chasse, Paris; p. 471-475
16. BRASS, E. (1925): *Aus dem Reich der Pelze*. Verlag Neue Pelzwaren- und Kürschner Zeitung, Berlin; p. 42
17. BUTYNSKI, T.M. (1973): Life History and Economic Value of the Springhare (*Pedetes capensis forster*) in Botswana. *Botswana Notes and Records*, Vol. 5, p. 209-213
18. DE VOS, A. (1978): Game as food. A report on its significance in Africa and Latin America. *Unasylva*, 29(116), p. 2-12
19. FALCONER, J. (1992): *Non-timber forest products in Southern Ghana. A Summary Report*. Overseas Development Administration, Forestry Series No. 2; London; 23 pp.
20. FRAGUGLIONE, D. (1973): Les problèmes posés par les entreprises de repeuplement avec le lièvre commun (*Lepus europaeus* Pallas 1778). In: *Transactions of the Xth Congress of the International Union of Game Biologists*; Office National de la Chasse, Paris; p 491-505
21. FREYE, H.-A. (1967): Ordnung Nagetiere. In: *Grzimeks Tierleben*, Vol. 11, p. 210; Kindler, Zürich
22. FREYE, H.-A. (1967): Andere Murmeltiere. In: *Grzimeks Tierleben*, Vol. 11, p. 231-233 and 238; Kindler, Zürich
23. FUJITA, H.O. and J.O. CALVO (1981): Las exportaciones de productos y sub-productos de la fauna silvestre en el quinquenio 1976-1980. *IDIA No. 397-400: Jan. - Apr. 1981*; p. 1-26; Instituto Nacional de Tecnologia Agropecuaria (INTA), Buenos Aires, Argentina

24. GONZALES-JIMENEZ, E. (1977): The capybara: an indigenous source of meat in tropical America. *World Animal Review*, 21, p. 24-30
25. GOSLING, L.M. and J.R. SKINNER (1984): 31. Coypu. In: *Evolution of domesticated animals*, ed. MASON, I.L.; p. 246-251; Longman Group, London and New York
26. GRAKOV, N.N. (1986): Game resources and their role in economical development. In: *Game resources of Siberia* (in Russ.), ed. JUDIN, B.B.; p. 5-11; Nauka, Novosibirsk, Russia
27. GRAU, J. (1984): 34. Chinchilla. In: *Evolution of domesticated animals*, ed. MASON, I. L.; p. 260-263; Longman Group, London and New York
28. GRZIMEK, B. (1972): Myxomatose der Kaninchen. In: *Grzimeks Tierleben*, Vol. 12, p. 451; Kindler, Zürich
29. HEINZ, G. (1981): Fleischexporte aus den wichtisgsten Produktionsländern Südamerikas. *Fleischwirtschaft*, 61(1), p. 14-22
30. HOFFMAN, M. (1958): *Die Bisamratte - ihre Lebensgewohnheiten, Verbreitung, Bekämpfung und wirtschaftliche Bedeutung*. Geest and Portig, Leipzig
31. IGUE, J.O. (1981): cited after GANMAYO, E. (1992): Elévage de l'Aulacode et sa contribution au développement du monde rural - cas du Sud Bénin. In: *Proceedings of the International Conference on Grass-cutter Production: Achievements and Prospects; Cotonou, Bénin*; p. 34; German Office for Technical Cooperation (GTZ), Eschborn, Germany
32. JACKSON, J.E. (1986): The Hare Trade in Argentina. *TRAFFIC Bulletin*, 7(5 ), p. 72
33. KÜSTERS, G. and R. GEURSEN (1993): Gesundheit für Mensch und Tier. Warum Tierversuche noch nötig sind. Piper, München und Zürich
34. JARMAN, M.R. (1972): European deer economics and the advent of the Neolithic. In: *Papers in economic prehistory*, ed. HIGGS, E.S.; p. 125-149; Cambridge University Press, Cambridge
35. KINLER, N.W., LINSCOMBE, G. and P.R. RAMSEY (1987): 27. Nutria. In: *Wild Furbearer Management and Conservation*, ed. NOVAK, M., BAKER, J.A., OBBARD, M.E. and B. MALLOCH; p. 326-343; Ministry of Natural Resources, Ontario, Canada
36. LEVER, C. (1985): Naturalised mammals of the world. Longman Group, London and New York
37. LUXMOORE, R. (ed.) (1992): *Directory of Crocodilian Farming Operations. Second Edition, Part 2*. WCMC/IUCN, Cambridge and Gland, Switzerland
38. McGEE, H.F. (1972): 2. The Use of Furbearers by Native North Americans After 1500. In: *Wild Furbearer Management and Conservation in North America*, ed. NOVAK, M., BAKER, J.A., OBBARD, M.E. and B. MALLOCH; p. 13-20; Ministry of Natural Resources, Ontario, Canada
39. MONK, C.E. (1981): History and present status of fur management in Ontario. In: *Proceedings of the Worldwide Furbearer Conference*, ed. CHAPMAN, J.A. and D. PURSLEY; p. 1501-1523; Frostburg, Md., USA
40. NACHTSHEIM, H. (1949): *Vom Wildtier zum Haustier*. Parey, Berlin
41. NOVAK, M. (1987): 25. Beaver. In: *Wild Furbearer Management and Conservation in North America*, ed. NOVAK, M., BAKER J.A., OBBARD, M.E. and B. MALLOCH; p. 282-313; Ministry of Natural Resources, Ontario, Canada
42. OJASTI, J. (1991): Human Exploitation of Capybara. In: *Neotropical Wildlife Use and Conservation*, ed. ROBINSON, J.G. and K.H. REDFORD; p. 236-256; University of Chicago Press, Chicago and London
43. PIECHOCKI, R. (1972): Familie Wühler. In. *Grzimeks Tierleben*. Vol. 11, p. 210, Kindler, Zürich
44. PIECHOCKI, R. (1972): Uberfamilie Bilchartige. In: *Grzimeks Tierleben*, Vol. 11, p. 384; Kindler, Zürich
45. POLETSKIY, V. (1986): Results of hunting during the last three years (in Russ.). *Okhota i okhotnichje khozjaistvo*, 10, p. 6

46. RAMSAY, B.J. (1994): *Commercial use of wild animals in Australia*. Bureau of Resource Sciences, Australian Government Publishing Service, Canberra, Australia
47. RAY, A.J. (1987): 3. The Fur Trade in North America: An Overview from a Historical Perspective. In: *Wild Furbearer Management and Conservation*, ed. NOVAK, M., BAKER, J.A., OBBARD, M.E. and B. MALLOCH; p. 21-30; Ministry of Natural Resources, Ontario, Canada
48. ROTH, H.H. and B. HOPPE (1996): Le marché de viande de brousse de Toumodi en Côte d'Ivoire (publication in preparation)
49. SAFONOV, V.G (1981): The status and establishment of fur resources in the USSR. In: *Proceedings of the Worldwide Furbearer Conference*, ed. CHAPMAN, J.A. and D. PURSLEY; p. 95-110; Frostburg, Md., USA
50. SAMKOV, J.A. and G.W. TRUBETZKOIJ (1974): Nutriazucht ausserhalb der USSR. *Deutcher Pelztierzüchter*, 48, p. 130-132
51. SHIRJAEV, V. (1989): Muskrat (in Russ.). *Okhota i okhotnichje khozjaistvo*, 13, p. 14-17
52. SMYTHE, N. (1991): Steps Towards Domesticating the Paca (*Agouti = Cuniculus paka*), and Prospects for the Future. In: *Neotropical Wildlife Use and Conservation*, ed. ROBINSON, J.G. and K.H. REDFORD; p. 202-216; University of Chicago Press, Chicago and London
53. STIEFF, A. and J.A. BAKER (1987): 56. Marketing and International Fur Markets. In: *Wild Furbearer Management and Conservation in North America*, ed. NOVAK, M., BAKER, J.A., OBBARD, M.E. and B. MALLOCH; p. 862-877; Ministry of Natural Resources, Ontario, Canada
54. STUTTARD, R.M. (1979): The hare as an object of sport. In: *Proceedings of the World Lagomorph Conference, Aug. 1979, Guelph, Ontario*, ed. MYERS, K.; p. 907-916; University of Guelph, Canada
55. TURTON, J.D. (1984): 36. Mongolian gerbil. In: *Evolution of domesticated animals*, ed. MASON, I.L.; p. 266-267; Longman Group, London and New York
56. WILLNER, G.R. (1984): 37. Muskrat. In: *Evolution of domesticated animals*, ed. MASON, I.L.; p. 268-272; Longman Group, London and New York
57. WRIGHT, J.V. (1987): 1. Archaeological Evidence for the Use of Furbearers in North America. In: *Wild Furbearer Management and Conservation in North America*, ed. NOVAK, M., BAKER, J.A., OBBARD, M.E. and B. MALLOCH; p. 3-12; Ministry of Natural Resources, Ontario, Canada

## 2.3.4 TERRESTRIAL CARNIVORES (*Fissipedia: CARNIVORA*)

Terrestrial carnivores comprise mustelids (**Mustelidae**), raccoons and allies (**Procyonidae**), bears (**Ursidae**), viverrids (**Viverridae**), Dog-like (**Canidae**) and Cat-like carnivores (**Felidae**). They are distributed in numerous species over all continents and include the majority of the so-called "furbearing" animals apart from some furproducing marsupial, rodent and seal species. Tab. 28 shows however, that terrestrial carnivores are not only utilised for fur production. The Gray wolf and Wild cat have been domesticated for companionship and the polecat for hunting. Some of the furbearing carnivores have been bred in captivity for more than a hundred years and are thus undergoing domestication. Civet cats produce musk and bears are also commercially utilised for their gall bladders as highly valued medicine in Asia.

### Historical background

Archaeological evidence in Europe and North America shows that some of the carnivorous furbearers, particularly *foxes*, were extensively hunted in prehistoric times, probably as much for food as for pelts (18, 42). However, compared with ungulates they were of rather limited nutritional importance. Neolithic man learnt to tan their skins and to utilise these for the making of warm clothing (12), and trade in furs developed from the early cultures of the 4th millennium BC onwards. Raw pelts were bartered by the hunters in Northern and Eastern Europe against luxury goods available from the more developed Mediterranean cultures. *Sable* pelts were traded as early as in 300 - 200 BC as "Skythian marten" via the Black Sea. In the 11th century trading of furskins from Western Siberia developed via the Novgorod merchants who sold them by the ship load to Norwegian vessels sailing up the northern Dwina (29). During the Middle Ages furskins from *squirrel, beaver, ermine, marten* and *sable* became one of the most important commodities in the trade of the Hanseatic League (12). In North America the arrival of the Europeans in the 16th and 17th centuries had a profound impact on the use of furbearers in that the subsistence hunting and trapping of the indigenous people was changed into a market dominated activity (31). The ever increasing European demand for furskins led to the establishment of Muskovy in Russia in 1555 and of the Hudson's Bay company in Canada in 1670. The fur trade, largely monopolised and constantly promoted by these powerful agencies, became a major force in the economic and political development of North America as well as Siberia. It served to stimulate continental exploration and provided financial incentives for early colonial ventures (31). By the 17th century furskins had become one of the most important export goods in North America and in the emerging Russian Empire, and the development of both depended to a large extent financially on the returns from this export (29). The Russian exploration and colonisation extended eastwards as far as Alaska where fur trapping interests clashed with those of the American companies until Alaska was sold to the USA in 1867 (27). Furskins were widely used as currency and some of the peoples in Siberia paid their tributes to the Russian czars in sable pelts (29). It was also customary for tribal chiefs and rulers to make presents in the form of valuable furskins. Czar Boris Godunov, for example, sent 40,360 sable pelts to the Austrian emperor to promote the joining of the two countries in the war against the Osmanian Empire (29).

**Table 28:** Overview of the utilisation of terrestrial carnivores

| English name | Scientific name | Original geographical distribution | Body weight (kg) | Utilisation |
|---|---|---|---|---|
| **MUSTELIDS - MUSTELIDAE** | | | | |
| Ermine or stoat | *Mustela erminea* | Europe, W.Siberia, N.America | 0.2-0.5 | trapped for fur |
| American mink | *-vision* | N.America[1)] | 0.7-1.8 | trapped and farmed for fur |
| Polecat | *-putorius* | Eurasia | 0.5-1.5 | domesticated into ferret; farmed as "fitch" |
| Asian weasels | -spp. | C.-E.Asia | 0.5-1.0 | trapped for fur (*M. sibirica* produces "Kolinski"); exp. farming |
| Martens | *Martes* spp. | Europe, W.Siberia, W.Asia, N.America | 0.6-1.8 | trapped for fur (*M. americana* prod. "Can. sable"); *M.pennanti* called "fisher" |
| Sable | *-zibellina* | C. and E.Asia | 0.6-1.8 | trapped and farmed for fur |
| European badger | *Meles meles* | Eurasia | 10-20 | trapped for fat used in medicine |
| American badger | *Taxidea taxus* | N.America | 6-14 | trapped for fur; formerly farmed |
| Ferret badger | *Melogale moschata* | E.Asia | 1-1.8 | trapped for musk; exp. breeding for fur |
| Striped skunk | *Mephitis mephitis* | N.America | 1.8-4.5 | trapped as pest and for fur; formerly farmed |
| Zorrino | *Conepatus humboldti* | South. S.America | 1.5-3 | trapped for fur |
| Wolverine | *Gulo gulo* | North. Eurasia and N.America | 10-28 | hunted for fur; exp. farm. |
| River otter | *Lutra canadensis* | N.America | 4.5-15 | trapped for fur; exp. farm. |
| **RACCOONS and allies - PROCYONIDAE** | | | | |
| Raccoon | *Proycon lotor* | N.-C.America[2)] | 5-8 | trapped as pest; formerly farmed for fur |
| **BEARS - URSIDAE** | | | | |
| Black bear | *Ursus americanus* | N.America | 80-200 | hunted for sport |
| Brown bear | *-arctos* | N.America, Eurasia | 120-350 t | hunted for sport |
| Polar bear | *-maritimus* | Arctic circumpolar | 180-500 | hunted for subsistence and sport |
| **VIVERRIDS - VIVERRIDAE** | | | | |
| Mongooses | *Herpestes* spp. | S.Europe, Africa, S.Asia[3)] | < 1 | semi-domesticated for control of mice, rats, snakes |
| Civet cat | *Viverra civetta* | Africa | 12-18 | farmed for musk |

| DOG-LIKE CARNIVORES - CANIDAE | | | | |
|---|---|---|---|---|
| Gray wolf | *Canis lupus* | N.Eurasia, N.America | 25-40 | domesticated into dog; farmed for fur |
| Coyote | *-latrans* | N.America | 9-16 | hunted as pest; farmed |
| Red fox | *Vulpes vulpes* | Eurasia, N.America, Australia | 3.5-7 | farmed in several colour phases for fur; hunted as pest |
| Arctic fox | *Alopex lagopus* | Arctic regions circumpolar | 2.5-5 | hunted for fur; farmed or ranched as "Blue fox" |
| S.Americ. foxes | *Dusicyon* spp. | S.America | 4-10 | hunted and trapped for fur |
| Raccoon dog | *Nyctereutes procyonoides* | E.Asia[4)] | 4-7 | farmed for fur |
| Gray fox | *Urocyon cinereoargenteus* | N.America | 3-6.5 | hunted and trapped for fur |
| **CAT-LIKE CARNIVORES - FELIDAE** | | | | |
| Lion and | *Panthera leo* | Africa s.o. Sahara | 160-200 | hunted for sport |
| leopard | *-pardus* | Africa, W.-C.Asia | 120-180 | hunted for sport |
| Mountain lion | *-concolor* | N.-S.America | 35-67 | hunted for sport |
| Canad. lynx and bobcat | *Lynx canadensis, -rufus.* | N.America | 7-13 | trapped for fur; experimentally farmed |
| Cheetah | *Acinonyx jubatus* | Africa, S.W.Asia | 30-60 | tamed for hunting |

1) introduced into Sweden, Norway, Finland and former USSR (established in European part of USSR, Far East and Altai mountains); 2) introduced into former USSR (established in western Russia, Tian Shan Region and Trans-Caucasia); small feral population in West Europe; 3) introduced onto several islands (see text); 4) introduced into former USSR and Europe (see text)

As of the 19th century the use of furskins in Europe became more and more sophisticated as described in Chapt. 3.2.4. Furs became a fashion article and thus the whims of fashion started to dominate the fur trade. This led on the one hand to the over-exploitation of some furbearer species such as *nutria* in Argentina, *beaver* in North America, *sable* in Siberia and *Raccoon dogs* in China and Japan. On the other hand it stimulated the production of furskins from captive-held furbearers at the end of that century. The main goals of the fur farming ventures were to produce rare and highly priced colour phases, like the melanistic mutant of the *Red fox* and various different colour strains of the *mink*, and large quantities of furskins of the same quality. The first mink farm was established during the American Secession War 1861-65 by T. D. Phillips and W. Woodcock in the state of New York, and breeding of "Silver" foxes in captivity commenced in 1892 when Charles Dalton obtained offspring from a wild-caught couple of "silver" mutants of the Canadian (standard) type (11).

Farming of furbearers gained importance in North America well before the First World War; and in 1913/14 the first *Silver fox* breeding stock was exported to Norway. This was the beginning of the Scandinavian fur farming industry which later overtook the USA and Canada becoming the most important producer of fur pelts next to the former USSR. In the 1920s a strong "hausse" for furs of any type led to the farming of a wide variety of other species, such as the *Striped skunk, American badger, raccoon, martens* and *polecat*, and to the expansion of fur farming over most of Europe and to the former USSR. In the 1930s as many as 11,000 Silver fox farms existed. However, reduced demand and marginal profitability forced farming of most of the other species, except mink, out of business. Silver fox farming reached its peak in 1940 when 1.35 million pelts were sold on the international markets (11).

The only fur farming which has developed historically outside North America and Europe is the breeding of *Raccoon dogs*. These were managed and kept in captivity in the Far East since ancient times (40). At the beginning of the 20th century there were thousands of small state-supported Raccoon dog farms in operation in Japan, and farming was introduced to Russia in 1928. Concurrently Raccoon dogs were liberated by the thousands in Siberia, Kasakhstan, the Caucasus and Western Russia (17) from where feral populations gradually expanded their range into Eastern Europe, Scandinavia and Romania.

During and after the Second World War the demand for *Silver fox* pelts declined and *mink* farming, particularly in the Scandinavian countries, emerged as by far the most economically important industry. Fox farming regained some importance and together with the breeding of *Arctic foxes* now lies in second place. The farming of the *polecat*, called "fitch", was first undertaken in Great Britain in the 1920s, but has been promoted into economic importance only since 1979 in Finland. It did not find access to the USA and Canada because of import prohibition for fear of the establishment of feral populations. It was however introduced into the USSR where it has gained greatest importance. Only in 1969 was the farming of Raccoon dogs taken up again in Finland using feral animals from Russia, and it has since spread from there to Denmark, Holland, France and Poland.

Other uses of carnivores also have a long history. Tamed, *semi-domesticated polecats* named *ferrets*, were used for rabbit hunting by the Berbers in North-West Africa around 20 AD, and probably long before that by the Romans when they brought the rabbit to Italy. The semi-domesticated ferret, mostly used in the form of a red-eyed albino race, was distributed together with the rabbit, and ferreting in Britain was first mentioned in 1223 and in France in 1421 (28).

Tamed *mongooses* have been used since early times in India, Java and Egypt for ridding human habitations of snakes, rats and mice. In Egypt the *ichneumon (Herpestes ichneumon)* even became a sacred animal and large numbers were kept there during the Ptolemaic era in 300-200 BC, as evidenced by the numerous mummified mongooses found in tombs (24). From Egypt the semi-domesticated Ichneumon mongoose was also imported by the Romans into Italy where ladies began to keep them as fashionable pets. A mosaic in Pompeii impressively depicts a mongoose fighting a snake (43). The domestic cat eventually replaced the mongoose in man's efforts to control rats and mice, and with the development of poultry mongooses became more important as pest animals. This status was greatly reinforced by ill-considered introductions of *Spotted mongooses (H. auropunctatus)* into Jamaica in 1872, Fiji in 1873 and 1883, and in the latter year also to the Hawaiian islands (24). The *Indian Grey mongoose (H. edwardsi)* has been introduced into Mauritius and the Ryukyu islands. All these deliberate introductions were made in the vain hope of controlling rats in the sugar plantations of these islands, but have led to the extermination of a number of indigenous taxa and the introduction of the rabies virus.

Another carnivore tamed for hunting purposes is the *cheetah*. This is one of the swiftest mammalian predators and was extensively kept by the ancient Egyptians as early as 1500 BC and later also by the Assyrians, to hunt down gazelles and other fast antelopes. The cheetah gained greatest popularity in India during the time of the Mogul emperors, between 1500 and 1800. One of them, Akbar, is said to have kept 9,000 cheetahs during his reign (13). Since cheetahs were not bred but had to be replenished from the wild, it is easy to imagine how the wild resource of these large cats in Asia was depleted to its present-day endangered status.

A rather unique historical use of a large carnivore was the keeping of *Striped hyenas*, also by the ancient Egyptians during the Old Kingdom, a period in which almost any manageable wild species was kept either for religious or utility purposes. Relief pictures from around 2500 BC show clearly how hyenas were shepherded and how they were restrained for being stuffed with roasted goose or duck, obviously in preparation of being eaten by man (43).

The musk of *Civet cats* has been a valued substance since biblical times and was later particularly appreciated in the Arab cultures. Although some civet musk also originated from the *Asian Civet cats (Viverra zibethica, -tangalunga* and *Viverricula indica)* and from the *Ferret badger (Melogale moschata)* the main centre of production has always been Ethiopia where the African species (*V. civetta*) is used. In the Middle Ages efforts were also made in European countries to keep Civet cats for producing musk (21), but were abandoned in favour of importing it through Nubia from Ethiopia. Later the trade of civet became a monopoly of the Franco-Abyssinian Railways, exports being routed through Djibouti. Attempts to establish Civet cat farming in German East Africa and French West Africa before the First World War failed.

The medical use of parts of *large cats* in Asia, particularly the *tiger*, is rooted deeply in Oriental traditions and myths which can be traced back to the Ch'in (221 BC) and Han dynasties (20 BC). Until today, the tiger symbolises strength and courage, and consumption of its flesh or parts is believed to confer these qualities to the consumer or patient (36). Similar to other providers of Oriental medicines, like rhinoceros, this use is seriously threatening not only *tiger (Panthera tigris)* and *Snow leopard (Uncia uncia)*, but also the *Asian bear species* (among others *Ursus thibetanus)*.

## Management and utilisation

Carnivorous mammals are predominantly utilised for the production of furskins and only to a much lesser extent for other purposes. Worldwide they are therefore less important in **subsistence hunting** than other animal groups which mainly provide food. In tropical regions *viverrids* and some *mustelids* are occasionally included in the diet of aboriginal or rural societies, and furskins, teeth and claws of some species, such as *Spotted cats* and *lions*, may be used for decorative purposes and as status symbols. In the cold regions, however, subsistence societies depend as much on furs from wild animals for clothing as on their meat for nutrition. There, trapping and hunting of furbearing carnivores, mainly *mink,* *martens, wolverines, raccoons, bears, wolves* and *foxes,* had and still has great importance in the local subsistence economies.

With the encroachment of market economies into the remotest areas of North America and Siberia, subsistence hunting of wild furbearers by the local peoples gradually turned into **commercial hunting and trapping**, furskins being taken for bartering or for sale to roving traders. The high value of pelts in the late 19th and early 20th centuries also enticed many non-indigenous adventurers and settlers to penetrate far into the Arctic north in search of furbearers. Commercial hunting and trapping of these animals has thus become a way of life in much of the northern regions and has developed there into a professional activity which persists widely today (39). Most trappers in North America are organised in professional associations which represent them at the local administrative level and are sometimes also concerned with the marketing of the harvested raw pelts (14). In the former USSR and now Russian Federation, hunters are members of State co-operative game management units or hunting societies to which specific hunting grounds are assigned. These co-operatives purvey the greater part of the harvested animals to the government. Game management is in general a strictly planned economic activity in which the hunters, through their management unit or society, participate with the Board for Game Management and Reserves of the local governments (35).

In North America the role of the trappers is gradually changing to one of being primarily managers of furbearing animals on State, Tribal or Private lands, acting as go-betweens for the public, the resource users, and local administrations. The co-operation of the trappers and private landowners with the local government agencies presupposes that the economic benefit from the harvesting of the furbearers accrues fully to them (26). The Provincial or State Governments assume thereby the responsibility for training the trappers, either directly or through their Associations, and for the education of the public. Furthermore they are providing the necessary control and the tools with which the trappers can work (26). This includes also the necessary research into the management of the different furbearers and the collection, analysis and dissemination of data.

Most of the *mustelid* furbearers are trapped and only the *larger carnivores* are hunted with guns and dogs. Most trappers have "trap lines" assigned to them which - in Canada - must be registered with the Administration. A great deal of research has gone into the most efficient, species-adapted, and humane trapping methods, but nevertheless the humaneness of trapping in general remains a major public controversy (7). Trapping is a strictly seasonal activity as the pelts must be harvested at the beginning of the winter when these are densest and of the best quality. The trapped animals are pelted locally and individual batches of raw furskins delivered to local collection points for grading. At these stations the trappers are usually also furnished with financial credit for supplying themselves with all necessary traps, ammunition and provisions for the whole season in advance. Management criteria and the regulation of harvesting have been comprehensively elaborated upon by NOVAK and coll. (1987) for all of the North American furbearer species on the basis of the available scientific knowledge and need not be further discussed here.

Apart from the trapping with the primary goal of furskin production, large numbers of raw pelts derive from the necessary **control of predators and nuisance species**, such as *rac-*

*coons, skunks, Red foxes*, and in more remote areas *coyotes, wolves* and *bears*; in Southern Africa *jackals* may also become pest animals in sheep farming areas. Fur production from these species is often only a secondary management goal and subject to quite different conditions. The first mentioned are all most successful species which adapt to many different habitats and penetrate into peri-urban and even urban areas (1). They raid refuse deposits and garbage containers, fruit gardens, pet food bowls and poultry houses. Raccoons invade attics, chimneys, garages and sheds where they may even den. All three species are considered to be carriers of the rabies virus and the control of this dangerous disease often involves their large-scale elimination either by shooting, poisoning or gassing of their dens. In North America responsibility for control rests with the wildlife or municipal authorities who in some jurisdictions employ specialist personnel for this task (1). In European countries fox control is routinely carried out by the respective "revier" hunters and in the case of rabies control in collaboration with the Veterinary authorities. In Australia, the most important exporter of Red fox pelts, it is largely commercialised (30).

Control of the larger carnivores, such as *Black* and *Brown bears*, and in some areas the *large cats*, like *Mountain lion, jaguar, leopard* and *lion*, is usually linked with **sports hunting**. This is based on license systems with hunting quotas established by the respective Wildlife or Forestry authorities. The pelts of these animals are utilised by the sports hunter as rugs or trophies. Hunting of *Polar bears* is strictly limited to the subsistence needs of the Inuit. Only in the North-West Territories of Canada are limited numbers sold to non-indigenous sports hunters, with returns going to the local Inuit communities (22).

As described above **farming of captive furbearers** has from an economic point of view become the dominant form of utilisation of terrestrial carnivores. The species most widely farmed at present are, in order of their importance, *American mink, Red fox*, mainly its color mutants, *Arctic fox, polecat (or fitch)* and *Raccoon dog*. Although mink and foxes have been propagated in captivity for more than hundred years in almost complete separation from the gene pool of the wild populations, they have not yet become truly domesticated (11). Their reproductive biology and behaviour have changed little and harvesting of pelts in farms therefore is as seasonal as the trapping of the wild furbearers; i.e. synchronised to the species natural reproductive and moulting cycles. This is mainly due to the fact that the overriding criteria for the selection of breeding stock was colour and quality of the furskins. In fact, over the decades a great deal of genetic knowledge has been gained concerning the heredity of these traits, enabling breeders to produce a large variety of different colour phases. Some selection was also done for high fertility and reproductive stability as these determine the productivity of fur farming. Good average litter sizes obtained with mink are 5 cubs and with foxes 5.5 (11, 33). Since there is a correlation between behaviour and fertility some behavioural selection has occurred, but only unconsciously and at too low an intensity to bring about greater docility in farm foxes (11). Farmed mink have generally become less shy and aggressive towards man although heavy gloves are always required for handling them (33).

Farmed furbearers are kept solitarily in cages above ground for hygienic reasons. Males are kept separately and mate on average 5-6 females only during the rutting season. Mink in particular require high levels of protein for nutrition which is provided in the form of fishery or poultry by-products, offals, and more recently dry pelleted rations based on plant

Inside of a typical mink farm in Denmark; each wire netting cage has a wooden nesting box attached for mother minks to give birth and to attend to their cubs, usually in litters of five. (Photo: Saga Furs of Scandinavia)

products and fortified with vitamins and minerals. Raccoon dogs are mainly vegetarian. Therefore, fur farming has developed near seas or large inland waters along the north-western coastlines and around the great lakes of the USA and Canada, in the Scandinavian and Baltic countries and Holland.

The only species which is also **ranched**, is the *Arctic fox*. In Yamal, Taimyr and Yakutia in the former USSR more than 600 permanent feeding stations have been used for more than 30 years to facilitate trapping and generally to increase the productivity of the trapping grounds from which the foxes would otherwise seasonally migrate away. Migrating foxes may also be taken alive and kept in cages until their fur is in prime condition (8).

The use of the semi-domesticated form of the *polecat* for rabbit hunting referred to as **ferreting**, diminished greatly when the *Myxoma* virus was introduced in rabbit control in the 1950s. It has become a mere sport and breeding of ferrets has dropped to a very low level. In Europe usually the red-eyed albino mutant form is preferred to the dark coloured natural phenotype as, in many habitats, these are seen and followed more easily by the hunter. Ferreting techniques vary: in Europe the ferret is usually muzzled before being sent into the rabbit's hole which is covered with a net. When the rabbit tries to escape from the ferret it bolts into the net where it is killed by the hunter. The *albino ferret* is also bred and kept in considerable numbers for physiological research. In the laboratory colonies ferrets are selectively bred for docility to ease handling and for maintenance of the colonies (28). In New Zealand where the ferret was introduced in large numbers it has become feral and now provides breeding stock for fitch farming (9).

**Civet cat farming** is limited to Ethiopia for the production of civet musk, the secretion of the species perineal glands and a highly valued substance in perfumery. The *Civet cats* are traditionally kept by farmers singly in wooden cages the design of which has not changed since the 19th century (41). Cages are placed in a row on trestles in darkened sheds and hygiene is usually very poor. A farmer may keep up to 10-15 animals which are fed with fruit, vegetables, maize porridge and occasionally meat and eggs. Formerly there were also bigger commercial farms keeping up to 300 animals under somewhat better conditions. Only males caught from the wild are used and no breeding has been attempted so-far. The musk is manually extracted from the restrained animal; production and marketing of civet musk are described in Chapt. 3.2.8.

## Economic importance

In contrast to most other animal groups the greatest economic importance of the terrestrial carnivores lies in the use of their skins. Only a few of the larger species, notably lion and leopard in Africa, Mountain lion in America and bears in America and Eurasia, are also important in commercialised sports hunting; the food value which is so dominant in rodents and ungulates, is insignificant and limited to some aboriginal societies. Other uses, such as the taming for hunting purposes, farming for musk or commercialisation of the organs of some carnivores in Oriental medicine, have only hobby or local importance, although they may be very wasteful.

In Tab. 29 an attempt is made to quantify the use of terrestrial carnivores for furskin production. The utilised hunting and trapping statistics have the usual deficiency of being incomplete and do not necessarily represent the numbers of pelts commercialised. Particularly in the former USSR they may comprise only minimum figures because only those pelts which were sold to government agencies may be included. Further, they do not reflect the considerable socio-economic importance of furskins obtained in subsistence hunting or from local predator control. All species which were formerly extensively trapped and hunted explicitly for the fur trade but which have now been placed on Appendix I or II of CITES, as described in Chapt. 3.2.4, have been omitted with the exception of the well-managed North American lynxes. The farming statistics usually give only the number of pelts which reach international markets. These may be lower than total production, as for example, in the former USSR where in 1984 only about 30% of the total production of mink farms was offered on western markets (34). In both forms of production there are considerable annual fluctuations mainly due to changing prices as a result of fashion demands or long-term ecological and economic trends. Although more recent farming statistics are available, figures of the 1980s have been used to make them more comparable to the latest published comprehensive trapping statistics.

Despite these shortcomings it can be clearly seen that the farming of captive-held stocks of furbearers has in most species a disproportionately greater significance for fur production than the harvesting of wild populations. Only about 1-2% of all mink pelts produced in 1982/84 originated from wild animals. Of the total of Arctic fox pelts only some 3% were obtained from wild populations. Only in a few furbearer species primarily managed and utilised for fur production, are the majority of traded pelts derived from the wild. Despite

**Table 29:** Pelt production from carnivorous furbearers[1)]

| Species | Region | Annual production (in thousands) by hunting (period) | by farming (period) | Literature references | |
|---|---|---|---|---|---|
| Ermine | form. USSR | 127.0 (1985) | - | (10) | |
| Mink | N. America | 430.6 (1982/83) | 6,200.0 (1984) | (34) | (4) |
| | Fennoscandia[2)] | 44.6 (1988/89) | 13,495.0 (1984) | (5) | (4) |
| | European Union | - | 3,310.0 (1984) | | (4) |
| | form. USSR | 50-75 (1960-80)* | 11,405.0 (1981) | (16) | (3) |
| | others[3)] | - | 3,240.0 (1984) | | (4) |
| Polecat (fitch) | N. America | - | 75.0 (1983/84) | | (9) |
| | Europe | insignificant | 377.5 (1983/84) | | (9) |
| | form. USSR | ? | 600.0 (1993) | | (32) |
| | China | ? | 5.0 (1983/84) | | (9) |
| Martens | N. America | 178.5 (1983/83) | - | (34) | |
| | Europe[4)] | 114.1 (1991) | insignificant | (5) | (9) |
| | form. USSR | ~78.0 (1950-80) | - | (10) | |
| Fisher | N. America | 25.1 (1982/83) | - | (34) | |
| Sable | form. USSR | 232.3 (1985) | 23.8 (1981/85) | (10) | (3) |
| Badgers | N. America | 43.5 (1982/83) | - | (34) | |
| | Europe[4)] | 62.4 (1988/89) | - | (5) | |
| Skunks | N. America | 279.5 (1982/83) | insignificant | (34) | (9) |
| Wolverine | N. America | 1.8 (1982/83) | - | (34) | |
| | form. USSR | ~1.8 (1960-80)* | experimental | (16) | (9) |
| River otter | N. America | 34.5 (1982/83) | - | (34) | |
| | form. USSR | ~7.5 (1960-80)* | experimental | (16) | (9) |
| Raccoon | N. America | 4,917.1 (1982/83) | insignificant | (34) | (9) |
| Gray wolf | N. America | 4.1 (1982/83) | insignificant | (34) | (9) |
| | form. USSR | 22-25 (1980-90)* | - | (20) | |
| Coyote | N. America | 612.5 (1982/83) | insignificant | (34) | (9) |
| Red fox (incl. Silver fox) | N. America | 534.4 (1982/83) | 52.0 (1984) | (34) | (9) |
| | Europe[4)] | 474.0 (1988/89) | 194.2 (1984) | (5) | (9) |
| | form. USSR | 487.3 (1924-59)* | 371.0 (1981/85) | (16) | (3) |
| | Australia | 351.0 (1982/83)[5)] | - | (30) | |
| Arctic fox | N. America | 15.7 (1982/83) | 2.0 (1984) | (34) | (9) |
| | Europe | insignificant | 2,915.0 (1984) | | (9) |
| | form. USSR | 108.0 (1985) | 1,168.2 (1985) | (25) | (37) |
| Raccoon dog | Japan | 35.0 (1980s) | insignificant | (40) | (40) |
| | form. USSR | n.a. | 2.0 (1984) | | (9) |
| | Europe[6)] | insignificant | 108.9 (1982/84) | | (9) |
| Gray fox | N. America | 315.5 (1982/83) | - | (34) | |
| Lynxes | N. America | 116.7 (1982/83) | insignificant | (34) | |
| Feral cat | Australia | 1.9 (1988/89) | - | (30) | |
| **Totals (in thousands)** | **Mustelids** | **>1,733.7** | **38,731.3** | *verified by BASKIN (1994) in litt. | |
| | **Procyonids** | **4,917.1** | **insignificant** | | |
| | **Canids** | **>2,951.0** | **4,813.3** | | |
| | **Felids** | **118.6** | **insignificant** | | |

1) excluded are species which have been placed on CITES Appendix I and some tropical species the pelts of which are occasionally commercialised, like Civet cats, jackals, etc.; 2) including Denmark; 3) mainly China, Japan and S.Korea; 4) Germany, Denmark, Finland, Austria, Switzerland, Sweden, Slovenia only; 5) refers only to exports; 6) mainly Finland, also Denmark, Holland and Poland

good progress in farming of sable, about 90% of their marketed pelts still originate from wild populations (3, 10, 32). Virtually all ermine, marten, fisher, skunk and lynx fur pelts are also derived from hunting because the farming of these species has proven to be either difficult or uneconomical. Other species of which pelts are commercialised, may be managed as predator or nuisance animals and therefore their pelts may be only a by-product, although this might be of great economic importance. This applies particularly to raccoons, coyotes, wolves and Red fox. The farming of the latter species only became economically important by breeding its rare melanistic mutant, the "Silver fox". Most of the hunted fox pelts, far exceeding in number those of farm-produced pelts, are from the Red fox which in many areas needs to be controlled. Only a small percentage of Red fox pelts are produced in farms and the great majority of traded fox pelts derive from the captive-bred Silver foxes. Altogether, about 50% of furs used in the manufacture of garments in 1982 were produced in farms (34), and this percentage has increased since.

***Fur production from wild as well as farmed furbearers*** has major economic importance in the USA, Canada, the former USSR, and also in Scandinavia and Holland. Production in the ***USA*** of wild furs from 1976 to 1983/84, averaged annually 15.9 million pelts and the total harvest in 1982/83 had a value of about Can.$ 203.1 mill. (approx. US$ 164.7 mill.) In ***Canada***, from 1964 to 1984, 3.5 million pelts were marketed annually and the total harvest in 1982/83 had a value of Can.$ 46.45 mill. (approx. US$ 37.67 mill.) (34). In volume the leading species in the fur trade were in Canada mainly rodents (*muskrat, Red squirrel* and *beaver*) and in the USA *muskrat, raccoon* and *opossum.* In monetary terms however, the *lynxes* produced in Canada alone a value of Can.$ 8.14 mill. exceeding that of any of the mentioned rodent harvests. *Raccoon* pelts in the USA and Canada together valued Can.$ 90.67 mill., thus exceeding by far the value of muskrats; *Red foxes* realised Can.$ 23.38 mill. and were thus more valuable than the *opossum* harvest, followed by *coyote* pelts with Can.$ 19.95 mill. (34). Interestingly Australia is the world's biggest Red fox exporter, having realised in 1982/83 Austr.$ 8.15 mill. (approx. US$ 7.6 mill.) from this species alone (30).

Although fur farming was historically initiated and first developed in North America it was economically overtaken by the fur industries in Europe and the former USSR during the 1930s and 1940s. Canada, for example, was the leading country in *Silver fox* farming until 1938 when about 8,000 farms produced up to 241,000 pelts. However, when fashion started preferring mink, Silver fox farming declined to only 500 pelts in 1964; it picked up again in the 1980s when the Canadian industry had an average output of some 22,000 Silver fox pelts (9). *Mink* farming is now by far the most important fur industry in both the USA and Canada and produced about 20% of the internationally traded pelts in 1983 with an approx. value of US$ 160 mill. Since then this share has further declined to about 16% (4).

The ***former USSR, now Russia***, is the world's biggest fur producer, but has in the past exported only about 15% of its wild and farm-produced pelts (2). In 1985 a total of 900,898 wild furskins were auctioned through V/O Sojuzpushnina in Leningrad and Hudson's Bay Company in London; further unrecorded outlets were through Leipzig, then East Germany, and private negotiators (2).

Fur farming was systematically developed in the former USSR only after the Second World War with emphasis on *mink, Arctic* and *Red foxes*. According to various Russian sources state mink farms increased their production from about 2 million/annum during 1961/65 to about 11.4 million/annum in 1981/85, reaching a peak of some 14.8 million pelts in 1994 of which 87% came from the former large state farms and 13% from smaller farms (6, 32). Annual production of Arctic foxes increased from about 395,000 in 1961/65 to a peak of 1.61 million in 1990, falling to 1.4 million in 1994 (500,000 originating from the large and 900,000 from small individual farms) (6, 32). Fox farming involving both red and silver colour phases, grew from about 387,000/annum in the 1961/65 to 505,000 in 1990. Only during the last ten years has Russia also become the leader in the farming of *polecats* or *fitch*, having imported breeding stock from Finland in the 1980s, with a reported production of 600,000 pelts in 1994 (32). Apart from some efforts at *sable* farming in China, Russia has also retained and further developed its monopoly in the breeding of this uniquely valuable species; the Zveroprom farm reportedly produced 31,000-32,000 sable pelts in 1994 (32). Since the greater part of fur production is consumed internally it is difficult to quantify its monetary value comparatively.

The ***Scandinavian countries*** together ***with Holland*** are the world's most important producers of farm pelts. In 1983 they marketed a total of 13.81 million mink pelts representing 50% of the internationally traded pelts (4). These were valued at about US$ 350 mill. when overproduction in 1986-89 caused a crash in prices and subsequently a world-wide decline in production, as discussed in Chapt. 3.2.4. Denmark is not only the world leader in producing *mink* but also in farming *Arctic foxes*, followed by Norway and Poland. Finland is the main producer of *Silver fox* pelts and has since 1969 pioneered the farming of *Raccoon dogs*. Production of so-called "finnraccoons" reached a volume of 105,000 pelts in 1982/83 (9), but has since decreased. It is also the major producer in Europe of fitch, the pelts of farmed *polecats* (250,000 pelts in 1982/83) (9).

During the last 20 years ***China*** has also developed into a fur exporting country. While traditional utilisation of wild animals for meat, skins and medical purposes has continued to flourish, it has adopted similarly pragmatic policies with regard to the introduction and use of wild species as neighbouring USSR. Commercial hunting is well established in the northern, western and southwestern provinces, where apart from *larger felids* and *bears, wolves, foxes* and *mustelids* are taken for their pelts. In the northwestern province of Chinghai, for example, 400,000 pelts were harvested from 1970 to 1972 (15). *Siberian weasels* are being processed in China into fur plates for export to Europe as "kolinski". Even in 1959 seventy fur farms were in operation breeding *mink, sable, martens, ermines*, and *badgers*. Mink farming has since developed into an export industry which marketed 5.0 million pelts in the peak year 1988, before numbers dropped to 0.5 million in 1992 (4).

Compared with wild furskins, farm-produced pelts are more consistent in quality and colour which is the most important factor in trade, allowing marketing of large batches of the same fur type and quality. Mink and fox pelts are produced in a great variety of colours which the wild harvest cannot provide. Therefore farm-produced pelts generally fetch higher prices than those of their wild counterparts. The development of fur farming, amongst other reasons, has led to a 50% decrease in wild fur production in Canada and a 36% decline in the USA since 1979/80 (34). As mentioned above farm production of mink had overtaken

Silver fox farming in economic importance by the early 1940s. At a level of about 3 million pelts/annum mink farming in North America competed favorably with wild fur production throughout the 1970s; but with annual farm pelt outputs increasing to more than 5 million in the early 1980s farming became much more important than wild fur production.

The ***economics of fur farming***, much like other wildlife farming, are generally dependent on cheap food resources, high reproductive rates and optimum numbers of breeding stock. However, in the case of fur farming rapidly changing market demands determine profitability. The history of fur farming shows how the farming of various species was either abandoned altogether, for example skunks and badgers in North America in the 1930s, or declined as in the case of Raccoon dogs in the former USSR in the 1930s, and Silver foxes world-wide in the 1970s, in response to fashion and market demands. The more recent development of finnraccoon and fitch farming show, however, that a species may also come into fashion again.

Fig. 11 shows that during the last decade the number of fur farms has generally decreased all over the world by more than 50%. This was mainly due to the economic reasons mentioned above: smaller farms were given up in favour of more productive large-scale operations, but the demand particularly for high-quality fur garments decreased as a result of the subjective campaigning of animal welfare groups against the fur industry, as discussed in Chapter 3.2.4. In 1993 the demand for fur apparel has revived again and has brought about price increases of 50% for quality pelts. This is likely to stimulate the regrowth of fur farming.

**Figure 11:** Development of fur farming[1)]

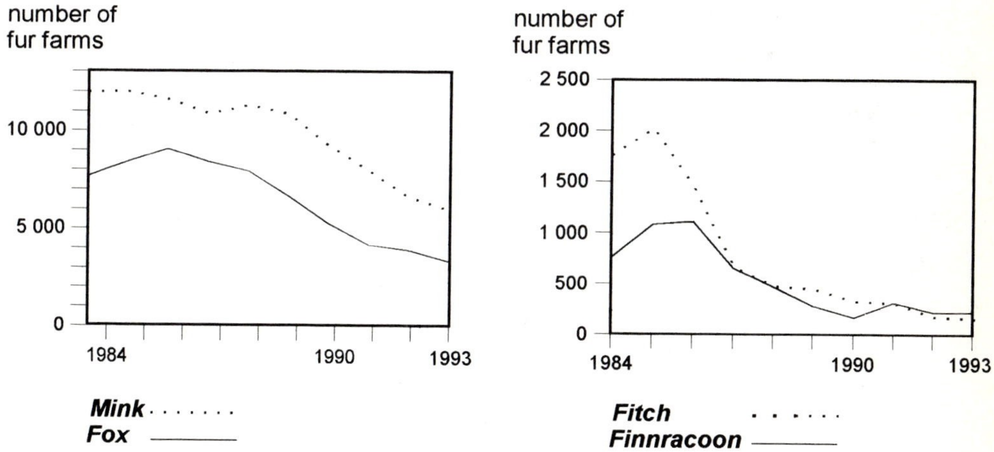


1) excluding the former USSR

Source: Oslo Fur Auctions (ANON. (4))

The larger ursid and felid carnivores have considerable ***importance in sports hunting***, although their pelts may also be marketed as rugs and command high prices. Such use presupposes efficient management in order not to become detrimental. This exists in the case of *Black* and *Brown bears* and *Mountain lions* in North America, and also for *lion* and *leopard* in some African countries. Studies in the USA and Canada (23, 19) have shown that the revenue from licence sales for hunting of bears alone can be substantial. If one considers the money spent by the hunters for equipment, lodging and food, the damage which bears cause may well be more than off-set. In Pennsylvania, for example, 250,000 hunters participated in 1979 in the harvest of 340 Black bears, each paying licence fees which were US$ 13.00 per hunter in 1983. In British Columbia the total estimated annual expenditure in 1981 of Brown bear hunters, 1,416 resident and 507 non-resident, was US$ 3.7 mill. of which US$ 1.3 mill. was spent locally. Brown bears are also offered for hunting in some Eastern European countries and in Russia where a tourist will pay up to US$ 2,000 for the trophy fee alone. Lion and leopard rank amongst the most demanded species in African safari hunting, valued for example in Zimbabwe at US$ 600-700 each in terms of just trophy fees. Purely commercial hunting of *Asian bears, tiger* and *Snow leopard* is based on the very high prices paid for their organs, used in Oriental medicine. Gall bladders and fat of bears, bones, claws, blood and flesh of tigers are much more valuable than their furskins, having a mystical importance to Chinese people (15). Similarly to the rhinoceros, this demand resulting in indiscriminate killing is seriously threatening the survival of the tiger all over Asia and the Snow leopard in China, and has started to affect even bear conservation in Canada.

The use of *ferrets* for rabbit hunting never had economic importance. Since myxomatosis has greatly reduced rabbits in most countries ferreting is merely a sport (28).

The use of *Civet cats* for ***production of musk***, which is still in high demand in the perfume industry, is economically important only in Ethiopia as described in Chapt. 3.2.8. The total number of farms was estimated in the 1980s at more than 200 with a total stock of 3,000-4,000 animals (41). It is not known how the most recent political upheavals in Ethiopia have affected civet farming.

## Implications for conservation

The commercial exploitation of carnivore species for furskins during the 19th century has reduced wild populations of some of these to low levels. However, it has never caused the extinction of any species because of the great degree of adaptability and the high reproductive efficiency of the mustelids and canids which are the main targets of the fur trade. The sable was near extermination at the turn of the century when its commercial exploitation was halted; but only 20 years later it had once again multiplied under human control into exploitable populations, constituting one of Russia's most valuable fur resources. Apart from the economic benefits from areas which are either marginal or unsuitable to agriculture, exploitation of wild furbearers has also engendered extensive research into most of the more important furbearer species and this enables us now to manage and to utilise them at sustainable levels. However, it is not only necessary to understand the biology and productivity of the different furbearers but also the ecological reasons for short or long-term population fluctuations of these to which harvesting must be adjusted. For example, martens in the USSR declined severely in the 1930s as a result of clearcutting of pristine forests. It took considerable time for them to adapt to replanted young forest before they could again be harvested (10).

More vulnerable than most mustelids are the felid carnivores as has been shown by the rapid decline of the Neotropical and Asian Spotted cats (*Felis* spp.) as a result of suddenly increasing fur fashion demands in the 1960s. Subjecting these and other vulnerable or endangered furbearers, such as local populations of the Asian ermines, weasels and martens (*Mustela altaica, -kathiah, -sibirica* and *Martes flavigula*), the Patagonian skunk or zorrino (*Conepatus humboldti*), River (*Lutra* spp.) and Sea otters (*Enhydra lutris*), the South American foxes (*Dasycyon* spp.) and lynxes (*Lynx* spp.) to trade restriction under CITES can only be seen as a welcome but temporary measure until knowledge and conditions are developed which will allow sustainable use of this resource to the full benefit of the local economies. The long-term conservation of these species, like so many others, depends more on the preservation of sufficient habitat than on protection against hunting. In the USA, into which the range of the neotropical cats once extended, these were extirpated not by hunting, but by the almost total conversion of the rich soils, which supported their habitat, into agricultural land like the Rio Grande valley (38).

It is important not to emotionalise the utilisation of furbearers unduly, although humaneness of fur trapping and hunting must be ensured. This bears the danger of jeopardising the legitimate rights of the producer nations and the people who live with and from the respective wildlife resources, which in turn will be counterproductive to their long-term conservation. It should be clearly understood that the vast majority of furs used for garments are produced from farmed animals, similar to the use of livestock, and that the greater part of wild furbearer pelts originate from nuisance animal species, like raccoons, Red foxes, and coyotes, whose numbers need to be necessarily controlled. Even the use of the relatively small proportion of other wild furbearers is today strictly controlled, fully legitimate and aids their conservation.

The large cats and bears, like other big game, will become less and less compatible with growing human populations and their agricultural activities. Given the special spiritual status which they enjoy not only with the local people but also in industrialised societies, their non-consumptive use for recreational purposes, be it in the form of viewing or well-controlled trophy hunting, seems to be the only way to generate economic returns which will off-set the damage they cause and give added justification for their conservation.

## Literature references

1. ALMEIDA, M.H. de (1987): 63. Nuisance Furbearer Control in Urban and Suburban Areas. In: *Wild Furbearer Management and Conservation in North America*, ed. NOVAK, M., BAKER, J.A., OBBARD, M.F. and B. MALLOCH; p. 996-1006; Ministry of Natural Resources, Ontario, Canada
2. ANONYMOUS (1985): Various notes from the *Fur Trade Journal* and *Fur Age Weekly*; *cited after* SHIEFF and BAKER (1987)

3. ANONYMOUS (1987): Advertising book for the farm-management of furbearing animals (in Russ.) (*cited after* BASKIN (1994): in litt.)
4. ANONYMOUS (1993a): World production for offering at the "western" markets. *Oslo Fur Auctions, official statistics*; Oslo, Norway
5. ANONYMOUS (1993b): *DJV-Handbuch 1993: Jagd aktuell*. Deutscher Jagdschutz-Verband e.V., Bonn
6. ANONYMOUS (1994): New Russia. Statistics. Moscow (*cited after* BASKIN (1994): in litt.)
7. BAKER, J.A. and P.M. DWYER (1987): 62. Techniques for Commercially Harvesting Furbearers. In: *Wild Furbearer Management and Conservation in North America*, ed. NOVAK, M., BAKER, J.A., OBBARD, M.F. and B. Malloch; p. 970-995; Ministry of Natural Resources, Ontario, Canada
8. BANNIKOV, A.G. (1970): Arctic fox in the USSR: biological premises of productivity. In: *Productivity and conservation in northern circumpolar lands*, ed. FULLER, W.A. and P.G. KEVAN; p. 121-127; IUCN, Morges, Switzerland
9. BARZDO J., LUXMOORE, R., BROAD, S. and D. JONES (1984): *A survey of wildlife farming operations*. WCMC/IUCN, Cambridge (unpubl. report)
10. BASKIN, L.M. (1990): Conflicts between wildlife and human activity in the forested zone of the USSR. *Transactions of the XIXth Congress of the International Union of Game Biologists*, ed. MYRBERGET, S.; p. 401-405; Norwegian Institute for Nature Research, Trondheim, Norway
11. BELYAEV, K.K. (1984): 23. Foxes. In: *Evolution of domesticated animals*, ed. MASON, I.L.; p. 211-214; Longman Group, London and New York
12. BRASS, E. (1925): *Aus dem Reich der Pelze*. Verlag Neue Pelzwaren- und Kürschner-Zeitung, Berlin
13. CLUTTON-BROCK, J. (1987): *A Natural History of Domesticated Mammals*. Cambridge University Press and British Museum (Natural History), Cambridge and London
14. GIROUX, A. (1987): 6. The Role of the Trapper Today. In: *Wild Furbearer Management and Conservation in North America*, ed. NOVAK, M., BAKER, J.A., OBBARD, M.F. and B. MALLOCH; p. 89-100; Ministry of Natural Resources, Ontario, Canada
15. GREER, E.C. and R.W. DOUGHTY (1976): Wildlife Utilisation in China. *Environmental Conservation*, 3(3), p. 200-208
16. HEPTNER, V.G., NAUMOV, N.P., JURGENSON, P.B., SLUDSKY, A.A., CHIRKOVA, A.F. and A.F. BANNIKOV (eds.) (1967): *Mammals of the Soviet Union* (in Russ.). Vol. 2; p. 380-383: Practical importance of red fox, and p. 491-507: Practical importance of mustelids; Vischaja schkola, Moscow
17. HEPTNER, V.G. and NAUMOV, N.P. (1974): *Die Säugetiere der Sowjetunion* (German translation of the Russian Edition 1976). Fischer, Jena, Germany
18. JARMAN, M.R. (1972): European deer economies and the advent of the Neolithic. In: *Papers in economic prehistory*, ed. HIGGS, E.S.; p. 83-97; Cambridge University Press, Cambridge
19. JONKEL, C.J. (1987): 36. Brown Bear. In: *Wild Furbearer Management and Conservation in North America*, ed. NOVAK, M., BAKER, J.A., OBBARD, M.F. and B. MALLOCH; p. 456-473; Ministry of Natural Resources, Ontario, Canada
20. KAPLIN, A.A. (1960): *Furs of the USSR* (in Russ.). Vneshtorgizdat, Moscow
21. KILLERMANN, S. (1931): Ein mittelalterlicher Bericht über die Zibethkatze. *Zeitschrift für Säugetierkunde*, 6, p. 235-236
22. KOLENOSKY, G.B. (1987): 37. Polar Bear. In: *Wild Furbearer Management and Conservation in North America*, ed. NOVAK, M., BAKER, J.A., OBBARD, M.F. and B. MALLOCH; p. 456-473; Ministry of Natural Resources, Ontario, Canada
23. KOLENOSKY, G.B. and S.M. STRATHEARN (1987): 35. Black Bear. In: *Wild Furbearer Management and Conservation in North America*, ed. NOVAK, M., BAKER, J.A., OBBARD, M.F. and B. MALLOCH; p. 442-455; Ministry of Natural Resources, Ontario, Canada

24. MASON, E. (1984): 28. Mongoose. In: *Evolution of domesticated animals*, ed. MASON, I.L.; p. 225-228; Longman Group, London and New York
25. NASIMOV, A.A. and J.A. ISAKOV (eds.) (1985): *Arctic Fox, red fox, wolverine* (in Russ.). Nauka, Moscow; p. 5-72
26. NOVAK, M. (1987): 10. The Future of Trapping. In: *Wild Furbearer Management and Conservation in North America*, ed. NOVAK, M., BAKER, J.A., OBBARD, M.F. and B. MALLOCH; p. 89-100; Ministry of Natural Resources, Ontario, Canada
27. OBBARD, M.E., JONES, J.G., NEWMAN, R., BOOTH, A., SATTERTHWAITE, A.J. and G. LINSCOMBE (1987): 64. Furbearer Harvests in North America. In: *Wild Furbearer Management and Conservation in North America*, ed. NOVAK, M., BAKER, J.A., OBBARD, M.F. and B. MALLOCH; p. 1007-1038; Ministry of Natural Resources, Ontario,Canada
28. OWEN, C. (1984): 26. Ferret. In: *Evolution of domesticated animals*, ed. MASON, I.L.; p. 225-228; Longman Group, London and New York
29. PAVLININ, W. (1966): *Der Zobel* (*Martes zibellina L.*). Neue Brehm-Bücherei No. 363, A. Ziemsen, Wittenberg-Lutherstadt, Germany
30. RAMSAY, B.J. (1994): *Commercial use of wild animals in Australia*. Bureau of Resource Sciences, Australian Government Publishing Service, Canberra, Australia
31. RAY, A.J. (1987): 3. The Fur Trade in North America: An Overview from a Historical Geographic Perspective. In: *Wild Furbearer Management and Conservation in North America*, ed. NOVAK, M., BAKER, J.A., OBBARD, M.F. and B. MALLOCH; p. 21-30; Ministry of Natural Resources, Ontario, Canada
32. ROSENBERG, I. (1994): *cited after* BASKIN (1994): in litt.
33. SHACKELFORD, R.M. (1984): 27. American mink. In: *Evolution of domesticated animals*, ed. MASON, I.L.; p. 229-233; Longman Group, London and New York
34. SHIEFF, A. and J.A. BAKER (1987): 56. Marketing and International Fur Markets. In: *Wild Furbearer Management and Conservation in North America*, ed. NOVAK, M., BAKER, J.A., OBBARD, M.F. and B. MALLOCH; p. 862-877; Ministry of Natural Resources, Ontario, Canada
35. SOKOLOV, V.E. and N.L. LEBEDEVA (1989): Commercial hunting in the Soviet Union. In: *Wildlife production systems*, ed. HUDSON, R.J., DREW, K.R. and L.M. BASKIN; p. 170-186; Cambridge University Press, Cambridge
36. SOWERBY, A.de (1933): The Tiger in China. *China Journal*, 18(2), p. 94-101 and 108-109
37. STOLBOV, S.G. (1991): Situation of furbearer production in farms (in Russ.). *Krolikovodstvo i zverovodstvo*, 2, p. 2-3.
38. TEWES, M.E. and D.J. SCHMIDLY (1987): 52. The Neotropical Felids: Jaguar, Ocelot, Margay, and Jaguarundi. In: *Wild Furbearer Management and Conservation in North America*, ed. NOVAK, M., BAKER, J.A., OBBARD, M.F. and B. MALLOCH; p. 696-716; Ministry of Natural Resources, Ontario, Canada
39. TODD, A.W. and E.K. BOGGES (1987): 7. Characteristics, Activities, Lifestyles, and Attitudes of Trappers in North America. In: *Wild Furbearer Management and Conservation in North America*, ed. NOVAK, M., BAKER, J.A., OBBARD, M.F. and B. MALLOCH; p. 59-76; Ministry of Natural Resources, Ontario, Canada
40. VALTONEN, M.H. (1984): 24. Raccoon dog. In: *Evolution of domesticated animals*, ed. MASON, I.L.; p. 215-217; Longman Group, London and New York
41. WOODFORD, J.D. (1990): *The status of wildlife in Ethiopia: Conservation and utilisation*. Agriconsulting, Rome (unpubl. report)
42. WRIGHT, J.V. (1987): 1. Archaeological Evidence for the Use of Furbearers in North America. In: *Wild Furbearer Management and Conservation in North America*, ed. NOVAK, M., BAKER, J.A., OBBARD, M.F. and B. MALLOCH; p. 3-12; Ministry of Natural Resources, Ontario, Canada
43. ZEUNER, F.E. (1963): A history of domesticated animals. Hutchinson, London

## 2.3.5 SEALS AND WALRUSES (*Pinnipedia: CARNIVORA*)

The pinnipeds comprise 3 distinct families: the Eared seals (**Otariidae**), the walruses (**Odobenidae**) and the True seals (**Phocidae**). The Eared seals (Fur seals and Sea lions), are distributed over the Southern Hemisphere and in the Pacific Ocean where they range as far north as the Bering Sea. Walruses are restricted to the most northerly circumpolar seas. Of the True seals the subfamily **Phocinae** is found only in northern seas, but the subfamily **Monachinae** is found in both hemispheres. The fur and subcutaneous blubber that pinnipeds use to retain their body heat in the much cooler sea water have both proved particularly attractive commodities to human hunters. Tab. 30 shows the commercial utilisation to which the different species of seal have been subjected mainly in the past.

### Historical background

Pinnipeds have been an attractive prey for human hunters wherever these animals were available to them. The earliest records of **subsistence hunting** of seals come from Northwestern Europe during the Stone Age (7). The area around the present-day Baltic Sea (*s. l.*) has yielded abundant remains associated with seal hunting. The tiny island of Hesselø in the Kattegat, for example, has been shown to have been a sealing station in the Stone Age (16). Hunting methods analogous to those used in the Stone Age existed almost up to the present in Eskimo or Inuit communities. Seals were particularly important to groups like the Netsilik of Northern Canada. A highly developed technology involving sophisticated harpoons and skin boats was developed by them to take seals in the water (4). Seals were also hunted in their breeding lairs in the spring, using dogs to scent out the occupied dens (21). All parts of the seal were utilised, but nothing was more important than the blubber. This could be eaten directly as it was, to provide an immediate energy source, but its most important function was as fuel for lamps. Eskimos in stone huts or snow houses could scarcely have survived northern winters without the heat and light provided by blubber lamps. The role of lighting by these lamps during the winter was particularly important, since the furskin clothing (often derived from seal) that the Inuit used to keep them warm required constant maintenance.

In Europe subsistence hunting for seal oil continued where seals were plentiful until very recently, but is now almost ended. LINNAEUS, in the diary of his journey to Öland and Gotland in 1745, described seal hunting practices in the Baltic. Netting and trapping were common methods. In the Orkney and Shetland Isles, in the Hebrides and in the west of Ireland, sealing continued, mainly for oil, until this century. Apart from lighting seal oil was used also as a medicine and a leather dressing (3, 15).

***Commercial hunting*** of seals differs in that the products of the hunt are sold on a cash basis. It usually involves some investment in hunting equipment and engaging crews with the object of securing the largest catch of seals possible. Commercial sealing hardly existed

**Table 30:** Overview of the utilisation of seals and walruses

| English name | Scientific name | Geographical distribution | Body weight (kg) | Utilisation[1)] |
|---|---|---|---|---|
| **FUR SEALS and SEA LIONS - OTARIIDAE** | | | | |
| Northern Fur seal | *Callorhinus ursinus* | north., but not arctic Pacific Ocean | 100-250[2)] | mainly for fur and leathers |
| Southern Fur seals | *Arctocephalus* spp. | South. Hemisphere, coastal and antarctic waters | 100-200 | mainly for fur, also meat and fat |
| Hooker's Sea lion | *Phocarctos hookeri* | N.Zealand, sub-antarctic islands | n.a. | mainly for oil |
| Australian Sea lion | *Neophoca cinerea* | S. and W.Australian coasts | 230-330 | mainly for oil |
| South American Sea lion | *Otaria bryonia* | South. and South-west. S.American coasts | 250-420 | mainly for oil |
| **WALRUSES - OBOBENIDAE** | | | | |
| Walrus | *Odobenus rosmarus* | shallow regions of circumpolar arctic coasts | 800-1,000 | mainly for ivory and hides |
| **TRUE SEALS - PHOCIDAE** | | | | |
| Common or Harbour seal | *Phoca vitulina* | N.Atlantic and N.Pacific | 80-100 | for subsistence, fur and leather |
| Spotted seal | *-largha* | N.Pacific, ice areas | 80-100 | for substistence and fur |
| Ringed seal | *Pusa hispida* | circumpolar arctic coasts, N.Atlantic | 60-90 | for subsistence; leather; meat and oil |
| Bearded seal | *Erignathus barbatus* | circumpolar arctic coasts, N.Atlantic | 250-380 | for subsistence; leather; meat and oil |
| Harp seal | *Pagophilus groenlandica* | arctic Atlantic Ocean | 80-150 | for fur ("white coats") and edible fat |
| Ribbon seal | *Histriophoca fasciata* | N.Pacific, seasonally ice areas | 70-95 | for subsistence and fur |
| Grey seal | *Halichoerus grypus* | N.Atl. coast. waters, N. and Baltic Seas | 120-200 | for subsistence |
| Hooded seal | *Cystophora cristata* | N.Atlantic and arctic waters | 200-350 | for furs ("blue-backs") and oil |
| South. Elephant seal | *Mirounga leonina* | south. coasts of S.America | 900-3,000 | for oil |
| North. Elephant seal | *-angustirostris* | Pacific coast of N.America | | for oil |
| Monk seals | *Monachus* spp. | tropical and sub-tropical coast. waters | 90-100 | for oil |

1) as described in the text, much of the utilisation of seals and walruses is now out-dated or incidental; it is here mentioned in an historical context only; 2) male Fur seals and Sea lions are much heavier than female

at all before the beginning of the 18th century when it developed mainly as an off-shoot of commercial whaling.

*Harp seals* were the first to be exploited commercially on a large scale. They breed on pack ice as three widely-separated populations; one at Jan Mayen in the Arctic Ocean, one in the White Sea, and the third in the North-West Atlantic around Newfoundland. The seals aggregate on the ice in huge colonies known as "whelping patches" where very large numbers of young and adults can be taken with the aid of suitable vessels. All three populations of Harp seals have been subjected to intense exploitation since about 1720. By the second quarter of the 19th century catches from around Newfoundland alone reached half a million seals yearly taken by 12,000 men from 300 schooners. Steamers were introduced in the 1860s and steel ships replaced wooden hulls in the early part of the present century (8, 9, 5, 19, 20). Where available, *Hooded seals* were also taken by the harp sealers. Their pups or 'blue-backs" provided a superior fur and the adults a larger quantity of blubber oil. *Walrus* hunting was at one time an important supplement to the Arctic Right whale fishery (and Harp sealing) in the 19th century. The chief products were ivory from the tusks and leather from the hides. This was in great demand for the manufacture of buffing wheels. The breeding grounds of the *Northern Fur seal* in the Pribilof Islands were discovered in 1786 and 1787 and sealing began immediately. It is estimated that 2.5 million seals were killed in the period 1786-1867 and the population was initially drastically reduced. However, restrictions on killing by the Russian and American authorities allowed recovery to something like the original population (12). Following the purchase of Alaska by the United States in 1867 the stock again declined. The North Pacific Fur Seal Convention, signed in 1911, banned pelagic sealing and subsequent careful management of the stock ashore allowed the herd to recover again. Catches rose from the 1920s to the 1940s, reaching a maximum of 117,164 in 1943. All species of *Southern Fur seals* have been exploited for their skins. The first records of hunting date from soon after the colonisation of Uruguay in 1515 when seals were killed at the mouth of La Plata. In South Africa sealing began about a hundred years later. However, it was in the 18th century that large scale fur sealing began, first in the Falkland Islands around 1775, the trade spreading throughout the range of the genus by the first decade of the 19th century. Fur seals were discovered in the South Shetland Islands in 1819 and three years later they had been almost exterminated in the Antarctic (4). Sealing reduced the numbers of Fur seals drastically. One ship sold 100,000 skins in Canton in 1798 and James Weddell calculated that 1.2 million skins had been taken from South Georgia alone by 1822 (22). No major cargoes were taken after 1850 but fur sealing lingered on in some remote localities until the early years of this century. Southern Fur seal skins were used for exactly the same purposes as those from the Pribilofs. However, many southern skins were used only for leather, being of inferior quality. Oil was not a valued product.

Besides Fur seals, sealers in the south, as in the north, would kill any available *Sea lions* for their oil, bringing them to near extinction in New Zealand and Australia.

When Fur seals began to become scarce in the Southern Hemisphere, the sealers turned their attention to killing *Southern Elephant seals* for their oil. They were both more difficult to harvest and produced a less valuable product, so that although they were locally reduced in many localities, they were never so near to extinction as the Fur seals. Elephant sealers were often itinerant whalers (17) and when this trade declined at the end of the century, so did elephant sealing. *Northern Elephant seals* were also heavily exploited off the coast of California and Mexico by sealers in the 19th century. Despite being reduced to only a few tens of individuals, this species has recovered well and is now once again numerous (14). All three species of *Monk seals* have been heavily exploited for their oil. The *Caribbean Monk seal (Monachus tropicalis)* has become extinct and the *Hawaiian* and *Mediterranean Monk seals (M. schauinslandi* and *-monachus)* were greatly reduced. The principal current threat to these species, however, seems to be disturbance.

## Recent utilisation of seals and walruses

Although the Second World War caused an interruption in some seal hunts, it did not have a profound effect on **commercial sealing** or the demand for seal products. *Harp seal* catches in the North-West Atlantic recovered rapidly because of the increasing demand for edible fats after the Second World War. Norway and the former USSR joined the North-west Atlantic hunt, but the latter withdrew in 1966 restricting commercial harp sealing to the White Sea from State farms around the coast (19, 20). Concurrently with the hunting of Harp seals, *Hooded seals* were also taken though in much smaller numbers. The commercial hunting of seals is carried out by parties of hunters who approach the large breeding groups on foot. Adult and sub-adult animals are shot and the young are clubbed. Owing to the behaviour of the breeding animals large numbers of seal can be killed in a short time. Harp seals were originally taken almost entirely for their blubber oil. In the postwar years the white furs of the youngest pups had become a main item of trade and to prevent the fur yellowing by oxidation of blubber oil in the hair fibres it was necessary to bring back the catch chilled in brine tanks. This resulted in the production of a much higher grade of oil, since no decomposition had taken place, and the product from Norwegian factories, particularly, was much used for the production of edible fats. Skins from older animals were also used as garment furs, and particularly in the leather trade and for providing fur trimmings in, for example, the luggage industry.

Although harvesting of *Northern Fur seals* started again after the war, the herd size and catches have declined, perhaps the combined results of adverse management practices, mortality arising from net entanglement, and changes in the trophic status of the seals' feeding grounds. Commercial sealing ceased in 1984 in the Pribilofs, but Russia continues to take skins up to a maximum of 5,000 per year from the Robben and Commander Islands. Northern Fur seals have always been taken almost solely for their skins, which when de-haired and dyed were in high demand in the furriers trade. Native communities utilised meat from the killed seals, and seal oil obtained when the skins were de-blubbered was used in the tanning process or as a cheap animal oil. Fur-seal oil has not been a valued product in its own right, nor has an international market for seal meat ever developed.

Commercial sealing of the *Southern Elephant seal* continued in South Georgia throughout the war. The catch of 6,000 adult bulls which produced 2,000 tons of oil a year was near the maximum sustainable yield of the stock which was in a very healthy condition at the time sealing ceased in 1965 as a result of the collapse of the whaling industry of which sealing was an off-shoot (4).

Elephant sealing in 1956 in the Antarctic. (Photo: N. BONNER)

Subsistence sealing continues to a very limited and decreasing extent in northern aboriginal communities. The collapse of the seal-skin market in Europe in 1982, described below, made sealing as a means of providing a cash crop uneconomic (13). This has affected the subsistence economies of northern aboriginal communities very negatively.

## Economic importance and trends

Similarly to the whaling industry, the exploitation of seals assumed great economic importance historically. Although there are many scattered records - some of which are cited above to document this -, there has been little attempt to systematically record international catches of seals as has been done with whales.

Between 1957 and 1961 catches of *Harp seals* averaged 316,000/year (6) declining to 284,000/year between 1960 and 1969. Harvest rates of *Hooded seals*, mostly taken with them, were around 30,000/year for Jan Mayen and 15,000/year for New Foundland, but have also since declined.

Until 1964 little public interest had been taken in the Harp-seal hunt. In that year, however, television shots of the hunting had a dramatic effect on public opinion. Pressure from animal welfare groups caused the Canadian government to introduce hunting quotas and apply regulations to sealing methods. A motion in the European Parliament to ban trade in products derived from young Harp and Hooded seals was adopted in 1982 and virtually destroyed the market. The sealers' catch in that year dropped to fewer than 167,000. In 1983 the European Economic Commission instituted a ban on the importation of products from

Harp and Hooded seal pups, causing the catch to still tumble further to an average of 36,000 animals per year between 1983 and 1987. Commercial killing of the white-coated Harp seal pups has also been banned by the Canadian government, but a small harvest continues for older seals, mostly by hunters operating from the shore. Canadian and Greenland Inuit take small numbers, as do the Norwegians. The Russian White Sea sealing continues at a higher rate.

In order to elucidate the effect of the anti-sealing campaign and European legislation on the European trade in seal skins an investigation was made in 1984 (11). Four species of seal were found to be commercially important to the European markets: the *Cape Fur seal, Harp* and *Hooded seals* and *Ringed seal*; another three species, the *Pribilof* and the *South American Fur seal* and *Common seal* are of marginal importance. Tab. 31 summarises the minimal catches of pinnipeds since 1982 as reported to FAO:

The *Northern Fur seal* was managed in terms of the Interim Convention on Conservation of North Pacific Fur Seals of 1957 jointly by USA and Russia until 1984. Currently, sealing of this species is managed under domestic legislation by the USA and Russia. The USA take has been limited to the native subsistence needs, between 3,713 and 1,241 seals a year in 1985-1992 (1). The *South American Fur seal* is still taken in Uruguay, but FAO Statistics show a rapid decline in catches (11).

The *Cape Fur seal* breeds along the coasts of South Africa and Namibia. Licenses to kill are granted to reduce pressure on fisheries, but the take has so far never achieved the established control quotas and the total population is on the increase. For 1994 a harvesting quota of 43,000 pups and 12,000 adults was allowed by the Namibian authorities (2). All skins are licensed under CITES for export, mainly to Germany.

The *Common or Harbour seal* is hunted in Iceland, Greenland and Russia. Only commercially insignificant numbers of skins are still traded internationally.

*Ringed seals* have been traditionally hunted in large numbers by arctic natives. In Canada there has been an annual kill of about 75,000, and in Greenland 40,000-50,000. Blubber, meat and other products were used locally, but the surplus skins traded to Europe. Because of the decrease in prices, however, very few skins still enter the European market. In fact in 1983 the Danish Royal Greenland Trade Department which buys the skins from the aboriginal population for auction had a stock pile of 200,000 skins which could not be sold at profit (11).

Harvests of *Harp and Hooded seals* have been regulated since 1980 by Canada and Norway on a quota basis which includes allocations to Arctic natives. They have remained, so far, below the established quotas every year.

Other commercially utilised seal species which, however, do not enter the European market are the *Ribbon and Spotted seals*. The former are hunted on a small scale in the Bering and Okhotsk seas. In Alaska only subsistence hunts of 100-250 animals a year are permitted, but in Russia catches of 3,000-3,500 have been permitted. The latter are hunted at a rate of about 10,000-15,000 a year by the Russians (18).

**Table 31:** Nominal catches of pinnipeds in 1982 - 1991[1)]

| Species | 1982 | 1983 | 1984 | 1985 | 1986 | 1987 | 1988 | 1989 | 1990 | 1991 |
|---|---|---|---|---|---|---|---|---|---|---|
| Northern Fur seal | 32,816 | 33,614 | 30,032 | 10,975 | 1,423 | 1,802 | 1,258 | 1,521 | 1,240 | 1,926 |
| South American Fur seal | 1,375 | | 7,060 | 6,946 | 6,172 | 6,720 | 6,549 | 6,709 | 5,439 | 5,375 |
| Cape Fur seal | 91,425 | 46,739 | 27,179 | 50,703 | 10,517 | 20,234 | 13,098 | 13,272 | 19,400 | 17,325 |
| South American Sea lion | - | - | - | - | - | - | 452 | 837 | 938 | 270 |
| Harbour seal[2)] | 71 | 56 | 48 | 1 | 32 | 12 | 97 | - | - | 5 |
| Ringed seal | 71,691 | 67,382 | 75,354 | 38,792 | 100[3)] | 37,919 | 45,120 | 34,673 | 40,168 | 47,221 |
| Bearded seal | 888 | 918 | 1,159 | 700 | - | 574 | 875 | 152 | 119 | 1,320 |
| Harp seal | 211,448 | 98,449 | 59,808 | 48,013 | 44,964 | 83,283 | 127,333 | 87,675 | 86,147 | 77,620 |
| Grey seal | 1,407 | 38 | 40 | 43 | 53 | 74 | 34 | 31 | 39 | 33 |
| Hooded seal | 30,105 | 5,699 | 5,996 | 2,837 | 2,932 | 11,006 | 6,354 | 1,765 | 2,939 | 11,984 |
| Seals not otherwise identified | 5,204 | 6,492 | 1,993 | 1,994 | 2,874 | 3,236 | 1,036 | 2,559 | 1,821 | 1,721 |
| **Total** | **446,430** | **259,387** | **208,669** | **160,004** | **69,067** | **164,860** | **202,206** | **149,194** | **158,250** | **164,850** |

1) includes only catches reported by FAO member countries, hence figures do not represent total yearly catches; 2) the returns for the number of Harbour seals caught are certainly too low; no catches are reported for Canada, the USA or Russia; 3) the very low return for Ringed seals in 1986 is because there was no return from Greenland in that year

Source: FAO Year Book of Fishing Statistics; Vol. 72 (1991)

The former rather important harvesting of *walruses* has also been reduced to the subsistence level though with the ultimate intention of producing ivory on a sustainable basis for commercial carving. Some 7% of the Pacific walrus population is taken each year and perhaps 8% or more of the Atlantic and Laptev Sea populations (10).

The European seal-skin market on which commercial sealing was always dependant virtually collapsed in 1984 with prices for seal skins and derived articles dropping by about 40% (11, 13). This was mainly due to the perceived unacceptability to the European and North American public of killing seals, especially baby animals, for their pelts which resulted in the EEC ban on import of products derived from young Harp and Hooded seals.

However, even before this ultimate development alterations in the demand for seal products had caused commercial sealing to decline. The expansion of industrial fishing and the substitution of vegetable oil in the production of margarine and cooking fats has destroyed the market for seal oil. The labour-intensive nature of sealing operations and wage levels in most potential sealing nations have reduced margins and thus rendered sealing more and more uneconomic. This trend is not likely to change in the near future.

## Impact on conservation

In view of the intensity of hunting it is surprising that seal species have suffered relatively little at Man's hands. Only one species, the Caribbean Monk seal (*Monachus tropicalis*), has become extinct, though several seal species (the other two Monk seals and most Fur seals) have been reduced to very low numbers by hunting. Some of these species have proved remarkably resilient. Antarctic Fur seals have recovered from a population perhaps as low as a few tens at the beginning of this century to around a million and a half to-day, a total which probably exceeds pre-exploitation stocks. This very rapid increase has been attributed to enhanced food resources consequent on the depletion of Antarctic whale stocks. Cape Fur seals as well as the Northern Elephant seal are also rapidly increasing in numbers due to the protection of their breeding sites and banning of commercial sealing. Grey seals were reduced in European waters mainly by subsistence hunting. With the disappearance of this type of life style these seals have also shown a great increase. While this may be due in part to the ending of hunting pressure it seems more probable that the prime cause may be, as in the case of the Antarctic Fur seal, an increase in food supply, in this case consequent on the selective removal of large fish predators (e.g. cod) leaving larger stocks of smaller fish to be exploited by the seals.

In conclusion, several seal stocks are presently sufficiently large and sufficiently well-documented for sustainable hunting quotas to be set on what in the past has been a commercial scale. These include the North-West Atlantic Harp seals, Elephant seals and Fur seals in South Georgia and South Africa. Some seal stocks, such as Ribbon and Spotted seals in the North Pacific and Harp seals in the White Sea, are still being commercially exploited without certainty of sustainability. Elsewhere, seal hunting currently and in the immediate future is likely to be mostly by indigenous peoples or conducted by governments to reduce damage to fisheries.

## Literature references


1. ANONYMOUS (1993): *Annual report to Congress*. Marine Mammal Commission; Washington, DC
2. ANONYMOUS (1995): Cape fur seal cull continues. *Oryx*, 29(1), p. 10
3. BERRY, R.J. (1985): *The natural history of Orkney*. Collins, London
4. BONNER, W.N. (1982): *Seals and Man: a study of interactions*. Washington University Press, Seattle, USA
5. BUSCH, B. C. (1985): *The war against the seals: a history of the North American seal fishery*. McGill-Queen's University Press, Kingston, Ontario, Canada
6. CAPSTICK, C.K., LAVIGNE, D.M., and K. RONALD (1976): *Population forecasts for North-West Atlantic harp seals, Pagophilus groenlandicus*. International Commission for North-West Atlantic Fisheries Research; Document 76/C/132; 22 pp.
7. CLARK, J.G.D. (1946): Seal-hunting in the Stone Age of Northwestern Europe: a study in economic prehistory. *Proceedings of the Prehistory Society*, 12(2), p. 12-48
8. COLMAN, J.S. (1937) : The present state of the Newfoundland seal fishery. *Journal of Animal Ecology*, 6, p. 145-159
9. COLMAN, J.S. (1949): The Newfoundland seal fishery and the second World War. *Journal of Animal Ecology*, 18, p. 40-46
10. FAY, F.H. (1981): Walrus *Odobenus rosmarus* (Linnaeus, 1758). In: *Handbook of Marine Mammals*, ed. RIDGWAY, S.H. and R.J. HARRISON, Vol. 1: The walrus, sea lions, fur seals and sea otter; p. 1-23; Academic Press, London
11. DIXON, A.M. (1984): The European Trade in Seal Skins. *TRAFFIC Bulletin*, 6 (3/4), p. 54-65
12. LANDER, R.H. and H. KAJIMURA (1982): Status of northern fur seals. In: *Mammals in the seas*. FAO Fisheries Series, No 5; Vol 4; p. 319-345; Food and Agriculture Organisation of the United Nations, Rome
13. LAVIGNE, D.M. and K.M. KOVACS (1988): *Harps and Hoods: ice-breeding seals of the North-West Atlantic*. University of Waterloo Press, Waterloo, Ontario, Canada
14. LE BOEUF, B.J. (1981): Mammals. In: *The natural history of Año Nuevo*, ed. LE BOEUF, B.J. and S. KAZA; p. 287-325; Boxwood Press, Pacific Grove, USA
15. LOCKLEY, R.M. (1966): *Grey seal, common seal*. André Deutsch, London
16. MØHL, U. (1970): *Fangstdyrene ved den Danske strande: den zoologiske baggrund for harpunerne*. KUML (Aarbog for Tysk Arkaeologiske Selskab) 1970; p. 297-329
17. MURPHY, R.C. (1947): *Logbook for Grace: Whaling brig "Daisy", 1912-1913*. MacMillan, New York
18. POPOV, L.A. (1982): Status of the main ice-living seals inhabiting inland waters and coastal marine areas of the USSR. In: *Mammals in the seas*. FAO Fisheries Series, No. 5; Vol 4; p. 361-381; Food and Agriculture Organisation of the United Nations, Rome
19. SERGEANT, D.E. (1976): *Studies on Harp seals of the western North Atlantic population in 1976*. International Commission for North-West Atlantic Fisheries; Research Document 76/X/124 (revised), Serial No. 4010; 32 pp.
20. SERGEANT, D.E. (1991): *Harp seals, man and ice*. Canadian Special Publications on Aquatic Sciences, 114; 153 pp.
21. SMITH, T.G. and I. STIRLING (1975): The breeding habitat of the Ringed seal (*Phoca hispida*): The birth lair and associated structures. *Canadian Journal of Zoology*, 53(9), p. 1297-1305
22. WEDDELL, J. (1825): *A voyage towards the South Pole, performed in the years 1822-24*. Longman Group, London and New York

## 2.3.6 WHALES, DOLPHINS and PORPOISES (*CETACEA*)

The cetaceans are aquatic mammals and may be divided into 2 suborders: the Baleen whales (**Mysticeti**) and the Toothed whales (**Odontoceti**). The former include Right whales (**Balaenidae**), Gray whales (**Eschrichtiidae**) and Finback whales or rorquals (**Balaenopteridae**), most of which have a world-wide distribution. The latter comprise Sperm whales (**Physeteridae**), Narwhal and White whales (**Monodontidae**), Beaked whales (**Ziphiidae**), dolphins (**Delphinidae**) and porpoises (**Phocaenidae**), of which only Sperm and Killer whales and some dolphins occur world-wide. The others are geographically and ecologically restricted to particular seas and coastal regions. Tab. 32 shows that much like the pinnipeds most species of this group of mammals were subjected to severe commercial exploitation, initially mainly for baleen and oil, but latterly particularly for meat. In recent years the learning capacity of some of the smaller species such as certain dolphins, has also been utilised by removing limited numbers of these from the wild and training them for display and even military purposes.

### Historical background

In much the same way as the terrestrial large mammals the marine cetaceans have been utilised for hundreds of years by aboriginal peoples for subsistence purposes. They still constitute an important resource to some indigenous people in the Arctic region, notably in Alaska, Northern Canada, Greenland and Siberia. With the spread of more advanced technology in the 19th century the consumptive utilisation of whales, or whaling, became a highly specialised branch of the fisheries industry in which a number of the industrialised nations participated.

#### Whaling

The advent of modern **large whale fisheries** can be dated to the mid-1860s when the Norwegian Sven Foyn first successfully used a harpoon fitted with an explosive grenade fired from a cannon mounted on a steam powered catcher vessel (14). The combination of these technologies allowed even the fastest swimming and less buoyant rorquals, including the largest species - the *Blue* and the *Fin whales* - to be successfully pursued and captured. Prior to this development the hand-harpoon whaling carried on from open boats under oars or sail was mainly limited to the slow swimming species which tend to float when dead, notably the *Bowhead* and *Right whales*, the *Humpback* and also the *Toothed Sperm whale*.

The ***North Atlantic*** stocks of *Blue, Fin, Sei* and *Humpback whales* were severely depleted in many areas by the early 20th century. The last fishery was off the west coast of Iceland which maintained a fairly constant level of catches following the Second World War. Early whaling off the ***Pacific coast*** of North America greatly reduced the numbers of *Gray whales*, which were also taken by traditional net and hand-harpoon fisheries off Japan. Modern whaling in the 20th century, from the coasts and pelagic factory ships, made great inroads into all the *Baleen* and *Sperm whale* stocks.

**Table 32:** Overview of the utilisation of cetaceans

| English name | Scientific name | Geographical distribution | Body length (m) | Utilisation[1)] |
|---|---|---|---|---|
| **BALEEN WHALES s.l. - MYSTICETI** | | | | |
| Right whales | *Eubalaena australis* | world-wide except tropics | 1-17.0 | for baleen, oil and meat |
| Bowhead whale | *Balaena mysticetus* | Arctic waters | 18.5 | for baleen, oil and meat |
| Gray whale | *Eschrichtius robustus* | North Pacific | 14.0 | for baleen, oil and meat |
| Blue, Fin, Sei and Minke whales | *Balaenoptera* spp. | world-wide | 11-30.5 | for baleen, oil and meat |
| Bryde's whale | *-edeni* | temperate waters only | 14.0 | for baleen, oil and meat |
| Humpback whale | *Megaptera novaeangliae* | world-wide | 16.0 | for baleen, oil and meat |
| Sperm whale | *Physeter macrocephalus* | world-wide | | |
| Narwhal | *Monodon monoceros* | Arctic waters | | |
| White whale or beluga | *Delphinapterus leucas* | Arctic waters | | |
| Baird's Beaked whale | *Berardius bairdii* | N.Pacific | | |
| Bottlenose whale | *Hyperoodon ampullatus* | N.Atlantic | | |
| Killer whale | *Orcinus orca* | world-wide | | |
| False Killer whale | *Pseudorca* spp. | world-wide | | |
| Long- and Short-finned Pilot whales | *Globicephala* spp. | N.Atlantic, S.Hemisphere, tropical/ temperate waters | | |
| Dusky and White-sided dolphins | *Lagenorhynchus* spp. | temp. S.Hemisphere coasts., N.Pacific | | |
| Common dolphin | *Delphinus delphis* | warm waters world-wide | | |
| Spotted and Striped dolphins | *Stenella* spp. | warm and temperate waters world-wide | | |
| Bottlenose dolphin | *Tursiops truncatus* | world-wide except polar seas | | |
| Northern Right Whale dolphin | *Lissodelphis borealis* | temperate N.Pacific | | |
| True porpoises | *Phocaena* spp. | cool N.Hemisphere coasts, temperate S.America | | |
| Dall's porpoise | *Phocaenoides dalli* | N.Pacific | | |

1) as described in the text, much of the utilisation of whales is now out-dated or incidental; it is here mentioned in an historical context only

Depletion of the Northern Hemisphere stocks of whales led to the development of major fisheries from the early 1900s from the coasts of ***Africa, Australia, New Zealand*** and ***South America***. *Humpback whales* were the early target. Antarctic whaling from ***South Georgia*** started in 1904 followed by a major expansion with the introduction of floating factory ships in 1925, which led to the over-exploitation of most large *Baleen whale* stocks in the Southern Hemisphere.

Modern whaling therefore has had a very large Norwegian input, spreading from the coasts of that country to the rest of the world, either directly by the activities of Norwegian companies and personnel, or through the influence of the Norwegian inspired technology. As a result, many whale stocks have been seriously depleted as they became successively the focus of catching activity. This last phase in the long history of whaling was effectively brought to an end in 1982 through the action by the member nations of the International Whaling Commission (IWC) in setting the catch limits for commercial whaling for all stocks to zero from the 1986 season. The final implementation of this decision was actually delayed until 1988, but apart from very small catches of whales for research purposes, commercial whaling is now in abeyance. The situation will be reviewed when a comprehensive assessment of the status of the major whales stocks has been accomplished, and a revised management scheme designed to provide adequate safeguards against further over-exploitation implemented.

Commercial whaling was directed against the large Baleen whales and the Sperm Whale. However, **small cetacean fisheries** also developed world wide until concern over their impact on the conservation status of these whales grew sharply in the 1970s (11, 12). Fisheries for *dolphins, porpoises* and *small whales* have been surveyed by the Small Cetacean Sub-Committee of the IWC's Scientific Committee (5). Catches of small cetaceans include not only ***directed fisheries*** for particular species in specific areas, but also some substantial ***incidental catches*** of small cetaceans taken in fisheries for other species.

More than 20 species of small cetaceans (including *Dall's porpoise, Short-finned Pilot whales, Baird's Beaked whale, Striped* and *Bottlenose dolphins*) are found in ***Japanese coastal waters*** and there have been local fisheries here since historical times. The catching operations have included drive fisheries, the use of hand harpoons, crossbows and now modern harpoon technology. Aboriginal hunting of the *Beluga* or *White whale* and the *Narwhal* has gone on for centuries in the ***Arctic regions*** of Greenland, Canada, Alaska and Siberia. Commercial whaling operations took large numbers more recently, but at the present time most of the hunting is again limited to catches by aboriginal people for subsistence use. A variety of hunting techniques are used including nets, drives, hand harpoons and increasingly, rifles.

The *Long-finned Pilot whale* has a long history of exploitation in the ***North Atlantic***. Today the only remaining fishery is the opportunistic drive at the Faroe Islands; occasional catches are taken by the small-type whalers in Greenland. The Faroes "grind" involves driving the whales, when they are seen passing the islands, onto the shore in suitable bays and usually entire schools are killed. The hunt began in the 9th century and records for the fishery go back to 1584, and are unbroken from 1709 to the present (9). Pilot whales are taken, together with *Killer, Beaked* and *other small whales, dolphins* and *porpoises* in various artisanal fisheries in the ***West Indies***. The Norwegian small-whale fishery in the North Atlantic took *Minke, Bottlenose, Killer* and *Pilot whales* in its small catcher boat operations before the introduction of the ban on commercial whaling. The very similar Greenland small-type whaling also took a mixed suite of these species in what is regarded as an aboriginal subsistence hunt for local consumption. *Common* and *Bottlenose dolphins* and *Harbour porpoises* have been hunted extensively since 1870 by fishermen in the ***Black Sea*** by netting and shooting. Although the fisheries have now been banned because of stock depletion there appears to have been a resumption of catching recently. There are many small fisheries taking small cetaceans around the world, largely for meat for local human consumption. Statistics on these fisheries are scarce, but it is known that the various artisanal fisheries in the ***coastal waters of Peru*** commonly take *Dusky dolphin (Lagenorhynchus obscurus)*, *Burmeister's porpoise (Phocaena spinipinnis)* and the *Common dolphin* (13). Similarly, although large catches of small cetaceans have been reported ***around Sri Lanka***, records are at best estimates and incomplete. At least 17 *small cetacean species* have been identified in the catches, as well as *Sperm, Fin* and *Humpback whales* (10).

Apart from the above described directed fisheries there has been a proliferation of gillnet, purse seine and other fishing gear in recent years that may take cetaceans incidentally (5). The widespread use of modern synthetic materials such as monofilament nylon for nets makes them particularly prone to promoting fatal incidental entanglement by cetaceans.

### Keeping of live cetaceans

Compared with the consumptive utilisation of cetaceans, the live keeping of some Toothed whale species, mainly of the dolphin family, in oceanaria or dolphinaria for research, display and entertainment is rather insignificant. However, due to their intellectual capabilities a special co-operative relationship has developed between dolphins and man which is worth mentioning.

Cases of social interaction between *dolphins* and man have been described by some of the early Greek writers, including Plutarch, Plinius and Herodot. Dolphins have allegedly saved people from drowning and have given rides to children. In modern times the playful behaviour of dolphins and their acceptance of being watched, touched and fed by tourists have greatly promoted tourism in some beach resorts, such as Monkey Mia in Australia. In the Carribbean, the Bahamas and Bermudas, commercially operated so-called "dolphin training camps" have been established in which dolphin-lovers are allowed to have physical contact with some of the smaller Toothed whales. In some regions dolphins "collaborate" with fishermen in fish drives into nets. In the coastal waters of Burma *Irrawaddy dolphins (Oracaella brevirostris)* are even individually cared for as valued "helpers" by village fishermen.

First attempts to capture and train live *Beluga whales* in large sea-water tanks were made in 1861 on the eastern coast of North America. Less successful were initial efforts in Europe to keep *Harbour porpoises* as this species is particularly difficult to maintain in captivity. The modern display of cetaceans started only in 1938 with the establishment of the Marineland oceanarium in Florida. During the 1960s dolphinaria mushroomed all over the world from Finland to Hongkong, attracting millions of visitors. The most frequently used species is the *Bottlenose dolphin*, but about 30 *other small Toothed whale species* are kept in dolphinaria.

### Utilisation of whales for oil and meat

Larger whales are taken with explosive gun harpoons whereas smaller species may be killed with hand harpoons or netted. They are either brought on board the catching vessels or towed to land for processing.

The ***processing*** of a *large whale carcass* starts with the removal of the blubber which encases the body. This is stripped off longitudinally, followed by the two dorsal meat fillets. The liver, viscera, ribs and intestines are next removed in turn, and finally the head is separated from the vertebral column and the bones dismembered. The lean meat is minced and heated in water to coagulate the protein. The flavouring substances are then filtered and concentrated before drying and mixing with the solid protein material from the other processes to produce protein meal for animal feed. The fat, meat, bone and viscera are pressure-cooked to remove the free oil. The remaining solid protein material is separated from the liquid phase and dried, while the liquid protein is concentrated and spray-dried. The blubber is minced and heated to remove the oil, the remaining solid part joining the other solid material in the pressure-cookers. All parts of the body of a whale yield some oil, the proportions varying by species. The blubber is the primary source providing 50-80% oil by weight, with the bones yielding 10-70% and the meat 2-8%. Whale livers are treated to recover the vitamin A, in which they are particularly rich, by mincing followed by digestion in caustic soda to liberate the vitamin-containing liver oil.

The processing of the ***smaller cetacean carcasses*** is, naturally, a physically simpler procedure. However, because the technique is governed fundamentally by the essentially similar anatomy of the dolphins and porpoises to that of the great whales, the separation of the constituent parts is much the same.

Whales were initially exploited mainly for oil (see Chap. 3.2.2). Only in the last phase of the whaling industry has whale meat become its primary product. There is still a ready market for the meat for human consumption in Japan, so that in addition to the whales caught by the Japanese whaling industry itself, much of the whale meat produced in recent years by other nations such as Iceland, Norway and the former USSR has been exported to Japan in addition to supplying their own domestic markets. As a result of a contraction in the catch available world-wide, the price of whale meat has risen sharply, so that from being a cheap foodstuff used in school and canteen meals immediately after the Second World War, it has become a gourmet item on the menu if and when available. Lower-grade meat is used as an animal feed, while whale-meat meal and the dried whale solubles are marketed as protein materials for use in the manufacture of balanced rations for livestock. Whale-meat extract, which is essentially the natural fluid present in the raw meat, is produced to exacting specifications for use as a meat flavouring for incorporation in a variety of food products designed for human consumption.

Dolphin meat is important as a food for human consumption, as an animal feed and as fertiliser. The exact use varies with the particular fishery concerned, and in many instances, this has changed over time with the availability of the dolphins and other competing sources of raw material for protein and oil.

A Sei whale on the slipway at the Iceland whaling station is started to be "flensed". (Photo: G. DONOVAN)

## Utilisation of small whales for research, display and entertainment

Compared with the consumptive utilisation of whales the removal of *small Toothed cetaceans* for non-consumptive purposes is rather limited. Any ***capture of live specimens*** for research and display is strictly controlled. It is usually undertaken in shallow coastal waters with nets up to 2 km in length. Bow-riding species are captured with so-called break-away-hoopnets. The netted animals are placed on a foam rubber stretcher in the water which is then transferred into a freight container. During transport the animals have to be kept cool and wet by constant sprinkling with water to avoid overheating and skin damage. They are then placed into large water basins in oceanaria or dolphinaria for training. If these are in coastal localities, carefully filtered and treated sea water is circulated directly through the facilities. Otherwise sea water has to be prepared artificially. Technological and biological knowledge of the management of dolphinaria and the care of dolphins has steadily increased, so that the life expectancy of dolphins in captivity may be as high as in the wild. Captive breeding is achieved which may make dolphinaria independent from the wild stocks in the foreseeable future.

Training of *dolphins* is based on the animals' natural behaviour, for example, their habits of jumping above the surface, balancing a piece of kelp on their backs or thrusting turtles and other objects out of the sea. If they carry out an exercise correctly they are rewarded with a fish, if not the food is withheld. In this way dolphins are capable of producing a remarkable range of behaviours on command which are presented as "shows". Special training exercises are devised for scientific purposes, for example, to learn their physiological or mental capabilities. *Dolphins, Pilot* and *Beluga whales* have also been trained in the USA and the former USSR for military purposes; for example to locate sea mines, to search for lost torpedoes, or to attack enemy frogmen. However the military benefit which may be expected from such training remains doubtful.

## Economic importance and trends

***Records of the catches*** of large whales were kept by the Bureau of International Whaling Statistics (BIWS) in Norway from 1930 to 1984; thereafter by the International Whaling Commission (IWC) (8). However, evidence has recently emerged (15) that the catch records submitted prior to 1972 (when the International Observer Scheme was introduced) by the former USSR were falsified, with massive under-reporting of some species such as *Blue, Humpback* and *Right whales*, but over-reporting of other species. The following figures should therefore be treated as reflecting general trends rather than precise data.

Fig. 12 shows the total numbers of large Whales killed for consumptive purposes since 1956. Chatches have declined steadily from 1963/64 on; *Blue whales* were already severely reduced at the start of this period, and the *Humpback whale* stocks also quickly declined, so that *Fin* and *Sperm whales* made up the bulk of the catches throughout the world. As the Fin whale stocks became depleted there was a greater emphasis on catching *Sei whales*. The statistics do not distinguish between *Sei* and *Bryde's whales* up to 1968, but it is evident that most of the increased catch was of Sei whales in the early 1960s. Humpback whales were given complete world-wide protection from commercial whaling from 1966, and Blue whales from 1968. Similar protection for Fin whales was introduced in the Southern Hemisphere in 1976, and for Sei whales in 1978.

**Figure 12:** Catch of large whales (by species) in 1956-1990[1)]

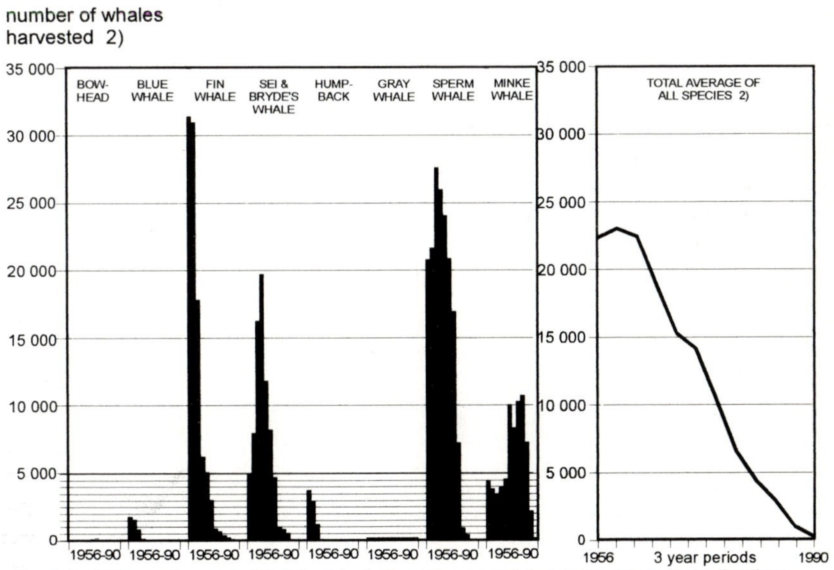


1) only in 1994 was it revealed that Soviet whaling ships caught during operations in the Southern Hemisphere, mainly in Antarctic waters, in 1956 - 1987 the following whales in excess of those reported to IWC: Blue ca 8 240, Sei & Bryde's 18 432,Humpback 45 767, Sperm ca 14 659, Right 3 211;
2) as annual means of 3-year catches recorded.

Source: Bureau of International Whaling Statistics and International Whaling Commission (8)

In Fig. 13 whaling activities are presented according to geographical regions. It will be seen that the Antarctic was by far the most important whaling ground during the peak period of the commercial industry, followed by the North Pacific. The pelagic factory ship operations also took *Sperm whales* in the South Atlantic, Pacific and Indian Oceans on their way to and from the Antarctic whaling grounds. The catches recorded for the most recent years in the North Atlantic and North Pacific reflect the aboriginal subsistence takes by Denmark (Greenland), the USA and the former USSR. A breakdown of these figures into countries shows the significant role of Japan in recent years; by 1980 it killed almost 3 times as many whales as any other nation. The former USSR took as many whales as Japan in the 1960s and early 1970s but stopped whaling in 1979.

*Minke whale* catches have assumed a much greater importance in the last few years as the stocks of the great whales, and the industry based upon them, have declined. The Antarctic fisheries of Japan and the USSR expanded dramatically in the 1970s to compensate in part for the lack of the larger species. A similar transition occurred in the fishery off Brazil which turned from the great whales to Minke whales in the mid-1960s.

**Figure 13:** Catch of large whales (by geographical areas) in 1956-1990[1)]

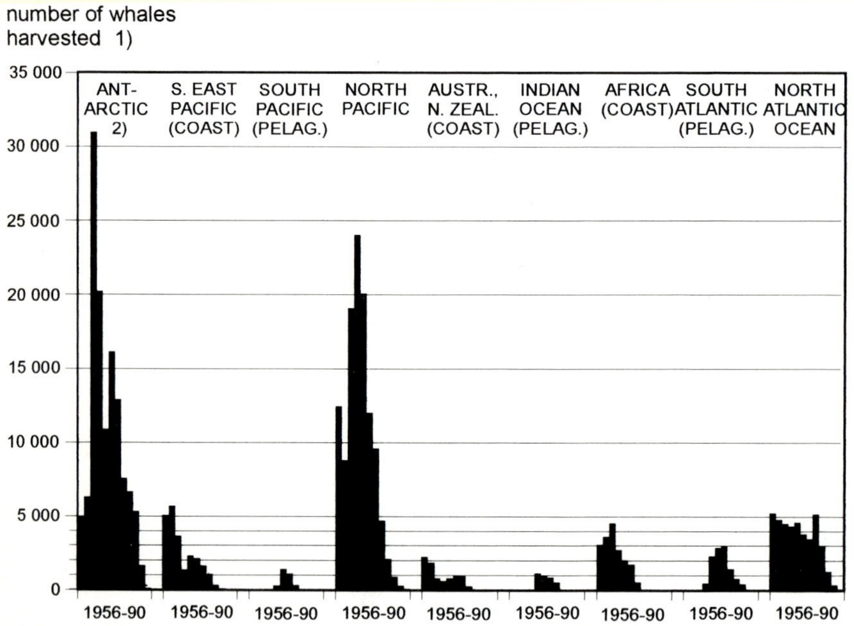


1) as annual means of 3-year total catches recorded; 2) only in 1994 it was revealed that Soviet whaling ships caught during operations in the Southern Hemisphere, mainly in Antarctic waters, in the 1960s and 1970s some 89 000 whales in excess of those reported to IWC; this implies correction of the whaling statistics of up to 3 000 additional whales in some years.

Source: Bureau of International Whaling Statistics and International Whaling Commission (8)

Norway had a major small whale fishery in the North Atlantic until the IWC ban on commercial whaling became effective in 1988. The small catches since then have been research takes under special permit. The similar Icelandic fishery ended in 1985. The catches from Greenland are from the aboriginal subsistence fishery. There had also been a consistent coastal fishery for Minke whales from Japan over the years, and the catch by the Republic of Korea was also important for two decades in the North Pacific.

Reliable catch statistics for the small whales and dolphins are generally not available. The data shown in Fig. 14. illustrate the main trends for a few species for which there are directed fisheries. The great bulk of small cetaceans killed were *"other species"* which constituted more than 90% of the total catch. These comprised mainly incidental catches of dolphins in in the eastern tropical Pacific tuna fishery, and other incidental takes in high sea drift net fisheries.

**Figure 14:** Catch of small whales (by species) in 1965-1990

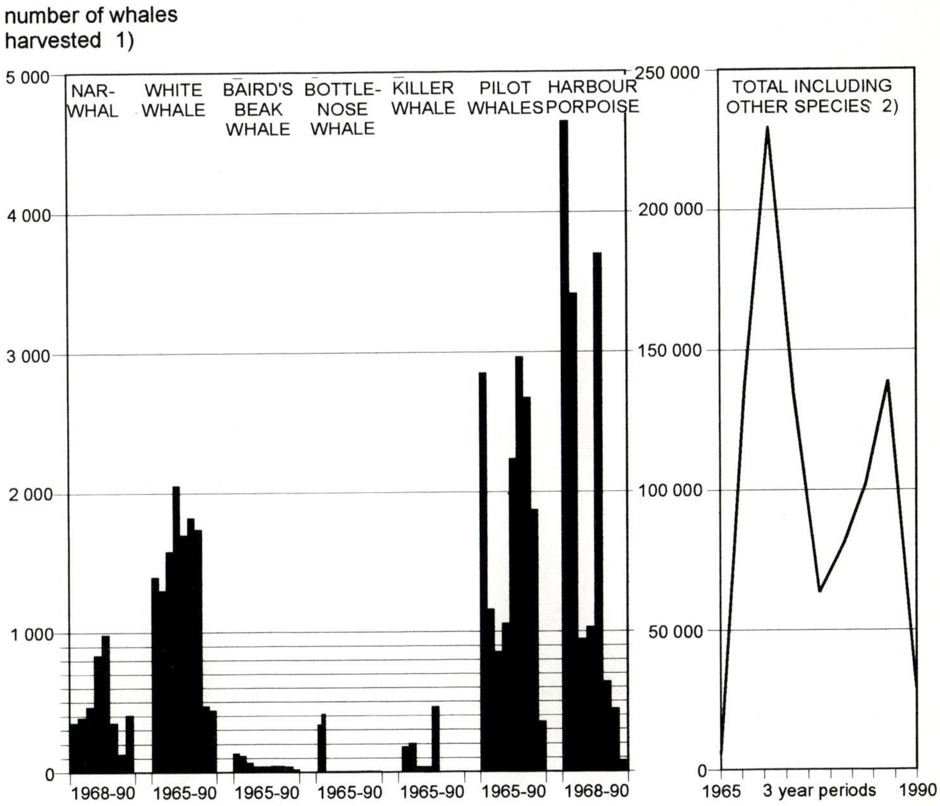


1) as annual means of 3-year total catches reported; 2) the number of whales of other species, incidentally caught, constituted between 88 and 90% of the total number of small whales caught.

Source: Bureau of International Whaling Statistics and International Whaling Commission (8)

In addition to the catch records, the BIWS has also assembled ***records of production*** by the whaling operations. The data derived from the International Whaling Statistics (1959-1988) (8) on production corresponding to the catch figures shown in Fig. 12-14 up to the end of commercial whaling under IWC control in 1986, are given in Fig. 15. In the 1950s the *primary product* from Baleen whales was still the *oil*, but it is obvious that as the catches of the large Baleen whales, and particularly the Fin whales, declined so too the oil production would follow a similar downward path. The oil yield was further depressed by the change over to *whale meat* as a preferred product, since the amount of oil obtained from a carcass is much reduced when the muscular tissue is prepared and frozen for marketing as meat. The tonnage of meat and other similar products for human consumption rose to a peak in the first half of the 1960s and then also fell with the progressively smaller catches. Some enhancement of the quantity of edible products was caused by the rapid growth of the Minke whale fishery in the Antarctic in the 1970s, but because of the small size of the Minke whales compared with the Fin and Sei whales, this only served to introduce a small delay in the inevitable decline in output.

There is a close correlation between the numbers of Sperm whales caught and the production of *Sperm whale* oil, both rising to a peak in the 1963-64 and 1964 seasons and subsequently declining to relative insignificance in the 1980s. Interestingly, the output of *whale liver* and *liver oil* does not seem to reflect the pattern of catches nearly so closely. This suggests that the production of these rather more specialised items was governed more by a consistent market demand than by the ability of the whaling industry to supply.

**Figure 15:** Average world output of products from whales in 1956-1985

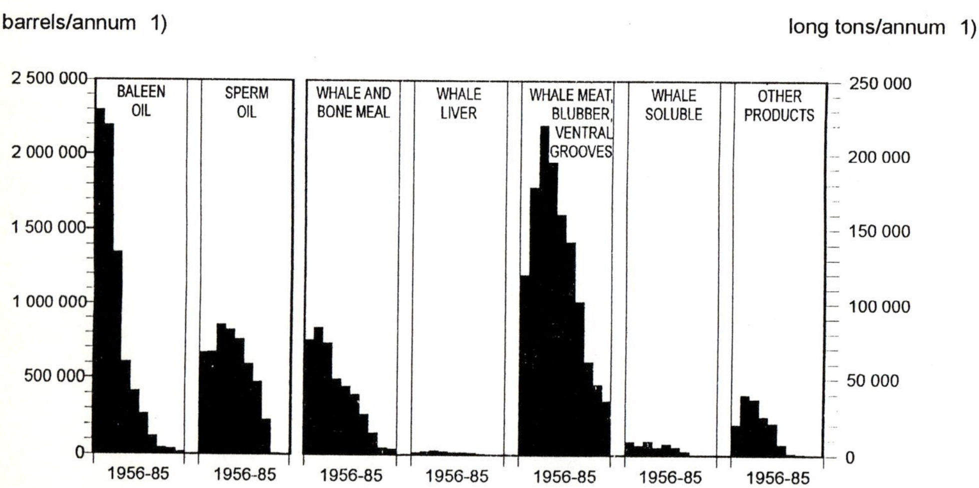


1) as annual means of 3-year total outputs

Source: Bureau of International Whaling Statistics (8)

The number of *small cetaceans* kept in ***oceanaria*** and ***dolphinaria*** around the world in 1990 did not exceed 1,200 animals (7). Compared with the consumptive use of cetaceans the economic importance of this type of utilisation is therefore clearly insignificant. However, the non-commercial benefits derived from these institutions were nevertheless most remarkable: in no other animal order has captive keeping produced as much scientific knowledge as with the cetaceans. The Bottlenose dolphin today counts among the best researched vertebrate animal species. Apart from this benefit, oceanaria and dolphinaria have and still are creating awareness of marine mammalian life in many millions of people.

## Impact on conservation

All species of Baleen whales (except the Pygmy Right whale, *Caperea marginata*) and most if not all of the stocks of each species are, or have been, exploited to some extent at various times (1). The more easily catchable species - the Right, Bowhead and Gray whales - were all reduced to very low levels of abundance by the beginning of the 20th century. No species has, however, been exterminated by whaling, although the Gray whale in the North Atlantic has been extinct since mediaeval times.

**Table 33:** Population estimates for some of the larger whale species[1)]

| Species and geographical area | Year | Population size | 95% Confidence interval |
|---|---|---|---|
| MINKE WHALES<br>Southern Hemisphere | 1982/83 - 1988/89 | 761,000 | 514,000 -1,138,000 |
| North Atlantic (Central and West Greenland areas) | 1987-91 | 31,300 | 23,400 - 37,100 |
| North Pacific (North-West Pacific and Okhotsk Sea) | 1989-90 | 25,000 | 12,700 - 48,600 |
| FIN WALES<br>North Atlantic | 1969-89 | 47,300 | 27,700 - 82,000 |
| GRAY WHALES<br>Eastern North Pacific | 1987/88 | 21,000 | 19,800 - 22,500 |
| BOWHEAD WHALES<br>Bering-Chukchi-Beaufort Seas | 1988 | 7,500 | 6,400 - 9,200 |
| HUMPBACK WHALES<br>Western North Atlantic | 1979-86 | 5,500 | 8,120 - 2,890 |
| PILOT WHALES<br>Central and Easternnorth Atlantic | 1989 | 780,000 | 440,000 -1,370,000 |

1) rounded to the third significant figure of the upper confidence limit

Source: Scientific Committee of the International Whaling Commission (8)

Modern whaling very dramatically reduced the stocks of Humpback and Blue whales by the middle of this century, and Fin and Sei whales have since been depleted throughout the world's oceans. Bryde's whales in the Southern Hemisphere are believed to be only lightly exploited so far, although there is a long history of catching of this species in the North Pacific. Minke whales have been exploited for decades in the Northern Hemisphere, but they have only recently been hunted in the Southern Hemisphere.

The Sperm whale has a more than 250 year history of commercial exploitation. The number of female Sperm whales may not have been so greatly affected by modern whaling except in the South-East Pacific. The males, which grow appreciably larger than the females, have been subjected to longer and more intense hunting, and have been significantly reduced in number in most areas.

Because of the considerable scientific uncertainties in determining the numbers of whales of different species and in different geographical stocks, the Scientific Committee of the IWC decided in 1989 that it could give whale population figures only for those species/stocks which have been assessed in detail and for which there is statistical certainty with respect to the numbers (3). Summarised results from the detailed assessments carried out up to 1991 are shown in Tab. 33.

The only stock which has demonstrably recovered under protection is the California Gray whale. Assuming a constant rate, the population was increasing at 3.2% per year (standard error 0.5%) over the period 1967/68 - 1987/88 with an average annual catch of 174 whales (4). Some Humpback whale stocks have increased, e.g. those in the Westernnorth Atlantic which show a rate of increase of 10.3% (95% confidence interval 2-23%) from photo-identified animals in the Gulf of Maine (2), and some Right whale stocks, such as those off Argentina and South Africa also show evidence of increasing at rates of 7.6% and 7.1% per annum respectively in recent years (3).

Even less information is available on the status of stocks of small cetaceans. The Scientific Committee of the IWC has concluded (5) that there is cause for concern over the severe reduction of certain stocks through both directed exploitation and incidental catches in fishing operations.

The overall picture is therefore one of considerable reductions of many of the stocks of cetaceans. For the great whales, this has led to the imposition of at least a temporary ban on commercial hunting of these species. Any resumption of commercial whaling would only be permitted on those stocks numerous enough to sustain carefully controlled and monitored catches. The IWC adopted a draft specification for the calculation of catch limits in a revised management scheme for Baleen whales in 1992 (6). This was the result of extensive computer simulations to give the best balance between the three objectives of the stability of catches; an acceptable risk that a stock not be depleted below some chosen level; and making possible the highest possible continuing yield. The model requires only a series of abundance estimates within a stock area, usually obtained by dedicated sightings surveys, and the past catches from that stock. Computer simulations tested for a variety of uncertainties in our knowledge of the whale stocks, including stock boundaries, the possibility of catastrophic natural disasters, as well as robustness to modifications of the mathematical model of stock's behaviour. Agreement has still to be reached on additional matters such as minimum standards for data, guidelines for conducting the sightings surveys and analysing the results, and a fully effective inspection and international observer scheme to monitor any catches. There is also concern over the humane killing aspect of the whaling operations. When all of these matters have been satisfactorily resolved and implemented, it should be possible to consider a resumption of commercial whaling on a sustainable basis.

However, Iceland left the IWC in 1992 because the Commission had failed to reach solutions acceptable to her own interests, while Norway announced that she will resume commercial whaling for Minke whales in the North-East Atlantic in 1993 under a legitimate objection procedure because of the delay in implementation of the Commission's revised management scheme.

For the smaller cetaceans there is much more cause for concern, with significant catches being taken in a number of directed fisheries of species which are poorly assessed but clearly depleted by the level of exploitation, and major kills also occurring in fisheries for which the dolphins are not the target species. There is no international body recognised as responsible for the assessment, regulation or monitoring of these species, since a number of governments do not accept the legal competence of the IWC for them.

As regards the capture of live small cetaceans for oceanaria and dolphinaria their number is too small to have any impact on the respective wild populations.

## Literature references


1. ANONYMOUS (1978): *Mammals in the Sea.* FAO Fisheries Series, No. 5(1), i-xxv; p. 1-264; Food and Agriculture Organisation of the United Nations, Rome
2. ANONYMOUS (1989): Report of the Scientific Committee. *Reports of the International Whaling Commission*, 39, p. 33-64
3. ANONYMOUS (1990): Report of the Scientific Committee. *Reports of the International Whaling Commission*, 40, p. 39-79
4. ANONYMOUS (1991): Report of the Scientific Committee. *Reports of the International Whaling Commission*, 41, p. 51-82.
5. ANONYMOUS (1992): Report of the Sub-Committee on Small Cetaceans. *Reports of the International Whaling Commission*, 42, p. 178-234
6. ANONYMOUS (1993): Chairman's Report of the 44th Annual Meeting. *Reports of the International Whaling Commission*, 43, p. 11-53
7. GEWALT, W. (1993): *Wale und Delphine - Spitzenkönner der Meere.* Springer, Heidelberg, Germany
8. INTERNATIONAL WHALING STATISTICS (1959-1988): Committee for Whaling Statistics, Oslo, and Secretariat of the International Whaling Committee, Cambridge
9. JOHNSEN, J. S. and P. ZACHARIASSEN (1982): Grindatol 1584-1964 of 1709-1978. *Fródskaparrit*, 30, p. 71-102
10. LEATHERWOOD, S. and R. R. REEVES (1989): *Marine mammal research and conservation in Sri Lanka 1985-86.* UNEP Marine Mammals Technical Report No. 1, i-vi; p. 1-138; UN Environmental Programme, Nairobi, Kenya
11. MITCHELL, E. (ed.) (1975a): Report of the Meeting on Smaller Cetaceans, Montreal, 1-11 April 1974. *Journal of Fisheries Research Bd.*, Canada, 32(7), p. 889-983
12. MITCHELL, E. (1974b): *Porpoise, Dolphin and Small Whale Fisheries of the World.* IUCN Monograph No 3; IUCN, Morges, Switzerland
13. READ, A. J., VAN WAEREBEEK, K., REYES, J. C., MCKINNON, J. S. and L. C. LEHMAN (1988): The exploitation of small cetaceans in coastal Peru. *Biological Conservation*, 46, p. 53-70
14. TØNNESSEN, J. N. and A. O. JOHNSEN (1982): *The History of Modern Whaling.* C. Hurst, London
15. YABLOKOV, A.V. (1994): Validity of whaling data. *Nature*, 367 (6459), p. 108

## 2.3.7 MANATEES AND DUGONGS (*SIRENIA*)

Manatees and dugongs are another group of large mammals which are fully adapted to an aquatic life. The manatees (**Trichechidae**) occur along the Eastern American coast from Florida to Brazil and in the Amazonas, in Western Africa from Senegal to Angola and in the Niger. They inhabit brackish shallow coastal waters, lagoons, river estuaries and rivers. The dugongs (**Dugongidae**) are marine animals of the Indian Ocean where they occur in shallow coastal waters from Madagascar and East Africa to Australia. Both live exclusively from submerged or floating aquatic plants and grow to several hundreds of kilogram weight. Tab. 34 shows that both manatees and dugongs have been relentlessly exploited in the past because of their tasty meat and fat; they are now considered to be endangered.

**Table 35:** Overview of the utilisation of sirenians

| English name | Scientific name | Geographical distribution | Body weight (kg) | Utilisation |
|---|---|---|---|---|
| **MANATEES - TRICHECHIDAE** | | | | |
| Florida manatee | *Trichechus manatus* | Florida, Carribean region, C.-North. S.American coast | 400-680 | for tourism, weed control |
| West African manatee | *-senegalensis* | W.African coast, Niger | 400-680 | for meat and fat only |
| Amazon manatee | *-inunguis* | Amazon river system | <200 | for meat, hides, fat |
| **DUGONGS - DUGONGIDAE** | | | | |
| Dugong | *Dugong dugong* | Madagascar, E.African, S.W., S.-S.E.Asian coasts | 230-900 | for meat, fat, oil, hides, and bones |
| Steller's sea cow (extinct) | *Rhytina gigas* | North. Pacific islands | up to 4,000 | for meat and hides |

### Historical background

The use of the American manatees in historical times throughout their range and the associated decline and extirpation of them in large parts of it has been well documented (1, 9, 6, 13). Their tender meat and oil was valued in Pre-Columbian time, and their bones used for ritual purposes. Mound pipes in the form of manatees were widely used by the Indians in Florida for ceremonial smoking. Surplus meat was sold by them to the first Spanish explorers. With the introduction of firearms by the European settlers extensive hunting of manatees commenced in the 18th century, particularly in Suriname from where some companies specialised in exporting manatee meat as slave food to the Antilles.

During the 19th century a commercially produced meat preparation from the *Amazon manatee* was widely marketed and during the 1930s and 1940s their hides were exported by the thousands for making water hoses and machine belts. The meat of the West Indian species was regularly offered on the markets of Belize until the 1930s and in Guayana into the 1960s. Although the *West Indian manatee* was never commercially harvested on a large scale, the hunting pressure diminished them severely over most of their range. There are reports that after the advent of the European "buccaneers", manatees in Florida were unscrupulously decimated. However, a more recent thorough synthesis of all literature came to the conclusion that *T. manatus* in the Southeastern USA was historically never further distributed and probably not much more abundant than at present (16). Within Florida manatees have become more widespread and possibly more abundant since the 1950s due to protection, greater availability of warm water in winter and in some areas an increase of exotic vegetation. In Africa the subsistence use of the *West African manatee* by the local fishermen intensified only during the last decades to such an extent that the West African species has now become rare throughout its range. This is not so much a result of more efficient trapping, but simply due to the enormous increase of the human population all along the coasts and the associated demand for more meat (17). However, the introduction in shallow waters of shark and other nylon nets in which they get entangled and drown, is also killing many manatees.

Similarly, the *dugong* has been over-exploited in its entire range of distribution and is also badly affected by shark netting and marine fisheries. Commercial dugong fisheries operated once from Ceylon where the meat is particularly appreciated by the Moslem population. In several cultures dugong products, particularly its oil, are prized for medicinal properties (10). Dugong meat was offered daily on the markets of Colombo until 1970 when the dugongs were given a legal protective status (11). Along the northeastern coast of Australia a dugong fisheries industry existed during the late 19th century which processed thousands of dugongs mainly for oil (11). Due to protection afforded by the Australian authorities dugongs have again increased in numbers there.

Only in 1741 a giant species of the Dugongidae was discovered on the Commander islands in the Bering Sea off the coast of Kamtchatka. This remarkable animal weighing up to 4,000 kg, was named after its discoverer *Steller's Sea cow (Rhytina gigas)*. In a similar way to the Giant doves on the island of Mauritius this species provides a particularly sad example of the results of indiscriminate exploitation of island fauna. After their discovery the Commander islands became the base for seal hunters who slaughtered the huge "sea cows" not only for meat and fat but utilised their special skins for making light boats. Within only twenty years the species was totally exterminated (12).

## Utilisation and conservation

Both *manatees* and *dugongs* are particularly easy to hunt and this has certainly contributed to their decline. In West Africa manatees are traditionally caught either in simple door traps built out of poles in shallow waters which are frequented by them, or they are fished with baited hooks to which floating drums are attached. They are then usually killed and eaten by the local fishermen (17). Dugongs are either speared, harpooned or netted while they graze algae in shallow sea waters. They also get entangled in nets set to catch Sea turtles in which they then drown (11). In Northern Australia where dugongs are fairly abundant, indigenous hunting in the Torres Strait at a level of about 1,000 animals per year seems to be sustainable. Habitat destruction appears to be a more important threat to them than exploitation (14).

Generally it is extremely difficult to control or regulate the hunting or trapping of the sirenians and in most of their range the only way to conserve them is to create especially protected reserves with a potential for tourist use. This, however, is also most difficult as both manatees and dugongs move around and the coasts on which they depend are invariably densely inhabited by fishermen. There is thus little scope to enhance their conservation through utilisation programmes.

Due to their docility and ability to convert otherwise not used plant material into favoured meat, sirenians would be potentially ideal amimals for farming which has been suggested for manatees in West Africa (5, 2, 17). However, their slow rate of reproduction and a daily food requirement of more than 25kg vegetable matter seems to make this rather impracticable.

Another possibility for utilising manatees for economic returns is to translocate them into certain water systems or canals which they could help to keep open by eating the floating vegetation. This has been attempted in Guyana, Suriname, Panama, Mexico and the USA with *Trichechus manatus* and in Panama and Brazil with *T. inungis* (2, 15, 8). In Guyana some 200 manatees were used for weed clearance during a period of 20 years. They were found to be well suited for this purpose. Under the local conditions each manatee required an area of at least 0.7 ha for sustenance, but was able to successfully clear up to about 1.4 ha. Prior to the introduction of the manatees annual expenditure on manual weed clearance had been US$ 750 for just one canal and this was saved during the subsequent fifteen years. The manatees also reproduced in this canal (7, 3). In Brazil, 42 manatees were introduced into the Curua-Una reservoir near Santarem and a captive breeding scheme to populate the large Tucurui reservoir with *Trichechus inungis* was planned in 1983 (4).

The Florida, wildlife authorities have probably furnished the best example for conserving *manatees* through intensive public awareness campaigns. These have created a high degree of consciousness and interest of the public for these unique mammals so that they receive much attention and are also being viewed for recreation in some of the overwintering localities. However, with the rapid increase of motor boats for recreational purposes a new threat to the manatees has arisen, as when submerged they are easily and inadvertently injured by the propellers of the boats.

## Literature references


1. ALLEN, G.M. (1942): *Extinct and vanishing mammals of the western hemisphere*. American Commission for International Wildlife Protection, Spec. Publ. 11; p. 528-552; Intelligence Print., Lancaster, Pa., USA
2. ALLSOPP, W.H.L. (1969): Aquatic Weed control by manatees - its prospects and problems. In: *Man-Made Lakes*, ed. OBENG, L.E.; p. 344-351; Ghana University Press, Accra, Ghana

3. ANONYMOUS (1974): *An international centre for manatee research. Report of a workshop held in Georgetown, 7-13 February 1974.* National Science Research Council, Georgetown, Guayana
4, BARZDO, J., LUXMOORE, R., BROAD, S. and D. JONES (1984): *A survey of wildlife farming operations.* WCMC/IUCN, Cambridge (unpubl. report)
5. BEAL, W.P. (1939): The Manatee as a food animal. *Nigerian Field*, 8, p. 124-126
6. BERTRAM, G.C.L. and C.K.R. (1968): *The Sirenians as aquatic meat producing herbivores.* Symposium of the Zoological Society of London, 21, p. 385-391
7. BERTRAM, G.C.L. and C.K.R. (1963): The Status of manatees in the Guyanas. *Oryx*, 7 (2/3), p. 90-93
8. ETHERIDGE, K., RATHBURN, G.B., POWELL, J.A. and H.I. KOCHMAN (1985): Consumption of Aquatic Plants by the West Indian Manatee. *Journal of Aquatic Plant Management*, 23, p. 21-25
9. HUSAR, S. (1977): The West Indian Manatee (*Trichechus manatus*). *Wildlife Research Reports*, 7; US Department of the Interior, Fish and Wildlife Service, Washington, DC
10. JONES, S. (1967): The dugong - its present status in the seas around India with observations on its behaviour in captivity. *International Zoo Yearbook*, 7, p. 215-220
11. KURT, F. (1972): Der Dugong. In: *Grzimeks Tierleben*, Vol. 12, p. 529-532; Kindler, Zürich
12. KURT, F. and H. WENDT (1972): Die Stellersche Sehkuh. In: *Grzimeks Tierleben*, Vol. 12, p. 532-535; Kindler, Zürich
13. LEFEBVRE, L.W., O'SHEA, T.J., RATHBUN, G.B. and R.C. BEST (1983): Distribution, Status, and Biogeography of the West Indian Manatee, In: *Biogeography of the West Indies: past, present and future*; ed. WOODS, C.A.; p. 567-610; Sandhill Crane Press, Gainesville, Fl., USA
14. MARSH, H. (1993): *in litt.*
15. McLAREN, J.P. (1967): Manatees as a naturalistic biological mosquito control method. *Mosquito News*, 27(3), p. 387-393
16. O'SHEA, T.J. (1988): The Past, Present and Future of Manatees in the Southeastern United States: Realities, Misunderstandings, and Enigmas. In: *Proceedings of the IIIrd Southeastern Nongame and Endangered Wildlife Symposium*, ed. ODOM, R.R., RIDDLE BURGER, K.A. and J.C. OZIER; p. 184-204; Georgia Department of Natural Resource, Social Circle, Ga., USA
17. ROTH, H.H. and E. WAITKUWAIT (1986): Répartition et status des éspéces de grands mammifères en Côte d'Ivoire. III. Les lamatins. *Mammalia*, 50(2), p. 227-242

## 2.3.8 ELEPHANTS (*PROBOSCIDEA*)

Elephants belong to a mammal order which was prehistorically very diverse and distributed all over Africa, Eurasia, North- and South America. Only 2 genera with 1 species each have survived into the Neolithicum, the Asian Elephant (*Elephas maximus*) and the African Elephant (*Loxodonta africana*). Both species were historically much more widely distributed than today. The Asian elephant occurred from Mesopotamia in West Asia through Persia all over the Indian subcontinent, the Malayan and Indochinese peninsulas far into China, and also on the Indonesian archipelago. Today there are only isolated populations of wild elephants left, mainly in India, Sri Lanka, Burma, Thailand and Indonesia. The African elephant inhabited nearly the entire African continent, including its northern part, and possibly ranged into Syria. Its distribution became limited in early historical times to Africa south of the Sahara; in West Africa only small relic populations have survived into the 1980s and in Eastern and Southern Africa the numbers of elephants have also dramatically decreased since 1976/78. Tab. 35 shows that both elephant species have been utilised by man since the earliest times for food and ivory and have also been semi-domesticated for various services.

**Table 35:** Overview of the utilisation of elephants

| English name | Scientific name | Geographical distribution | Shoulder height (m) | Utilisation |
|---|---|---|---|---|
| Indian elephant | *Elephas maximus* | S.-S.E.Asia | 2.40-3.00 | semi-domesticated for transport, riding, timber extraction; use of ivory |
| North African elephant (extinct) | *Loxodonta africana berbericus* | N.W.Africa | ~2.40 | semi-domesticated for warfare; hunted for ivory |
| African Forest elephant | -*a. cyclotis* | South. W.-C. Africa | 1.80-2.40 | exp. semi-domestication; hunted for ivory and meat |
| African Savanna elephant | -*a. africana* | North. W.-E. and S.Africa | 2.40-3.40 | hunted for ivory, meat, hides; used for tourism |

### Historical background

The association between elephants and man has been more varied and emotional than with any other animal group. It reaches back into the Middle Pleistocene some 400,000 years ago, when the ancestral forms of today's elephants, the *mammoth (Mammuthus primigenius)* and the *Straight-tusked elephant (Palaeoloxodon antiquus)* were already extensively hunted by hominids (8). Whereas the latter became extinct before the end of the Glacial Ages the mammoth survived into the Upper Palaeolithicum some 11,000 years ago, driven finally north with the retreat of the ice at the end of the last glaciation. Ice-age archaeological sites in the Ukraine have provided ample evidence for the hunting of the mammoth and the use of its bones and tusks for the construction of dwellings (17). Whole herds of perishing mammoths were preserved in the northern ice, and in 1821, for example, no less than

90,000 kg of mammoth tusk were recovered from the New Siberian islands of the Arctic ocean alone (8). The mammoth tusks have been utilised in ivory carving since the Middle Ages up to the present-day.

The use of ivory for making artefacts commenced in prehistoric time and by the 2nd millennium BC it was already well developed in the Mediterranean area, long before the procurer animal was known. In fact, to Homer the Greek word *Elephas* meant "ivory" and not "elephant". Ivory has remained one of the most important and valuable animal raw materials ever since. The art of ivory carving was further perfected in the Assyrian, Babylonian, Phoenician and Egyptian civilisations (8) but it is not known from where all the ivory which was used in those early times came from; some of it may have originated from an assumed *Syrian elephant subspecies*. During the Roman era the use of ivory reached a level of extreme extravagance. Calligula, for example, had the stables of his horses made out of ivory (28). The enormous quantities of ivory which were processed in those days have certainly originated to a large extent from the *northern subspecies of the African elephant*. This exploitation has probably contributed significantly to the extermination of this subspecies by the 6th century AD (16). Already before the time of Christ, from the 6th century BC, elephant tusks were exported from Africa via Ethiopia to Arabia and further east into Asian countries (31). In these the supply of ivory from the *Asian elephant species* has not sufficed to meet the growing demand considering that only males in the Asian species carry tusks, and that these were also valued for training into war elephants. Later on the Arabs organised a flourishing trade in "white gold" together with slaves from the interior of Africa to the ports along the eastern coast of Africa to Oman, Arabia, Egypt, Persia, India and China. This trade was essential for the expansion of major carving centres in South and Eastern Asia from the 18th century onwards, described in Chapt. 3.2.7. With the development of the British and French empires in the 19th century the trade in ivory eventually reached its all-time peak before the First World war, when about 1,000 t of ivory were leaving Africa annually. If one puts the average tusk weight at that time to 5-10 kg this quantity of ivory corresponded to the killing of about 50,000-100,000 elephants per year (26).

Whereas the prehistoric hunters probably viewed elephants simply as a source of meat and raw materials, depicting their hunting in many caves in Europe, Africa and Asia, the peoples of the Indus valley in the North-West of the Indian subcontinent learnt to capture and to tame the *Indian elephant* for services probably as early as the 3rd millennium BC (25). The skill to train captive elephants for work was introduced from there to Southern India by the Dravidian people when they were displaced from the Indus valley by the invading Aryans in the course of the 3rd millennium BC (29). The semi-domesticated elephants were used for the agricultural colonisation of wet valleys such as the Ganga and Bramaputra. Apart from captive stocks having to be replenished from the wild, the semi-domesticated elephant has thus contributed significantly to the decline of the wild elephant populations by enhancing conflict between these and man's agricultural interests. With agricultural development proceeding all over the Indian subcontinent this conflict with its multiple facets has become the major cause for the early diminution of the wild elephants not only in India but all over Asia (29).

However, apart from the agricultural conflict, the capture of elephants for training as war machines must have constituted an unsustainable drain on the wild elephant resource. This use was already mentioned in a script of about 1100 BC and has been subsequently reported for many battles in West Asia and the Indian subcontinent (29). Elephants were used as a sort of heavy cavalry for frontal attacks, but also had great logistic importance by carrying baggage. The possession of large elephant stables and of war elephants remained a matter of prestige in India and the neighbouring countries well into the Middle Ages. The Delhi Sultanate around 1340 for example, maintained 3,000 elephants of which 750-1,000 were trained for warfare (29); the Moghul emperors later on are reported to have kept as many as tens of thousands of elephants (21). The use of trained elephants in warfare became known to the western world after the defeat by Alexander the Great of the Persians in 331 BC and of King Porus of the Punjab in 326 BC (33, 8). He took over the captured war elephants and within a century the use of trained elephants was extended via Syria and Egypt throughout the Mediterranean theatre.

In Egypt knowledge of tamed Asian elephants had existed as early as 1500 BC as evidenced by an engraving of a domesticated Asian Elephant on the tomb of vizier Rehkmir (12). However, only during the confrontation of the Ptolemy dynasty with the Syrians from about 300 BC did efforts commence to also tame and train *African elephants* which were well-known in Egypt from its southern regions and the neighbouring Kingdom of Meroë, where African elephants might have been semi-domesticated earlier (28). Being cut off from the supply of Indian-trained elephants through Syria, they were shipped in on specially designed boats via Red Sea ports near Massowa; however, detachments of soldiers were also sent to capture local elephants (33). These were trained initially with the help of hired Indian "mahouts", but the training capability was soon acquired by the locals and this enabled Ptolemy III and IV to fight successfully against their Asiatic opponents. In the battle of Raphia in 217 BC African elephants were pitted against Indian war elephants but according to contemporary accounts the African elephants proved inferior to the Indian in size and training (33). The example of Egypt must have inspired the Carthaginians into training local elephants, enabling them to invade Sicily with sixty African war elephants in 262 BC. This event and the subsequent attack by Hannibal who crossed the Pyrenees and Alps with 37 war elephants in 218 BC, brought the Romans in contact with trained elephants which they then used themselves for more than 200 years for warfare and shows. The last recorded use in warfare was in 43 AD by Marc Anthony (28). As in India the capture of large numbers of elephants in Northwestern Africa for domestication combined with ivory hunting and desiccation of their habitat, certainly contributed to the early decline and extirpation of the Northern African Elephant subspecies.

On the Indian subcontinent mainly the spread of settled agriculture, but also capture and hunting had already eliminated wild elephants from the larger part of their range by the 10th century (29). However, by then elephants had passed indelibly into the culture of Asian peoples, through art, legend and religion. As local stocks became insufficient to meet the demand for trained elephants and no captive breeding was attempted, trade in live elephants from areas where they were still abundant, developed. Such trade had in fact already been well established early on; Ceylon, for example, exported elephants in the 6th century BC. Although the utilisation of elephants for warfare ceased with the introduction of gun powder and cannons, the demand for trained elephants has persisted over the centuries into our time. Traditional use for transport in difficult terrain continued and the possession of trained elephants still signifies wealth and prestige sought by princes, landlords and temples. With the colonial era in the 18th and 19th century semi-domesticated elephants assumed another economically important role in forestry, extracting timber from swampy areas and assisting in the transport of the logs. Thus the importance of live elephant trade has continued through the centuries until today (21).

The drain of *Asian elephants* through capture and control of crop raiders accelerated in the 17th and 18th centuries and reached a peak when the British colonisers introduced sport hunting. Records from 1868 to 1980 suggest that about 30,000-50,000 elephants were captured or killed within a century on the Indian subcontinent alone, and over 100,000 in the whole of Asia, exceeding by far the number of presently living wild elephants (20, 28). Although administrative and legislative attempts to protect wild elephants and to regulate their capture for training commenced in India in 1873, they did not halt further decline until 1975 when the Asian elephant was placed on Appendix I of CITES as endangered species.

The *African elephants* remained a plentiful resource until excessive commercial hunting for ivory in the 19th century started to diminish elephant populations locally, first in South and West Africa (27). In contrast to the historical situation in Asia, serious conflict with agricultural intensification arose only after the Second World War, but by 1977/80 there were still about 1.2 million elephants inhabiting the greater part of Eastern, Central and Southern Africa (9). However, this number was reduced to only 600,000 within the following ten years, some countries like Kenya, Tanzania and Zambia, losing almost three quarters of their elephant populations (3). Due to this spectacular decline the African elephant was also placed on Appendix I of CITES in 1989. The controversies resulting from this decision are discussed in Chapt. 1.4.

Also in contrast to the situation in Asia, the ancient skill of taming *African elephants* for use was lost and only remembered by the Belgian colonisers at the end of the 19th century. King Leopold I of Belgium himself took an interest in the possibility of domesticating the African elephant in the Belgian Congo and ordered the establishment of an elephant domestication station at Api, which was later transferred to Gangala-na-Bodio in the far North-West of Congo (now Zaire) (14). It was thought that similarly to its Asian cousin the African elephant could be profitably employed in forestry work. Various capture methods were experimented with (19) and Indian "mahouts" were imported to train local "carnaks", the name applied with pride to the elephant handlers of the Azande tribe. History has shown that the African elephant may be trained as well as the Asian. However, the elephants at the Gangala-na-Bodio elephant station never attained the same degree of docility as their Indian cousins. Injuries and fatal accidents to the elephant drivers at Gangala-na-Bodio occurred quite frequently. By 1930 there were some 70 trained elephants in use in the Congo (15), but other African countries did not follow suit (7). With the mechanisation of forestry work as of the 1950s the Belgian initiative began to loose its economic justification and the Gangala-na-Bodio station became a mere curiosity, supplying zoos and circuses in Europe with elephants. It barely survived the post-independence turmoil with only 9 trained elephants left in 1974 (12) and it is now administered by the Institut National Zairois pour la Conservation de la Nature for research purposes.

## Management and utilisation

Only in exceptional cases do wild elephants still inhabit permanently unprotected agricultural or forest lands. In many countries, however, elephant populations disperse seasonally from wildlife or forest reserves far into adjacent agriculturally utilised areas where they are perceived as a hazard and pest. **Management of wild elephants** in Asia as well as in Africa must therefore aim at minimising the conflict situation, but at the same time enhance non-consumptive uses of elephants which will give economic justification for maintaining the elephant herds. This can best be achieved by physical separation of the *wild elephants* from agricultural lands by means of ditches, elephant proof or electric fences. Other methods are deterrence in regularly patrolled buffer zones around the elephants permanent range combined with planning of agricultural development that takes the elephants into account (29). The perfection of the drug immobilisation technique in the 1960s has also made it possible to capture and translocate relic or stray elephants into wildlife conservation areas. In some instances whole herds have been driven by helicopter and ground crews from agricultural lands into sanctuaries.

In Africa as well as Asia the most promising use of elephants is ***to incorporate them into wildlife tourism,*** but this presupposes non-disturbance of the elephant population for long periods with resultant easy observability of them. Strictly controlled trophy hunting can be the most lucrative form of utilisation and may also be applied to shy elephant herds outside of conservation areas (5). ***Consumptive use*** of *African elephants*, except for illegal poach-

A killed African elephant provides up to 1,000 kg of valuable meat to the indigenous people. (Photo: V. RADTKE; New York Zoological Society)

ing for meat and tusks, is now mainly limited to conservation areas in Southern Africa in which elephant populations exceed the carrying capacity of the range. There, management is not only concerned with the protection of elephants and prevention of crop damage by them, but must also aim at protecting the habitat and landscape against the elephants. To this end elephant populations are selectively "culled" on the basis of computerised population models, either by dart immobilisation from helicopters and subsequent transport to central slaughter and processing plants, or by shooting of whole herds with automatic weapons on the ground to minimise the disturbance factor. Although meat, bones and carcasses are economically utilised, this is only a secondary aspect of the necessity to regulate populations according to ecological needs (24, 11, 18, 10).

In the 1960s, prior to the spectacular decline of elephants in East Africa due to commercial ivory poaching in the 1970s, meat production from elephant herds also had economic importance in Uganda, Tanzania and Zambia (22, 23). In the Murchison Falls National Park in Uganda some 2,000 elephants were culled in 1965-67 and 600 in the Nkomasi Reserve in Tanzania in 1967-68. The average dressed out carcass weight was 850 kg (50% of liveweights) (13), but a fully grown elephant will yield more than a ton of meat. The culling operations were contracted out to private enterprises which would market the meat locally. In a UNDP/FAO project in the Luangwa Valley of Zambia in 1970-74 processed elephant meat was deep-frozen into 25 kg blocks for sale in the 600 km distant industrialised "copper belt" (1). In Southern Africa elephant meat is converted into the popular "biltong" or canned as pet food (11). Technologies were developed to slice elephant hides for better tanning, enabling the production of high-prized quality leather articles from them.

An alternative to culling is to reduce elephant populations by planned and controlled capture of young animals for sale or translocation into other conservation areas. This use is also applicable to the *Asian elephants*, considering the continued demand for trained male elephants (29).

The **use of elephants in a tamed, semi-domesticated state** applies only to the *Asian elephant* and has a long tradition as discussed above. Until recently provision of elephants for training depended entirely on capture; breeding within the semi-domesticated herds has always been discouraged for economic and practical reasons. The *khedda* (or *kraal*) capture method was historically developed in Northern India where it is still preferred. This allows multiple captures by driving whole herds into a stockade in which suitable animals are selected and removed with the aid of tamed adult females. In Southern India individual elephants are either *pitted* or captured by different versions of the *mela-shikar* method. This involves the noosing of the head or foot of a selected young elephant from the back of a tame riding elephant. In some of the countries in which trained elephants are still extensively used, capture is now regulated in terms of a quota system in accordance with estimates of the wild populations. In Burma, for example, the annual quota from 1973 to 1980 was 200 and in 1981-82 150 animals, but only a total of 1,121, i.e. an average of about 125/year, were captured during this period (2). The newly captured are then "broken" by first tying and starving them. Later they are led between already trained adult females and learn by reward and punishment to understand verbal commands and tactile signs from the *mahout* who usually remains attached to the animal for life. Full training takes 2-3 years, but the optimum for forestry work may only be attained at the age of 25-30 years (2).

Asian elephants at forestry work in Thailand. (Photo: C. KANTAPANIT; by courtesy of the Forest Industries Organisation, Thailand)

Semi-domesticated elephants are still extensively used in forestry work in Burma, Thailand and some Indian states, as well as for ceremonial, transport and tourist uses. In forestry work elephants are kept in camps from which they are employed for timber extraction only for a few hours per day during a season of 7 months. After work they are left to forage by themselves in the semi-natural surroundings of the camps (2). The majority of semi-domesticated elephants are owned by the forestry departments or timber corporations, but there is also private ownership.

The demand for trained elephants continues but cannot be met anymore from the wild. Therefore captive breeding, although very costly, is being promoted. The average breeding success in Burma during 1972-82 was only 2.3% of the total stock of semi-domesticated elephants as compared to 4.2% recruitment by purchase or capture (2). With the knowledge of the elephant's reproductive physiology now available and the possibility of chemical restraint, artificial insemination seems to be feasible and is being experimented with in zoos.

## Economic importance

Elephants have had great economic importance throughout history as ***suppliers of ivory*** which has been highly valued by man since the earliest times as described above. Fig. 16 shows the volume of ivory exports from Africa since 1950 until all trade in ivory was banned under CITES in 1989. The annual exports increased steadily from about 200 t to a peak of more than 1,000 t in the early 1980s. Thereafter quantities of exported raw ivory diminished drastically but the ivory price increased threefold. Elephants generated annual monetary values between US$ 64 and US$ 49 mill. until this dropped to US$ 21.3 mill. in 1988 despite the price increase (3). The international trade of ivory is further discussed in Chapt. 3.2.7.

**Figure 16:** African raw ivory exports in 1950-1989 and their value in 1979-1989[1)] (after BARBIER and coll. (3))

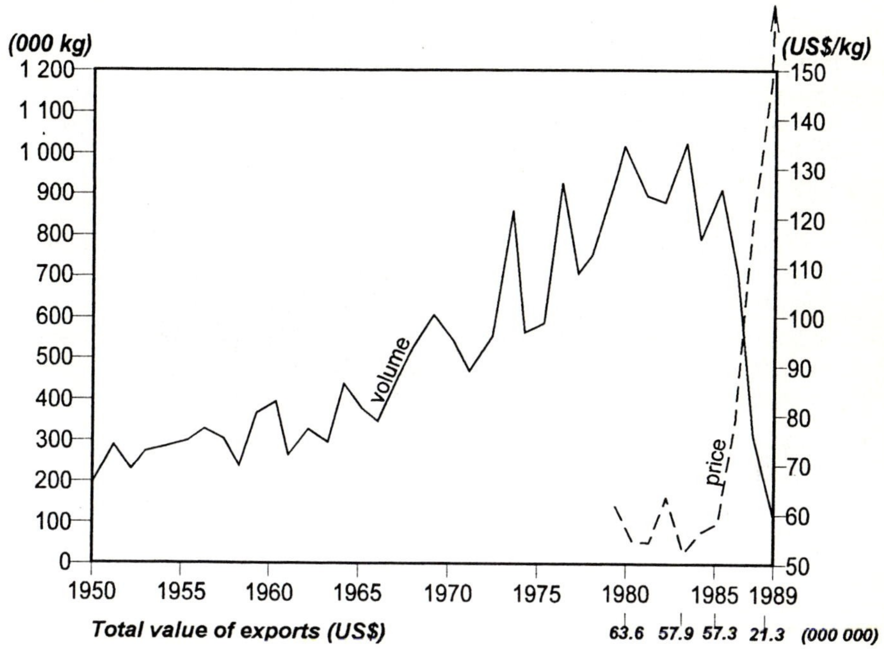


There is no doubt that the documented exploitation of the African elephant for ivory was unsustainable and would have lead to a dangerous depletion of this species on the continent as a whole. However, the wisdom of a total ban of any commerce in elephant products as the most appropriate means for elephant conservation is being strongly contested by countries in which elephant populations have been maintained at high levels due to rational and effective management. It is argued that a deprivation of the right to gain financial returns from elephants lays an unwarranted financial burden on the management authorities of these countries at a time when budgets for conservation are less adequate than ever. In Zimbabwe, for example, where the elephant population has doubled since and despite the introduction of culling in 1965, it is estimated that protection against determined illegal hunting entails an annual recurrent expenditure of no less than US$ 200 per km$^2$. National governments can hardly be seen to invest such moneys unless the elephants together with other wildlife earn it (18). It has been shown in Zimbabwe that elephant populations under appropriate management are capable of returning US$ 378/km$^2$. Sports hunting earned 48.16%, culling operations 23.39%, elephant control 10.45% and natural deaths 18% of this return. In terms of commodities, 43% of the total return came from ivory, 7% from meat and 2.5% from hide sales. Management, however, is not primarily aimed at producing such returns but at the conservation of elephants of which the bulk live in wildlife reserves at ecologically acceptable levels; the inherent consumptive use is a strictly secondary aspect, as discussed above.

In this context, the economics of the much abused *"elephant control"* complex also need to be reviewed. This term has been used in the past for any apparently justified elimination of individual or entire herds of elephants which were - often subjectively - deemed to constitute a hazard to humans, agriculture or forestry activities. Like any animal control it should, however, be based on objective economic considerations. Only where irrigated or industrial plantations with high cash returns are at stake, like sugar, oil palms, rubber, etc., are high investments into protective barriers justified. Electric fences in Asia as well as Africa have shown an 80% efficiency depending on maintenance, and are much less expensive than trenches or elephant proof fences (29). In Malaysia electric fences at an investment of US$ 1,200/km plus recurrent maintenance cost of about US$ 1,600/km over five years proved to be well cost-effective if each km of fencing protected 200 ha of oil palm plantation at 80% efficiency (4). However, in sparsely settled areas with subsistence crops it is most economical to kill crop raiders and to utilise their carcasses and ivory. The resulting financial returns will certainly offset the crop losses. In the CAMPFIRE programme in Zimbabwe, referred to in Chapt. 1.4., commercial trophy hunting of elephants has yielded 64% of the total income from trophy fees levied by the rural districts concerned (5). It is being successfully incorporated into necessary elephant control whereby the monetary return from a "problem" elephant may be increased tenfold (30).

Even before the international banning of ivory trade, studies were conducted to analyse the indirect economic importance of ***elephants for tourism***, which is not reflected by the above, but is discussed in Chapt. 1.7. They suggest that in countries with important wildlife tourism the indirect use value of elephants far exceeds that of any potential sustained ivory harvests (32). In Kenya, for example, elephants were estimated to be worth US$ 25 mill. per annum for tourism, which in terms of numbers of foreign visitors increased by 10-15% per year. Almost 50% of questioned tourists considered Kenya no longer worth visiting, if elephants were significantly reduced in number (6). This does not yet include the important indirect use value of elephants derived from their natural function in diversifying savannas and forest ecosystems (32).

The ***use of semi-domesticated Asian elephants*** has previously had a much greater importance than today. Tab. 36 shows that the stocks held and mainly employed in forestry have diminished considerably in all countries for which statistics are available. However, if compared with the most recent estimates for the still-existing wild populations they are relatively important. In fact, in some countries there were almost as many elephants in captivity as in the wild at the time of the survey; and of the still existing 50,000-70,000 Asian elephants, ±33% at an average were semi-domesticated. The decrease of semi-domesticated elephants was mainly due to the decline of the wild populations and the resultant difficulties and costs of recruiting young elephants for training in the absence of any significant breeding (21). Only to a lesser degree, has mechanisation in forestry also made semi-domesticated elephants redundant. However, in Burma, Thailand and Laos they are still extensively used and are considered to have definite advantages over machines. They are much less destructive and particularly useful in clearing jams of rafted logs during river transport. In 1982 the State Timber Corporation in Burma logged 1.2 mill. t of which only 2% was solely extracted by machines (2). Apart from forestry, semi-domesticated elephants are part of Asian culture and have an important traditional role in religious and economic life. They are also becoming increasingly important in tourism as ideal vehicles for game viewing and

Use of semi-domestic Asian elephants for tourist visits to inaccessible conservation areas in India. (Photo: WWF-GERMANY)

other transport in national parks. The relocation of Indian rhinos into various wildlife reserves in Assam would not have been possible without trained elephants pulling the heavy transport crates over the swampy ground.

## Implications for conservation

Elephants in Africa provide an interesting example of the consequences of different approaches to the management of wildlife as a natural resource and of the varying levels of technical capabilities in this respect. These have resulted in three different situations for the elephants in Africa as a species that is totally incompatible with man's agricultural aspirations:

- Failure to protect and to purposefully manage elephant populations in the course of colonisation and agricultural development in most of West Africa has resulted in loss or early depletion of the resource below the level of any possible utilisation;
- Purely preservation-minded protection of elephants in East Africa was fairly successful in maintaining the resource until suddenly increasing demand for ivory has created a very high illegal hunting pressure on the resource which in turn has diminished it to a low but still potentially utilisable level;

- Management on an ecological basis with an open mind to economic utilisation of the re-source in Southern Africa has enhanced this into a relative abundance which does not only allow for profitable use, but necessarily calls for regulatory intervention to maintain its viability in future.

**Table 36:** Estimated numbers of elephants in Asian countries

| Country | Semi-domesticated past[1)] | Semi-domesticated 1980s[2)] | Wild elephants 1989[3)] | Total 1980s |
|---|---|---|---|---|
| India | n.a. | 2,200-2,800 | 16,455-21,970 | 18,655-24,770 |
| Burma | ~7,000 (1940) | | | |
| | 5,973 (1978) | 5,400 | 6,000-10,000 | 11,400-15,400 |
| Thailand | 13,392 (1950) | | | |
| | 8,438 (1972) | 3,500-5,000 | 2,925-4,550 | 6,425-9,550 |
| Indonesia | | 50-100 | 2,800-4,800 | 2,850-4,900 |
| Sri Lanka | 670 (1955) | 400-500 | 2,000-4,000 | 2,400-4,500 |
| Laos | 902 (1975) | 1,000-1,300 | 1,500-2,000 | 2,500-3,300 |
| Cambodia | 582 (1975) | 500-600 | 1,500-2,000 | 2,000-2,600 |
| Vietnam | n.a. | 600 | 500-1,000 | 1,100-1,600 |
| Malaysia | n.a. | <50 | 1,300-5,000 | 1,350-5,050 |
| Bangladesh | n.a. | 50 | 200-350 | 250-400 |
| Nepal | n.a. | 60-80 | 50-85 | 110-165 |
| China | n.a. | 15 | 100-230 | 115-245 |
| Bhutan | n.a. | 0 | 60 | 60 |
| All Countries | | 13,825-16,495 | 35,390-56,045 | 49,215-72,540 |

1) according to OLIVIER (21); 2) according to IUCN/SSC Specialist Group on Asian elephants and C. SANTIAPILLAI, WWF (years not specified); 3) according to SUKUMAR (29) (reference years vary)

The reasons for the differences in the three regions are deeply rooted in colonial history and the varying mentalities of the colonisers. This different background is also reflected in many other areas of public life, administration and conservation in general. It is thus an over-simplification to merely hold ivory hunting responsible for Africa's elephant crisis: the causes for the elephant decline are complex and differ even within the mentioned regions. Accordingly there is no simple solution to elephant conservation and remedies must be ad-justed to the three different conservation situations. Whereas the total ban n ivory trade will certainly aid the survival of the elephants in West and also in Central Africa, it is equally certain that it will harm elephant conservation in Southern Africa by removing an important economic incentive for the maintenance and protection of elephants at present levels. In East Africa, of course, one can live with the ban and profit from non-consumptive use of elephants but only as long as no ecological need for population reductions arises.

In Asia the situation of the elephant is different again. Ivory hunting has never been the major reason for the elephant decline, this having been caused mainly by the early clash of elephants with agricultural development and human settlement and by the excessive capture of young elephants for replenishment of semi-domesticated elephant stocks. The conservation requirements there are to create as much economic justification as possible for protective management of the few remaining wild populations by development of tourist game viewing and as a source of well controlled capture of young males for sale. Since the demand for elephants for training presently exceeds sustainability of capture it is essential to promote, even if costly, more captive breeding within the semi-domesticated elephant stock.

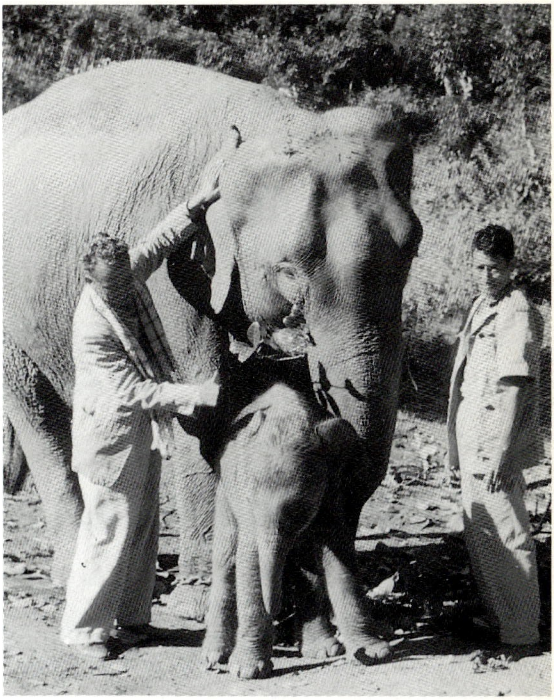

Breeding of privately-owned domestic elephants in India is still rare. (Photo: H.H. ROTH)

## Literature references


1. ALBL, P. (1971): Studies on assessment of physical condition in African elephants. *Biological Conservation*, 3, p. 134-140
2. ANONYMOUS (1992): Elephants, distribution, status and conservation in Burma. Ministry of Agriculture and Forests, Rangoon (*cited after* BARZDO, J., LUXMOORE, R., BROAD, S. and D. JONES (1984): *A survey of wildlife farming operations*, WCMC/IUCN, Cambridge (unpubl. report))
3. BARBIER, E.B., BURGESS, J.C., SWANSON, T.M. and D.W. PEARCE (1990): *Elephants, Economics and Ivory*. Earthscan, London
4. BLAIR, J.A.S., BOON, G.G. and N.M. NOOR (1979): Conservation or cultivation: The confrontation between the Asian elephant and land development in peninsular Malaysia. *Land Development Digest*, 2, p. 27-59
5. BOND, I. (1994): The Importance of Sport-hunted African Elehants to CAMPFIRE in Zimbabwe. *TRAFFIC Bulletin*, 14(3), p. 117-119
6. BROWN, G. and W. HENRY (1989): *The Economic Value of elephants. LEED Discussion Paper 89-12*; London Environmental Economic Centre, London
7. CALDWELL, K. (1925): Elephant domestication in the Belgian Congo. *Journal of the Society for the Preservation of Fauna of the Empire*, Part 7, p. 71-82
8. CLUTTON-BROCK, J. (1987): *A Natural History of Domesticated Mammals*. Cambridge University Press and British Museum (Natural History), Cambridge and London
9. CUMMING, D.H.M. and P. JACKSON (eds.) (1984): The Status and Conservation of Africa's Elephants and Rhinos. *Proceedings of a Joint Meeting of the IUCN/SSC African Ele-*

*phants and Rhinos Specialist Groups at Hwange Safari Lodge, Zimbabwe, 30 July - 7 August 1981*; IUCN Gland, Switzerland

10. CUMMING, D.H.M. (1983): The decision-making framework with regard to the culling of large mammals in Zimbabwe. In: *Management of large mammals in African conservation areas*, ed. OWEN-SMITH, R.N.; p. 173-186; Haum Educational, Pretoria, South Africa
11. DE VOS, V., BENGIS, R.C. and H.J. COETZEE (1983): Population control of large mammals in the Kruger National Park. In: *Management of large mammals in African conservation areas*, ed. OWEN-SMITH, R.N.; p. 213-231; Haum Educational, Pretoria, South Africa
12. DOUGLAS-HAMILTON, I. (1984): 21. African elephant. In: *Evolution of domesticated animals*, ed. MASON, I.L.; p. 193-198; Longman Group, London and New York
13. FIELD, C.R. (1974): Scientific utilisation of wildlife for meat in East Africa: A review. *Journal of the South African Wildlife Management Association*, 4(3), p. 177-183
14. GRZIMEK, B. (1956): Die belgische Elefanten-Zähmungsstation Gangala-na-Bodio. *Säugetierkundliche Mitteilungen*, 4, p. 1-10
15. HUFFMANN, C. (1931): La domestication de l'elephant au Congo Belge. *Bulletin Agricole du Congo Belge*, 22(1), p. 7-22
16. JOELEAUD, L. (1932): Etudes de géographie zoologique sur la Bérbérie. Les Proboscidiens. I. L'Eléphant d'Afrique. *Bulletin de la Societé de Zoologie de France*, 56, p. 489-527
17. KLEIN, R.G. (1974): Ice-Age hunters in the Ukraine. *Scientific American*, 230(6), p. 96-105
18. MARTIN, R.B., CRAIG, G.C. and V.R. BOOTH (eds.) (1992): *Elephant Management in Zimbabwe. Second Edition*; Department of National Parks and Wildlife Management, Harare, Zimbabwe
19. OFFERMANN, P. (1930): La capture et la domestication des éléphants dans la province Orientale. *Bulletin Agricole du Congo Belge*, 21, p. 384-387
20. OLIVIER, R.C.D. (1978): Distribution and status of the Asian elephant. *Oryx*, 14, p. 379-424
21. OLIVIER, R.C.D. (1984): 20. Asian Elephant. In: *Evolution of domesticated animals*, ed. MASON, I.L.; p. 185-192; Longman Group, London and New York
22. PARKER, I.S.C. (1985): Perspectives on wildlife cropping or culling. In: *Conservation and wildlife management in Africa*, ed. BELL, R.H.V. and E. McSHANE-CALUZI; p. 233-253; Peace Corps, Washington, DC
23. PARKER, I.S.C. (1987): Game cropping in the Serenget region. *Parks*, 12(2), p. 12-13
24. PIENAAR, U de V. (1969): Why elephant culling is necessary. *African Wildlife*, 23, p. 181-194
25. RAO, H.S. (1956): History of our knowledge of the Indian fauna through the ages. *Journal of the Bombay Natural History Society*, 54, p. 151-280
26. RICCIUTI, E.R. (1980): The ivory wars. *Animal Kingdom*, 83 (1), p. 6-58
27. ROTH, H.H. and I. DOUGLAS-HAMILTON (1991): Distribution and status of elephants in West Africa. *Mammalia*, 55(4), p. 489-527
28. SCULLARD, H.H. (1974): *The elephant in the Greek and Roman world*. Thames and Hudson, London
29 SUKUMAR, R. (1989): *The Asian Elephant: Ecology and Management*. Cambridge University Press, Cambridge
30. TAYLOR, R.D. (1993): Elephant management in Nyaminyami District, Zimbabwe: turning a liability into an asset. *Pachyderm*, 17, p. 19-29 (*cited in Oryx*, 28(3), p. 145-146)
31. WARMINGTON, E.H. (1974): *The Commerce between the Roman Empire and India*. Vikas, New Delhi, India
32. WESTERN, D. (1989): The Ecological Value of Elephants: A Keystone Role in African Ecosystems. In: *The Ivory Trade and the Future of the African Elephant*; ITRG report prepared for the Second Meeting of the CITES African Elephant Working Group, Gaberone, Botswana
33. ZEUNER, F.E. (1963): *A History of domesticated animals*. Hutchinson, London

## 2.3.9 ODD-TOED UNGULATES (*PERISSODACTYLA*)

Ungulate mammals constitute the most important wildlife resource worldwide. They have evolved into two distinct orders: the Odd-toed (**Perissodactyla**) and Even-toed or Cloven-hooved ungulates (**Artiodactyla**). Perissodactyles comprise tapirs (**Tapiridae**), rhinos (**Rhinocerotidae**) and Horse-like ungulates (**Equidae**). Tapirs are animals of the dense evergreen forests and occur in three species (*Tapirus* spp.) in Central and South America and in one species (*Acrocodia indica*) in South-East Asia. Rhinoceros were historically distributed in 2 species over most of the savanna zones of Africa (*Diceros bicornis* and *Ceratotherium simum*) and in 3 species in the swamps and dense forests of South and Southeastern Asia (*Rhinoceros* spp. and *Dicerorhinus sumatrensis*). All members of these animal groups have become rare and are endangered. Of the Horse-like ungulates only some of the subspecies of the Savanna zebra (*Equus quagga* sspp.) are still abundant over large parts of East and Southern Africa; the progenitor species of the Domestic horse (*E. przewalskii*) which occurred historically from Western Europe to Central Asia, and that of the Domestic ass (*E. africanus*) in North-eastern Africa have both survived only in very small relic populations and are on the verge of becoming extinct. Similarly the Asian Hemione equids (*E. hemionus* sspp.), historically much utilised, have all become very rare. Tab. 37 shows that the main use of tapirs has been for meat and hides whereas the rhinos have always been persecuted for their horn. Zebras are ranched for meat and hides and hunted for sport.

**Table 37:** Overview of the utilisation of perissodactyles

| English name | Scientific name | Geographical distribution | Body weight (kg) | Utilisation |
|---|---|---|---|---|
| **TAPIRS - TAPIRIDAE** | | | | |
| American tapir | *Tapirus* spp. | C. and S.America | 230-300 | for subsistence hunting |
| Asian tapir | *Acrocodia indica* | S.E.Asia | 250-350 | for subsistence hunting |
| **RHINOS - RHINOCEROTIDAE** | | | | |
| Black rhino | *Diceros bicornis* | C., E. and S.Africa | 1600-2000 | all rhinos are killed for their |
| White rhino | *Ceratotherium simum* | S. and C.Africa | 3200-3600 | horn and other |
| Indian rhino | *Rhinocerus unicornis* | S.Asia | 1600-2100 | body parts which |
| Javan rhino | *-sondaicus* | S.E.Asia | n.a. | are highly esteemed |
| Sumatran rhino | *Dicerorhinus sumatrensis* | S.E.Asia | 900-2000? | in Asian medicine |
| **HORSE-LIKE UNGULATES - EQUIDAE** | | | | |
| Wild horse | *Equus przewalskii* | originally Eurasia | 200-300 | domest. into dom. horse |
| African ass | *-africanus* | N.E.Africa | 200-300 | domestication into dom. donkey |
| Savanna zebra | *-quagga* sspp. | E. and S.Africa | 150-280 | ranched for hides and meat, hunted for sport |
| Kiang, onager, kulan | *-hemionus* sspp. | W. and C.Asia | 250-350 | hunted for sport and meat; onager was historically semi-domestic. |

## Historical background

Perissodactyles comprise mammal groups of rather different ecological and socio-cultural background and these are therefore best discussed separately. The one thing they have in common, however, is that - apart from the Savanna zebras - all of them have become rare and most of them are now endangered species with little or no potential for economic utilisation. Although exploitation for specific products, like rhino horn, has certainly been a most significant factor for their decline, perissodactyles in general are considered phylogenetically a less successful order than other ungulates. Restricted to specific habitats and with limited productivity they comprise vulnerable species which are prone to extermination. In fact, the majority of the equids had already become extinct in prehistorical times due to factors unrelated to excessive hunting pressure.

*Tapirs,* as animals of the humid tropical forest zones, have probably always been an important food resource for the local hunting and gathering forest dwellers just like any other forest fauna. Since tapirs are not very productive they were over-exploited as soon as human settlement and local population densities increased and impinged on their habitat (7). All tapir species with the exception of the *Lowland tapir (Tapirus terrestris)* in the Amazonian region have thus been reduced to non-utilisable, relic populations.

The *rhinoceros* have suffered a similar fate although the underlying socio-cultural reasons for their persecution and consequent decline were different. The rhinoceros was at the root of the myth of the "*unicorn*" of the Greeks and Romans, and the legendary powers attributed to the terrible weapon of this imaginary beast have been sought ever since in the horn of its living prototype. By the Middle Ages the identity of the rhinoceros and the "unicorn" was taken for granted and one of the goals of Marco Polo's travel was to find this most treasured animal (24). A result of the mediaeval mystical beliefs was also the use by royalty and politicians of beakers carved out of rhino horn to detect the presence of poison in drinks (21). Physicians prescribed powdered rhino horn up to the 16th century "as cure for all poisons, for fevers, for bites of mad dogs and scorpions, for falling sickness, worms, flukes, loss of memory, and the plague". The Chinese started to use rhino horn as a medicine over a thousand years ago, and traditional Chinese doctors still believe it to be unparalleled. It was mainly this medical demand for horn, and any other parts of the rhino, including even blood and urine, coupled with unrestricted hunting for sport during the colonial time (2) that led to a rapid decline in numbers of *Asian rhinos* during the 19th century until finally hunting of the Indian species was prohibited in 1910 (12). Although the Asian rhino species were preferred, the Asian market for rhino horn also caused large-scale commercial hunting of the *White rhino* in Southern Africa. This was a commercial supplement to ivory hunting until this species was nearly exterminated over most of its southern range at the turn of the century (20). Another reason for the slaughter of the rhino during the colonial era in Africa was simply to furnish meat for labourers and to make highly prized ox-wagon whips out of their hides (21). The *Black rhino* has survived the hunting pressure of that time in greater numbers due to its more wary behaviour and retreat into remote inaccessible bush areas. In the 1967 there were still an estimated 11,000-13,500 Black rhinos in Africa (9). This was, however, most probably an underestimate owing to the difficulties of accurate census of this species in those days.

The greatest historical importance of the *equids* was that two species, the *Eurasian Wild horse* and the *African Nubian ass* were domesticated early on for riding and draught purposes. More than any other animals these two domestic species have greatly influenced man's history, rendering possible continent-wide migrations and conquests, but also becoming an essential economic factor in agriculture, trade and communication. However, prior to and concurrently with the domestication of the horse and ass, the *onager* played an important role in the ancient cultures of Western Asia. It was formerly widespread throughout Persia and has been much hunted for food since prehistoric times. The onager is the most frequently depicted equid in Mesopotamian art of the 3rd and 2nd millennia BC and it was assumed for a long time that it had been domesticated by the Sumerians in the 3rd millennium BC. However, it is now held that most of the commonly found onager remains were from wild onagers and that this equid was never truly domesticated, probably because of its inherent refractory behaviour (3). It was used in a semi-domesticated state, possibly even in harness, and also interbred with the introduced domestic horses and donkeys (26). Until recent times hunting of the tough *Hemione asses* was considered to be a royal sport, which has certainly contributed to their decline or even extirpation.

In a few localities of their world-wide distribution by man, horses and asses became feral. In North and South America the so-called "mustangs" were such *feral horses* abandoned by early European settlers and later broken in for riding by the Indians. In Australia herds of feral horses, known as brumbies, were reported as of the late 1860s as "pest" animals (14). They are now common and widely distributed in Central and Northern Australia where they are shot in conjunction with feral goats, dromedaries and kangaroos as noxious animals and used for pet meat (18). Asses have also become feral in Australia and in many of the habitable parts of the Sahara (3).

The *zebras* in Africa were never utilised to the same extent as antelopes as a source of food, although their skins served various purposes. The early European settlers in South Africa, the Boers, shot the local zebra subspecies, the *quagga*, by the thousands to make grain bags out of their skins. This resulted in the extermination of this zebra subspecies in 1850-70 (11). Zebra have also been perceived as particularly undesirable livestock competitors as they graze very close to the ground and may thus become destructive to the grass cover. In arid areas, such as Botswana, they have been killed by the tens of thousands by fencing them off from their natural seasonal water supplies. With the development in the 1960s of game ranching in Southern Africa, *zebra* - as an important ecological component of the local savanna ungulate community - were included in the commercial utilisation of wildlife for meat and skins.

Domestication of the zebra was suggested during the early European colonisation with a view to their assumed resistance to trypanosomiasis; but this was never seriously attempted. Like many other species, zebras have been tamed occasionally to draw carts or carriages mainly for reasons of prestige. For example, one of the Berlin bakeries used to operate zebra-drawn carts for deliveries in the 1920s (11).

## Management and economic importance

Of the *tapirs* only the *Lowland tapir (Tapirus terrestris)* of the Amazonian region still has some economic importance for the subsistence of the local indios. However, since these swamp living animals are sensitive to human activities (7) even this species has declined and is now also considered to be endangered. Surveys of subsistence hunting in Amazonia have shown how the use of tapirs, usually speared or hunted with dogs for food and hides, has diminished over the past decade (23, 1). Although tapir meat is not particularly favoured it constituted, together with *peccaries*, between 75% and 85% of the total weight of game animals taken for food in village samples in 1973/74 (22) and it featured among the ten most popular dishes in restaurants sampled in 1975 in Manaus (25).

Management of *rhino* during the last fifty years has been strictly aimed at protection, but their illegal commercial exploitation exclusively for horn has continued to-date. Although the efforts to preserve the *Great Indian rhino* in a number of game reserves and national parks in Assam and Nepal have prevented the extermination of this species, which was imminent in the 1920s, all of the *Asian rhino species* are barely surviving the continuing pressure for their horn. Their only future lies in the enhancement of their non-consumptive use in tourism as exemplified with the *White rhino* in South Africa. This species was spectacularly brought back by propagation under strict human control of only half a dozen survivors in the two small game reserves Umfolosi and Hluhluwe in Natal at the turn of the century to more than 700 in the 1960s (17). To avoid destruction of habitat in these reserves a programme of systematic re-distribution into protected areas in and outside the species

White rhinos are captured at Umfolosi Game Reserve in South Africa for redistribution into conservation areas and onto private land for tourist viewing. (Photo: WILDLIFE SALISBURY)

former range was initiated in 1961, with a view to diversifying the wildlife resources of these areas for tourism development. In 1981 the total number of the *Southern White rhino* over all of its former range had increased to 2,801 animals (4). An important prerequisite for the redistribution was the development of chemical restraint and darting techniques in the early 1960s which made capture and translocation of rhinos possible. The amenable and docile White rhino is now the only rhino species which can be easily observed in public reserves and on well protected private lands. The *northern race of the species*, however, discovered only in 1908, has not enjoyed such protection and is thus now near extinction.

In the 1960s and 1970s procurement of rhino horn focused on the *Black rhino* as the only species which was still roaming the savanna areas in large numbers. It became the principal source of rhino horn for the Asian markets until all trade in rhino horn was banned under CITES in 1977, as described in Chapt. 3.2.8. In 1981 there were still some 12,000 Black rhino left (4) but illegal commercial hunting continued despite the trade ban. It is estimated that from 1980 to 1986 this has realised no less than US$ 8 mill. (10); by 1993 the species had been reduced to only about 2,400 animals. To further discourage and curb the poaching of Black rhino the wildlife authorities in Namibia and Zimbabwe have embarked on systematic de-horning of remaining rhinos (15) and their concentration into a few particularly well-patrolled areas. At the same time captive breeding programmes in Zimbabwe and the USA were initiated.

Only the *Savanna zebras* of East, Southern and South-West Africa (*E. quagga granti, --boehmi, --chapmani,* and *--antiquorum*) are still abundant enough to be managed for commercial returns. They are either ranched together with antelope, sold as trophy animals or serve game viewing (19, 16, 5). With the exception of elephants, the zebra has become one of the most valuable herbivores because of the extraordinary prices, up to US$ 450, paid for well-tanned zebra skins used for decorative purposes. Although their meat was also converted in quantities into the popular "biltong" for local consumption the return from their skins in game ranching and cropping was already in the 1970s much greater than from their meat. In Tanzania for example, the returns from zebra skins in cropping operations amounted to 95% as compared with only 5% from the sale of their meat (6).

The number of *feral horses* in Australia exceeded 112,000 in 1984 (8), but this fluctuates significantly in the arid zone with environmental conditions (13). They are managed and utilised like other noxious animals by field shooting, usually at watering points. Carcasses are either deboned in the field and transported in refrigerated boxes to processing plants or are quartered for subsequent deboning at the plants. More than 10,000 feral horses are killed annually for producing pet food, meat for human consumption, hides and hair for domestic as well as export markets (18). Equally several thousand *feral donkeys* are shot for pet meat in the Northern Territory and Western Australia each year. However, the commercial harvesting has largely been supplanted by helicopter shooting for pest control. In 1989/90 no less than 50,000 feral donkeys were thus killed and left to waste (18).

## Implications for conservation

All perissodactyles have suffered badly from man's persecution and most of their species are depleted to non-utilisable population levels.

The rhinos provide one of the saddest examples of wanton destruction of a most valuable unique wildlife resource for purely commercial, misdirected motives. At the same time, however, the Southern White rhino shows what wise management and conservation can do not only to save a species but to bring it back to yield sustainable benefit to man. Had the principles and possibilities of natural resource management been understood 30 years ago they could have been applied to the Black rhino which at that time still occurred in some countries in manageable and utilisable populations. Instead of trying the impossible, i.e. to prevent the killing of animals which to the poacher are each worth a multiple of his annual income, in often almost uncontrollable areas, it might have been wiser to respond to the market forces with well-controlled and financed programmes of sustainable utilisation. However, in contrast to the situation with the elephant, it is now too late for consumptive forms of utilisation and the only option is to try and propagate the Black rhino under human control, as is being done with the other rhino species, for tourist viewing, attributing thus a non-consumptive value to the species.

The Savanna zebras are the only perissodactyles which still have a resource value that can and should be consumptively utilised outside conservation areas. With this group of animals, only economic benefit accruing from them to the landowners can offset the prejudicial and factual competition with livestock and this is a prerequisite for conserving zebras in ranching areas where they still occur in large numbers.

## Literature references

1. AYRES, J.M, MAGALHAES LIMA, D. de, SOUZA MARTINS, E. de and J.L.K. BARREIROS (1991): On the Track of the Road: Changes in subsistence hunting in a Brazilian Amazonian Village. In: *Neotropical Wildlife Use and Conservation*, ed. ROBINSON, J.G. and K.H. REDFORD; p. 83 - 92; University of Chicago Press, Chicago and London
2. BUTLER, J. (1846): A sketch of Assam with an account of the hill tribes (*cited after* LAURIE and coll. 1983)
3. CLUTTON-BROCK, J. (1987): *A Natural History of Domesticated Mammals*. Cambridge University Press and British Museum (Natural History), Cambridge and London
4. CUMMING, D.H.M. and P. JACKSON (eds.) (1984): The Status and Conservation of Africa's Elephants and Rhinos. *Proceedings of the Joint Meeting of IUCN/SSC African Elephant and African Rhino Specialist Groups at Hwange Safari Lodge, Zimbabwe, 30 July - 7 August 1981*; IUCN, Gland, Switzerland
5. CUMMING, D.H.M. (1989): Commercial and safari hunting in Zimbabwe. In: *Wildlife production systems*, ed. HUDSON, R.J., DREW, K.R. and L.M. BASKIN; p. 147-169; Cambridge University Press, Cambridge
6. ELTRINGHAM, S.K. (1984): *Wildlife Resources and Economic Development*. John Wiley, New York

7. FRAGOSO, J.M.V. (1991): The Effect of Hunting on Tapirs in Belize. In: *Neotropical Wildlife Use and Conservation*, ed. ROBINSON, J.G. and K.H. REDFORD; p. 154-162; University of Chicago Press, Chicago and London
8. GRAHAM, A., JOHNSON, K. and P. GRAHAM (1986): An aerial survey of horses and other large animals in the Alice Springs and Gulf regions (*cited after* RAMSAY 1994)
9. GRZIMEK, B. (1968): Wieviel Spitzmaulnashörner gibt es noch? (*cited after* Grzimeks Tierleben, Vol. 13, p. 53; Kindler, Zürich)
10. HUDSON, R.J. and D.H.M. CUMMING (1989): Recreational and commercial hunting. In: *Wildlife production systems*, ed. HUDSON, R.J., DREW, K.R. and L.M. BASKIN; p. 133-114; Cambridge University Press, Cambridge
11. KRUMBIEGEL, I. (1958): *Einhufer*. Neue Brehm-Bücherei No. 208; A. Ziemsen, Wittenberg-Lutherstadt, Germany
12. LAURIE, W.A., LANG, E.M. and C.P. GROVES (1983): Rhinoceros unicornis. *Mammalian Species*, No. 211; American Society of Mammalogists; Department of Zoology, Brigham Young University, Provo, Ut., USA
13. LETTS, G.A., BASSINGTHWAITHTE, A. and W.E.L. De VOS (1979): Feral animals in the Northern Territory (*cited after* RAMSAY 1994)
14. LONG, J.L. (1988): Introduced birds and mammals in Western Australia (*cited after* RAMSAY 1994)
15. MILLNER-GULLAND, E.J., LEADER-WILLIAMS, N. and J.R. BEDDINGTON (1993): Is dehorning African rhinos worth-while? *Pachyderm*, 27, p. 52-58 (*cited in Oryx*, 28(3), p. 147-148)
16. MOSSMAN, S.L. and A.S. (1976): *Wildlife utilisation and game ranching*. IUCN Occasional Paper No. 17; IUCN, Morges, Switzerland
17. PLAYER, I. (1967): Translocation of white rhinoceros in Africa. Oryx, 9, p. 137-150.
18. RAMSAY, B.J. (1994): *Commercial use of wild animals in Australia*. Bureau of Resource Sciences, Australian Government Publishing Service, Canberra, Australia
19. ROTH, H.H. (1966): Game utilisation in Rhodesia in 1964. *Mammalia*, 30(3), p. 397-424
20. ROTH, H.H. (1967): White and black rhinoceros in Rhodesia. *Oryx*, 9, p. 217-231
21. ROTH, H.H. (1969): *The Present Situation of the Rhinoceros*. World Wildlife Yearbook 1969; p. 307-312; WWF, Morges, Switzerland
22. SMITH, J.H. (1976): Utilisation of game along Brazil's transamazon highway. *Acta Amazonica*, 6 (4), p. 455-466
23. VICKERS, W.T. (1991): Hunting yields and game composition over ten years in an Amazon Indian Territory. In: *Neotropical Wildlife Use and Conservation*, ed. ROBINSON, J.G. and K.H. REDFORD; p. 53-81; University of Chicago Press, Chicago and London
24. WENDT, H. (1956): *Auf Noahs Spuren. Die Entdeckung der Tiere*. Grote, Hamm, Germany
25. WETTERBERG, G.B. (1976): *Amazon fauna preferred as food*. FAO/FO: DP/BRA/71/454: Working Document 23; 17 pp.; Food and Agriculture Organisation of the United Nations, Rome
26. ZARINS, J. (1976): *The domestication of Equidae in third millenium BC Mesopotamia*. Thesis No. T-26263; Joseph Regenstein Library, University of Chicago, Chicago, Ill, USA

## 2.3.10 SUIFORM UNGULATES (*Nonruminantia: ARTIODACTYLA*)

The large order of Even-toed or Cloven-hooved ungulates (**Artiodactyla**) is usually broken down into three distinct groups: the non-ruminating suiform mammals (**Nonruminantia**), the tylopodes (**Tylopoda**) and the ruminant ungulates (**Ruminantia**). The suiform ungulates comprise pigs (**Suidae**), peccaries (**Tayassuidae**) and hippopotami (**Hippopotamidae**). Wild pigs are distributed in several species (*Sus* spp.) and range from North-West Africa and Western Europe to Eastern Asia, including Japan, and over India and South-East Asia; sub-saharan Africa is inhabited by another 3 species (*Potamochoerus porcus, Phacochoerus aethiopicus, Hylochoerus meinertzhageni*). Peccaries occur in 2 species only in South and Central America (*Tayassus* spp.). Hippopotami (*Hippopotamus amphibius*) were formerly abundant in all river systems south of the Sahara. Tab. 38 summarises the utilisation of these animals for meat and hides.

**Table 38:** Overview of the utilisation of suiform ungulates

| English name | Scientific name | Geographical distribution | Body weight (kg) | Utilisation |
|---|---|---|---|---|
| **PIGS - SUIDAE** | | | | |
| European Wild boar | *Sus scrofa scrofa* | W. Europe to E. Asia[1], N.W. Africa | 150-250 | domesticated into dom. pig; hunted for sport, meat |
| Asian Wild boar | *-s. vittatus* | S., S.E., E. Asia[2] | | and control; farmed for meat |
| Bearded pig | *-barbatus* | S.E. Asia | 70-150 | hunted for meat |
| Warty pigs | *-verrucosus* sspp. | S.E.Asia | | semi-domesticated |
| Babirusa | *Babyrousa babyrussa* | S.E. Asia | 80-100 | hunted for meat |
| Warthog | *Phacochoerus aethiopicus* | Africa south of Sahara | 50-120 | hunted for trophies and meat |
| Bush pig | *Potamochoerus porcus* | Africa south of Sahara | 50-80 | hunted for meat, control |
| **PECCARIES - TAYASSUIDAE** | | | | |
| Peccaries | *Tayassus* spp. | S. and C. America | 18-30 | hunt. for meat and hides; exp. farming |
| **HIPPOPOTAMI - HIPPOPOTAMIDAE** | | | | |
| Hippopotamus | *Hippopotamus amphibius* | African rivers/lakes | 2000-3000 | for meat where contr.; sport hunt. |

1) introduced to North and Central America, Argentina (HALTENORTH and TRENSE (10))
2) domestic pigs have become feral in New Guinea, Melanesia, Australia and New Zealand

### Historical background

Palaeolithic rock paintings document that the *European Wild boar* is one of the oldest game animals and was, next to the Red deer, the most important source of food for Stone Age man (14, 17). Apart from its material value the Wild boar also assumed great spiritual importance as reflected the Germanic mythology. It was probably domesticated simultaneously in several parts of its wide range of distribution, domestication involving the European (*Sus scrofa scrofa*) as well as its Asian forms (*S.s. vitattus*). In contrast to the ancestors of most other domestic animals, it has not only survived but has become abundant in many areas of its distribution coexisting with the domestic pig. This is due to its great adaptability to diverse habitats, its omnivorous feeding habits and its remarkable reproductive capabilities. All through the mediaeval era boar constituted "royal game" and its meat was relished. However, due to the elimination of all predators and the reduction of closed forests in West and Central Europe, Wild boar populations proliferated and became a serious nuisance to the developing agriculture in many areas. As of the 18th century this led in some European countries to repeated efforts to eradicate them. Only in Great Britain and Scandinavia, however, were these efforts successful. In Central Europe boar populations were merely diminished during the 18th and early 19th centuries, when populations started to increase again. Boar were reintroduced to Holland and Denmark where they had been extirpated previously, and by the second half of the 19th century the species reached a new peak of abundance (21). In Germany, for excample, this led in 1883 to the classification of boar as "vermin" with the legal obligation of landowners to eradicate them. Nevertheless boar populations further increased in numbers during the first decades of the 20th century. In the late 19th and early 20th centuries Wild boar were also transplanted from Europe into North and Central America as well as Argentina for hunting purposes (10). Whereas the total number of boar hunted in Germany in 1885/86 was estimated at about 11,900, this number had risen to 31,000 by 1935 although large parts of Germany had been lost to France and Poland at the end of the First World War (21). At that time the need to manage boar populations more rationally was recognised by inclusion of the species as "game animal" in the revised German Hunting Law of 1934. Lack of management and control during the Second World War resulted in a new upsurge of boar populations resulting in the most serious infliction of damage to agriculture in the postwar years when hunting was still prohibited. By that time, however, the ecological and material benefits derived from boar were gradually understood and weighed against agricultural damage.

In South-East Asia various other wild pigs, such as the *Bearded pig (S. barbatus)*, the *Sulawesi* and *Javan Warty pigs (-verrucosus)*, and the *babirusa (Babyrousa babyrussa)* have also been hunted for meat since the earliest times together with the *Asian Wild boar (S. srofa vittatus)*. Domestic pig breeds derived from this latter were introduced early on to New Guinea, Melanesia, Australia and New Zealand. There, they reverted in many areas to a feral life becoming major pest animals and being hunted as such (22, 13). Historical records show that in Australia escaped pigs were a problem around Sydney even in 1795 and *feral pig* populations had formed in New South Wales by the mid-1800s (22). Today there are large populations of wild pig established throughout New South Wales, Queensland and the northern portion of the Northern Territory.

In Africa the *Bush pig* is as adaptable and widespread as the Wild boar. As in Europe, the removal of its major predator, the leopard, has in some areas caused increases of this species into nuisance populations. Both *Bush pigs* and *warthogs* produce delicate meat which, however, cannot be commercialised because they are considered to be a reservoir of the highly contagious African Swine Fever virus. Warthogs are also one of the major hosts of the blood-sucking Tsetse flies (*Glossina* spp.) and their numbers have been reduced in some parts of Africa in the control of trypanosomiasis.

*Peccaries* have always been one of the most important sources of meat for the indigenous peoples in the forest zone of South and Central America. They have, however, also long been hunted commercially for their skins from which a particularly soft and durable leather is produced, which is used, for example, in glove making. After the Second World War, in 1946-66 a total of nearly 3 million peccary skins were exported only through Iquitos (8).

Due to their weight and aquatic life *hippopotami* are less hunted for subsistence by local populations than other, smaller ungulates. However, as in the case of elephants, hippo populations in conservation areas sometimes have to be reduced for ecological reasons. Such control operations were carried out temporarily in the 1960s in Uganda and have been regularly conducted since the 1970s in the Kruger National Park in South Africa (6).

## Management and utilisation

Wild pig populations are utilised through **subsistence, sports and commercial hunting**. Most of them are difficult to hunt because of their volatile behaviour, hiding during the daytime in dense under-wood or in swampy areas. They are usually tracked down with the aid of dogs and then killed at close range by spearing or shooting. *Peccaries* are also trapped in pits. An exceptional method for hunting *Bearded pigs* is customary with the indigenous peoples in Malaysia, Sumatra and Borneo. This species use to migrate seasonally in large numbers along distinct and conspicuous forest trails. During these migrations the hunters awaited the pigs at rivers which they had to cross by swimming. There they speared them from dugouts in the river (11, 7). Fully grown *Wild boar* can be extremely aggressive and therefore boar hunting is considered to be quite "sportive". *Warthogs* in Africa inhabit open savanna and have strictly diurnal habits. They are therefore easy to hunt and are now often included in commercial safari hunting tourism as a "fill-in" species or are offered in countries which have few big game, like Senegal (1).

In European countries boar hunting is now well regulated in terms of management plans which specify hunting quotas. These aim at regulating the size and sex and age structure of

Hunting of Wild boar is still most popular in Germany; the carcasses are sold to restaurants. (Photo: B. WINSMANN-STEINS)

the hunted populations in accordance with local ecological conditions. Depending on the nutritional conditions the rate of reproduction may vary from 50 to 150% of the spring population (21). This explains their pest status in many parts of their range and the sustainability of harvesting rates in excess of 50% (12). Management of Wild boar is often primarily directed at controlling population growth and the utilisation of the meat is only a secondary aspect; it is either sold directly to restaurants or commercially marketed.

In South and South-East Asia and in Australasia, *feral pigs*, originating either from *Sus scrofa vitattus*-breeds, or from abandoned stock of *European domestic pigs*, are widely used in a semi-domesticated state. The pigs live and reproduce in dense vegetation near human habitations, associate with the domestic buffalo herds, raid the fields and at times appear in the villages. They are hunted like wild pigs for meat, but in some countries piglets are caught and raised in the villages. In New Guinea women even breast suckle them, and the possession of numerous pigs in Melanesia, for example, is a matter of social prestige. In Melanesia, New Guinea and some of the Sunda Islands, like Bali, pigs have always played an important role in spiritual life and the myths of the people. Feral pigs are thus kept to have them ready for certain festivals at which they are then sacrificed (22, 17). An exceptional form of utilising Wild boar has been reported from Rajestan in India. There, hand-reared and tamed boar were trained to follow the blood trail of wounded leopards enabling the hunters to finish them off (9).

Feral pigs in Australia are estimated to number well over one million (26) and are regarded as a major pest, causing extensive environmental and agricultural damage. Legislation requires landowners to suppress wild pig populations on their land. Commercial harvesting therefore is welcome as a means of control and has developed since 1980 into a Wild boar meat industry. Pigs are killed by part-time or professional shooters, partly eviscerated in the field and must be delivered within two hours to a field chiller. The rural chiller operators pay the shooters according to the weight of the carcasses and supply them to the licensed game meat establishements in Sydney and Brisbane. There the boar carcasses are inspected and if passed, processed exclusively for export (23).

With the increasing demand for boar meat, growing numbers of Wild boar in Europe are subjected to **more intensive management regimes**. They are kept in closed-off hunting reserves or ranches of at least 300-400 ha where they receive supplementary feed and are selectively culled to enhance off-take. In some European countries, but also in Japan and the USA, *Wild boar* are bred in captivity and farmed like domestic pigs, in order to produce boar meat and live hoglets for restocking purposes. Animals from such farms may be slaughtered in abattoirs after *antemortem* inspection and marketed like domestic pigs (3).

Where the management of conservation areas calls for the reduction of *hippopotami* in certain river courses, or where these animals clash with fisheries development, selective **culling** of this species may be required. This may, as for example in Zimbabwe during drought years, be combined with license hunting by residents or non-residents who apart from the sport, value the meat in the form of small strips of sun-dried salted "biltong". Hippopotami are usually shot with a heavy calibre rifle in the brain when they emerge for respiration. The immediately dead animals sink to the ground, but after 1-2 hours float up to the surface due to the development of methane gas in their digestive tract. They are then pulled on land with a tractor. In the Ugandan operations in the 1960s the culled animals were sold on the spot to local traders who would then retail the meat in surrounding villages and townships. The meat was well accepted and considered to have a high nutritional value (15). In the Kruger National Park the dead animals are transported to a central game abattoir. There they are either processed into fresh refrigerated meat and biltong or are canned (6).

## Economic importance

**Subsistence utilisation** of wild pigs still has some importance in Brazil where *peccaries* are extensively hunted in the Amazon region for food and hides (4). In a 12 month survey in 1973/74 of game meat consumption in two localities along the transamazonian highway, peccary meat constituted 54.6 and 39.0% respectively of the total weight of game taken (24). The skins are mainly a by-product of the meat hunting, but nevertheless they are widely commercialised for export mainly to Europe, Japan and North America (5, 4). During the last three years prior to the export ban on wildlife products in Brazil, a total of 1,932,496 peccary skins were exported (24), and they have remained thereafter the most commonly confiscated wild mammal product. In Peru exports declined in 1982, also after controls were imposed, to 101,000. Paraguay and Argentina have been other major exporters in the past, with annual exports of peccary skins valued in Argentina at US$ 256,250 (16). In South-East Asia *wild pigs* were also extensively used for subsistence in the past, but there are no records to show the present importance of this use.

**Sports hunting** of *Wild boar* in Europe is very popular and of considerable economic dimension. During 1969-1972 about 140,000 boar were killed annually in European countries, excluding the European part of the former USSR. This represented a total value at the time of about DM 28 mill. (approx. US$ 9.3 mill.) for the meat alone, i.e. 6% of the value of all game production including those countries in which boar do not exist anymore (18). The percentage value of hunted boar varied greatly according to the different ecological and sociological conditions, from 36% in the former German Democratic Republic to less than 5% in Austria and former Czechoslovakia. In 1991/92, however, the recorded numbers of hunted boar in just Germany, Holland, Austria, Hungary, Slovenia and Switzerland had risen to 375,989 (2); the total of the European boar harvest was estimated at 525,000 (12). In Germany alone, which accounted for 83% of hunted boar, these yielded about 16,000 t of meat with a value of DM 87.7 mill. (approx. US$ 51.59 mill.), which represented 32.6% of the value of total game production (2).

Next in economic importance is the **commercial hunting** of *feral wild pigs* in Australia, where 271,133 of these animals were processed through export game meat establishments in 1992. The total value of meat derived from wild pigs has varied between Austr.$ 10 and 20 mill. (approx. US$ 7.3-14.6 mill.) per annum (23). Apart from this value, game meat companies handling the carcasses paid at least Austr.$ 4-5 mill. to shooters and chillers. The wild pig industry provides significant employment and is an important part of the rural economy in parts of Australia. The commercial harvesting of wild pigs in Australia has also an important, though largely unrecognised, role in reducing the damage which these animals cause to agriculture. As is the case with other introduced wild animal species which have become noxious, both their resource and pest values have to be weighed against each other with a view to establishing multi-use management of the feral pigs, incorporating commercial use, recreational hunting and pest control (26, 19, 20).

Fig. 17 shows that the international ***demand for boar meat***, as reflected from export figures of producer countries, has greatly increased world-wide during the last decade. It is favoured mainly in Europe, but also increasingly consumed in Japan and the USA. Partly because of a long-lasting European tradition to eat boar meat and partly because of its specific flavour, it clearly fetches higher prices than pork. The total international trade in fresh and frozen boar meat amounts to at least 10,000 t, valued in 1990 at Austr.$ 40.4 mill. (approx. US$ 31.7 mill.), of which Australia supplied 20-30%, as the world's biggest exporter after Poland (23). The actual consumption of boar meat is, of course, much greater than these trade figures suggest. Germany as one of the most important consumer countries, for example, produced, according to hunting statistics, about 15,298 t of boar meat in 1990 (2) of which it has exported 1,472 t to other EU countries. In the same year, however, it has also imported 3,966 t (23) which suggests an internal consumption of about 17,800 t. Two types of boar meat are traded, boar game meat and meat of farmed wild-type pigs. Whereas the former is derived from hunting the latter originates from farm animals which are slaughtered at an abattoir after *ante mortem* inspection.

**Figure 17:** Export of fresh and frozen wild boar meat by the major suppliers to all countries in the European Union in 1981-1990 (after RAMSAY (20))

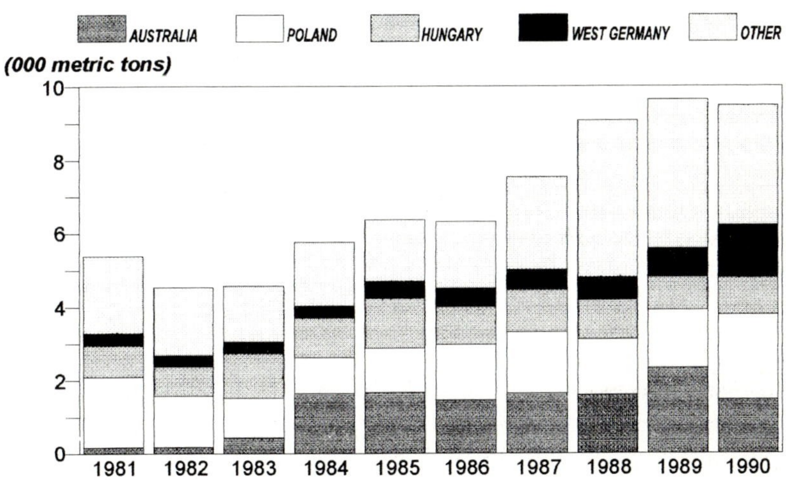


***Total value of boar meat imported into EU and Japan (in Austr.$ 000 000)***

***12.2 15.0 21.7 38.5 50.6 27.7 24.1 30.5 40.6 40.4***

Recognising the commercial advantage inherent in the consumption trend towards lean meats, **farming** of *Wild boar* has developed since the 1970s in some European countries, Japan and the USA. Some of the farms specialise in the production of meat and cross-breed boar with domestic pigs to facilitate their keeping and to increase their fecundity. Economically more important is the breeding of boar for supplying hunting estates with live boar for stocking, which is usually combined with meat production. In France it was estimated that in 1980 about 100,000 boar were bred on more than 800 farms (3). In Japan where hunting of boar has also been a tradition since at least the 17th and 18th century, growing demand for boar meat has increased the hunting pressure to a point of depleting the wild resource. As a result boar farming was initiated and in 1980 there were 1,300 breeding boar established on 160 farms (3). In the USA Wild boar introduced from Europe for hunting were estimated in 1975 to number about 10,000 on more than 100 ranches (25). To a lesser degree peccaries are also bred in captivity in the USA and also experimentally in a few South American countries, for hunting purposes (3).

The commercial utilisation of *hippopotami* for meat and skins is coincidental with the need to control this species in certain localities for ecological reasons. It has therefore no economic motive or implication. In the Kruger National Park an average of 80 hippo were killed per year in 1977-1981 yielding about 144,000 kg of meat. Of this a mass of 97,648 kg was canned for staff rations or sale; hides and by-products were also commercially utilised (6). In Zimbabwe the wildlife authorities were earning Z$ 3,000 (approx. US$ 360 in 1993) from any hippo that needed to be culled, through the issue of hunting licenses.

## Implications for conservation

Utilisation of Wild boar and feral pigs follows from the necessity to control their populations as with many other nuisance animals. Attempts to exterminate them - and also Bush pigs and warthogs in some parts of Africa for the control of Tsetse fly - have generally failed. Management of these species at sustainable population levels with a view to minimising damage to agriculture but optimising the ecological and material benefits from them, is thus the most rational approach to this animal group. It is applied in large parts of the range of distribution of these species. They are also partly protected against over-exploitation in all those countries in which moslems predominate in the human population, as these strictly reject any pig meat for food. However, some of the more localised Asian pig species, such as *Sus verrucosus, -barbatus, Porcula salvania* and *Babirussa babyrussa,* have certainly been over-exploited in the past to dangerously low numbers and now require the fullest possible protection. The status of peccaries in South America needs to be investigated.

Hippopotami populations have steadily decreased in the past everywhere in Africa. However, this was not so much due to subsistence hunting as to the incompatibility of this large species with agricultural development and human settlements. As with elephants it will only survive in protected areas. In these areas, occasional and well supervised population control for ecological reasons is a legitimate aspect of their management and will only contribute to the conservation of the species.

## Literature references


1. ANONYMOUS (1986): *La Chasse en Senegal.* Direction des Eaux, Forêts et Chasse; Dakar-Hann, Senegal (Information pamphlet)
2. ANONYMOUS (1993): *DJV-Handbuch 1993: Jagd aktuell.* Deutscher Jagdschutz-Verband e.v., Bonn, Germany
3. BARZDO, J., LUXMOORE, R., BROAD, S., and D. JONES (1984): *A survery of wildlife farming operations.* WCMC/IUCN, Cambridge (unpubl. report)

4. BODMER, R.E., PUERTAS, P., MOYA, I.L. and R. GILL (1994): Managing wildlife to conserve Amazonian forests: population biology and economic considerations of game hunting. *Biological Conservation*, 67, p. 29-35
5. BROAD, S. (1984): The peccary skin trade. *TRAFFIC Bulletin*, 6(2), p. 27-28
6. DE VOS, V., BENGIS, R.G. and H.J. COETZEE (1983): Population control of large mammals in the Kruger National Park. In: *Management of large mammals in African conservation areas*, ed. OWEN-SMITH, R.N.; p. 213-231; Haum Educational, Pretoria, South Africa
7. FRÄDRICH, H. (1968): Familie Altweltliche Schweine und Familie Pekaris. In: *Grzimeks Tierleben*, Vol. 13; p. 74 - 103; Kindler, Zürich
8. GRIMWOOD, I.R. (1968): *Conservation of Wildlife and the Establishment of National Parks and Reserves in Peru*. Appendix III of the Recommendations to the Ministry of Overseas Development, London
9. GUMAN SINGH, K. (1956): Wild boars being used as blood hounds. *Journal of the Bombay Natural History Society*, 53, p. 258-259
10. HALTENORTH, T. and W. TRENSE (1956): *Das Grosswild der Erde und seine Trophäen*. Bayerischer Landwirtschaftsverlag, Bonn-München-Wien
11. HISLOP, J.A. (1949): Some field notes on the bearded pig. *Malayan Nature Journal*, 4, p. 7-23
12. HOFFMAN, R.R. (1991): Wild game in Europe: a multi-billion dollar industry. *The Deer Farmer*, 75, p. 67-72; Wellington, New Zealand
13. HOWARD, W.E. (1965): *Control of introduced mammals in New Zealand*. New Zealand Department of Scientific and Industrial Research, Information Series No. 45, Wellington, New Zealand
14. JARMAN, M.R. (1972): European deer economies and the advent of the Neolithic. In: *Papers in economic prehistory*, ed. HIGGS, E.S.; p. 125-149; Cambridge University Press, Cambridge
15. MANN, I. and B. GRZIMEK (1962): Vom Fleisch der Flusspferde und Antilopen. *Fleischwirtschaft*, 6, p. 502-504
16. MARES, M.A. and R.A. OJEDA (1984): Fauna Commercialisation and conservation in South America. *Bioscience*, 34, p. 580-584
17. MOHR, E. (1960): *Wilde Schweine*. Neue Brehm Bücherei No 247; A. Ziemsen, Wittenberg-Lutherstadt, Germany
18. NÜSSLEIN, F. (1972): Der ökonomische Wert der Wildproduktion in Europa (ohne Sowejetuntion). *Zeitschrift für Jagdwissenschaften*, 20, p. 85-95
19. O'BRIEN, P.H. (1987): Socio-economic and biological impact of the feral pig in New South Wales: an overview and alternative management plan. *Australian Rangelands Journal*, 9, p. 96-101
20. O'BRIEN, P.H. and P. MEEK (1992): Feral pig management. Final Report to the Rural Industries Research and Development Corporation (*cited after* Ramsay 1994)
21. OLOFF, H.-B. (1951): *Zur Biologie und Ökologie des Wildschweines*. Beiträge zur Tierkunde und Tierzucht, Vol. 2, Frankfurt/Main, Germany; 95 pp.
22. PULLAR, E.M. (1953): *The wild (feral) pigs of Australia: their origin, distribution and economic importance*. Memoirs of the Natural History Museum, Melbourne, No. 18, p. 7-23
23. RAMSAY, B.J. (1994): *Commercial use of wild animals in Australia*. Bureau of Resource Sciences; Australian Government Publishing Service, Canberra, Australia
24. SMITH, J.H. (1976): Utilisation of game along Brazil's transamazon highway. *Acta Amazonica*, 6(4), p. 455-466
25. TEER, J.G. (1975): Commercial uses of game animals on range lands of Texas. *Journal of Animal Science*, 40(5), p. 1000-1008
26. TISDELL, C.A. (1982): *Wild pigs: environmental pest or economic resource?* Pergamon, Sydney

## 2.3.11 CAMELIDS (*Tylopoda: ARTIODACTYLA*)

Camelids are primitive ruminating Even-toed ungulates (**Artiodactyla**) constituting a suborder (**Tylopoda**) of these with only one family, the camels (**Camelidae**). All tylopodes are extremely well adapted to harsh desert or high altitude conditions, the 2 *Camelus* species occurring from Central Asia to North Africa and the 2 *Lama* species in the western and southern parts of South America. Tab. 39 shows that four domestic animal species derive from them, and that the two South American wild species are hunted and utilised for wool and meat.

**Table 39:** Overview of the utilisation of camelids

| English name | Scientific name | Geographical distribution | Body weight (kg) | Utilisation |
|---|---|---|---|---|
| **CAMELIDS - CAMELIDAE** | | | | |
| Wild camel | *Camelus ferus* | C.Asia | 450-550[1)] | domesticated into Two-humped camel |
| Wild dromedary (extinct) | *-dromedarius* | West. Asia | 350-500[1)] | domesticated into One-humped camel |
| Guanaco | *Lama guanicoë* | West. and South. S.America | 60-100 | domesticated into Llama and alpaca; exp. farming |
| Vicuña | *-vicugna* | Andean region of S.America | 40-60 | hunted for meat and wool; exp. farming |

1) approx. weight of domestic form only

### Historical background

Wild camels in Central and Western Asia were domesticated long before the time of Christ into the Two-humped Bactrian and the One-humped Arabian camel or dromedary. These have become an essential basis for human life in the arid desert zones of these regions, providing mainly transport but also important life-sustaining raw-materials such as wool, meat, milk and dung for fire. Whereas the progenitor of the *Bactrian camel* has survived in small numbers in Central Asia, the wild ancestor of the *dromedary* became extinct early on and was never known. It was supposed to have occurred in Arabia until the beginning of the Christian era (31). Domestication of the camels was never rigid and they were allowed to more or less continue their original lifestyle, never ever being stabled or penned. Therefore domestication has not modified the camel species much. Only the dromedary was gradually distributed far beyond its original range, all over Northern and Eastern Africa. Even in 1405 they had been introduced to the Canary Islands and they were temporarily transplanted to various localities in South America, the USA and Southern Africa (18). Only in Northern Australia, however, to where it was first brought in 1840 did the dromedary become feral and is now inhabiting all arid zones of the continent like a wild animal species (20). Historically the dromedary also had a significant military function. The Arab tribes used it for raids and dromedary racing is still a most popular sport in Arabia. The Romans and later on the European colonial powers maintained dromedary units for warfare in the arid parts of Africa.

According to recent research (22) it is possible that the two South American domestic camelids, the llama and alpaca, existed before the Asian species were domesticated. Previously it was thought that the *guanaco* was the progenitor species of both, but this may have been descended from a now extinct camelid species which might have been the progenitor of all three (3). The use of the two domestic forms, the llama mainly as an animal of burden and the alpaca as wool producer, was well developed in the Andean region of Pre-Incan times, long before the Spaniards introduced sheep, cattle and mules. However, these have never been able to fully replace the local llamas and alpacas, and since the 1960s concerted, scientifically-based breeding and husbandry efforts were made to improve the productivity of the South American domestic camelids.

At the same time an appreciation of the resource value of the wild camelid species developed, although an escalation of market-dominated hunting after the Second World War had brought the more vulnerable and valuable *vicuña* to the verge of extinction by the end of the 1960s (16) and reduced the *guanaco* to some 600,000 by the 1980s (8, 30). Historically, the guanaco was the dominant herbivore on the southern grasslands and steppes of South America numbering many millions in Patagonia the centre of its distribution. As were the bisons in North America, guanacos were the life-sustaining resource for the aboriginal Indians in Patagonia and Tierra de Fuego until the Europeans arrived in the 16th century (12). By the mid-1900s most guanaco had been all but eliminated from the Patagonian pampas as they were perceived as undesirable competitors of the introduced sheep (15), much as kangaroos are in Australia. Also they were indiscriminately hunted for their meat and the valuable pelts of the newly born young, the *chulengos*. In fact, despite the dramatic reduction throughout their former range, legal harvesting of chulengos remained an economically important industry in Argentina into the late 1970s when the species was placed onto Appendix II of the CITES and international trade in "guanaquito", the pelts of the chulengos, was curbed.

The *vicuña* always seem to have been ecologically restricted to the Puna, the high-altitude pastoral plateau of the Central Andes (8). Archaeological evidence from preceramic sites in the Puna region suggests that vicuña, together with other equid and cervid ungulates, were already extensively utilised in 7000-5000 BC. As the latter became scarcer, some taxa being exterminated, the percentage ratio of camelid to cervid remains increased from 55:42 to 82:17 in 5500 BC (22). This is interpreted as an indication of the increasing use of camelids, possibly in a semi-domesticated state. However, in contrast to the guanaco, the strongly territorial behaviour of the vicuña precluded this from domestication. It was, nevertheless extensively utilised for its wool and meat. The Incas managed it through well-regulated periodic hunts, called "chacos" (17). These were carried out in each district at 3-5 year intervals, under strict royal supervision, rotating the hunting annually through alternate areas and thus allowing the fleece of the vicuñas to regrow. For a chaco the Incas ordered out 20,000-30,000 Indian beaters who encircled an area of 4,000-5,000 km$^2$, closing in on the vicuña herds and other game animals until these would be either killed or taken by hand for shearing. The sheared animals would be released. Later the chaco-method was improved by driving the animals with the aid of staked out ropes with attached coloured rags into large stone corrals where animals could be sorted out, killed or sheared. Except in these hunts the Indians were not permitted to take game. In fact, the wool of the vicuña was strictly reserved for weaving cloth for royalty and ornamentation of temples, a task of the sacred women priests. With the subjugation of the Inca empire by the Spanish conquistadores this well-controlled utilisation system broke down and was replaced by indiscriminate persecution of the vicuña with firearms and dogs by anyone. Excess market hunting for the extremely fine wool continued into the early part of this century. Between 1902 and 1907, Peru alone exported for example 19,409 kg wool and 14,554 dried skins, representing a kill of more than 50,000 animals (19, 17). Only in the 1920s was protective legislation prohibiting hunting and commercialisation of the vicuña enacted in all three Andean countries, but it had little or no effect (17). By the early 1950s there were still about 400,000 vicuñas left (7), but by the late 1960s these numbers had declined to an estimated 5,000-10,000 in Peru and fewer than 2,000 in Bolivia, Chile and Argentina (16).

## Management, utilisation and economic importance

Compared with the great economic importance of the domestic camelids, present subsistence as well as commerical use of wild camelids is negligible. Both, guanaco and vicuna have become endangered species and need to be primarily managed for conservation and multiplication until they can be sustainably harvested.

The only feral population of the *Arabian camel* or *dromedary* in Australia is utilised by **commercial shooting** for processing the carcasses into pet food, or by **live capture for slaughter** at abattoirs. Domestic trade in camel meat has only developed since 1988 and it mainly supplies restaurants which cater for tourists (25). The dromedary population, distributed over all the arid regions of the continent, was estimated at more than 43,000 animals in 1988 (29).

The **harvesting** of *guanacos* in Argentina until the early 1980s was uncontrolled and probably unsustainable as it removed mainly the chulengos, the youngest age class, without knowing how this would effect the productivity of the guanaco population. The numbers of guanaquito, the pelts of the young born guanaco, which were exported from Argentina increased from 22,397 in 1976 to 86,324 in 1979, before diminishing as of 1980 due to the imposed CITES trade restriction to an annual 11,452 skins and 11,890 cloth items (2). With an average export value of US$ 1.65 mill. per annum it had ranked sixth in the value of utilised wildlife species (10). In Southern Chile and Tierra de Fuego a conservation programme established in the 1970s included studies of the possibilities for sustainable harvesting of guanaco (4). Experimental cropping in Tierra de Fuego where guanaco were most abundant, was initiated in 1980 by the Chilean Forestry and Wildlife Service, but was soon terminated when it was discovered that the cropped population had a high incidence of sarcocystosis (9). Although *Sarcocystis*, a protozoan muscle parasite, is not pathogenic to humans the lesions it causes are disagreable to the consumer. Sarcocystosis occurs locally depending on the presence of undefined carnivores which constitute the definite host of the parasite; it has also diminished the economic viability of utilisation programmes in certain localities of East Africa (27). Generally, however, guanaco meat - like other game meats - fetches much higher prices than beef (21) and its sale together with wool and skin production should provide a viable basis for the economics of guanaco utilisation. Later studies in Southern Chile were aimed at providing biological base knowledge for sustained harvesting of guanaco populations (9). More recent aerial surveys have shown that the number of guanaco just in Argentine Patagonia have increased again to about 1.5 million (11) and that therefore sustainable commercial utilisation of guanaco as an alternative or supplement to the ailing sheep industry in Patagonia is of immediate concern.

A conservation programme for the *vicuña*, established with international aid in the Pampa Galeras of Peru in 1967, had from the start the ultimate goal of rendering the species economically beneficial to the local people. The present vicuña population has been managed on behalf and with the co-operation of the local peasant communities, and under this control it increased rapidly from about 1,200 animals in 1967 to more than 42,000 in 1980 (13), when the carrying capacity of the area obviously became exceeded.

In addition to the successful propagation of the species, the Pampa Galeras project produced most detailed biologically-based guidelines for the extensive agricultural utilisation of *vicuña* (14). This was initiated after the vicuña population had reached an exploitable level of some 20,000 animals in the 1970s (13). Practical utilisation comprised **capture of live animals** in mobile pens **for shearing and/or for translocation** into adjacent ranges, as well as limited cropping of young males for meat. Wool and skins were stored until legal trade would be permitted and returns could be channelled back to the local peasant communities. Nevertheless, approach and methods of utilisation remained a controversial subject (6). Various studies have been made to determine a sustainable yield of vicuña populations (26) and to analyse the potential economic benefit to the local, very poor peasants from vicuña utilisation (23, 28, 24). Upon request by the three Andean countries in 1987, distinct populations of the vicuña were removed from Appendix I to Appendix II to allow for international trade in cloth made from the wool of these populations. *Vicuña* wool is in fact by far the finest existing fibre used in wool production. Its average diameter is only 13.2 microns, compared with 22.8 microns in the finest merino sheep wool (17), but wool productivity in the vicuña is low, varying between 300 and 400 g wool/animal every 2-3 years during a productive lifespan of 8 years; i.e. a vicuña does not produce more than about 1.2-1.8 kg of wool during its lifetime (17, 13). However, in 1982 wool fetched more than US$ 100/kg on the illegal market and vicuña had thus a 20times and 10times greater economic value than sheep and alpaca respectively under similar ecological conditions (13).

Vicuñas under strictly controlled management at Pampa Galeras in Peru, utilised for wool, meat and skins. (Photo: K. HOFMANN)

More intensive utilisation of the South American camelids by **farming** has been suggested for both species. In Peru, Don Francisco Paredes Perez had started breeding *vicuña* on his Cala Cala ranch in 1919. In 1957 the Cala Cala herd comprised about 400 animals which were regularly driven into funnel-like stone corrals for shearing under restraint (17). In order to improve wool productivity vicuña were cross-bred with alpaca, and in 1969 there were about 300 *Paco-vicuña* hybrids on the property. Although yielding longer and therefore more wool, the hybrids are often sterile but no less difficult to handle. After the death of Don Paredes, management of Cala Cala deteriorated in the 1970's and productivity declined. Since no breeding records were ever kept, no scientific conclusions could be drawn from the Cala Cala breeding and husbandry efforts. The only other captive herd of vicuña was established by the Instituto Nacional de Tecnologia Agropecuaria (INTA) at Abra Pampa in Argentina in 1958 and comprised 400 animals in 1985 (1). It is solely run for agricultural research, cross-breeding with *llama* and *alpaca*, and shearing of wool. Selective breeding for tameness has been unsucessful so far.

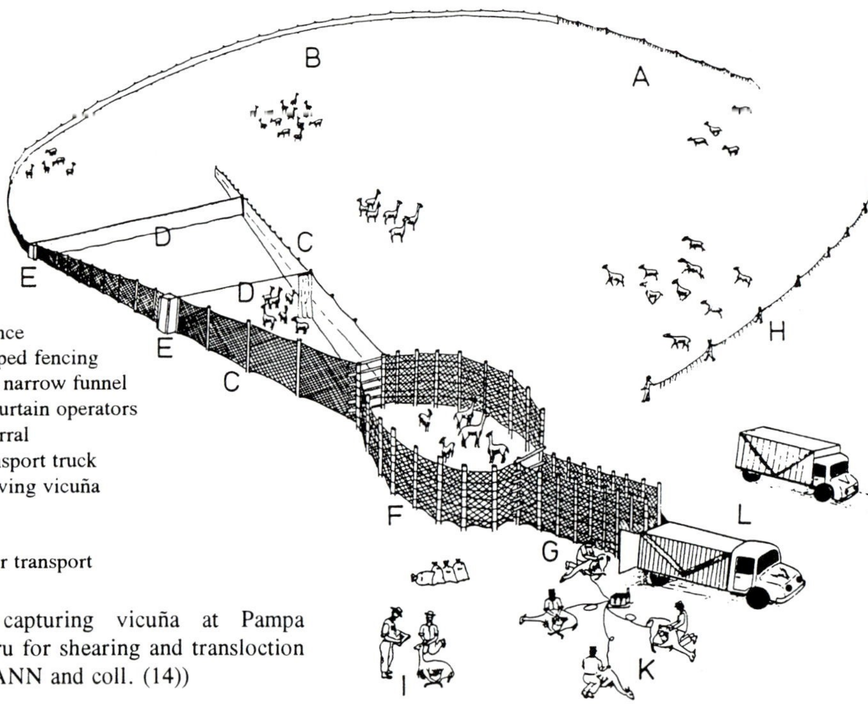


A - access
B - guiding fence
C - funnel-shaped fencing
D - curtains to narrow funnel
E - hides for curtain operators
F - holding corral
G - exit to transport truck
H - beaters driving vicuña
I - recorders
K - shearers
L - vehicles for transport

System for capturing vicuña at Pampa Galeras in Peru for shearing and transloction (after HOFMANN and coll. (14))

Since 1977 INTA has also been experimenting with captive *guanaco* at Trelew with a view to farming them for meat and wool (5).

## Implications for conservation

Over-exploitation in the past has severely affected both of the South American wild camelids, the guanacos and vicuñas. Whereas the former were decimated below exploitable population levels mainly because they were perceived as grazing competitors to sheep, the latter were brought to the verge of extinction purely for reasons of monetary gain. Fortunately, protective measures since the 1970s have halted the excessive and uncontrolled exploitation and have resulted in the return of both species to utilisable population levels. Decisive for this success was the fact that the conservation efforts included considerations as to how the two species could and should contribute sustainably to the economies of the local human populations. Consequent upon the realisation that these species could only survive in the future if managed on the basis of ecological and biological knowledge for economic returns, much research went into them. Furthermore, the conservation efforts were also aimed at actively involving the local communities through educational programmes and economic incentives. With this foundation, programmes of sustainable utilisation should proceed in the future and will aid conservation as long as the marketing of the products, particularly the exceptionally valuable vicuña wool, remains under effective international control. Conservation of the South American camelids is linked to economic utilisation more than in other animal groups because of their unique adaptation to extreme environmental conditions and their important resource potential for the very poor local peasant populations.

## Literature references


1. BARZDO, J., LUXMOORE, R., BROAD, S. and D. JONES (1984): *A survey of wildlife farming operations*. WCMC/IUCN, Cambridge (unpubl. report)
2. BROAD, S., LUXMOORE, R. and M. JENKINS (1988): *Significant Trade in Wildlife. A review of Selected Species Listed in CITES Appendix II*; Vol. 1, Mammals; p. 158-167; IUCN/WCMC, Cambridge
3. CLUTTON-BROCK, J. (1987): *A Natural History of Domesticated Mammals*. Cambridge University Press and British Museum (Natural History), Cambridge and London
4. CUNAZZA, P.C. (1980): *El guanaco - importante resorso natural renovable de magallañes*. Secunda edición. Publ. No. 17; Corporación National Forestal, Ministerio de Agricultura; Santiago, Chile
5. DEFOSSE, A.M., GARRIDO, J.L., LAPORTE, O.J. and L. DUGA (1981): *Cría de guanocos en cautividad. Variacion de su crescimiento y calidad de su lana*. Centro Nacional Patagónico, Secretaría de Estado de Ciencia y Tecnología; Contribucion no. 45; Puerto Madryn, Chubut, Argentina
6. ELTRINGHAM, S.K. and W.J. JORDAN (1981): The vicuña of the Pampa Galeras National Reserve. The conservation issue. In: *Problems in management of locally abundant wild animals*, ed. PEWELL, P.A. and S. HOLT; p. 277-289; Academic Press, New York
7. FRANKLIN, W.L. (1973): *High, wild world of the vicuña*. National Geographic Magazine, 143, p. 76-91
8. FRANKLIN, W.L.(1982): Biology, ecology and relationship to man of the South American camelids. In: *Mammalian biology in South America*, ed. MARES, M.A. and H.H. GENOWAYS; p. 457-489; Special Publications Series, Vol. 6; Pymatuning Laboratory of Ecology and University of Pittsburgh; Pittsburg, Pa., USA
9. FRANKLIN, W.L. and M.A. FRITZ (1991): Sustained Harvesting of the Patagonia Guanaco: Is it Possible or Too Late? *In: Neotropical Wildlife Use and Conservation*, ed. ROBINSON, J.G. and K.H. REDFORD; p. 317-336; University of Chicago Press, Chicago and London

10. FUJITA, H.O. and J.O. CALVO (1981): Las exportationes de productos y subproductos de la fauna silvestre en el quinquenio 1976/1980. *IDIA* No. 397-400, Jan. - April 1981; Instituto Nacional de Tecnologia Apropecuaria, Bariloche, Argentina
11. GARRIDO, J.L., MAZZANTI, R. and D. A. GARRIDO (1988): Distributión y densidades de guanaco en la Patagonia Argentina. *VI. Convención Internacional de Especialistas en Camélidos*; Oruro, Bolivia
12. GILMORE, R.M. (1950): Fauna and ethnozoology of South America. In: *Handbook of South American Indians*, ed. STEWARD, J.; Vol. 6, p. 345-464; Bulletin of the Bureau of American Ethnology, 143; 715 pp.
13. HOFMANN, R. (1982): Vikunjabewirtschaftung als Ergänzung der Haustierhaltung in den Hochanden. *Giessener Beiträge zur Entwicklungsforschung*, Reihe, I Band 8; University of Giessen; Giessen, Germany
14. HOFMANN, R.K., OTTE, K.C., PONCE C.F. and M. A. RIOS (1993): *El manejo de la vicuña silvestre*. Sociedad Alemana de Cooperacion Tecnica (GTZ); Eschborn, Germany
15. HOWARD, W.E. (1970): *Relationship of wildlife to sheep husbandry in Patagonia, Argentina*. FAO Report 14, p. 1-31; Food and Agriculture Organisation of the United Nations, Rome
16. JUNGIUS, H. (1971): The vicuña in Bolivia: the status of an endangered species, and recommendations for its conservation. *Zeitschrift für Säugetierkunde*, 36, p. 129-246
17. KOFORD, E.B. (1957): The Vicuña and the Puna. *Ecological Monographs*, 27, p. 153-219
18. KÖHLER-ROLLEFSON, I.U. (1991): Camelus dromedarius. *Mammalian Species* No. 375; p. 1-8; American Society of Mammalogists
19. MADUEÑO, C.G. (1912): Monografia de la vicuña (*cited after* KOFORD 1957)
20. McKNIGHT, T.L. (1969): *The Camel in Australia*. Melbourne University Press, Carlton, Australia
21. MILLER, S.D., ROTTMANN, J. and R.D. TABER (1973): Dwindling and endangered mammals of Chile: Vicuña, llama, hippocamelus, and pudu. *North American Wildlife Natural Resources Conference*, 38, p. 55-68
22. PIRES-FERREIRA, J.W. and E., and P. KAULICKE (1976): Preceramic animal utilisation in the Central Peruvian Andes. *Science*, 194, p. 483-490
23. RABINOVICH, J.E., HERNANDEZ, M.J. and J.L. CAJAL (1985): A simulation model of the management of vicuña populations. *Ecological Modelling*, 30, p. 275-295
24 RABINOVICH, J.E., CAPURRO, A.F. and L.L. PESSINA (1991): Vicuña Use and the Bioeconomics of an Andean Peasant Community in Catamarca, Argentina. In: *Neotropical Wildlife Use and Conservation*, ed. ROBINSON, J.G. and K.H. REDFORD; p. 337-357; University of Chicago Press, Chicago and London
25. RAMSAY, B.J. (1994): *Commercial use of wild animals in Australia*. Bureau of Resource Sciences, Australian Government Publishing Service, Canberra, Australia
26 RODRIGUEZ, R., GLADE, A. and E. NUÑEZ (1983): *Bases para el manejo de la vicuña en la Provincia de Parinacota, I. Region, Chile*; Documento Trabajo No. 3; Corporacion Nacional Forestal; Arica, Chile
27. SACHS, R. and C. (1968): Survey of parasitic infestations of wild herbivores in the Serengeti Region in Northern Tanzania and the Rukwa Region in Southern Tanzania. *Bulletin of Epizootic Diseases in Afrika*, 16(4), p. 455-472
28. SANDI, A., SANCHEZ, E. and M. YARIGANO (1987): *Plan nacional Vicuña*. Analisis del plan nacional Vicuña 1965-2013. Plan alternativo. Ministerio de Agricultura; Lima, Peru
29. SHORT, J., CAUGHLEY, G., GRICE, D. and B. BROWN (1988): The distribution and relative abundance of camels in Australia. *Journal of Arid Environments*, 15, p. 91-97
30. TORRES, H. (1985): *Distribution and conservation of the guanaco*. IUCN/SSC South American Camelid Specialist Group. Special Report No. 2; IUCN Morges, Switzerland
31. ZEUNER, F.E. (1963): *A History of Domesticated Animals*. Hutchinson, London

## 2.3.12 DEER (*Cervidae: Ruminantia: ARTIODACTYLA*)

The two most important groups of ruminant Cloven-hooved ungulates (**Artiodactyla**) in the context of utilisation are the deer (**Cervidae**) and the Horned ungulates (**Bovidae**). The term "deer" is here used to include the more primitive Musk deer (**Moschinae**) and Barking deer (**Muntiacinae**) with the true Eurasian deer (**Cervinae**), American and Roe deer (**Odocoileinae**), moose (**Alcinae**) and reindeer (**Rangiferinae**). Deer are indigenous to all continents with the exception of Africa south of the Sahara. They inhabit very diverse climatic zones from the Arctic to the equator in large numbers. Their habitat varies accordingly from boreal tundra, swampy temperate forest and cultivated forest edges, semi-arid mountains to tropical rain forest. Although the significance of deer as a prime source of food diminished with the spread of the domesticated bovids, their commercial utilisation has become particularly important since the Second World War. Tab. 40 summarises which species of deer are being commercially utilised, and Tab. 41 shows how they were distributed by man all over the world, initially for sports hunting and more recently for farming. The purpose for which they are used varies according to culture and socio-economic conditions and Tab. 42 lists products which are being derived from deer.

## Historical background

Deer were the most important prey animals of the prehistoric hunter-gatherers for hundreds of thousands of years (18). Only when agriculture and livestock husbandry developed in the Neolithic era did domesticated animals gradually become the main source of meat. In the inhospitable northern regions, however, which were unsuitable for any of the emerging domestic livestock, man remained entirely dependent on *reindeer*, a species uniquely adapted to those regions. Apart from hunting them he learnt also to utilise reindeer more intensively. He started to follow the large herds of reindeer on their seasonal migrations, gaining more and more control over them. A special symbiotic relationship developed between the nomadic herdsmen and reindeer from which both derived significant benefits. As early as the Stone Age this process led to the semi-domestication of this species. First evidence of reindeer husbandry in Norway dates back to AD 890 (56).

But also in all other regions of the world, deer retained a special importance to man throughout history as a major economic resource. In Europe *Red deer* resisted the climatic changes of the Glacial Ages and became the most important game animal for hunting purposes. *Fallow deer*, however, survived only in Asia minor and Mesopotamia. They never regained their former distribution naturally, but the Assyrians, Babylonians and Egyptians had already started to keep them in large enclosures, probably to facilitate their killing for religious purposes (24). The Phoenicians distributed Fallow deer within the Mediterranean region and the Romans introduced them eventually to Western and Central Europe. There Fallow deer became, together with the indigenous Red deer, an important park animal of the noble class during the Middle Ages. In England, for example, there existed no less than

2,000 deer parks during the 12th and 14th century (17). As of the 16th century Fallow deer were released more and more into the forests for hunting purposes (40). Thus, the species became re-established all over Western, Central and Southern Europe by the 18th century.

**Table 40:** Overview of the utilisation of deer

| English name | Scientific name | Original geographical distribution | Body weight (kg) | Utilisation |
|---|---|---|---|---|
| **MUSK DEER - MOSCHINAE** | | | | |
| Siberian musk deer | *Moschus moschiferus* | C.-E.Asia | 10-12 | hunted and farmed for musk |
| Dwarf musk deer | *-berezowski* | E.-S.E.Asia | 8-10 | |
| Himalayan musk deer | *-chrysogaster* | Himalayan region | 12-15 | |
| **BARKING DEER or MUNTJACS - MUNTIACINAE** | | | | |
| Barking deer | *Muntjak muntjak* sspp. | S.-S.E.-E.Asia[1)] | 20-30 | hunted for meat |
| **TRUE EURASIAN DEER - CERVINAE** | | | | |
| Fallow deer | *Dama dama* | W.Asia, N.Africa | 60-120 | hunted and farmed for meat |
| Axis deer or chital | *Axix axis* | S.Asia | 60-120 | hunted for trophies/meat |
| Hog deer | *-porcinus* | S. and S.E.Asia | 40-60 | hunted for subsistence |
| Sambar deer | *Cervus unicolor* | S. and S.E.Asia | 150-250 | hunted for trophies; farmed |
| Javan or Rusa deer | *-timorensis* sspp. | S.E.Asia | 60-120 | hunted and farmed for meat |
| Sika deer | *-nippon* sspp. | E.Asia | 80-130 | hunted and farm. for velvet |
| European Red deer | *-elaphus* | W., S., C.Europe | 120-300 | hunted for trophies and meat; farmed for meat and velvet |
| Siberian Red deer or maral | *-e. sibiricus* | C. and E.Asia | 160-240 | |
| Asian Red deer or malu | *-e. xanthopygus* | E.Asia | 250-350 | |
| Wapiti or elk | *-e. canadensis* | N.America | 280-380 | hunted, ranched and farmed for meat |
| Thorold's or White-lipped deer | *-albirostris* | C.Asia | 100-140 | farmed in China only |
| Père David's deer or milu | *Elaphurus davidianus* | C.Asia | 120-200 | bred for propagation |
| **AMERICAN and ROE DEER - ODOCOILEINAE** | | | | |
| Roe deer | *Capreolus capreolus* | W.Europe, C.-E.Asia | 20-50 | hunted for trophies, meat |
| White-tailed deer | *Odocoileus virginianus* | N.America | 60-120 | hunted for trophies, meat |
| Mule deer | *-hemionus* | West. N.America | 60-120 | hunted for trophies |
| Brocket deer | *Mazama* spp. | C. and S.America | 18-25 | hunted for subsistence |
| **MOOSE - ALCINAE** | | | | |
| Moose | *Alces alces* sspp. | North. Eurasia and N.America | 350-480 | hunted for trophies; ranched for meat, semi-domestication |
| **REINDEER - RANGIFERINAE** | | | | |
| Eurasian reindeer | *Rangifer tarandus* | North. Eurasia | 60-160 | domesticated |
| Caribou | *-t. caribou* | North. N.America | 80-180 | hunted for meat, skin, velvet |

1) *M. m. reevesi*, introduced to England, has become a semi-domesticated garden pet in some places (see (1) in Chapt. 2.3.15)

In the Far East of Asia, *Père David's deer* were probably kept thousands of years ago in enclosed areas both for hunting and for providing the especially favoured antler velvet for medical purposes. The last herd of this species was kept in the large imperial park of Nan Hai-Tze south of Peking from where they were first scientifically reported in 1865 by Père David, a French Jesuit priest (24). After the imperial herd perished in 1895 due to natural flooding of the Nan Hai-Tze park, the species was bred and propagated at Woburn Park in England by the Duke of Bedford. By 1946 the Worburn herd numbered 300 animals and in 1956 Père David's deer were repatriated to China (24, 8). Today there is considerable interest, not only in China, to use this species commercially (8).

The inhabitants of Central Asia had traded antler velvet, originating from the *local Red and Sika deer subspecies*, for hundreds of years with China where there was great demand for this as medicine. As wild deer became scarcer, herds were kept and propagated in enclosures for regular harvesting of antler velvet, in Mongolia as early as the 1830s and in China certainly by the 1880s (64). In Japan the local Sika deer were also kept and tamed from the Middle Ages, mainly for religious purposes, rendering them gradually semi-domesticated (24).

With the development of hunting as a privilege of the upper classes in Europe during the 19th century and strict control over game, Red as well as Fallow deer populations increased rapidly in many European countries. This resulted in many areas in an over-population of deer during the last century. To off-set the resulting habitat damage and to prevent large die-offs it became a standard practice of European deer managers, and in Germany even a legal requirement, to provide for supplementary feeding during winter time. This implied the establishment of feeding structures and in many cases also some fencing. In order to recover some of the costs of deer management, hunting rights were leased out and thus, deer hunting in many European countries developed into a commercially oriented form of wildlife utilisation.

To provide for deer hunting in overseas territories, Red and Fallow deer were introduced from Europe to many countries in the Southern Hemisphere, particularly New Zealand where no deer occurred. This had the most undesirable ecological effects and by the 1920s and 1930s it became urgent to reduce the deer in order to prevent further destruction of vegetation and soil erosion (31). The need to drastically cull deer led to the development of planned hunting of deer for commercial production of venison. This utilisation was eventually intensified into the farming of various deer species. The first license for a deer farm was granted in 1970. Due to lack of breeding stock, growth of deer farming was initially slow and by 1975 there were only 25 farms with a total of about 5,000 deer. This changed with the introduction of helicopter capture and by 1985 there were already 374,000 deer farmed of which about 84% were *Red deer*, 8% *Fallow deer* and the remainder *Wapiti hybrids* and other species such as *Sika, Sambar, Rusa, Axis, Hog* and *White-tailed deer* (22). The most recent survey in 1993 indicated that deer stock on 2,040 registered farms had increased to 1,300,000 animals of which 96% were Red Deer, 3% Fallow deer and 1% others (3). The success of deer farming in New Zealand has led to introduction of deer species for farming purposes to many countries all over the world (see Tab. 43 (b)).

The need to provide a source of meat for seafarers on their way to India had prompted as early as 1639 the introduction of *Rusa deer* to the island of Mauritius. They thrived in the moist forest which originally covered the island. With the development of large sugar estates during the 19th century they provided sport hunting for the feudal landowners and meat for the mainly Indian labour force. As the deer population became gradually split up into distinct herds bound to fenced estates, a more intensive use of the herds commenced in the late 1960s as a more productive alternative to livestock development (53). By the early 1980s about 10,000 deer, almost a third of the small islands total deer population, was under farming regimes of varying intensity (39).

**Table 41:** Introduction of deer species for release, extensive breeding or farming

| Deer species | from | to |
|---|---|---|
| Fallow deer | Asia Minor | initially to mediterranean countries, later by Romans, perhaps also by Normans, to West Europe; in 16-17th century further spread by nobles all over Europe |
| | Europe | New Zealand (1864), Australia (1840), Taiwan, South Korea (1956/70), Fiji, Canada (1851), USA (1878), Argentina (1905), Chile (1887), South Africa (1869) |
| Europ. Red deer | Europe | New Zealand (1851), Australia, Argentina, Reunion |
| | Australasia | South Korea (1956/70), Taiwan, South Africa |
| Malu | East Asia | England |
| Wapiti | North America | New Zealand (1905), Australia |
| Hog deer | South Asia | New Zealand, Australia, Europa |
| Sambar | South Asia | New Zealand (1875), Australia, China, Taiwan, South Korea (1956/70) |
| Rusa deer | South Asia | Mauritius (1639), New Zealand (1907), Australia, New Caledonia, Papua New Guinea (1928), Madagascar, Comores |
| Sika deer | East Asia | England, Ireland (1860), Germany, Denmark, Sweden, New Zealand (1885), Australia, USA, Russia (1950), Madagascar |
| Axis deer | South Asia | New Zealand, Australia, South Korea (1956/70), USA, Argentina, Chile, Germany, Sweden, England |
| Père David's deer | East Asia | England |
| | England | China, South Korea (1956/70), New Zealand |
| Domestic reindeer | Siberia | Alaska (1891-1902), Mongolia, China, South Korea (1956/70) |
| | North Europe | Canada (1908, 1921, 1929) |
| White-tailed deer | North America | New Zealand (1901), Czechoslovakia (1852), Finland (1970), South Korea (1956/70) |
| Moose | Siberia | New Zealand (1900), China |

Sources: HOWARD (31), WHITEHEAD (64), CHAPMAN and CHAPMAN (17), BARZDO and coll. (8), SCOTTER (54), RAMSAY (47)

A similar situation existed in Scotland where the *Red deer* utilised the rather unproductive highlands together with sheep. In 1970 the Rowett Research Institute together with the Hill Farming Research Organisation in Scotland started to compare the productivity and profitability of sheep and deer farming. As a result a first private deer farm was established in 1974; by 1984 60 fenced deer farms were in operation with a total stock of about 3,000 -

5,000 animals. 90% of these were Red deer, and 10% other species (Fallow, Sika and Père David's deer) (25). First efforts to farm Fallow deer in Europe were initiated in West Germany in 1974 by the Landwirtschaftskammer of Rheinland (49). They were so successful that Fallow deer farming for purely agricultural purposes spread within only the last decade to most other West European countries.

In North America the indigenous deer was a most important resource for subsistence hunting by the Indians until the White settlers started to deplete it in the 19th century by excessive hunting. Thus wapiti were extirpated from large parts of their original eastern range which extended as far as Pennsylvania by the turn of the century. The development of more responsible and controlled management of wildlife in the 1930s had a similar effect to that previously observed in Europe: both *White-tailed* and *Mule deer* populations started to proliferate in many areas inducing utilisation of the deer resources, both on State and Private land (67). In the USA this is mainly oriented towards trophy hunting, allowing also the use of the various non-indigenous deer species which have been introduced on private estates since the 1950s. In Canada considerable interest has developed in the ranching or farming of *wapiti* in conjunction with *moose* and bison (*Bison bison*) mainly for venison production. This has led since 1978 to the establishment of some large game ranches, and also to the fencing of large tribal wildlife areas, including building on-site game abattoirs (51).

## Management and utilisation

The spectrum of deer utilisation ranges from different types of hunting or keeping and breeding deer for recreational purposes, to farming them as an agricultural activity. Tab. 40 shows that it involves almost all species.

**Hunting** of deer **for subsistence** is still important in the northern regions, where mainly *caribou* and *Wild reindeer, moose, wapiti* and other *Red deer, White-tailed, Mule deer* and in Siberia also *Roe deer*, provide food and raw materials for the native people. However, some of the tropical forest species, such as *Axis, Hog, Sambar* and *Rusa deer*, and also the small muntjaks or *Barking deer* in South, South-East and East Asia, and *White-tailed* and *Brocket deer* in Latin America are also utilised for subsistence wherever they still occur. As with other animal groups, hunting of deer became more and more commercialised with the development of cash economies. Today the Indian and Innuit tribes in Alaska and Canada and some 22 different ethnic minorities in the northern and central parts of the former USSR depend, also economically, to a large extent on hunting (34, 57, 11). There are vast areas of relatively low productivity in these regions, such as tundra, taiga, deserts and mountains, where livestock production is not possible or only marginal. Almost all males of the native population in these areas are involved in **commercial hunting** of *Wild reindeer, moose, Roe* and *Red deer* mainly for meat production and *Musk deer* for musk (33, 68). Administrative control of the hunting activities is complex and reflects the differences in social conditions. Whereas in Canada native people have retained special aboriginal hunting rights, in Alaska hunting is regulated by government-imposed harvesting quotas and restricted seasons. Generally traditional hunting in both of these countries receives more and more legal protection and preference over other hunting (34). Also in the former USSR retention of hunting traditions by the involved ethnic minorities as well as the local Russians is an important aspect. Commercial hunting is planned and executed as an economic activity on the basis of a hunting license system. It is organised through state-controlled hunting co-operatives ("Game management units") which purvey most of the game products to state marketing organisations (57). Depending on the different species of deer, the hunting methods comprise stalking, driving, battue, baying with dogs, calling during the rut, and in the case of reindeer spearing or shooting at water crossings during mass migrations. Hunting is usually a team activity.

Control mechanisms have been relatively effective both in North America and the former USSR at reversing the decline of some deer species, such as *moose, Wild reindeer* and *caribou* (34, 57). They seek to respond to the major natural and anthropogenetic fluctuations of Wild reindeer and caribou populations which in the past have locally threatened the survival of native people. On both continents well-managed commercial hunting of deer is considered to have a good potential for further development (34, 57). However the possible negative effect on hunting of the most recent political changes in Russia will have to be observed.

**Table 42:** Utilisation of deer products

| From live animals | Processing/marketing methods |
|---|---|
| Cropped antlers with velvet | dried or frozen and cut into thin slices, or ground into powder for medicinal purposes |
| Shed antlers | ground into powder for medicinal purposes |
| Musk | scooped out on paper, or into jars, preserverd with antiseptic for the perfume industry |
| From slaughtered animals | |
| Venison | frozen or cooked for human consumption |
| Sinews | cooked for medicinal purposes |
| Skin | salted and tanned for leather goods |
| Pizzle<br>Velvet | sectioned into pieces, dried, cooked or preservered in drinking alcohol for medicinal purposes |
| Antlers | in subsistence economies used for various utility purposes; mounted as trophies; ground into powder for medicinal purposes |
| Bones | cut into pieces, ground for medicinal purposes or for bone meal production |
| Tail | sectioned and dried for medicinal purposes |
| Heart/liver | cooked for medicinal purposes |
| Placenta/blood | jellied or extracted from velvet for medicinal purposes |

A special aspect of commercial hunting in Russia and Asian countries is the production of antler velvet and musk, both highly priced commodities in Asian medicine and the perfume industry. Hunting for antler velvet has formerly resulted in the over-exploitation of the *Asian Red deer subspecies* and the *Sika deer* in East Asia and has probably also contributed to the decline of other Asian deer species. Although both antler velvet and musk, are now more and more produced on farms, these commodities are still also economically important in commercial hunting of *maral, Wild reindeer* and *Musk deer* in Russia. Protective management is considered to have increased local densities of *Musk deer*, allowing an animal harvest of about 5,000 animals, which represents about 5% of the estimated total population (57). Processing and trade in antler velvet and musk are described in Chap. 3.2.8.

**Recreational or sports hunting** of deer has evolved in Europe since mediaeval times into either the "revier" hunting system or free-hunt. A "revier" constitutes a designated hunting ground of a minimum of 75-150 ha which is either kept by the landowner or leased out to registered hunters for long periods (15). Whereas the revier hunting system implies co-responsibility and mutual law enforcement by state agencies and the revier owner, free-hunting is based on a license system. The latter concept was adopted and further developed in North America and has increasingly assumed commercial features during the last decades. Landowners with hunting rights may either lease these rights, charge fees for access to hunting grounds or sell identified types and numbers of deer to sports hunters. This may pertain to state agencies, including hunting co-operatives in the former USSR, as well as to private landowners including farmers, ranchers or tribal communities. The fact that large numbers of middle and upper class people mainly in the industrialised countries are prepared to pay substantially for hunting opportunities and trophies from game animals has constituted a major financial incentive for the development of commercialised trophy hunting in all continents. Particularly important as trophy animals are the different *Red deer subspecies*, including the wapiti or "elk" and the maral, *White-tailed, Mule* and *Roe deer*, and *moose*. The size and beauty of the deer's antlers is the most important aspect of trophy hunting and the influence on antler development by supplementary feeding has thus become a much disputed subject. Similarly, a lot of introductions of particularly strong Red deer stags, and the crossbreeding of different Red deer subspecies were undertaken with a view to improving antler quality genetically. European Red deer, introduced into Argentina, for example, grow particularly impressive antlers and have thus become favourite with trophy hunters (12). However, some of the *exotic deer species* which were introduced onto ranches in the USA as of the 1950s first for curiosity, are now also being commercially utilised for trophy hunting. The use of the meat and other products of trophy animals was formerly a purely secondary aspect, although the quantities of venison resulting from sports hunting have always been substantial (41, 2). Only with the consumer trend to lean meats and technological progress in game management are combinations of commercial trophy hunting and venison production are beginning to be favoured, particularly in the USA (67).

The production and sale of live animals for replenishment of the hunted populations or stocking of other hunting estates or deer parks, has become an important sideline of this type of deer utilisation. In France, for example, *Roe deer* are bred in large numbers solely for hunting purposes; and in England *Père David's deer* have been successfully bred for re-establishment of the species. In the USA live sales of exotic deer species like *Axis, Fallow* and *Sika deer*, and in Canada export of *wapiti* command increasingly high prices (67, 51). Such utilisation relates to the more intensive use of deer, either kept and managed in fenced areas or herded on natural range. Both of these have a long history as shown above.

**Keeping of deer in enclosed areas** might have historically been motivated by religious needs for animal sacrifices, but later it became an important feature of the mediaeval recreational hunting of the noble class (15). In Western Europe, and also Japan, ***deer parks*** have retained some socio-cultural importance to-date as non-commercial recreational establishments. *Fallow, Sika* and *Père David's deer* have been kept and bred in parks for hundreds of years bringing about varying degrees of semi-domestication of these species. Also *Axis deer* were introduced to Europe by the Romans and established in some parks in England and continental Europe (24). Attempts have been made to introduce *White-tailed deer*, but have failed except in Czechoslovakia and Finland. The Chinese subspecies of the *Barking deer (Muntiacus m. reevesi)* was introduced to various deer parks in England a century ago from where it has become well established in several localities. In places it is a semi-domesticated garden pet ((1) in Chapt. 2.3.15).

As of the late 1960s economic and ecological considerations inspired more intensive use of deer for purely agricultural purposes. This led to the concept of "deer farming", first in New Zealand and then in Europe, and "deer ranching" in North America, as described below. ***Farming of deer*** implies the keeping and breeding of semi-domesticated deer stock in fenced enclosures of varying sizes usually with supplementary feeding for the express purpose of producing venison and other deer products (see Tab. 42). Deer farming has developed very rapidly all over the world and is still expanding. Tab. 43 (b) shows numbers of deer farms and stock held in 1992. In New Zealand, and Great Britain mainly *Red deer* are utilised whereas in continental Europe, Australia and North America Fallow deer are preferred; *Sika deer* are farmed in Russia, China, Taiwan and South Korea and *Rusa deer* in Mauritius. Intensive research, mainly in New Zealand, Scotland and Germany, has produced a sound foundation of knowledge for the farming practice of Red and Fallow deer. In particular, stock size in relation to space and available pasture; sex ratio for maximum productivity; reproductive behaviour; nutritional requirements and health under different husbandry regimes; and also practical issues such as fencing, feeding, slaughter and carcass processing have been investigated. Present-day deer farming and the regulatory procedures

Modern feeding facility on a Fallow deer farm in Germany. (Photo: G. REINKEN)

pertaining to it are based on these investigations, the results of which are published in various manuals, textbooks and official guidelines (66). Generally speaking, deer farming is most intensive in New Zealand whereas in the UK and continental Europe less-intensive farming is favoured (50). Of greatest importance is optimal feeding. Natural pastures have to be fertilised and well managed. All calves must receive supplementary feed early on and during autumn and winter food for the adult stock must also be supplemanted. Whereas in New Zealand animals have to be brought to so-called "Deer Slaughter Premises" for slaughter, in Europe and the USA deer are predominantly killed humanely on the farms. Fallow deer are considered to produce a superior quality of venison (48) and this contains, indeed, particularly low levels of lipids as shown in Chapt. 3.2.1. In Europe and the USA most venison is marketed directly to the consumers which is most profitable. In New Zealand however, the quantities of produced venison are much greater and this implies organisation of marketing. There, venison is thus exported through the New Zealand Game Exporters Association and the Game Industry Board.

These bodies also organise the profitable sale of antler velvet to South Korea and Hong Kong. This is produced by sawing the antlers off before the velvet starts drying and by processing it according to consumer needs. Although the removal of antlers can be carried out quite painlessly under anaesthesia it is prohibited in Europe for humane reasons which constitutes a serious economic handicap to European deer farmers.

A special development is the ***farming of Musk deer*** for production of musk. This species is an ancestor of the true cervids and does not grow antlers. Instead, the males grow tusk-like canines and have a pod on the penis in which a strongly smelling secretion or "musk", collects. *Musk deer* occur all over Central Asia from the Himalayas through China and into Siberia. For hundreds of years Musk deer were hunted specifically for winning its musk as a highly valued ingredient of perfumes (43). As with other animals the high value of musk has induced the keeping of Musk deer in enclosures for periodic extraction of musk. Attempts to produce musk from captive animals were made in Russia even during the last century (7, 8). In response to the dwindling wild stocks Musk deer farming as a commercial activity was developed in China since 1958 (70). Four state farms which started extracting musk under the authority of the State Bureau of Pharmacy by 1979 were established in the Sichuan Province. Due to the high demand for musk, increasing numbers of rural communes have since taken up musk farming on a smaller, co-operative scale. The total number of Musk deer kept in 1984 in these farms was estimated at some 2,000 or more (28).

Musk deer are solitary living animals and therefore difficult to breed. They are best kept in large enclosures at a ratio of 3-7 females to one male. In some farms the males are placed into wooden cages for easier handling (13). They produce musk from the age of 2.5 years and on average 6-8 g can be extracted from them per year. In India experimental musk deer farms were established in Himachal Pradesh in 1965 and in Utar Pradesh in 1981 (27). They have, however, not yet started any commercial production. Plans for Musk deer farms have also been reported from Nepal and Bhutan (8).

Only in North America is consideration also being given to a holistic approach to deer utilisation in the form of ***ranching***. This involves the management of multi-species indigenous ungulates contained on large tracts of private more-or-less wildland, for the purpose of meat production. Compared with the more intensive deer farming, multiple species ranching has the advantage of a better utilisation of the full spectrum of the natural food resources and thus higher biological productivity. It is also preferable from a point of view of landscape conservation. However, the development and successful application of this concept has been constrained by legal and administrative problems and a general animosity of the public against the commercialisation of indigenous wildlife (67). In Canada ranching involves mainly *wapiti, moose* and also *Mule* and *White-tailed deer* in association with Plains bufalo. Relevant biomass and production data have been collected since 1907 at the Elk Island National Park in Alberta which is stocked with these species. Despite the mentioned problems there were 251 ranches operating in 1987 just in Alberta, Saskatchewan and Manitoba, holding 5,100 bison and 2,900 wapiti (51). Numbers of meat-production-oriented game ranches in the USA are also limited. Most of them are situated in Texas, some others in New Mexico, Colorado, Wyoming and Utah. They mainly utilise populations of re-introduced wapiti, native White-tailed and Mule deer in association with introduced exotic ungulates, mainly *Axis, Fallow* and *Sika deer, blackbuck, mufflon* and *Barbary sheep*. More recent operations with introduced European Red and Fallow deer in New York State should rather be classified as deer farming.

Similar to the experience in Africa, game ranching in the USA has been more and more integrated with trophy hunting, utilising particularly the introduced exotic deer species (61, 67). Such combination seems to be the only economically profitable form of semi-intensive commercial utilisation of mixed deer populations.

The only deer species which can be utilised by **herding** is the *reindeer*. This is explained by the fact that it differs from all other deer by being a highly sociable and gregarious species. Because of the great importance of reindeer for the survival of man in the inhospitable northern regions, he learnt early on to exploit their special behavioural characteristics. In reaction to danger or disturbances, social ungulates have a tendency to gather into compact groups or herds and then to follow a "leader". "Herding" implies the knowledge to predict the behaviour and movements of the respective herds (and to control these by providing artificial leadership (10). Herdsmen have to adapt, to a large extent, to the biological needs and the seasonal migrations of the herded animals which implies a nomadic lifestyle with all its hardships. Therefore herding and semi-domestication of reindeer initially only developed in some parts of Northern Asia and possibly in Fennoscandia, where the local tribes were mentally prepared for this. During the last centuries, however, reindeer husbandry was adopted by most of the ethnic minorities in Northern Russia, although hunting of *Wild reindeer* is still preferred by some (11). Today the reindeer constitutes a fully domesticated species with different breeds which show genetic characteristics of domestication. As in the case of deer farming, there is a wealth of literature on all aspects of reindeer husbandry and their products. In Russia there is a total stock of about 2.3 million domesticated reindeer as compared with about 1 million wild ones, mainly in the Taimyr and Yakutiya regions (57). In Fennoscandia there are about 781,000 domestic reindeer, but only an insignificant number of wild ones have survived in the mountains of Southern Norway (55).

Despite its potential the *caribou* of North America was never domesticated. The native people of Alaska and Canada had vast wildlife resources available to them and the utilisation of these by hunting evidently offered greater incentives and independence. In order to supplement the food basis of the indigenous population, domesticated reindeer were first introduced from Siberia to Alaska in 1891 and 1902. By 1932 the reindeer population had increased to 641,000 animals, but then crashed. Since 1950 their total number has fluctuated around 20,000 (59). The initial success of this operation prompted similar introductions from Norway into Canada, in 1908 into Newfoundland and in 1921 onto Baffin Island (54). Both of these efforts to introduce reindeer husbandry to the Canadian Inuit failed completely. Nevertheless another attempt was made in 1929 by driving 3,000 reindeer over 2,600 ks of Arctic tundra from Alaska to the MacKenzie River Delta in the North-West Territories of Canada which took 5 years. Similar to the other introductions this unique operation was intended to improve the economic basis of the native Inuit after the wildlife resources on which they depended had been severely depleted, partly because of the commercialisation of hunting. The reindeer herd was established in a Reindeer Preserve of 47,000 km$^2$ under Government control, and the intention was to turn over smaller herds to suitably trained Inuit herdsmen, each assigned herding ranges. None of the Inuit-owned herds succeeded in the long run and by 1964 the last one of these reverted back into Government custody (54). It had obviously not been possible to achieve the difficult psychological task of changing hunters into herdsmen. In 1974 the entire reindeer herd, numbering 5,200 animals, was sold to private enterprise. This was able to gradually improve the economics of reindeer production by making use of the growing demand for reindeer meat in southern towns and the oriental market for antler velvet, and also by introducing innovative husbandry and management techniques, such as helicopters and snow-mobiles. The herd increased to about 15,000 animals by 1986. In 1978, 60 animals of this herd were flown to the Belcher Islands, also in the North-West Territories. The rapid increase of this herd in the following eight years to about 500-700 animals indicates that it has been successfully established. This herd is owned and managed by the local people. (54).

The only other deer species which has a ***potential for domestication like reindeer*** is the *moose*. It is the largest deer species and has a similar distribution to the forest-type reindeer, but ranges much further to the south. Like reindeer it is well adapted to the northern taiga and tundra. Captive moose are particularly easy to tame and according to prehistoric rock paintings were probably herded and used for draught at an early time in Siberia (18). Later they seem to have been replaced by reindeer and horses, but moose were sporadically used as riding animals in Scandinavia as late as the beginning of the 19th century (69). Between 1934 and 1957 several moose nurseries were set up in the former USSR. But only after the Second World War in 1946 was an Experimental Moose Farm established at the Pechora-Ilichsky State Reserve in the foot-hills of the Ural mountains (35, 65). There it was intended to investigate the potential of moose for meat and milk production, and also for transport. Another experimental project was started in 1963 at the Kostroma Agricultural Research Station 330 km North-East of Moscow. By 1969 a total of 174 calves had been raised at Pechora-Ilichsky and successive breeding has reached the fifth generation. The total moose stock in the two experimental stations, and as of 1980 in a few others, is around 70-90 animals. To establish semi-domesticated herds of moose it is essential to raise calves artificially and this has been to some extent mechanised. The moose are kept in a free-grazing system, returning voluntarily to the station for supplementary feed. This implies that the females are mated by wild moose bulls and full genetic domestication will thus not be achieved (9). Meat yields at 5-6 years of age range between 175 and 230 kg carcass weight. Dairy moose may produce at an average up to 6 l of milk daily during a lactation period of six months (36, 35); special properties for the treatment of gastrointestinal disorders are attributed to it (20).

In Canada the only evaluation of the agricultural potential of moose has been conducted at Ministik Research Station of the University of Alberta (60). Interest in Canada focused mainly on the ranching of moose as discussed above. The main use of captive bred moose is for live sale, with individual animals valued at US$ 5,250-7,400 in the early 1980s (60).

## Economic importance

Due to the world-wide distribution of deer and their multiple use they have greater economic importance than any other group of wild animals. This is reflected both from the quantities of venison and other products derived from them and the monetary returns which they generate mainly through sports or trophy hunting. Tab. 43 (a) attempts to compare the importance of the different types of their utilisation, hunting, herding and farming or ranching, in terms of **meat production**. Most of the compiled data are only minimum figures, referring to different years, or rough estimates based on assumed harvesting rates and average dressed carcass weights. Nevertheless they clearly show that ***hunting*** produces far greater quantities of meat than herding, farming and ranching together. The most important species in this respect are *White-tailed* and *Mule deer, moose, Red deer*, inclusive of the *wapiti, malu* and *maral*, and the *Wild reindeer*. The quantities of venison which these species produce are even greater if subsistence hunting by indigenous people in the northern regions and in South America is included. *Moose*, for example, provide more than 25% of the food of indigenous people in Northern Canada (29) and *Brocket deer* are one of the most important meat resources for the forest-dwelling people of South America (62). Sports hunting records for *Red, Fallow* and *Roe deer* tend to be incomplete, because many countries where these species are hunted, are excluded, as for example Argentina, where Red deer have been imported for sport hunting (45). Also, the wide-spread hunting of the *Asian deer* species for sport and for meat, is underrepresented. Rusa deer, for example, have been introduced to New Guinea, the Philippines and New Caledonia where they have built up to large hunted populations (14). In Irian Jaya, for example, local communities harvested 2,250 deer in 1993 which provided them with an annual income of more than US$ 51,000 (4).

The next most important use of deer for meat production is undoubtedly the ***herding*** of *semi-domesticated reindeer* which assumes particular importance as the only means of meat production in Arctic regions, apart from hunting. However, it is important to note that this form of utilisation implies a psychological preparedness and capability of local tribes to involve themselves in reindeer husbandry and indulge in the associated nomadic life-style. With the socio-economic changes which have also occurred in the remote northern regions, there is a tendency to transform herding into ranching activities, controllable by fencing and modern techniques (11, 54, 56).

Compared with hunting and herding the more intensive ***farming*** and ***ranching*** of deer which have only emerged during the last two decades, produce much less venison.

**Table 43 (a):** Estimated venison production - from hunting and herding (1980/92)

| Species[1] | Geographical region | Estimated total population (in numbers) | Mean annual harvest (in numbers) | Dress. carc. meat yield (in t) | Literature references |
|---|---|---|---|---|---|
| White-tailed and | North America | ~ 15,000,000 | 2,982,000 | ~ 100,000 | (44) |
| Mule deer | | ~ 1,500,000 | 500,000 | ~20,000 | (44) |
| Moose | Fennoscandia | 600,000 (1988) | 215,000 (1983/91) | 28,000[2] | (2) (15) (60) |
| | former USSR | 831,000 | 95,800 (1988) | ~ 12,454[2] | (38) |
| | Poland | 6,000 | 1,600 | (~ 208)[2] | (60) |
| | North America | 900,000 | 54-70,000 | 10,000 | (44) |
| Red deer (inclusive | Europe[3] | (~ 540,000)[4] | 162,477 (1989/92) | (~ 10,600)[5] | (2) (15) |
| wapiti, malu, maral) | former USSR | ~ 208,600 | 5,220[6] | (~ 626)[6] | (38) (19) |
| | North America | ~500,000 | 85,000[7] | ~ 11,000[7] | (44) |
| Roe deer | Europe[3] | (~ 5,600,000)[4] | 1,666,636 (1989/92) | ~ 21,666[8] | (2) (15) |
| | former USSR | 738,500 | 41,200 (1988) | 824[9] | (38) |
| Wild reindeer | former USSR | ~ 1,000,000 (1991) | 134,250 (1988) | ~ 6,041[10] | (11) (38) |
| (incl. caribou) | Norway | ~ 35,000 | 8,859 (1991) | (~ 443)[10] | (56) (2) |
| | North America | ~ 2,652,000 (1980s) | 50,000 | ~ 2,300 | (44) |
| Domestic reindeer | former USSR | 2,300,000 | (523,750)[11] | (~ 20,950)[12] | (11) |
| | Fennoscandia | 781,000 (1985) | 234,000 (1985) | ~ 7,230 (1985) | (56) |
| | North America | 35,000 (1980) | ~ 3,500 (1980) | (~ 140)[12] | (54) |
| | Mongolia and China | 1,000+ (1983) | insignificant | unknown | (8) (56) |
| Fallow deer | Europe[3] | (~ 151,000)[4] | 45,240 (1990/92) | (~ 1,583)[13] | (2) (15) |
| Asian deer species | Europe[14] | (~ 4,130)[4] | 1.239 (1990/92) | (~ 43)[13] | (2) (15) |
| | North America[15] | 45,991 (1984) | ~ 2,000[16] | (~ 70)[13] | (67) |
| | Mauritius[17] | 30,000 | 5,000 | (~ 225)[13] | (19) |
| **Total** | **wild populations** | **(30,342,221+)** | **5,571,521** | **(~ 226,083+)** | |
| | **herded populations** | **3,117,000+** | **761,250+**[18] | **28,320+** | |

Figures in brackets are only extrapolations - years in brackets denote specifically reported periods, otherwise these were not indicated.
1) in sequence of importance for venison production; 2) based on medium dressed carcass weight of 130 kg; 3) only Germany, Denmark, Norway, Sweden, Finland, Holland, Switzerland, Austria, Hungary and Slovenia; 4) extrapolated from the harvest, assuming this to represent at least 30% of total population; 5) based on medium d.c.w. of 65 kg; 6) only Asian wapiti and maral with medium d.c.w. of 120 kg; 7) refers only to wapiti; 8) based on medium d.c.w. of 13 kg; 9) based on medium d.c.w. of 20 kg; 10) based on medium d.c.w. of 45 kg; 11) extrapolated from a production of 41,900 t liveweight, assuming average liveweight of 80 kg; 12) based on an average d.c.w. of 40 kg; 13) based on medium d.c.w. of 35 kg; 14) Sika deer in Germany and Denmark only; 15) Axis (38,035) and Sika deer (7,956) on private land; 16) low off-take because of live sales; 17) Rusa deer only (see Table (b)); 18) total reindeer harvest according to SKJENNEBERG (55) was 789,000 animals

Being entirely commercially oriented this is, however, inevitably of a better quality and more easy to market. The data compiled on venison production from tamed or ranched deer are also incomplete, in so far that much of the farming, for example *Sika* farming in the former USSR, China and more recently Korea, is directed towards velvet production. This does not entail slaughtering, and venison production therefore is only an unrecorded secondary line. Considering the large numbers of farmed sika the quantities of venison thus obtained must nevertheless be substantial. Furthermore farming and/or ranching of deer are in a phase of rapid development and therefore there is a strong demand for breeding stock everywhere. This causes many farms and ranches to sell their female off-spring rather than to slaughter it for venison. The most important species for farming are *Red deer* and increasingly *Fallow deer*. Much interest and experimentation is being focused on hybridisation of wapiti with other Red deer subspecies and, in China, also with sika in order to increase productivity (19, 22).

An analysis of **monetary returns** from deer utilisation not only confirms but reinforces the above economic rating of the different forms of utilisation. Although hunting produces the greatest quantities of venison, its real economic importance results from the willingness of sportsmen and trophy hunters to invest significant amounts of money in hunting and the trophy cult. In European ***sports hunting***, for example, anything up to US$ 10,000 may be spent for a top-class deer trophy (15) and hunters in the European Union paid in 1993 an estimated 3.5 millard ECU (equivalent to US$ 4.1 milliard) for their sport (2). In North America hunters (17 mill. in the USA in 1980 and 1.8 mill. in Canada in 1981) were estimated to have spent about US$ 3.5 bill. on big game hunting (US$ 2.5 bill. in USA (1) and US$ 1.0 bill. in Canada (23)). Whereas governments and wildlife authorities profit in terms of land leases and license fees, landowners may realise greater monetary returns per land unit from hunting than from any other form of land use. In Texas, for example, the total of hunters fees paid to hunt native deer on private lands amounted to US$ 110 mill. in 1974 (61). In a survey in Wyoming more than half of the landowners indicated that fee hunting formed an important part of their income (46). Mainly in the western industrialised countries sports hunting with all its associated subsidiary commercial services has thus become a major industry, in economic terms comparable with livestock production.

However, large-scale ***commercial hunting*** which involves not only deer, but hare, wild pigs and fur animals as well, can also generate in some countries monetary returns of overall national importance. The total gross value of annual game production in the former USSR, for example, was estimated at Rb 40 mill. (equivalent to US$ 62.8 mill. in 1986) and the commercial value of the harvested 550,000 to 600,000 ungulates at Rb 25 mill. (equivalent to US$ 39.25 mill. in 1986) (57). Profits from commercial hunting amounted to 52-58% of household incomes of the local population. Although not hunted solely for commercial purposes moose venison in Sweden earned about US$ 150 mill. annually in the early 1980s (30), but harvests have since decreased.

Because of the limitations of reindeer ***herding*** this is seen economically as a special case. It has led to the domestication of this species, but the economic importance of reindeer husbandry is restricted to the northern parts of Fennoscandia and Russia. Although it is the only form of production in Northern Fennoscandia it has only limited national economic importance there. No more than 10,000 families own reindeer from which no more than

3,000 earn most of their income (56). In Russia, however, reindeer husbandry is still the economic basis of millions of people inhabiting its northern and central regions. Next to commercial hunting, which increasingly involves Wild reindeer too, it is the second most important branch of the economy in Central Siberia (11).

Deer ***farming*** started off as an agricultural activity in areas which are ecologically or for other reasons only marginal to conventional agricultural production, such as the Scottish Highlands, the scrublands in Mauritius and others. In such areas it assumed particular local economic importance not only by producing meat from otherwise rather unproductive land, but also by generating employment and income. Then, however, the superior capability of deer for efficient food conversion was recognised and deer farming was found to be competitive to livestock production also in agriculturally more productive areas. The development of intensified deer farming was economically enhanced by the consumer trend to lean meats and the significant supplementary profit which could be gained from the sale of antler velvet. Nevertheless, deer farming in most countries, although of local importance and profitable to the individual farmer, remains economically rather insignificant in comparison with livestock production. Only in New Zealand has deer farming gained an overall national economic dimension which is comparable with other branches of livestock production. Total export earnings from the deer farming industry there reached a record of NZ$ 200 mill. (equivalent to US$ 100 mill.) in 1992/93 representing 2.5% of New Zealand's overall exports of pastoral products. At some NZ$ 9,450 per ton (equivalent to US$ 5,100) the average export return was approximately double the level of beef and lamb prices (37).

The **economics of deer farming** and ranching depend strongly on the local conditions of production and marketing. Cost of establishment may be as high as in extensive livestock production, the greatest costs being fencing and initial breeding stock. Compared with intensive livestock production systems in Germany they are, however, considerably lower (48, 6). Returns on investments have been good to very good (between 11% (63), 18% (58) and 28% (25)). In Canada recurrent production costs on wapiti/bison ranches were found to be lower than for beef production (51). Fallow deer farming in Germany was rated the second most efficient long-term pasture use after dairying (50). There, Fallow deer are preferred because of their greater physical resistance, greater availability and higher prices paid for fresh venison for this species. Generally, Fallow deer are, however, considered to be less docile than Red deer, less meat productive, and produce only a lower quality of velvet. Quality and the possibility to market antler velvet to the Asian consumers, however, are a most important economic aspect of deer farming. Some of the New Zealand deer farmers gain major profits from it (22), and in a representative wapiti ranch in Canada vension earned US$ 86,880 and velvet US$ 32,800, i.e. it produced no less than an additional 42% return (51). During the last decade, deer farming profited like most other game farming enterprises, from an ever-increasing demand for breeding stock which has enhanced their profitability greatly. Considering the rapid increase of deer farming activities all over the world this demand is bound to diminish in future and will render deer farming less profitable and more competitive. On the other hand, however, deer farming is also likely to gain a greater share in the international game meat trade, replacing other sources of game meat, because of the higher standard of hygiene and easier marketability of farmed venison as discussed in Chapt. 3.2.1.

**Table 43 (b):** Estimated venison production - from farming and ranching (1992)

| Countries | | Number of farms | Total deer stock held (numbers) | | | Venison production (t) | | Literature |
|---|---|---|---|---|---|---|---|---|
| | | | Red | Fallow | Other | Recorded | Estimated | references |
| New Zealand | | 5,500 | 1,100,000 | 70,000 | 300,000 | ~ 15,000[1)] | | (22) (32) |
| Australia | | 1,451 | 65,000 | 52,000 | 13,000 | | (~ 819)[2)] | (52) (47) |
| Germany | | 4,500 | 4,000 | 73,000 | | ~ 1,000 | | (50) (2) |
| Great Britain | | 300 | 43,200 | 7,500 | some sika | | (~ 371)[2)] | (25) (26) |
| Ireland | | 120 | 3,000 | 1,000 | some sika | | (~ 35)[4)] | (50) |
| Other Europe[3)] | | 2,071 | 12,620 | 93,600 | insignificant | | (~ 685)[4)] | (21) |
| Canada | (fa) | 758 | 300 | 28,000 | - | | (~ 285)[5)] | (16) |
| | (ra) | 251+(1987) | 2,900 (1987) | - | some moose | | ~ 418[6)] | (51) |
| USA | (fa) | 400 | 300 | 12,000 | 6,800 | | (~ 185)[5)] | (42) |
| | (ra) | few | 1,529+(1979) | 10,507 (1984) | 45,991 (1984) | | | (67) |
| Argentina | | few[7)] | 8,000 | | | | ~ 62[5)] | (45) |
| Mauritius[8)] | | ~ 20 | - | - | ~ 4,000 | | ~ 54[9)] | (19) |
| China[10)] | | >130 | 65,000 (1981) | - | 195,000 (1981) | in most Asian countries venison is only a by-product of velvet production, the oldest goal of deer farming | | (19) (8) |
| South Korea[11)] | | 1,235 (1981) | 5,000 (1985) | - | 52,000 (1985) | | | (19) (8) |
| Taiwan | | large number | 800 (1988) | 1,500 | 22,020 (1983) | | | (8) |
| former USSR | | unknown | ~ 70,000 (1983) | - | ~ 65,000 | | | (19) (8) |
| Mongolia | | unknown | insignificant | - | - | | | (8) |
| **Total** | | **16,736+** | **1,381,449** | **349,107** | **703,011** | **~ 19,000+ +** | | |

(1) venison export only; (2) based on 65 and 40 kg d.c.w. respectively at 12% annual stag off-take; (3) Sweden, Denmark, France, Switzerland, Austria, Italy, Yugoslavia; establishment of further farms in Belgium, Holland and Spain has only commenced; (4) based on 65 and 40 kg d.c.w. respectively at 15% annual harvest; (5) based on 65 and 40 kg d.c.w. respectively at 25% annual harvest; (6) based on 100 kg d.c.w. at 12% annual stag off-take; (7) some hold more than 2,500 animals; (8) only Rusa deer farmed or ranched (probably some overlap with Table (a)); (9) based on 45 kg d.c.w. at 30% annual harvest; (10) mainly Asian wapiti and Sika deer are farmed, but also Sambar, White-lipped deer and Reindeer; (11) other countries in which deer farming has been initiated or is contemplated are Malaysia, Thailand, Philippines, Indonesia, Papua New Guinea, Coledonia and Japan

## Implications for conservation

Deer provide some good examples of the beneficial effects which management may have on wildlife populations. Although they have been utilised for hundreds of years most of the commercially hunted or farmed deer species are still quite common. Only in parts of their original wide ranges have some of the deer subspecies become rare or even endangered. This might have been partly due to over-exploitation during the 19th and early 20th century, mainly for antler velvet in Central and East Asia, or trophy hunting in other areas. Generally, however, more intensive management has resulted in an enormous propagation of deer in many countries. In Sweden, for example, there are now probably more moose than ever before, and in the former USSR several species of deer, including Wild reindeer, have multiplied into utilisable populations after over-exploitation was checked by planned management. Although management was initially prompted by immaterial aesthetic reasons, commercial considerations became an important aspect of it. Some deer species have, in fact survived only because of being managed, bred and farmed by man for hundreds of years, like Père David's and Fallow deer, Dybowski-, North Chinese, Japanese and Formosan sika deer. Today it is much easier to obtain live deer, even of the rarer Asian species, from captive herds or farms than to capture specimens in their wild ranges. Therefore, procurement of stock for commercial farming does not constitute a threat to wild populations anymore. In future this will replace more and more venison traded from wild populations and thus eventually alleviate hunting pressures on these, as it has already done in respect of antler velvet.

Seriously endangered species, such as the barasingha (*Cervus duvauceli*), Eldi's (*C.eldi*) and White-lipped deer (*C. albirostris*) in Asia, and guemal (*Hippocamelus* spp.), pudu (*Pudu* spp.), Pampas (*Odocoileus bezoarcticus*) and Swamp deer (*O. dichotomus*) in South America, are hardly affected by commercial utilisation. Their decline had mainly ecological reasons and more recent efforts to breed some of these species in captivity will only enhance their conservation.

Musk deer have certainly been seriously depleted for reasons of commercial use, and because of the exceptional value of musk this might further endanger wild stocks. Here transition from hunting to farming might aid their conservation.

## Literature references

1. ANONYMOUS (1982): *1980 National Survey of Fishing, Hunting and Wildlife - associated Recreation (Initial Findings).* US Bureau of Sport Fisheries and Wildlife, Washington, DC
2. ANONYMOUS (1993): *DJV-Handbuch 1993: Jagd aktuell.* Deutscher Jagdschutz-Verband e.V., Bonn, Germany
3. ANONYMOUS (1994a): Farmed Deer Census. *The Deer Farmer* (Wellington, New Zealand), 118, p. 10
4. ANONYMOUS (1994b): Using rusa. *IUCN/SSC Deer Specialist Group News, June 1994*, No 11, p. 11
5. ASHER, G.W. (ed.) (1993): *Proceedings of the Ist World Forum on Fallow Deer Farming,* 10-13 March 1993 in Mudgee, NSW, *Australia.* Australian and New Zealand Fallow Deer Societies, Sydney and Wellington; 244 pp.

6. BACH, P. (1991): Ökonomik der Dam-und Rotwildhaltung in Gehegen. In: *Damwild und Rotwild in landwirtschaftlichen Gehegen*, ed. BOGNER, H.; p. 150-172; Paul Parey, Hamburg and Berlin
7. BANNIKOV, A.G., USTINOV, S.K. and P.N. LOBANOV (1980): *The Musk Deer* (*Moschus moschiferus*) *in the USSR*. IUCN, Gland (unpubl. report)
8. BARZDO, J., LUXMOORE, R., BROAD, S. and D. JONES (1984): *A survey of wildlife farming operations*. WCMC/IUCN, Cambridge (unpubl. report)
9. BASKIN, L.M. (1987): Moose domestication experience. *Swedish Wildlife Research*, Suppl. 1, p. 741-743
10. BASKIN, L.M. (1989a): Herding. In: *Wildlife production systems*, ed. HUDSON, R.J., DREW, K.R. and L.M. BASKIN; p. 187-196; Cambridge University Press, Cambridge
11. BASKIN, L.M. (1989b): Reindeer husbandry in the Soviet Union. In: *Wildlife production systems*, ed. HUDSON, R.J., DREW, K.R. and L.M. BASKIN; p. 197-206; Cambridge University Press, Cambridge
12. BELLATI, J. (1993): *in litt.*
13. BISTA, R.B., SHRESTA, M.N. and B. KATTEL (1979): Domestication of the Dwarf musk deer (*Moschus berezovskii*) in China (*cited after* GREEN 1989)
14. BORGUE, J. le (1964): *Géographie de la Nouvelle Calédonie et des Iles Loyauté*. Nouméa, Nouvelle Caledonie
15. BUBENIK, A.B. (1989): Sport hunting in continental Europe. In: *Wildlife production systems*, ed. HUDSON, R.J., DREW, K.R. and L.M. BASKIN; p. 115-133; Cambridge University Press, Cambridge
16. BURTON, B. (1993): The Current Status of Fallow Deer Farming in British Columbia (*cited from* ASHER 1993)
17. CHAPMAN, N.G. and D.I. (1980): The distribution of Fallow deer: a worldwide review. *Mammal Review*, 10(2/3), p. 62-138
18. CLUTTON-BROCK, J. (1987): *A Natural History of Domesticated Mammals*. Cambridge University Press and British Museum (Natural History), Cambridge and London
19. DREW, K.R., BAI, Q. and E.V. FADEEV (1989): Deer farming in Asia. In: *Wildlife production systems*, ed. HUDSON, R.J., DREW, K.R. and L.M. BASKIN; p. 334-345; Cambridge University Press, Cambridge
20. DUDIN, V.A. and A.P. MIKHAILOV (1990): Moose domestication, husbandry and management at the Kostroma moose farm; In: *Transactions of the XIXth Congress of the International Union of Game Biologists*, ed. MYRBERGER, S.; Vol 1, p. 332-333; Norwegian Institute for Nature Research, Trondheim,Norway
21. EKLUND, L. (1993): Fallow Deer Farming in Scandinavia (*cited from* ASHER 1993)
22. FENNESSY, P.F. and P.G. TAYLOR (1989): Deer farming in Oceania. In: *Wildlife production systems*, ed. HUDSON, R.J., DREW, K.R. and L.M. BASKIN; p. 309-332; Cambridge University Press, Cambridge
23. FILION, F.L., JAMES, S.L., DUCHARME, J.L., PEPPER, W., REID, R., BOXALL, P. and D. TAILLET (1983): *The importance of wildlife to Canadians: Highlights of the 1981 National Survey*, Canadian Wildlife Service, Ottawa, Canada
24. FLETCHER, T.J. (1984): 16. Other deer. In: *Evolution of domesticated animals*, ed. MASON, I.L.; p. 138-145; Longman Group, London and New York
25. FLETCHER, T.J. (1989): Deer farming in Europe. In: *Wildlife production systems*, ed. HUDSON, R.J., DREW, K.R. and L.M. BASKIN; p. 323-333; Cambridge University Press, Cambridge
26. GLICKSTEN, M. (1993): Fallow Deer Farming in Britain (*cited from* ASHER 1993)
27. GREEN, M.J.B. (1986): The distribution, status and conservation of the Himalayan musk deer (*Moschus chrysogaster*). *Biological Conservation*, 35, p. 347-375

28. GREEN, M.J.B. (1989): Musk production from Musk deer. In: *Wildlife production systems*, ed. HUDSON, R.J., DREW, K.R. and L.M. BASKIN; p. 401-409; Cambridge University Press, Cambridge
29. HAMILTON, G.D. (1981): Practical importance of moose and other wild foods to natives in a remote Northern Ontario community. *Alces*, 17, p. 44-55
30. HAWLEY, A.W.L., SYLVÉN, S. and M. WILHELMSON (1983): Commercial moose meat production. *Livestock Production Science*, 10, p. 507-516
31. HOWARD, W.E.. (1965): *Control of introduced mammals in New Zealand*. New Zealand Department of Scientific and Industrial Research, Information Series No. 45; Wellington, New Zealand
32. JACOBSEN, B. (1993): Fallow Deer Farming in New Zealand (*cited from* ASHER 1993)
33. KARELOV A.M. and S.G. SEMKIN (1981): *cited after* SOKOLOV and LEBEDEVA (1989)
34. KLEIN, D.R. (1989): Northern subsistence economies. In: *Wildlife production systems*, ed. HUDSON, R.J., DREW, K.R. and L.M. BASKIN; p. 96-111; Cambridge University Press, Cambridge
35. KNORRE, E.P. (1973): History and advances in experimental studies on moose domestication in the Pechora Ilychsky Reserve (in Russian). In: *Odomashnivaniye Losya*, ed. SABLINA, T.B.; Nauka, Moscow
36. KOZHUKHOV, M.V. (1973): Results of the 20-year experimental studies on moose domestication (in Russian). In: *Odomashnivaniye Losya*, ed. SABLINA, T.B.; Nauka, Moscow
37. KROSTITZ, W. (1994): Game meat: production, consumption and trade - Recent developments and prospects (unpubl. report)
38. KUZYAKIN, V.A. (1991): Status of ungulate resources and hunting systems in the Soviet Union. In: *Wildlife Production: Conservation and Sustainable Development*, ed. RENEKER, L.A. and R.J. HUDSON; p. 89-95; University of Fairbanks, Fairbanks, Alaska
39. LALOUETTE, J.A. (1985): Development of deer farming in Mauritius (*cited after* DREW and coll. 1989)
40. LINDNER, K., (1940): *Die Jagd im frühen Mittelalter. Geschichte des deutschen Waidwerks*. Walter De Gruyter, Berlin
41. NÜSSLEIN, F. (1974): Der ökonomische Wert der Wildproduktion in Europa (ohne Sowjetunion). *Zeitschrift für Jagdwissenschaften*, 20, p. 85-95
42. O'HARA, D. (1993): Fallow Deer Farming in the USA (*cited from* ASHER 1993)
43. PARRY, E.J. (1925): *Parry's Cyclopaedia of Perfumes*; Vol. 2, p. 473-486, J. and A. Churchill, London
44. PAYNE, H.C. (1989): Sport hunting in North America. In: *Wildlife production systems*, ed. HUDSON, R.J., DREW, K.R. and L.M. BASKIN; p. 134-146; Cambridge University Press, Cambridge
45. PORDOMINGO, A.J. (1993): *in litt.*
46. POWELL, J. and S. BAHR (1987): Preliminary Survey: Wyoming Farm and Ranch Outdoor Recreation Enterprises (*cited after* YORKS 1989)
47. RAMSAY, B.J. (1994): *Commercial use of wild animals in Australia*. Bureau of Resource Sciences, Australian Government Publishing Service, Canberra, Australia
48. REINKEN, G. (1990): *Deer farming*. Farming Press, Ipswich, UK
49. REINKEN, G. (1963): Grünlandnutzung mit Damtieren. *Landwirtschaftliche Zeitschrift Rheinland*, 45, p. 25-26
50. REINKEN; G. (1993): *in litt.*
51. RENECKER, L.A., BLYTH, C.B. and C.C. GATES (1989): Game production in Western Canada. In: *Wildife production systems*, ed. HUDSON, R.J., DREW, K.R. and L.M. BASKIN; p. 248-267; Cambridge University Press, Cambridge
52. ROBERTS, J. (1993): The Status of Fallow Deer Farming in Australia (*cited from* ASHER 1993)

53. ROTH, H.H. (1969): *Deer management and utilisation in Mauritius*. Working Doc. 11-69(105), FAO/UNSF-348-MAR4; Food and Agriculture Organisation of the United Nations, Rome; 10 pp.
54. SCOTTER, G.W. (1989): Reindeer industry in North America. In: *Wildife production systems*, ed. HUDSON, R.J., DREW, K.R. and L.M. BASKIN; p. 223-242; Cambridge University Press, Cambridge
55. SKJENNEBERG, S. (1984): 15. Reindeer. In: *Evolution of domesticated animals*, ed. MASON, I.L.; p. 128-137; Longman Group, London and New York
56. SKJENNEBERG, S. (1989): Reindeer husbandry in Fennoscandia. In: *Wildife production systems*, ed. HUDSON, R.J., DREW, K.R. and L.M. BASKIN; p. 207-242; Cambridge University Press, Cambridge
57. SOKOLOV, V.E. and N.L. LEBEDEVA (1989): Commercial hunting in the Soviet Union. In: *Wildife production systems*, ed. HUDSON, R.J., DREW, K.R. and L.M. BASKIN; p. 170-185; Cambridge University Press, Cambridge
58. STANTIAL, J. (1987): Deer Farming - an investment to add to your portfolio (*cited after* REINKEN 1990)
59. STERN, R.O., AROBIO, E.L., NAYLOR, L.L. and W.C. THOMAS (1980): Eskimos, Reindeer and Land (*cited after* SCOTTER 1989)
60. SYROECHKOVSKY, E.E., ROGACHEVA, E.V. and L.A. RENEKER (1989): Moose husbandry. In: *Wildife production systems*, ed. HUDSON, R.J., DREW, K.R. and L.M. BASKIN; p. 369-386; Cambridge University Press, Cambridge
61. TEER, J. (1984): Strategies and techniques for production of wildlife and livestock on western ranches (*cited after* YORKS 1989)
62. VICKERS, W.T. (1991): Hunting Yields and Game Composition over Ten Years in an Amazon Indian Territory. In: *Neotropical Wildlife Use and Conservation*, ed. ROBINSON, J.G. and K.H. REDFORD; p. 53-81; University of Chicago Press, Chicago and London
63. WALTON, T. (1991): Adding things up the American way (*cited after* REINKEN 1990)
64. WHITEHEAD, G.K. (1972): *Deer of the world*. Constable, London
65. YAZAN, Y. and E.P. KNORRE (1964): Domesticating elk in a Russian national park. *Oryx*, 7, p. 301-304
66. YEREX, D. and I. SPIERS (1987): *Modern Deer Farm Management*. Ampersand, Wellington, New Zealand
67. YORKS, T.P. (1989): Ranching native and exotic ungulates in the United States. In: *Wildife production systems*, ed. HUDSON, R.J., DREW, K.R. and L.M. BASKIN; p. 268-306; Cambridge University Press, Cambridge
68. ZABRODIN, V.A., KARELOV, A.M., DRAGAN, A.V., SEMKIN, S.T. and G.R. POPOV (1983): *cited after* SOKOLOV and LEBEDEVA (1989)
69. ZEUNER, F.E. (1963): *A history of domesticated animals*. Hutchinson, London
70. ZHANG, B. (1983): Musk deer: their capture, domestication and care according to Chinese experience and methods. *Unasylva*, 35, p. 16-24

## 2.3.13 ANTELOPES (*Bovidae: Ruminantia: ARTIODACTYLA*)

Antelopes form several different subfamilies of the Horned ungulates (**Bovidae**): the duikers (**Cephalophinae**), the Pigmy antelopes, dik-diks, and oribi (**Neotraginae**), the Spiral-horned antelopes (**Tragelaphinae**), the blesbok, harte- and wildebeests (**Alcelaphinae**), the roan, sable and oryx (**Hippotraginae**), the kobs, water- and reedbucks (**Reduncinae**), the impalas (**Aepycerotinae**), the gazelles, springbok and blackbuck (**Antilopinae**) and the saiga (**Saiginae**); the North American Pronghorn antelope is considered to constitute a separate family (**Antilocapridae**). Members of these groups differ greatly in external appearance and size, their weight ranging from only 1.5 to over 500 kg. They are distributed in 73 species all over Africa, the range of oryx and gazelles extending as far as Western Asia. Only a few species, such as the nilgai and the blackbuck are indigenous to the Indian subcontinent, and the saiga is specifically adapted to the salt steppes of Central Asia. During recent decades a number of African and Indian antelope species have been established successfully on ranches in the USA and Argentina. All species of antelope are utilised for food and many provide important raw materials or trophies. Tab. 44 lists the more important species, and shows which are commercially ranched or farmed.

### Historical background

Just as deer are in the temperate and cold zones, antelope have always been a major source of food and raw materials for the indigenous peoples of Africa. In Western Asia *gazelles* also played a most important role as a source of meat before the development of settled agriculture, i.e. in pre-neolithic times. They were not only hunted for meat but also herded, tamed and kept in large enclosures for utilitarian purposes (14). With the domestication of goats and sheep their importance as food animals gradually diminished. The reasons for this shift are not well understood as discussed in Chapt. 1.2. On the one hand the wild gazelle resources might have been over-exploited to such an extent that the breeding of domesticated goat and sheep species was easier. On the other, however, the gazelles certainly did not lend themselves biologically to domestication as well as the goats and sheep. Nevertheless, gazelles have continued to be favourite hunting animals in the arid countries of Western Asia to-date, but have now become extinct or rare in most of Western Asia.

In Ancient Egypt a variety of antelope species such as *gazelles, oryx* and *addax* were tamed and kept in captivity as depicted in many tombs from as early as 2000 BC (14). There is no evidence, however, to show that these were truly domesticated for utility purposes. It is more likely that they were merely kept as pets or for religious sacrifices.

With the colonisation of Africa by the Europeans, the continent's extraordinary antelope resources which had supported the local population for hundreds of years, soon became locally over-exploited. Mainly in Southern Africa antelopes together with the zebras were hunted to the point of extinction, in order to gain extra cash returns from the sale of their

**Table 44:** Overview of the utilisation of the most important antelopes

| English name | Scientific name | Geographical distribution | Live weight (kg)[1] | Utilisation |
|---|---|---|---|---|
| **DUIKERS - CEPHALOPHINAE** | | | | |
| Grey duiker<br>Forest duikers | *Silvicapra grimmia*<br>*Cephalophus* spp. | savannas s.o. Sahara<br>moist forests s.o. Sahara | 10-14<br>8-24 | for subsist. hunting<br>major source of meat in W.-C.Africa |
| **PIGMY ANTELOPES, DIK-DIKS and ORIBIS - NEOTRAGINAE** | | | | |
| Royal antelope<br>Oribi<br>Small buck<br>Dik-diks | *Neotragus pygmaeus*<br>*Ourebia ourebi*<br>*Raphicerus* spp.<br>*Madoqua* spp. | moist forests W.Afr.<br>savannas s.o. Sahara<br>South. Africa<br>E.Africa and Namibia | 2-6<br>10-20<br>7-14<br>3-6 | for subsist. hunting in West Africa<br>for ranching in S.Africa<br>commercial hunting in E.Africa for meat, skins |
| **SPIRAL-HORNED ANTELOPES - TRAGELAPHINAE** | | | | |
| Kudus<br>Bushbuck<br>Nyala<br>Eland<br><br>Nilgai | *Tragelaphus* spp.<br>*-scriptus*<br>*-angasi*<br>*-oryx*<br><br>*Boselaphus tragocamelus* | E.-South. Africa<br>thickets s.o. Sahara<br>South. Africa<br>E.-South. Africa<br><br>India[2] | 90-270<br>30-60<br>100-120<br>430-600<br><br>150-180 | for ranch., trophy hunt.<br>for subsist. in W.Africa<br>for farming in Natal<br>for ranching, semi-domestication<br>for trophy hunting in USA |
| **HARTEBEESTS, WILDEBEESTS and BLESBOK - ALCEPHALINAE** | | | | |
| Hartebeest<br>Wildebeest or gnou<br>Tiang, topi, sassaby<br><br>Bles/bontebok | *Alcephalus buselaphus*<br>*Connochaetes* spp.<br>*Damaliscus* spp.<br><br>*-dorcas* | savannas s.o. Sahara<br>E.-South. Africa<br>savannas<br><br>South. Africa | 120-180<br>90-140<br>90-160<br><br>60-100 | for cropping in E.Africa, farming and ranching in South. Africa<br>for farming in S.Africa |
| **ROAN, SABLE and ORYX ANTELOPES - HIPPOTRAGINAE** | | | | |
| Roan antelope<br><br>Sable antelope<br><br>Oryx antelope | *Hippotragus equinus*<br><br>*-niger*<br><br>*Oryx* spp. | savannas s.o. Sahara<br><br>South. Africa<br><br>South. and E.Africa | 180-270<br><br>180-270<br><br>150-220 | for trophy hunting in W.-E.Africa<br>for trophy hunting, ranching<br>for ranch., semi-domest. |
| **WATER- and REEDBUCKS, KOBS - REDUNCINAE** | | | | |
| Waterbuck<br>Kob, lechwe, puku<br>Reedbucks | *Kobus defassa*<br>*Kobus* spp.<br>*Redunca* spp. | savannas s.o. Sahara<br>savannas s.o. Sahara<br>savannas s.o. Sahara | 170-230<br>70-100<br>30-60 | for ranch., trophy hunt.<br>for subsist., cropping<br>for ranching |
| **GAZELLES, SPRINGBOK, BLACKBUCK and IMPALAS - ANTILOPINAE and AEPYCEROTINAE** | | | | |
| Impala<br>Gazelles<br><br>Springbok<br>Blackbuck | *Aepyceros melampus*<br>*Gazella* spp.<br><br>*Antidorcas marsupialis*<br>*Antilope cervicapra* | E.-South. Africa<br>N.-E.Africa, W.Asia<br>South. Africa<br>India[2] | 40-60<br>20-50<br><br>20-30<br>15-25 | major species for commerical hunting, farming, ranching; some trophy hunting<br>for trophy hunting in USA, Argentina |
| **SAIGA ANTELOPE - SAIGINAE** | | | | |
| Saiga antelope | *Saiga tatarica* | C.Asia | 25-45 | commercial hunting in Russia and Kasakhstan |
| **PRONGHORN ANTELOPE - ANTILOCAPRIDAE** | | | | |
| Pronghorn antelope | *Antilocapra americana* | N.America | 15-25 | for trophy hunting |

1) live weights vary widely according to sex, different subspecies, localities and management; in commerical use for meat antelopes dress out at about 55-60% of the live weights (cold dressed weight); in traditional rural use all parts except hooves are consumed i.e. about 90% of the live weights; 2) introduced to USA and Argentina (only blackbuck)

skins. Records show that hundreds of thousands of skins were exported every year during the 1860s and 1870s resulting in the *decimation of the plains ungulates* in South Africa. Towards the end of the 19th century antelopes started to assume a special importance as trophy animals for the "White sports hunters". ROWLAND WARD defined criteria for measuring and evaluating trophies, and periodically issued records of the biggest trophies acquired. During the last decades, trophy hunting in Africa, involving mainly antelope species, has become an economically important aspect of controlled utilisation of wildlife, as discussed below.

The early European colonisation of the southern part of the African continent has led to special conditions of land tenure there; i.e. an estimated 69% of the total area of today's South Africa is privately owned (16). Game on this land was traditionally used by the European farmer families by holding periodic shoots together with friends. The resulting game meat was mostly used for home consumption or dried and sold as "biltong", sun-dried and salted strips of meat (48). Only with the development of biological and ecological knowledge of Africa's game animals in the 1950s and 1960s was the commercial concept of game farming and ranching developed, first in Transvaal (40) and in former Rhodesia, now Zimbabwe (17).

Due to the inherent ecological advantages and the apparent economic potential of game ranching and farming, considerable international interest developed in the 1960s and 1970s in this form of wildlife utilisation as a possible way to put agriculturally marginal or tsetse-fly-infested lands to better use. However, it was often not understood that the *concept of ranching or farming of African ungulates* had been successfully developed in the framework of private landownership prevailing in only very few African countries. Only since 1984 were efforts initiated in Zambia, Zimbabwe and Botswana to extend wildlife utilisation concepts to communal lands and thus to adapt them to the needs and aspirations of the rural communities, as discussed in Chapt. 1.4. Specific programmes have developed which involve rural communities with wildlife-rich lands in the management and controlled utilisation of wildlife, such as the CAMPFIRE, LIRDP, ADMADE, and CBNRM Projects (6). These are probably the most promising approach to the sustainable utilisation of the vast antelope resources of Africa.

As of the 1920s, but mainly since 1945, a large number of antelope species were introduced on private ranches in the USA. Originally more a hobby, stocks of some of these are now being commercially utilised (59). The *Indian blackbuck* was first introduced to Argentina in 1918 and has since been propagated into several commercially utilised populations on Private land (26).

## Management and utilisation

As with other animal groups the utilisation of antelopes ranges from extensive to highly intensive depending mainly on the existing systems of land tenure and socio-political conditions. The distinction of different types of utilisation, as referred to below is, of course, rather artificial and some of these may be practised together in order to maximise the benefit from the antelope resources.

The most widely practised use of antelopes is **subsistence hunting** by rural people all over Africa for the procurement of meat for home consumption. This is carried out either by full-time village hunters or by other villagers as a side line to their normal farming activities. In some countries seasonal hunting in large groups is also customary. This traditional form of wildlife utilisation still constitutes a major socio-economic factor in many parts of Africa, particularly in the remoter tsetse-fly-infested areas where the use of domestic animals is limited. It involves virtually *all species of antelope*, although the smaller species are usually of greatest importance. However, in many forested areas of West and Central Africa, subsistence hunting, carried out by traditional means, has gradually become more and more commercialised. Animals are no longer hunted solely for village consumption, but also for earning cash by marketing them along the roads or on special "bush meat" markets in the urban centres (4, 3). Larger species are cut into pieces and smaller ones marketed *in toto*. Depending on the distances involved they are marketed either smoked or fresh. Distinct chains of trading in bush meat have evolved during the last decades in Ivory Coast, Ghana, Nigeria, Congo and Zaire. In these countries there are major urban "bush meat" markets on which hundreds of tons of meat are sold every year (24, 58, 44). Apart from rodents, monkeys and other small mammals the bush meat comprises mainly *duiker, Royal antelope, bushbuck*, and locally also some *oribi* and *kob*.

Red-flanked duiker (*Cephalophus rufilatus*) for sale at the Toumodi bush meat market in Côte d'Ivoire. (Photo: T. HOFMANN)

In the course of European colonisation, mainly in East and Southern Africa the use of antelopes intensified into different types of **commercial hunting for meat.** On most of the large European estates, but also on State and Communal lands antelopes were hunted to provide meat rations for labourers in lieu of cash, in forestry, roads or other European supervised work. Planned programmes of systematic cropping or culling of antelopes on State lands developed in Uganda, Tanzania, Zambia, Rhodesia (now Zimbabwe) and South Africa. They aimed primarily at producing meat, but more often than not this was only a secondary aspect of the primary goal of reducing and indeed eliminating certain game species. In attempts to eliminate Tsetse fly and control cattle diseases, approx. 1.3 million game animals were shot from the 1920s to the 1960s (36). These programmes also involved the larger antelope species, such as *hartebeest, wildebeest, kudu, roan, waterbuck* and *reedbuck*.

In Central Asia *Saiga antelope* have been extensively hunted throughout the 19th century, mainly for export of their horns to China (5). These are used like antler velvet, for medicinal purposes and still command high prices. From the turn of the century through the 1920s saigas were almost completely extirpated from most of their historical range extending from Kalmykia, West of the Wolga and the Caspian Sea, through Kasakhstan to Western Mongolia. In 1919 the authorities instituted a hunting ban for Kalmykia and in 1923 also for Kasakhstan. However, the recovery of the saiga population was mainly due to the events of the Russian Revolution when huge areas of the arid zones were vacated creating an ecological vacuum in which the species could develop its full reproductive potential (61). In the 1950s the species had recovered its entire range of about 2.5 mill. $km^2$ arid steppe land and numbered about 2 million head (800,000 in Kalmykia and 1.2 million in Kasakhstan) (5). In 1951 controlled hunting of the Kalmykian sub-population and in 1955 of the Kasakhstan populations was re-introduced and organised through state-controlled hunting cooperatives (SIWs). Cropping is carried out by hunters' brigades consisting of 4-5 professional hunters, mainly during night-time shooting from vehicles with mounted spot lights. Such a brigade would crop and process 100-500 saiga within 5-6 hours and be able to harvest 10,000-20,000 antelope in one season, i.e. from October until December when rutting starts (5). Since 1979-1981 drive hunts into nets have been introduced as a more efficient harvesting method. Herds are driven by motor cyclists and netted animals are either killed on the spot or taken alive to centralised processing facilities (61). The SIWs were given full responsibility for all aspects of management, such as censusing, poaching control and application of research results. This system of co-responsibility of the wildlife authorities and the resource users allowed cropping to be adapted to the marked ecological fluctuations of population productivity. In a species which utilises such vast areas of arid steppe this is essential. During periods of population increase harvesting was at 20-25% of the total population, and during periods of natural depression and population decrease only 8 - 10% (61). The live weight of saiga varies between 30 and 50 kg in males and 20 and 40 kg in females with a dressing out percentage of 50-55%. All parts of the animals are commercially utilised and a harvest of 100,000 animals produces on average 1,217.7 t of fresh meat and 7.8 t of tongues for local marketing and consumption, 80.5 t of first grade and 161.0 t of second grade offal for feeding of farm animals, and 8.3 t of fresh horn yielding about 4.3 t of cut-off dried horn for export (39). The skins are processed into chromic and suede leather and extracted fat is used for industrial purposes.

The latest politico-economic developments in the former USSR have had rather negative repercussions on the management and sustainable utilisation of saiga. The possibility of earning much wanted foreign exchange has become a very strong incentive for killing saiga indiscriminately for their horns, and this has given rise to illegal uncontrolled hunting and private marketing of products. Moreover, the predominant off-take of adult males has resulted in a marked shift of the sex ratio of populations with a consequent decrease of productivity (61).

Antelope are, next to deer, the most favourite group of animals for **sport and trophy hunting**. As mentioned above this has developed in Africa in the course of European colonisation, and much of the wildlife legislation introduced by colonial administrations (and still widely in force) was mainly oriented towards such use. However, as wildlife became generally scarcer, as of the 1960s, trophy hunting was more and more restricted to specific areas and in many countries it was completely banned. With the increasing affluence of the industrialised nations the demand for trophy hunting nevertheless rose steadily, rendering it a specific form of tourism for a privileged class. The economic potential of such tourism was clearly recognised in East and Southern Africa and led to the development of a lucrative "hunting safari" business as described in Chapt. 1.7. All of the *Spiral-horned antelopes, sable, oryx* and *waterbuck* but also smaller species such as *springbok* and *impala* are particularly important species in this context. Trophy hunting is a most lucrative form of wildlife utilisation but requires a high level of management and rather efficient administrative control which exists so far in only a few African countries. There is still great potential for its development, particularly also in the framework of community-based wildlife management programmes, such as for example, the CAMPFIRE in Zimbabwe. It is based either on hunting "concessions" on State or Communal lands leased out to safari operators; or it may be a supplementary activity on privately owned game ranches or farms. On these, hunting tourism has rapidly developed, often as a supplement to game farming or ranching.

Clients are accommodated either in special hotels or luxury camps for periods of up to 10 days during which they are given the opportunity to hunt certain animal species. They are usually accompanied all the time by trained hunting guides who ensure that only the allowed animals are killed in a professional and humane manner. For each animal killed a substantial trophy fee is charged which varies in accordance with the rarity of the trophy animal and the demand for it. As most trophy hunters wish to get a variety of trophies during one hunt it is vital to maintain a diversity of species. Where this is not possible tour operators or game farmers in South Africa often have co-operative arrangements for exchange hunts. Thus a hunt may start with a bontebok in the Cape Province and end with a sable in Transvaal. The trophies are either directly exported or, where skilled taxidermist services are available, processed locally; the meat and skins are of secondary importance. The trophy hunter by definition is mainly interested in large adults, usually male animals, which means that only a small percentage of the game population is suitable for this type of utilisation. Trophy hunting is thus complementary to the more intensive utilisation for meat. However, it makes the animals shy, and is thus rather incompatible with game viewing in adjacent national parks and other conservation areas.

Only in recent decades have antelopes become an important economic resource also in the USA. Tab. 45 shows the wide range of antelope species which have been imported, mostly since 1945, and propagated on private ranches mainly in Texas, probably initially for non-commercial purposes (53). As some species, particularly the *Indian blackbuck* and *nilgai* multiplied rapidly, commercial use of the increasing herds became obvious. This is achieved both through sale of live animals to other landowners and trophy hunting. Meat production from them as a primary management goal is rarely practised. As more self-sustaining antelope populations are building up, other species such as *wildebeest, kudu, oryx, sable* and *waterbuck* might become important trophy animals in the USA. However, the exotic species may prove to be too competitive with indigenous game which will certainly diminish interest in their propagation.

In South Africa and Namibia, a special type of commercialised hunting is practised, the so-called ***"biltong" hunting***, strictly in the framework of game farming, as discussed below. This is offered by game farms to local residents who must cater for themselves and are allowed a fixed bag of the commoner species of antelope, such as *springbok*, *blesbok* and *impala*. They are not interested in the trophy, only the sport and the venison. This is taken home for fresh consumption or turned into the popular "biltong", sun-dried salted strips of meat. The game farmer can thus double the return from his game crop, but must tolerate considerable disturbance of this.

**Table 45:** Antilope species introduced and propagated on ranches in the USA since 1945

| Scientific name | numbers of ranches holding the species 1982/83 | Scientific name | numbers of ranches holding the species 1982/83 |
|---|---|---|---|
| *Neotragus batesi* | 1 | *Hippotragus equinus* | 5 |
| *Cephalophus* spp. | 1 | *-niger* | 7 |
| *Tragelaphus euryceros* | 3 | *Oryx leucoryx* | 1 |
| *-strepsiceros/imberbis* | 7 | *-dammah* | 10 |
| *-oryx* | 13 | *-gazella* | 18 |
| *-spekei* | 5 | *Addax nasomaculatus* | 14 |
| *-angasi* | 4 | *Kobus lechwe* | 3 |
| *Boselaphus tragocamelus* | 8 | *-ellipsiprymnus* | 7 |
| *Alcephalus buselaphus* | 4 | *Gazella dorcas* | 3 |
| *Connochaetes taurinus* | 7 | *-granti* | 3 |
| *-gnou* | 2 | *-gazella* | 2 |
| *Damaliscus dorcas phillipsi* | 3 | *-thomsoni* | 4 |
| *-dorcas dorcas* | 1 | *Antidorcas marsupialis* | 4 |
| *-lunatus* | 3 | *Antilope cervicapra* | 28 |
| | | *Litocranius walleri* | 1 |

Total of 30 species recorded by 45 members of the Exotic Wildlife Breeding Association (ANONYMOUS (2))

**Ranching and farming** of antelopes has developed mainly in Southern Africa as both depend largely on private ownership of extensive tracts of land which exists in South Africa, Namibia, Botswana and Zimbabwe. "Ranching" is a rather extensive activity by which segments of a natural game community are exploited for production of meat and skins, and is usually only possible on estates of more than 10,000 ha. These may remain unfenced and carry up to 18 different species of antelope which constitute the bulk of the animal biomass. The most common and abundant species are *impala, Blue wildebeest, Greater kudu, eland, waterbuck, Grey Duiker*, *steenbok*, and in the southwestern ranges also *springbok* and *oryx*. Upon request, landowners are granted cropping quotas based on periodic censuses. The rate of sustainable annual off-take of antelope may vary between 20 and 30%, exceptionally up to 40%, of the total population, depending on the species and the range conditions (47, 60). The cropping is carried out by shooting from a vehicle at night with a spot light, sometimes also from hides. Usually telescoped rifles of 0.243" calibre or greater are used, but for the smaller antelopes 0.22" is favoured as it causes less damage. This technique has the advantage of the least disturbance to the surviving population, allowing the hunter to shoot cleanly at short distance and keeping the carcasses cool until they arrive at the processing plants. The dressed carcasses are either refrigerated or processed into biltong. The skins are salted and carefully dried and stretched for bulk marketing. In South Africa private holdings are generally smaller and mostly fenced; they lend themselves rather to the more intensified "farming". The size of game farms may vary between 20 and more than 20,000 ha, depending on the ecological conditions. In the Cape Province 74% of game farms are 1,000-10,000 ha; in Transvaal 70% comprise only 1,000-4,000 ha, and the minimum area permitted for a registered game farm is 50 ha for each species kept (7). The most important antelope species farmed are *springbok, blesbok* and *impala; nyala* are locally important in Natal and Eastern Transvaal.

In all of the Southern African countries specialised private enterprises offer services to game ranchers or farmers. One of the biggest, Kovisco Ltd., is based at Harrisburg in the Orange Free State and in 1984 operated five helicopters in conjunction with ground crews and 25 t freezer trucks (7). These carried annually carcasses of 50,000-60,000 springbok and blesbok to a centralised abattoir for processing in compliance with veterinary export requirements. Other contractors in Namibia operate with sophisticated mobile processing plants (31).

The only other country in Africa where private ownership of extensive game-rich tracts of savanna lands exist, is Kenya. However, the official attitude in this country was never favourable towards commercial meat production from wildlife. In 1977 the sale of all wildlife products was banned in favour of more efficient poaching control. Only the Athi River Ranch which has pioneered game ranching since the early 1970s (1), managed to circumvent the adverse legislation. The ranch comprises 20,000 ha and is fully fenced. Major species for meat production there are *Thomson's* and *Grant's gazelle, hartebeest, wildebeest, impala* and *eland.*

Intensive game farming implies management at relatively high stocking rates. On some farms it includes husbandry, rotation of stock over different fenced paddocks, and the driving of selected animals into funnel-shaped pens either for live sale, treatments or slaughter. To facilitate these operations a maximum of docility is desirable. As a logical consequence efforts were initiated in the 1960s at domestication of some of the most suitable antelope species, *eland* and *oryx*. The biological implications and the desirability of domesticating

Mobile field abattoir with refrigerator truck for processing mainly springbuck in Namibia. (Photo: E. JOUBERT and C.P. VISAGIE)

further wild ungulates has been discussed in general in Chapts. 1.2 and 1.4. In addition to individual farmers in South Africa and former South-West Africa who bred tamed antelope more or less as a hobby, large herds of semi-domesticated eland were established for experimental purposes in the Drakensberg Reserve in Natal (29) and on the Manyoli Ranch in former Rhodesia (42). Later on in 1972 a comprehensive project to investigate the domestication of eland, oryx and buffalo was launched on the Galana Game Ranch in Kenya (32). All of these experiments proved the suitability of these species for domestication. The herds were started with hand-raised calves and herded in large fenced paddocks. They were regularly penned and put through races for weighing, milking or treatments and remained totally docile. It was found that night feeding was essential for eland to achieve maximum growth rates. Herds in Natal and Rhodesia were thus left to graze all night and were rounded up by herdsmen during the day. On the Galana Ranch it was felt that the nocturnal feeding habit of eland and its tendency to wander in search of food was incompatible with the established management regime and experiments with eland were discontinued in favour of oryx and buffalo. Economic comparisons of eland production on Manyoli Ranch and oryx on Galana Ranch with beef production on the same semi-arid lands proved generally favourable (46, 33, 43, 55). Nonetheless, the externally funded experimental breeding of neither species ever reached a commercial phase for purely political subjective reasons. In former Rhodesia it became necessary to move the herd in 1966, when it already comprised more than 50 animals of five successive generations, to State land on the ecologically quite different and clearly unsuitable high plateau where it gradually deteriorated. In Kenya a law was passed in 1977 banning the sale of game meat which induced the supporting African Wildlife Leadership Foundation to withdraw their funding. Only on the Baobab farm near Mombasa is a large herd of semi-domesticated eland still propagated (7).

The only continuing long-term effort at domestication of eland was made outside Africa on the Askanya Nova State Farm in Southern Ukraine which is now part of the Ukrainian Institute of Animal Husbandry. Four pairs of eland were introduced in 1892 by the then private owners of the Askania Nova Estate and they have ever since been propagated and bred under semi-captive conditions; i.e. animals are free ranging, but have food and shelter provided during the winter months. In 1963 the herd comprised 400-500 descendants of the original stock and have been subjected to selective breeding mainly for milk production (56). Up to 1990 72 eland have been milked, some for more than 5 lactations and had produced a total of 60,000 l of milk which is believed to have special curative properties against certain ailments. Average duration of lactation is 200 days with average daily yields of 1.8 kg of milk which contains at an average 10.43% fat (57).

The early postulation that wildlife ranching or farming could and should replace cattle production under certain ecological conditions has been found to be somewhat euphoric. There have been a number of studies attempting to determine the biological productivity of game animals in comparison with livestock. In South Africa biological production parameters for *blesbok* (12), *impala* (22), *eland* (46) and *springbok* (47) were investigated. In Kenya the productivity of *gazelles* and of a mixed *hartebeest-wildebeest-impala-eland* game population were studied on the only private game ranch at Athi River (25, 10, 50). It was generally found that antelope have a higher energy requirement than cattle under similar ecological conditions. Thus it is not possible to stock antelope species according to a hypothetical total biomass without causing serious range deterioration (37).

Only a wide spectrum of different ungulate species with different physio-ethological adaptations and varying feeding habits may attain a higher biomass without range deterioration. Such biodiversity is usually not present on smaller to medium size game farms, but only on larger game ranches. On game farms it is often not even wanted because the management of multi-species game stock for meat production is generally much more difficult than that of single species antelope populations. However, the overall biological productivity of game farms may well be increased without overstocking by introduction of some of the natural bulk grazers, such as White rhino, buffalo and zebra that will supplement the selective feeding of the antelope species. A more diverse game stock can then be utilised profitably by supplementing the game cropping with trophy hunting. The most common approach to the problem of achieving optimum productivity is, of course, to combine antelope farming with cattle ranching. The cattle function just as well as complementary bulk and roughage feeders. Contrary to early conceptions "game farming will never be a way of obtaining food by unsophisticated means" (23); nevertheless, it has become increasingly popular in South Africa as a supplementary agricultural activity with only a few farmers deriving their income solely from game.

Even in the earliest stages of game farming in Transvaal in the 1950s, the **sale of live antelope for stocking of private land** all over South Africa became a major source of income for many farmers (40). With the steady increase in popularity of game farming the demand for live animals has continued to-date. It constitutes an important aspect in the economic utilisation of antelope populations in Southern Africa, as well as in the USA. The original capture methods of running the animals down by vehicle or darting them with anaesthetics have been replaced with more efficient netting or driving of antelopes into funnel-shaped enclosures with a crush at the end. For funnelling, coloured plastic sheeting is used and the driving is usually carried out by helicopter. An average of 8-10 animals may be captured in this way during an hours flying time, but for species with strong herding instincts, such as wildebeest and zebra, this may increase to a hundred under favourable conditions. Captured antelope are usually sold by auction and most commonly comprise *duiker, springbok, blesbok, Blue wildebeest, kudu* and *eland*. Usually, the demand for females is greater than for males. Therefore it is common practice to combine live sales with cropping operations whereby the surplus males are utilised for meat.

**Game viewing** as a form of commercial game utilisation on Private land has only developed more recently. It is not yet widely spread and usually combined with hotel tourism. In Eastern Transvaal and also in Zimbabwe a number of special luxury game lodges have been built on Private land from where clients are taken individually or in small groups on photographic safaris. In Botswana where no commercial killing of game animals on Private land is permitted several owners have embarked on this form of game utilisation (7). It depends entirely on the presence of a wide variety of game species and is rather incompatible with consumptive wildlife utilisation and conventional agriculture. However, some game farms do manage to combine it with other utilisation programmes by allowing game viewing drives at certain times in certain areas by arrangement with nearby hotels.

## Economic importance

The major significance of ***subsistence hunting*** lies in the provision of animal protein for the nutrition of rural populations as discussed in Chapt. 1.5. Antelope represent a major component of the wild animals eaten by local people all over Africa. However, the volume of meat derived from antelopes varies greatly from country to country. In many of them there are indications that the levels of subsistence hunting are no longer sustainable. Village hunters have to go farther and farther to locate their prey, and in many areas of Africa commercial hunting for meat is abandoned because it has become too time consuming and strenuous. In some areas a switch from the *favoured antelope species* to other animal groups, such as monkeys, small rodents and finally bats has been observed. From this it may be concluded that antelope populations are being depleted. However, in other areas, particularly in West Africa, market studies over many years have shown a remarkable continuity of offered species, quantities and sex and age composition. Here it may be assumed that the level of exploitation of antelopes is, in fact, sustainable (24, 58, 44).

***Game ranching and farming*** on Private lands in Southern Africa has steadily expanded since it was first initiated in the 1960s and has gained relative economic importance. In former Rhodesia (now Zimbabwe) in 1964 game animals, *mainly antelopes*, already contributed as much meat to the nutrition of the population as goats and pigs together (41). The average total meat yield from game per ha increased from 0.84 kg in 1964 to 1.2 kg in 1974 (38), in addition to cattle ranching. In the following decade the area devoted to game utilisation increased by 62%, from 59,700 to 197,000 ha (13), some game ranches producing as much as 4.5 kg game meat/ha/annum (51).

In South Africa provincial records show that in the Cape Province a total of 41,376 large game animals were harvested in 1981/82 on 1,148 registered game farms yielding 520 t of meat and involving about 26% of the total area of private land (30). In Transvaal, 528 registered game farmers operated in 1983 although the number of farmers utilising game commercially as a side line was probably much greater; in fact the total number had been estimated in 1975 at over 3,000 (60). In the Orange Free State in 1983 an estimated 6,000 landowners utilised game which had been largely reintroduced since the 1950s. This stock produced about 50% of South Africa's game meat exports, mentioned below (35). In Natal there were 30 game ranches registered by 1983, covering an area of 53,625 ha, i.e. 1% of all farmland; but a much greater number derive some income from mixed farms (45). The total number of landowners in South Africa who derive some income from game was estimated in 1984 at 10,000-20,000 (7). Surveys have shown, however, that only a small percentage of them earn more than 50% of their farm income from game. In Namibia about 77% of all private landowners utilise game in one form or another; 17% offer trophy and 15% biltong hunting. Only 5%, however, have completely abandoned livestock in favour of wildlife utilisation, predominantly as wildlife-based tourism (34). The overall potential for meat production from the more common antelope species in South Africa is estimated in Tab. 46. At an average price of R 2/kg (based on R 1/kg offered by wholesale contractors and R 3/kg obtained directly by farmers at retail in 1983) the overall potential financial value of such production was about R 17.76 mill. (approx. US$ 19.89 mill.).

**Table 46:** Estimated potential for meat production from antelope species in South Africa in 1981 (according to CONROY and GAIGHER (16))

| Species | Average live weight (kg) | Dressed carcass yield (%) | Estimated total population | Estimated annual cropping rate (%) | **Potential annual meat yield (t)** |
|---|---|---|---|---|---|
| Impala | 41 | 58 | 300,000 | 30 | **2,140.2** |
| Springbok | 33 | 58 | 470,000 | 30 | **2,698.7** |
| Blesbok | 55 | 55 | 100,000 | 20 | **605.0** |
| Wildebeest | 182 | 58 | 25,000 | 20 | **527.8** |
| Oryx | 135 | 57 | 10,000 | 25 | **192.4** |
| Great Kudu | 170 | 59 | 100,000 | 25 | **2,507.5** |
| Eland | 300 | 61 | 5,000 | 15 | **137.3** |
| **Total** | | | | | **8,808.9** |

The only private game ranch in Kenya, at Athi River, produced over some years an average of 4 kg game meat/ha from a *gazelle-hartebeest-wildebeest-impala-eland* community. Favourable returns of US$ 4.5/kg induced a 66% reduction of cattle in favour of creating additional wildlife habitat (50).

Apart from the biological productivity of the farmed species, as discussed above, profitability of meat production from antelopes depends mainly on the efficiency of the cropping and processing operations, the endemic presence or absence of major livestock diseases, and marketing opportunities. Efficient night hunting from vehicles may reduce harvesting costs to only 7% of the gross return, whereas use of helicopters may increase them to 43% (18). Optimum use of the skins and other by-products, in addition to the meat, is another critical factor. However, the harvesting must not only be technically cost-efficient, but be rigorously adapted to climatic fluctuations. In wet years game stock tends to build up rather quickly so that large natural die-offs may occur when drought strikes. To prevent such losses cropping has to be stepped up ruthlessly and, if possible, well in advance.

The tolerance of antelope to tsetse-fly-borne trypanosomiasis is hypothetically one of the most positive factors that argue for the promotion of game utilisation in Africa. However, it does not apply to most of the commercial game ranches and farms in Southern Africa as these are largely located outside of the tsetse belt. There, the endemic presence of bovine Foot and Mouth Disease is a major factor as this prohibits exporting game meat to Europe. Only South Africa with the exception of parts of Northern and Northeastern Transvaal, and Namibia South of the Etosha Pan are considered FMD-free and these countries have thus been able to develop an important export of game meat to Europe. In order to protect meat exports into the EU the governments have built very costly veterinary fences across the entire countries which prevent any wild animals from penetrating into the FMD-free zone. Game meat exports commenced in 1972 (16) and during 1980/81 1,446 t of antelope meat were exported from South Africa earning an estimated R 8 mill. in foreign currency (approx. US$ 9.4 mill.) (30). The gross income from "venison" exports from Namibia in 1980 amounted to R 8.77 (approx. US$ 11.4 mill.) (16). However, this export has since decreased considerably due to trade problems discussed in Chapt. 3.2.1.

In Zimbabwe large parts of the wildlife-rich western and northern regions are FMD-endemic. Although the veterinary authorities have been trying to contain the disease by fencing off these regions and by systematic eradication of buffalo which are believed to be a reservoir of the FMD virus, import of frozen antelope meat into Europe was never permitted. This has limited the sale of meat to local markets and has thus greatly reduced the profitability and financial security of game ranches in this country. Cattle diseases may, of course, also affect game ranching and farming directly by their transmission to the wild ungulate populations (60). For example, in West and East Africa Bovine Rinderpest is highly contagious to most antelope species and has regularly caused large die-offs of antelope populations. Heartwater Fever is considered a main reason for the absence of springbok in much of Transvaal.

In most cases game ranching and farming are economically viable and may generate maximum income, only if different types of utilisation are employed together. In particular ***trophy hunting*** is producing high levels of extra return, and the sale of live animals is a very important economic supplement as long as market demand continues (8).

Tab. 47 compares the value of *different antelope species* in South Africa in terms of different usages. If one takes a mean value between gross and retail sale of meat, the same animal sold alive may produce up to 70% more return; hunting for sport and producing biltong doubles the wholesale value of animals; if hunted for its trophy between 78 and 138% more financial return may be generated from an individual animal, according to the species. Fig. 18 shows game production in Namibia since 1978 and the evolution of the different forms of wildlife utilisation. Game cropping depended on the export market which collapsed

**Table 47:** Comparison of approximate financial values (in SA Rand) of four common species of antelope under different systems of commercial utilisation in South Africa in 1982 (after BARZDO and coll. (7))

| Utilisation system | Financial value of individual dressed out carcasses | | | |
|---|---|---|---|---|
| | Springbok DCW 20 kg | Impala DCW 24 kg | Eland DCW 183 kg | Gr. Kudu DCW 100 kg[1)] |
| **Cropping** | | | | |
| Wholesale of meat[2)] | 20 | 24 | 183 | 100 |
| Retail sale of meat[3)] | 60 | 72 | 549 | 300 |
| Mean value | 40 | 48 | 360 | 200 |
| **Live animal sale[4)]** | 50-80 | 50-90 | 300-480 | 200-300 |
| Mean value | 65 | 70 | 390 | 250 |
| Increase of value | **+62%** | **+46%** | **+7%** | **+25%** |
| **"Biltong" hunting[5)]** | 40 | 48 | 363 | 200 |
| Approx. increase of value | **+ 100% if compared with whole sale meat prices** | | | |
| **Trophy hunting[6)]** | 95 | 95 | 653 | 455 |
| Approx. increase of value | **+138%** | **+98%** | **+78%** | **+128%** |

1) dressed carcass weights from CONROY and GAIGHER (16); 2) R 1.0/kg, as offered by one contractor; 3) R 3.0/kg; 4) lower price refers to purchase from a provincial park, upper price to sale by private farm; 5) calculated from tariffs of private farms offering "biltong" hunting; 6) according to Natal Parks Board

**Figure 18:** Game production on Private land in Namibia in 1979-1992 (after KRUG (34))

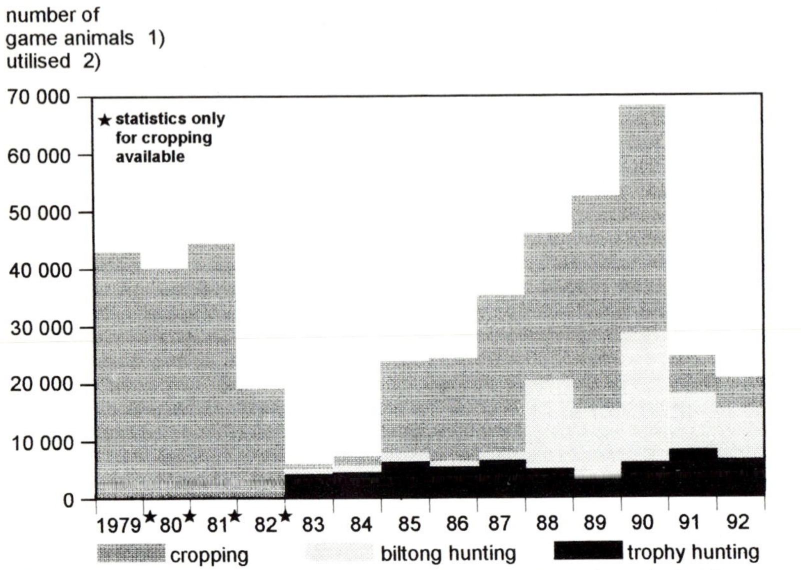


1) in game meat production: 92% springbok, 5% oryx, 3% kudu; in trophy hunting (in 1991/2 only): 24% warthog, 20% springbok,16% oryx, 12% kudu, 24% other antelope species, 3% zebra, 1% others; 2) animals used for home consumption are excluded.

twice, first in 1983 and then in 1991 as discussed in Chapt. 3.2.1; however, trophy hunting developed steadily (34). In Namibia the total value of game production in 1980 was estimated at R 12.5 mill. (approx. US$ 13.75 mill.) (30). Of this R 4.2 mill. were generated by commercial cropping, R 4.7 mill. by trophy hunting and R 1.2 mill. by live animal sales; the value of home consumption was estimated at R 2.4 mill. In Zimbabwe on two of the largest game ranches, Lone Star and Buffalo Range, the equivalent of US$ 150,000 and US$ 200,000 respectively were earned from trophy hunting as compared with only US$ 25,000 and US$ 40,000 respectively from regular game cropping plus some returns from skins and curios (51). It is therefore obvious that diversification of wildlife utilisation is a most important economic aspect of game ranching and farming. It renders it not only most profitable but also makes it less prone to crises such as drought, disease outbreaks and sudden marketing restrictions.

The common ***combination of game utilisation with livestock production*** in Southern Africa is not only ecologically advantageous as discussed above, but should also be seen in the financial context. In 1982 a comparative economic study under savanna conditions showed that a cattle farm almost doubled net profit to R 5/ha/annum if *impala* cropping was included (15). Another study compared merino sheep and *springbok* production in the arid Karoo region in 1978-1985 (49). The utilisation of springbok increased the net profit of the merino farm by 11.3%. In both cases stocking rates of livestock and antelopes were independent of each other.

The most comprehensive comparative economic analysis of cattle and game ranching was carried out by the WWF Multispecies Animal Production Systems Project in 1989 on randomly selected ranches in the southwestern semi-arid zone of Zimbabwe (28). The results of this study, summarised in Tab. 48, show that the mean returns on capital investments were significantly higher on game ranches. Mean net returns per ha and annum were also highest in game ranching, but if the export value of beef was considered, cattle ranching was more profitable.

**Table 48:** Comparative economic analysis of different multi-species animal production systems in Southwestern Zimbabwe (after JANSEN and coll. (28))

| Sampled ranches | | Financial analysis mean returns | | Economic analysis mean returns | |
|---|---|---|---|---|---|
| type | no. | on capital | per ha/ann. | on capital | per ha/ann. |
| Cattle | 45 | 1.8% | 2.52 Z$ | 12.4% | **18.53 Z$** |
| Game | 12 | **10.5%** | **5.47 Z$** | **20.9%** | 12.01 Z$ |
| Mixed | 32 | 3.6% | 3.02 Z$ | 14.8% | 15.59 Z$ |

***Commercial cropping*** of antelope populations on State land as initiated in East Africa in the early 1970s has little or no economic importance anymore. In Uganda the annual production from two very large tracts of savanna land was 0.8 kg and 0.2 kg per ha. As these amounts constituted only a small fraction of the present potential, and only meat was utilised, the net gain was only 10-60% of the total cost of the operation (27). In Tanzania the skins of harvested animals were also utilised and operations therefore more profitable. Meat marketing, however, was found to be a major economic constraint which eventually led to the discontinuation of the project (9). In Kenya an FAO project investigated the practical problems and the productivity of game cropping on grass savanna in the Kajiado District in 1970-1978. Up to 50 animals, mostly *wildebeests* and *hartebeests* were harvested per day, processed in a mobile plant and marketed fresh in Nairobi. Net economic return per wildebeest cropped was calculated at 60% of its market value (52).

In contrast to ranching and farming of game on Private land, the economic analysis of cropping schemes on State land has generally lacked serious and long-term study of sustainability as the basis of any productivity assessments. Many of them, of course, were not intended to be sustained over long periods.

The commercial cropping of *Saiga antelopes* in the former USSR since the 1950s has yielded probably greater economic returns than the use of any African antelopes. Fig. 19 shows the number of animals harvested and the quantities of meat produced, with peaks of 3,500-5,980 t per annum. Apart from the meat, skins and fat, the use of the horns is a major economic factor. In 1975-1981 about 42-52 t of horn was exported annually, but this has since, in 1991-93, increased to about 75 t at a price of US$ 50/kg (61). The total value of saiga production in Kasakhstan from 1955 to 1981 was estimated at Rb 25.2 mill. (21) and the Kalmykian subpopulation alone produced about Rb 1.2-1.3 mill. per annum (11). Meat productivity of the Kalmykian saigas in the 1970s was at an average 0.5 kg/ha (of semi-desert) and thus higher than in conventional livestock (mainly sheep) husbandry (19).

However, anthropogenic influences, such as intensification of livestock production with consequent desertification, and infrastructural development interfering with the migration routes of saiga, has gradually reduced the Kalmykian subpopulation to only 150,000-160,000 animals, i.e. to less than 20% of the 1950 population size (54). In 1987 cropping was suspended in Kalmykia, part of the Russian Federation, for five years, except for limited sport hunting. In Kasakhstan, which is now an independent state, ecological degradation of the saiga habitat was much less and the observed population fluctuations were mainly due to natural causes such as severe winters and epizootics. The saiga continue to constitute one of the country's most important resources, comprising in 1991-93 some 825,000-975,000 animals (61). In contrast to the situation in Kalmykia the saiga in Kasakhstan are still capable of evading negative anthropogenic influences which are only localised. However, the administrative problems of management and trade control resulting from the recent political changes, need to be addressed in order to ensure the sustainability of saiga utilisation.

**Figure 19:** Harvests and meat yields of Saiga antelope in 1951-1990 (after ZHIRNOV (61))

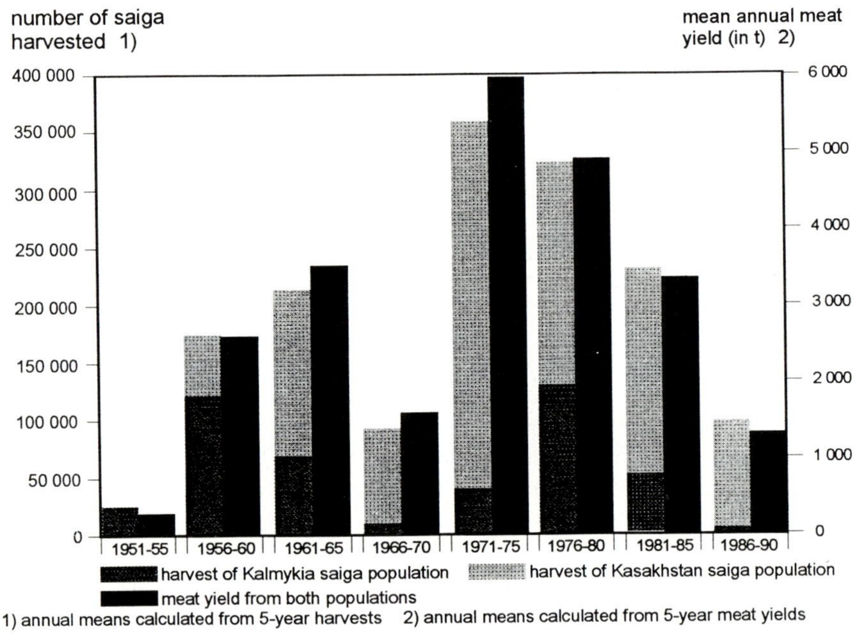


## Implications for conservation

Antelopes provide some of the best examples of the beneficial influence which controlled management and economic utilisation may have on the conservation of game species. Game farming in Southern Africa has not only generally enhanced the conservation of most antelope species in that region, but has resulted in their redistribution over wide areas in which they had been either depleted or eliminated. Nowhere has it threatened a species with extermination; on the contrary, it has saved several rare taxa, such as the bontebok and the Black wildebeest from extinction. Asian species, such as blackbuck and nilgai are now more numerous in America where they have been introduced and managed for use, than in their original ranges in India where they have become very rare.

The controlled hunting of the Saiga antelopes in the former USSR has provided a positive example as to how a species which had been at the brink of extermination through over-exploitation, can be sustainably utilised in the framework of planned and supervised management. However, the special irrational monetary value of saiga horns on the Asian markets combined with the difficult transition from the previous centralised administration to a market oriented economic system bears great dangers for the sustainability of saiga utilisation in the future. It is hoped that the lesson of the last century is remembered and that the devastating trade in saiga horn will be curbed through the CITES mechanism.

Controlled trophy hunting is rarely intensive enough to seriously threaten a wild population. By generating particularly high economic returns from particular species it gives, however, a strong incentive to management and conservation of these. Only uncontrolled commercial persecution of antelopes for meat with sophisticated fire arms as has developed in parts of Africa from the original subsistence hunting, as well as the random killing for "sport" as occurs in some of the arid countries, may deplete antelope populations to the point of extirpation, and thus have very negative implications for conservation.

The IUCN Global Antelope Survey (20) showed that antelope populations have generally declined greatly in large parts of Africa mainly due to the rapid growth of the human population and the associated loss of wildlife habitat and mounting hunting pressure. However, of the 57 species occurring in East and North-East Africa, 60% were still classed in the 1980s as "satisfactory/not threatened", of the 35 species occurring in Southern and South-Central Africa 80%, and of the 44 species occurring in West and Central Africa 50%. A total of 10 antelope species and 7 localised subspecies are considered to be at immediate threat and another 11 species and 14 subspecies are potentially at risk of being extirpated. In outlining strategies for conservation of antelopes the IUCN Antelope Survey discusses the following options:

- strict protection in core conservation areas where no compromise with short-term economic motives is possible if conservation goals are to be attained;
- tourism as a form of non-consumptive utilisation of the antelope resources;
- sustained yield consumptive utilisation, and
- development and implementation of multiple resource use programmes.

It concludes that consumptive utilisation is likely to play a decisive role in the conservation of antelopes outside conservation areas. The long-term success of this approach entirely depends, however, on the integration of antelope utilisation into development programmes which achieve an acceptable compromise between the needs of the rural people and the requirements of wildlife.

## Literature references


1. ANONYMOUS (undated): *Wildlife ranching, a new concept of land use*. Brochure produced by Wildlife Ranching and Research, Athi River, Kenya; 18 pp.
2. ANONYMOUS (1983): *Breeder's Directory 1982-83*. Exotic Wildlife Breeders Association, Kerriville, Ta., USA
3. AJAYI, S.S. (1979): *Utilization of Forest Wildlife in West Africa*. FO/MISC/79/26; Food and Agriculture Organisation of the United Nations, Rome
4. ASIBEY, E.O.A. (1974): Wildlife as a source of protein south of the Sahara. *Biological Conservation*, 6(1), p. 32-39
5. BANNIKOW, A.G. (1963): *Die Saiga Antelope (Saiga tatarcia L.)*. Neue Brehm Bücherei No. 320; A. Ziemsen, Wittenberg-Lutherstadt, Germany
6. BARBIER, E.B. (1992): Community-based development in Africa. In: *Economics for the Wilds*, ed. SWANSON, T.M. and E.B. BARBIER; p. 103-135; Earthscan, London
7. BARZDO, J., LUXMOORE, R., BROAD, S. and D. JONES (1984): *A survey of wildlife farming operations*. WCMC/IUCN, Cambridge (unpubl. report)
8. BERRY, M.P.S. (1986): A comparison of different wildlife production enterprises in the Northern Cape Provinces, South Africa. *South African Journal of Wildlife Research*, 16(4), p. 124-128
9. BINDERNAGEL, J.A. (1975): *Wildlife utilisation in Tanzania: The ecology of three wildlife areas in Tanzania with special reference to wildlife utilisation*. FAO Project URT/72/011; Working Document No. 1; Food and Agriculture Organisation of the United Nations, Rome; 121 pp.
10. BLANKENSHIP, L.H. and S.A. QUORTRUP (1974): Resource management on a Kenya ranch. *Journal of the South African Wildlife Management Association*, 4 (3), p. 135-190
11. BLISNJUK, A.I. (1982): Ecology and rational utilisation of the saiga antelope population North-West of the Caspian Sea (in Russian; *cited after* ZHIRNOV 1993)
12. CATTO, G.G. (1976): Optimal production from a Blesbok Herd. *Journal of the Environmental Management*; 4 (2), p. 95-121
13. CHILD, G. (1985): *A preliminary investigation of game ranching in Zimbabwe*. Department of National Parks and Wildlife Management, Harare, Zimbabwe (unpubl. report)
14. CLUTTON-BROCK, J. (1987): *A Natural History of Domesticated Mammals*. Cambridge University Press and British Museum (Natural History), Cambridge and London
15. COLLINSON, R.F.H. (1979): Production economics of impala. In: *Proceedings of a Symposium on Beef and Game Management*, ed. ABBOTT, M.A.; p. 90-103; Cedora, Pietermaritzburg, South Africa
16. CONROY, A.M. and I.G. GAIGHER (1982): Venison, aquaculture and ostrich meat production: action 2003. *South African Journal of Animal Science*, 12, p. 219-233
17. DASMANN, R.F. and A.S. MOSSMAN (1961): Commercial utilisation of game animals on a Rhodesian Ranch. *Wildlife*, 3 (3), p. 7-14
18. DENSHAM, W.D. and A.J. TOMKINSON (1979): Economics of game cropping. *Lämmergeyer*, 27, p. 19-24
19. DESHKIN, W.W. and N.W. MENKOWA (1981): Comparative aspects of the utilisation of wild and agricultural ungulates (in Russian; *cited after* ZHIRNOV 1993)
20. EAST, R. (comp.) (1988): *Antelopes. Global Survey and Regional Action Plans*. Parts 1-3. IUCN/SSC Antelope Specialist Group; IUCN, Gland, Switzerland
21. FADEEV, W.A. and A.A. SLUDSKY (1982): The Saiga antelope in Kasakhstan (Ecology and economic importance) (in Russian; *cited after* ZHIRNOV 1993)
22. FAIRALL, N. (1983): Production parameters of the impala (Aepyceros melampus). *South African Journal of Animal Science*, 13 (3), p. 176-179

23. FAIRALL, N. (1995). The use of non-domesticated African mammals for game farming. Acta Zoological Fennica (*cited after* BARZDO and coll. 1984)
24. FALCONER, J. (1992): *Non-timber forest products in Southern Ghana. A Summary Report.* Overseas Development Administration; Forestry Series No. 2; London; 23 pp.
25. HOPCRAFT, D. (1975): *Productivity comparison between Thomson's gazelle and cattle and their relation to the ecosystem in Kenya.* Thesis; Cornell University, New York
26. JACKSON, J. (1985): La distribución de la población de antilopa cervicapra en la Argentina. *IDIA* No. 429-432 Sept.-Dec. 1984; p. 95-98; Instituto de Tecnologia Agropecuaria; Buenos Aires, Argentina
27. JAHNKE, H.E. (1975): *Conservation and utilisation of wildlife in Uganda. A study in environmental economics.* IFO München; Afrika Forschungsberichte No. 54; 132 pp.
28. JANSEN, D., BOND, I. and B. CHILD (1992): Cattle, Wildlife, Both or Neither - Results of a Financial and Economic Survey of Commercial Ranches in Southern Zimbabwe. WWF-Multispecies Animal Production Systems Project, Project Paper No. 27; Harare, Zimbabwe
29. JEFFERY, R.C.V. and J. HANKS (1981): Body growth of captive eland *Taurotragus oryx* in Natal. *South African Journal of Zoology*, 16, p. 183-184
30. JOOSTE, J.F. (1983): Game farming as a supplementary farming activity in the Karoo (*cited after* BARZDO and coll. 1984)
31. JOUBERT, E., BRAND, P.O.J. and G.P. VISAGIE (1983): An appraisal of the utilisation of game on private land in South West Africa. *Madoqua*, 11 (3), p. 197-219
32. KING, J.M. and B.R. HEATH (1975): Game domestication for animal production in Africa. *World Animal Review*, 16, p. 23-30
33. KING, J.M., HEATH, B.R. and R.B. HILL (1977): Game domestication for animal production in Kenya. 1: Theory and practice. *Journal of Agricultural Science*, 89(2), p. 445-457
34. KRUG, W. (1994): *Wildlife Management in Namibia. Nachhaltige Wildtierbewirtschaftung als landwirtschaftlicher Betriebszweig und integrierter Bestandteil des Naturschutzes.* Dipl. Thesis; University of Giessen, Giessen, Germany
35. LE ROUX, P. (1983): *Cited after* BARZDO and coll. (1984)
36. MATTHIESSEN, P. and B. DOUTHWAITE (1985): The impact of tsetse fly control campaigns on African wildlife. *Oryx*, 19(4), p. 202-209
37. MENTIS, M.T. (1977): Stocking rates and carrying capacities for ungulates on African rangelands. *South African Journal of Wildlife Research*, 7(2), p. 89-98
38. MOSSMAN, S.L. and A.S. MOSSMAN (1976): *Wildlife utilisation and game ranching.* Occasional Paper No. 17; IUCN, Morges, Switzerland
39. PRISSJASHNJUK, W.E. and A.W. PRONJAEV (1982): The production in hunting of Saiga antelope (in Russian; *cited after* ZHIRNOV 1993)
40. RINEY, T. and W.L. KETTLITZ, (1964): Management of large mammals in the Transvaal. *Mammalia*, 28, p. 189-248
41. ROTH, H.H. (1966): Game utilisation in Rhodesia in 1964. *Mammalia*, 30 (3), p. 387-423
42. ROTH, H.H. (1970): Studies on the agricultural utilisation of semi-domesticated eland (*Taurotragus oryx*) in Rhodesia. I. Introduction. *Rhodesian Journal of Agricultural Research*, 8, p. 67-70
43. ROTH, H.H., KERR, M.A. and J. POSSELT (1972): Studies on the utilisation of semi-domesticated eland (*Taurotragus oryx*) in Rhodesia. 5. Reproduction and herd increase. *Zeitschrift für Tierzucht und Züchtungsbiologie*, 89 (1), p. 69-83
44. ROTH, H.H. and B. HOPPE-DOMINIK (1994): Le marché de viande de brousse à Toumodi en Côte d'Ivoire (publication in preparation)
45. ROWE-ROWE, D.T. (1984): *Game utilization on private land in Natal.* Natal Parks, Game and Fisheries Preservation Board; Pietermaritzburg, South Africa; 8 pp.
46. SKINNER, J.D. (1971): Productivity of the eland: An appraisal of the last five years research. *South African Journal of Science*, 67(12), p. 534-539

47. SKINNER, J.D. (1975): Technological aspects of domestication and harvesting of certain species of game in South Africa. In: *Proceedings of the IIIrd World Conference on Animal Production*, ed. REID, R.L.; p. 119-125; Sydney University Press, Sydney
48. SKINNER, J.D., von LA CHEVALLERIE, M. and J.H.M. VAN ZYL (1971): An appraisal of the springbok for diversifying animal production in Africa. *Animal Breeding Abstracts*, 39(2), p. 215-224
49. SKINNER, J.D., DAVIES, R.A.G., CONROY, A.M. and H.M. DOTT (1986): Productivity of springbok *Antidorcas marsupialis* and merino sheep *Ovis aries* during a Karoo drought. *Transactions of the Royal Society of South Africa*, 46, p. 149-164
50. STELLFOX, J.B. (1985): *Mixed-species game ranching in Kenya*. Ph.D. Thesis; University of Alberta, Edmonton, Canada
51. STYLE, C. (1987): The economics of game ranching in Zimbabwe. In: *Proceedings of the International Symposium and Conference on Wildlife Management in Sub-saharan Africa*; 6-12 Oct. 1987, Harare, Zimbabwe. FAO/CIC/GGF; Paris; p. 179-189
52. SWANK, W.G., CASEBEER, R.L., THRESHER, P.B. and P.B., WOODFORD, (1974): *Cropping, processing and marketing of wildlife in Kajiado District, Kenya*. FAO Project KEN/526, Working Document No. 6; Food and Agriculture Organisation of the United Nations, Rome; 77 pp.
53. TEER, J.G. (1975): Commercial uses of game animals on rangelands of Texas. *Journal of Animal Science*, 40(5), p. 1000 1008
54. TEER, J.G. (1992): Conservation Status of the Saiga Antelope in Kalmykia and Kazakhstan. *Species (Journal of the Species Survival Commission of IUCN, Gland, Switzerland)*, 17, p. 35-38
55. THRESHER, P. (1980): The economics of domesticated oryx compared with that of cattle. *World Animal Review*, 36, p. 37-43
56. TREUS, D.B. and D. KRAVCHENKO (1968): *Methods of rearing and economic utilisation of eland in Askanya-Nova Zoological Park*. Symposium of the Zoological Society of London, 21, p. 395-411
57. TREUS, M.Y., VITAKOVA, P.N. and V.M. DZHUROVICH (1992): Comparative characteristics of lactation in moose and eland. *Alces*, Suppl. No. 1; p. 238; Lakehead University, Thunder Bay, Ontario, Canada
58. WILSON, V.J. and B.L.P. (1991): *La chasse traditionelle et commerciale dans le sud ouest du Congo*. TURACO; Research Report No. 4, Chapt. 16, p. 279-389; Turaco Press, Brussels; in association with CONOCO OIL Co.
59. YORKS, R.T.P. (1989): Ranching native and exotic ungulates in the United States. In: *Wildlife production systems*, ed. HUDSON, R.J., DREW, K.R. and L.M. BASKIN; p. 268-285; Cambridge University Press,Cambridge
60. YOUNG, E. (1975): Technological and economic aspects of game management and utilisation in Africa. In: *Proceedings of the IIIrd World Conference on Animal Production*, ed. REID, R.L.; p. 132-141; Sydney University Press, Sydney
61. ZHIRNOV, L. (1993): *Ressourcen der Saiga Antilopen: Nutzung, Bewirtschaftung und Schutz* (unpubl. draft paper; 25 pp.)

## 2.3.14 WILD CATTLE, BUFFALOES, BISONS, AND WILD SHEEP AND GOATS (*Bovidae: Ruminantia: ARTIODACTYLA*)

Wild cattle, buffaloes and bisons all constitute a subfamily (**Bovinae**) of the Horned ungulates (**Bovidae**). Only the muskox is considered to be more closely related to sheep and goats which are another subfamily (**Caprinae**). Both have worldwide distribution. Tab. 49 shows that five bovine species and two caprine species were either fully or partly domesticated in early times and have become the material basis of many societies around the world. Others, like the African buffalo, the North American bison and the Arctic muskox have played a most important role in the subsistence of indigenous peoples; more recently they have also been ranched for meat or sold as trophy animals.

**Table 49:** Overview of the utilisation of bovines and caprines

| English name | Scientific name | Geographical distribution | Body weight (kg) | Utilisation |
|---|---|---|---|---|
| **WILD CATTLE, BUFFALOES and BISONS - BOVINAE** | | | | |
| Aurochs (extinct) | *Bos primigenius* | Eurasia | 350-600[1)] | domest. into dom. cattle |
| Yak | *-mutus* | C.Asia | 265-400[1)] | domest. for transport, meat and milk |
| Gaur | *-gaurus* | S.Asia | 625-770 | semi-domest. into gayal |
| Banteng | *-javanicus* | S.E.Asia | 400-730 | domest. into Bali cattle |
| Indian Buffalo | *Bubalus arnee* | S.Asia | 350-650[1)] | domest. into House and Water buffalo |
| American buffalo | *Bison bison* | N.America | 430-700 | farmed for meat and hides |
| African buffalo | *Syncerus caffer* | Africa s.o. Sahara | | hunted for trophies and meat |
| **SHEEP and GOATS - CAPRINAE** | | | | |
| Muskox | *Ovibos moschatus* | Arctic Canada, Greenland[2)] | 230-350 | hunted for meat and fur, farmed for hair |
| Chamois | *Rupicapra rupicapra* | Europe[3)] | 20-60 | hunted for trophies |
| Markhor | *Capra falconeri* | C.Asia | 80-110 | hunted for trophies |
| Bezoar goat | *-aegagrus* | W.Asia | 25-40 | domest. into dom. goat |
| Himalayan tahr | *Hemitragus jemlahicus* | Himalayan Asia[3)] | 75-100 | hunted for trophies |
| Bharal | *Pseudois nayaur* | Himalayan Asia | 25-70 | hunted for trophies |
| Asiat. Wild sheep | *Ovis ammon* sspp. | W.-C.Asia | 80-160 | domest. into dom. sheep; hunted for trophies |
| Mouflon | *-a. musimon* | Europe, W.Asia | 30-50 | hunted for trophies |
| Snow, Dall's and Bighorn sheep | *-canadensis* sspp. | E.Siberia, N.America | 100-150 | hunted for trophies |

1) approx. weight of domestic form only; 2) transplanted to Spitzbergen, Alaska and Siberia; 3) introduced into New Zealand

## Historical background

No other animal groups have been as important historically as the bovines and caprines for the development and sustenance of mankind. This is mainly due to the fact that these have proved to be particularly suitable for domestication, yielding no less than seven domestic animal species. These were already distributed in the earliest historical times over all continents and have gradually replaced the utilisation of their wild progenitors, but also to a large extent that of other ungulates. Nevertheless, wild bovines have continued to-date to be an important resource for subsistence societies and various economic uses.

In temperate zones of ***Eurasia*** the *aurochs*, next to *deer* and *Wild boar*, was the most important prey of the prehistoric hunters (16), but became extinct in the Middle Ages probably mainly due to the habitat changes in the course of agricultural development. In ***North America***, however, buffaloes roamed the prairies by the millions until the 19th century, being utilised sustainably by the indigenous hunter societies as their most important resource for meat, hides and other animal raw materials. Only when the European settlers on their westward quest for land and riches started to exploit the buffalo resource ruthlessly was this almost destroyed. The major commercial use of the *American buffalo*, or *bison*, in those days was for hides and bones only, being thus very wasteful. By the late 19th century less than 100 Plains bison remained in private herds in the USA and a remnant population of about 300 Wood bison survived in Northern Canada (4, 10). At that time preservation efforts by the US Government commenced and the few remaining captive animals provided the parent stock for several herds in federal protected areas. Under human control the number of bison increased from 800 in 1895 to 30,000 in 1972 (10).

In contrast to all other ungulates in North America, bison were always perceived more as an economic asset than wildlife, perhaps due to their extirpation from virtually all of their original range. A market for live bison had developed in the early 1900s and the preservation and propagation of this species was largely achieved through private efforts, confining and propagating bisons predominantly on privately owned lands. The concept of wildlife in the USA, however, presupposes public ownership and free access. Due to the special status of bison their commercial production was greatly facilitated, although in general the economic use of indigenous wild animals has never been fully accepted in the USA. As an exception, therefore, surplus animals from the federal reserves were routinely sold to private producers and these have become the main source of breeding stock for a rapidly developing bison farming industry, both in the USA and Canada (13).

In ***Central Asia*** with its extreme climatic conditions it was obvious that local people should tame and use the extremely well-adapted local wild cattle species, the *yak*, in early historical times for transportation, for use of their dung for fire, and for milk and wool for subsistence. Over thousands of years this traditional use has not changed and although the species was never selectively bred for special traits it gradually became domesticated. In fact the domestic yak differs only in size from the wild species which it has replaced in most of its original range of distribution. Domestic yak are also crossbred with zebu cattle into so-called "yakows", and are now used far beyond the species original range in Afghanistan, the former USSR, Mongolia and China (3). In ***South Asia***, however, the *Indian buffalo* or *arni*, inhabiting open swamp lands, was turned into a true domestic animal with different breeds for different uses. As House buffalo it was spread since the Middle Ages to Egypt and East Africa and as far as Southeastern Europe and the Ukraine; as Water buffalo it has become indispensable in Asian riziculture-based agricultural life. Because of its adaptation to tropical swampy habitats it was successfully introduced to most East Asian countries, the Philippines, New Guinea, Polynesia and also to Brazil, the USA and Australia (2, 7). In many areas the buffalo is not a fully domesticated animal and lives under semi-feral conditions, managed and used in a semi-domesticated state. Particularly in Australia where it was brought as early as 1824, the species became feral in the northern coastal flood plains by 1843 (23). Its resource value was recognised over a century ago and an industry based on field shooting of buffalo for hides developed at the end of the 19th century. Until the Second World War an average of 7,000 bull hides were exported annually, but in 1956 the market collapsed due to competition from other hide exporters (25). The other Asian bovines have either been only partially domesticated, like the forest-dwelling *gaur*, or domestication was geographically limited with no selective breeding, like the *banteng*.

In ***Africa*** the *African buffalo* has remained abundant, both in savanna and forest zones, to-date and has constituted since time immemorial, together with many other bovid species, a most important resource for indigenous peoples, particularly in those areas which are infested with Tsetse fly and are thus unsuitable for livestock production. Only with the intensified agricultural development since the 1960s have buffalo come into conflict with human activities, not only because of competition for grazing and water but in Southern Africa also because they were implicated in the occurrence and spread of Foot and Mouth Disease in domestic livestock. As a result of these conflicts Savanna buffalo have been eliminated from many areas indiscriminately and wastefully. Locally the Savanna buffalo has also been ranched together with antelope, or was subjected to culling or cropping schemes and commercial exploitation. Although the species is without doubt biologically well suited for domestication, no serious efforts were ever made to breed and farm it in confinement like other bovids. Only in the framework of the Galana Ranch Research Project in Kenya in the 1970s were buffalo included in the investigations into the feasibility of farming various other African bovids (18). As regards the buffalo it was found that its rather coarse meat did not compete with that of antelope or beef, and this might have been the reason why this bovine was never subjected to domestication efforts.

In the ***Arctic***, archaeological evidence shows that in some areas, for example on the Ellesmere Island and in North Greenland, prehistoric subsistence societies existed for hundreds of years BC in total dependence on *muskoxen* (21). Also in Alaska, Northern Canada and in Siberia archaeological sites have revealed that muskoxen constituted a major food item in northern subsistence economies in historical times until the populations of this species started to decline severely over much of its circumpolar range, becoming extirpated in Siberia a few hundred years ago and in Alaska by the late 1800s (31, 6). Only in Northern Canada and East Greenland did muskoxen continue to be of some subsistence importance into the 19th century until intensification of hunting for fur eliminated the species also over most of its remaining Canadian range. The behavioural and ecological characteristics of muskoxen renders this species particularly vulnerable to hunting and this fact combined with climatic changes has probably contributed significantly to their local extirpation in much of its original range (6).

Several attempts have been made since 1900 to transplant muskoxen to localities outside its range in Sweden, Iceland, Spitzbergen and Norway. Only on Spitzbergen were these successful and a herd of some hundred animals survived the Second World War occupation of this island (24). Within its historical range it was reintroduced successfully to West Greenland in the 1960s and the Taimyir Peninsula and the Wrangel Island of Siberia in 1975 (22). In the 1930s interest emerged in utilising muskoxen agriculturally. A captive herd originating from East Greenland, was first established in 1930 at College in Alaska; however, after the project there was abandoned the whole herd was liberated onto the Nunivak Island in the Bering Sea where it multiplied to an exploitable level and from where the species was then transplanted to other localities in Alaska (19, 20). In the early 1960s new attempts were made in Norway, Canada and Alaska to utilise muskoxen by breeding them in captivity mainly for production of qiviut, the particularly fine underwool of muskoxen. (33). Only in Alaska have these been pursued to-date, at Palmer, based on a herd of 33 animals obtained from Nunivak Island in 1964 and initially kept at the University of Alaska at Fairbanks (32).

Compared with the bovines, wild *caprines* never had a similar historical importance in subsistence or commercial hunting. Some species, however, have always been favourite trophy animals for the engaged sports hunter. The Domestic goat, carried around the world by man, has become feral in many countries and on remote islands causing tremendous damage; the most impressive example is New Zealand where they have contributed significantly to the erosion of whole mountain ranges (15).

## Management and utilisation

Only muskoxen and African buffalo are utilised to a limited extent in the form of **subsistence hunting**. *Muskox* populations have generally increased in East Greenland and Arctic Canada during the last decades (22), and several transplanted herds in Alaska have multiplied to exploitable levels (19). However, in Canada the Inuit prefer to hunt caribou and the indigenous people in Alaska have no tradition of hunting muskoxen, resulting in somewhat divergent views on their use. Some prefer to offer surplus animals to non-resident trophy hunters with the inherent cash benefits accruing to them, others wish to use them for their own subsistence needs (20). Subsistence hunting of the *African buffalo* with traditional means, both in the savanna and forest zones, was formerly widely practised , but with the increasing availability of firearms after the Second World War **hunting** became more and more **commercialised** with meat being sold partly to urban areas. However, agricultural development coupled with hunting and in some areas enzootics, have reduced buffalo significantly over most of Africa, so that today they are only still of some importance as a source of meat in the less accessible parts of the forest zone (1, 28).

In Northern Australia feral *Water buffalo* have also been commercially hunted for their meat and hides since 1959 (25). Since most of the buffalo meat is being exported, the original field shooting had to be abandoned in 1968 in favour of live capture with subsequent transport of the animals to export approved abattoirs. Herds of feral buffalo are driven by helicopter and four-wheel-drive vehicles into large capture yards. Young animals that are suitable for domestication are selected for live sale and the adults slaughtered like domestic cattle. Like bison, buffalo meat is leaner than beef (17) and has therefore found more and more export markets in Europe. Live animals are exported into Asian countries for use as domesticated animals in agriculture. Since 1985, however, feral buffalo became subject to extensive culling in the framework of the National Brucellosis and Tuberculosis Eradication Programme. In 1988/89 this accounted for at least 36% of all water buffalo killed and the total population of feral buffalo was diminished from an estimated 341,000 head in 1985 to 122,000 in 1989 (5), jeopardising the future of the Water buffalo's export industry.

Some hill tribes of Northern India and Burma, mainly the Nagas are utilising *mithans* or *gayals* which are believed to derive from the gaur, in a semi-domesticated state (3). Herds are allowed to browse freely in the woods, some animals returning to the villages for protection at night, others remaining in the forests. They are managed by provision of salt licks and mating by wild gaur bulls is encouraged. They remain fairly docile and animals are collected from the forest only when needed. The hill tribes use mithans mainly as sacrificial animals in religious ceremonies and as a kind of currency for buying goods and brides or for paying penalties (29). However, lately mithans are increasingly crossbred with cattle and then also used outside their original range for fieldwork, as a meat supply and for milk production (3).

**Ranching and farming** of wild bovines has become the most important form of their economic utilisation and is practised with bison and muskoxen in North America and with African buffalo in Southern Africa. In Canada the Elk Island National Park in Alberta served initially as an important model for ranching of game animals (27). However, few individual farmers controlled large enough tracts of land to be able to fully exploit the productivity of multi-species game communities. This fact and legal as well as economic constraints forced game production into more intensive forms, such as wapiti and bison farming. Herds of farmed *bison* range in size from a few animals to 3,000 head on some of the major enterprises. They are usually confined by especially strong fencing and receive supplementary feed similar to that of cattle. Bison are superior to cattle in their ability to graze native range throughout the year, to forage in deep snow and resist extreme cold. However, they are generally less productive than cattle although their dressing out percentage compares favourably with cattle. Cross-breeding with cattle, producing either "cattalo" or "beefalo" hybrids, has so far failed to give satisfactory results. In those states in which bison are legally defined as "domestic" animals (see below) finished steers are usually processed through cattle abattoirs with *ante-* and *post-mortem* inspections. Only very large enterprises can afford their own slaughter facilities. Bison meat is so far predominantly sold on domestic markets in preference to beef because of being typically leaner (8).

Despite decades of research into the farming of *muskoxen*, first at Fairbanks University and then at Unalakleet in Alaska, this is still in an experimental stage, since the late 1970s at Palmer, a former dairy farm in Southwestern Alaska (32). Domestication projects in Norway and Canada were abandoned because of the continuing necessity of taming the young by early weaning and bottle feeding with associated high costs (33, 34). At Palmer, calves are weaned at 2-3 months of age, early enough to ensure taming and late enough to prevent neo-natal mortality and stress. They are kept on fenced pastures with supplementary feeding of hay. The main product of farming is qiviut, the extremely fine underwool of muskoxen.

Muskoxen are farmed in Alaska for their particularly fine wool, the qiviut. (Photo: .Anthony Verlag, Starnberg, Germany)

It is harvested in spring by combing the animals in crushes, and processed into a fibre of uniform quality (32).

*African buffalo* were included in the multi-species game ranching which developed in Rhodesia, now Zimbabwe, in the 1960s and 1970s as described in Chapts. 1.4 and 2.3.13. However, buffalo meat did not compete with that of antelope and cattle, and in 1974 all wild buffalo in the livestock ranching areas were eliminated as suspected reservoir of the Foot and Mouth Disease virus. These areas were fenced off and classified as FMD-Free Zone from which the European Community accepted imports. Similarly, remaining wild buffalo populations in South Africa, Botswana and Namibia are restricted to remote non-agricultural areas by veterinary fences.

During the last two decades, many game ranches or farms have incorporated **trophy hunting** into their enterprises as an important economic supplement. Hunting clients, however, often wish to include an African buffalo in their bag and this necessitates flying them at extra cost to areas where buffalo still occur. Therefore, live young buffalo have come into great demand for re-introduction on Private lands, and the breeding of "FMD-free" buffalo stock is being initiated.

### Economic importance

Compared with smaller species of ungulates and rodents, wild bovines are of limited importance in ***subsistence economies***, except for the *muskoxen* in some areas of Arctic Canada and East Greenland, and the *African forest buffalo* in parts of the Congolese rainforest block. Generally the African buffalo is too large an animal for subsistence hunters, and smaller animals which can be trapped or killed with shot guns are preferred for ease of transport and preservation. Only where illegal ***meat hunting becomes commercially organised*** as for example in some of the West African countries (28), may buffalo constitute a fairly important resource economically, with meat being smoked for marketing to urban areas. In West Africa this and other bush meat is of considerable national economic importance as it substitutes for otherwise necessary beef imports for the dense coastal human populations as discussed in Chapt. 1.4.

The number of feral *Water buffalo* taken for commercial utilisation in Australia fluctuated in 1980-88 between 30,000 and 56,000 animals, including slaughter at export abattoirs, culling under BTEC and live sales, but decreased to only 12,384 in 1991. Accordingly, the buffalo meat export diminished from more than 3,000 t per year to 1,206 t in 1991 (25). With the decline of the feral buffalo population there is a tendency towards farming of disease-free domesticated Water buffalo herds which is supported by government with a view to maintaining the profitable buffalo meat exports. Similarly, *feral goats* are utilised in Australia by commercial hunting or trapping with subsequent abattoir slaughter. Their numbers are estimated in the hundreds of thousands and they constitute an economically important meat resource. Exports of goat meat from these feral populations mainly into Asian countries has increased from 4,514 t in 1981/82 to 11,441 t in 1991/92, constituting a value of more than Austr.$ 20 mill. (approx. US$ 15 mill.) (25).

***Ranching*** or ***farming*** of *bison* in North America has developed into a well-established industry of considerable economic importance. The total number of privately kept bison in 1985 was about 76,500 head, Plains buffalo only. It was estimated in 1989 that about 10,000 were slaughtered annually in the USA and another 1,000 in Canada (13). The envisioned availability of Wood bison stock, which was at that time still considered an endangered subspecies, for commercial production is likely to boost bison farming further as the demand for bison meat in 1989 exceeded the supply. Live bison sold at between US$ 590 and US$ 1,000 according to sex and age, and the price of bison meat at that time was 1.3-1.5 times that of equivalent beef cuts. By-products, such as hides, skulls, horns, tails, hooves and also some trophy hunting, contribute significantly to the profitability of bison farming (13). One of the most important factors in the economics of bison farming is the legal status of bison as "domestic animal" in most states of Canada and the USA. This is most exceptional for any farmed wild animals and has only recently found a parallel with the Fallow deer in Germany (26). It is explained by the special economic status which bison have always been accorded in North America as mentioned above. Being "domestic", although biologically still non-domesticated, slaughter and marketing of bison are greatly eased and the meat could qualify for export to the European Union.

The economics of farming captive *muskoxen* or of herding them in a semi-domesticated state for qiviut are still very difficult to evaluate because management conditions are too variable and markets for qiviut fibre not developed; no final system of husbandry and production of qiviut has emerged as yet. Considering the uniqueness of this potential resource in otherwise agriculturally almost unproductive areas, muskoxen farming may well become relatively important in the future when more breeding stock will become available. Adult muskoxen yield 2-3.5 kg qiviut per annum which sold in 1989 at over US$ 220/kg. Under the conditions of the Palmer experimental station the ratio between the value of produced qiviut and food consumed by a muskox was US$ 2.20-5.06 : US$ 0.42 (32).

The biggest economic potential of the African buffalo lies, presently, in its use in commercial ***trophy hunting***. As a meat producer it cannot compete commercially with antelope or cattle, although buffalo hides have a much higher market value for speciality leather production. In Zimbabwe *trophy buffalo bulls* were sold by safari outfitters at US$ 1,353 in 1986 (9) and at more than US$ 1,047 in 1993 (12). However, its value lies mainly in the ability of safari hunting outfitters to offer a larger spectrum of trophy animals. The buffalo ranks after elephant, lion, leopard and some antelope species amongst the most sought after trophy animals. As Australia's largest ungulate the *feral Water buffalo* is also a highly priced quarry for local and foreign hunters. In 1990 there were some nine safari operators established in the Northern Territory and trophy fees for Water buffalo bulls averaged about US$ 800 (25).

Similarly, wild caprines can produce very considerable returns through controlled trophy hunting. For example, in Mongolia a hunting quota of 50 *argali* rams at a trophy fee of US$ 20,000 generated a million dollars return in the 1970s, not considering indirect benefits such as job creation (30).

## Implications for conservation

The conservation of the remaining populations of endangered or rare wild bovines is largely unrelated to the present-day economic use of wild bovines. In fact, most of the endangered bovine species, such as the tamaraw (*Bubalus mindorensis*), the anoas (*Anoa depressicornis* and *-quarlesi*) and the kouprey (*Bos sauveli*) in South-East Asia and the European bison (*Bison bonasus*), as well as the wild progenitors of the yak, gayal, Bali cattle and Water buffalo, have become so rare that only rigid protection against illegal hunting might help them to survive in protected areas. The issue of legal and planned utilisation hardly applies to them, although they are considered a most valuable genetic resource and an important potential for tourism (3, 11). The farming of bison and muskoxen, as useful as it may be in propagating these species, has also little if any implication for conservation of the remaining wild stocks of these species. The only bovines which may be aided significantly by controlled utilisation are the two forms of African buffaloes, still abundant in many parts of the continent. Typical of other large ungulates and lions in Africa, but also of bears and large cats in North America, Europe and Asia, they are incompatible with agricultural development. Unless buffaloes in Africa produce sufficient economic returns they will without doubt be eliminated outside specially protected areas. Such returns are best obtained through commercialised trophy hunting which has already brought about the re-introduction of Savanna buffalo in areas where they had been previously eliminated. The Forest buffalo also has a tremendous potential for meat production in the humid forest zones of West and Central Africa in which domestic cattle succumb to trypanosomiasis. This subspecies is not as forest dependent as previously thought and could well be managed in secondary, otherwise unproductive forest areas (14, 28).

Wild caprines present a somewhat different case. Due to their rather inaccessible habitats they are less vulnerable to illegal hunting and usually do not conflict so much with agriculture. Well controlled trophy hunting in the framework of planned wildlife management can produce very considerable returns from them which could be channelled into their conservation.

## Literature references


1. AJAYI, S.S. (1979): *Utilisation of Forest Wildlife in West Africa.* FO:MISC/76/26; Food and Agriculture Organisation of the United Nations, Rome
2. ANONYMOUS (1981): *The Water Buffalo: New Propects for an Under-utilised Animal.* BOSTID, Report No. 32; National Research Council; National Academy Press, Washington, DC
3. ANONYMOUS (1983): *Little-known Asian Animals with a Promising Economic Future.* BOSTID, Report No. 46., National Research Council; National Academy Press, Washington, DC
4. BANFIELD, A.W.F. and N.S. NOVAKOWSKI (1960): The survival of the Wood Bison (*Bison bison athabascae* Rhoads) in the North-West Territories. *Natural History Papers*, 8; National Museum of Canada, Ottawa
5. BAYLISS, P. and K.M. YEOMANS (1989): Aerial Survey of buffalo, cattle and Bali cattle in the top end of the Northern Territory and adjacent areas (*cited from* RAMSAY 1994)
6. CAMPELL, J.M. (1978): Aboriginal human overkill of game populations: examples from interior North Alaska. In: *Archaeological Essays in Honour of Irving B. Rouse*, ed. DUNNELL, R.C. and E.S. HALL jun.; p. 179-208; Mouton, The Hague, Netherlands
7. COCKRILL ROSS, W. (ed.)(1971): *The husbandry and health of the domestic buffalo.* Food and Agriculture Organisation of the United Nations, Rome
8. COX, B.L. (1978): *Comparison of Meat Quality from Bison and Beef Cattle.* Thesis; University of Saskatchewan; Saskatchewan, Canada
9. CUMMING, D.H.M. (1989): Commercial and safari hunting in Zimbabwe. In: *Wildlife production systems*, ed. HUDSON, R.J., DREW, K.R. and L.M. BASKIN; p. 147-169; Cambridge University Press, Cambridge
10. DARY, D.A. (1974): *The Buffalo Book.* Avon Books/Swallow Press, Chicago
11. DE VOS, A. (1970): Research and Management Needs for South-East Asian Bovids. *Papers and Proceedings of the IUCN Eleventh Technical Meeting; New Delhi, India, 25-18 Nov. 1969*; Vol. I, p. 129-132; IUCN, Morges, Switzerland
12. FERGUSON, R.A. (1994): *Sport Hunting in Zimbabwe - 1993.* Zimbabwe Professional Hunters and Guides Association; Harare, Zimbabwe
13. HAWLEY, A.W.L. (1989): Bison farming in North America. In: *Wildlife Production Systems*, ed. HUDSON, R.J., DREW, K.R. and L.M. BASKIN; p. 346-362; Cambridge University Press, Cambridge
14. HOPPE-DOMINIK, B. (1992): Habitatnutzung von Waldbüffel, *Syncerus caffer nanus*, Bongo, *Tragelaphus eurycerus*, und Waldelefant *Loxodonta africana cyclotis*, auf Forst-Fahrwegen im Regenwaldgebiet der Elfenbeinküste. *Säugetierkundliche Mitteilungen*, 34, p. 31-41
15. HOWARD, W. (1965): Control of introduced mammals in New Zealand. New Zealand Department of Scientific and Industrial Research, Information Series No. 45; Wellington, New Zealand
16. JARMAN, M.R. (1972): European deer economies and the advent of the Neolithic. In: *Papers in economic prehistory*, ed. HIGGS, E.S.; p. 125-149; Cambridge University Press, Cambridge

17. JOHNSON, E.R. and D.D. CHARLES (1975): Comparison of live-weight gain and changes in carcass composition between buffalo (*Bubalus bubalis*) and *Bos taurus* steers. *Australian Journal of Agricultural Research*, 26, p. 415-422
18. KING, J.M. and B.R. HEATH (1975): Game domestication for animal production in Africa: Experiences at the Galana ranch. *World Animal Review*, 16, p. 1-8
19. KLEIN, D.R. (1988): The establishment of muskox populations by translocation. In: *Translocation of Wild Animals*, ed. NIELSEN, L. and R. BROWN; p. 298-318; Wisconsin Humane Society and Caesar Kleberg Wildlife Research Institute; Milwaukee, Wisconsin, USA
20. KLEIN, D.R. (1989): Northern subsistence hunting economies. In: *Wildlife production* systems, ed. HUDSON, R.J., DREW, K.R. and L.M. BASKIN; p. 96-112; Cambridge University Press, Cambridge
21. KNUTH, E. (1967): Archaeology of the muskox way. *Contributions du Centre d'etudes Arctiques et Finno-Scandinaves, No. 5*; Paris
22. LENT, P.C. (1988): Ovibos moschatus. *Mammalian Species*, No. 302; American Society of Mammalogists; Department of Zoology, Brigham Young University, Provo, Ut., USA
23. LETTS, G.A. (1962): Early livestock introductions to the "Top-End" of the Northern Territory. *Australian Veterinary Journal*, 40, p. 84-88
24. PEDERSON, A. (1958): *Der Moschusochs*. Neue Brehm Bücherei No. 215; A. Ziemsen, Wittenberg-Lutherstadt, Germany
25. RAMSAY, B.J. (1994): *Commercial use of wild animals in Australia*. Bureau of Resource Sciences, Australian Government Publishing Service, Canberra, Australia
26. REINKEIN, G. (1993): *in litt.*
27. RENEKER, L.A., BLYTH, C.B. and C.C. GATES (1989): Game production in Western Canada. In: *Wildlife proeduction systems*, ed. HUDSON, R.J., DREW, K.R. and L.M. BASKIN; p. 248-267; Cambridge University Press, Cambridge
28. ROTH, H.H. and B. HOPPE-DOMINIK (1987): Répartition et status des grandes espéces de mammifères en Cote d'Ivoire. IV. Buffles. *Mammalia*, 51 (1), p. 89-109
29. SIMOONS, F.J. and E.S. (1968): *A Ceremonial Ox of India: The Mithan in Nature, Culture and History*. University of Wisconsin Press, Milwaukee, Wisc., USA
30. SZANIAWSKI, A.(1990): *in litt.*
31. UPENSKI, S.M. (1984): Muskoxen in the USSR: some results of and perspectives on their introduction. In: *Proceedings of the Ist International Muskox Symposium*, ed. KLEIN, D.R., WHITE, R.G. and S. KELLER; p. 12-14; Biological Papers of the University of Alaska, Special Report No. 4; Fairbanks, Alaska
32. WHITE, R.G., TIPLADY, A.B. and P. GROVES (1989): Qiviut production from muskoxen. In: *Wildlife production systems*, ed. HUDSON, R.J., DREW, K.R. and L.M. BASKIN; p. 387-400; Cambridge University Press, Cambridge
33. WILKINSON, P.F. and P.N. TEAL (1984): The muskox domestication project: an overview and evaluation. In: *Proceedings of the Ist International Muskox Symposium*, ed. KLEIN, D.R., WHITE, R.G. and S. KELLER; p. 164-166; Biological Papers of the University of Alaska, Special Report No. 4; Fairbanks, Alaska
34. YOUNG, B.A. and L. GREER (1975): *Muskoxen: The Question of Domestication*. University of Alberta; Edmonton, Canada

## 2.3.15 OTHER MAMMALS

Virtually any wild animal may be utilised by aboriginal people for subsistence as discussed in Chapt. 1.4. The utilisation of a particular species depends on tradition, but also on its availability and on social factors. Whereas bigger species will be taken only by hunters, many other small animals are caught incidentally either as a side-line of the agricultural activities during the off-seasons or by children (3). Many of the incidentally utilised smaller mammals, particularly of the rodent order, have not been specifically mentioned in the preceding chapters, although all of the more commonly and commercially used groups have been discussed. There are, however, some species which are of local importance, but which do not belong to the mammal orders covered above. These are listed in Tab. 50 and briefly discussed in the following.

**Table 50:** Overview of the utilisation of some other mammals

| Zoological Order | English and scientific names | Utilisation |
|---|---|---|
| INSECTIVORA<br>Insectivores | Tailless tenrec (*Tenrec ecaudatus*), Russian desman (*Desmana moschata*), moles (*Talpa* spp.) | for food, musk and fur |
| CHIROPTERA<br>Bats | Fruit bats (*Eidolon* spp., *Pteropus* spp.) | for food |
| EDENTATA<br>Edentates | Armadillos (*Euphractus* spp., *Dasypus* spp.) | for food, laboratory use |
| PHOLIDOTA<br>Pangolins | Pangolins (*Manis* spp.) | for food |
| ARTIODACTYLA<br>Even-toed ungulates | Mouse deer (*Tragulus* spp.), Water deer (*Hydropotes inermis*), Giraffe (*Giraffa* spp.) | for food, laboratory use, pets, hunting, ranching |

Larger species of *insectivores*, although usually smelling rather unpleasantly, may also be utilised for food, for example the Tailless tenrecs in Madagascar. They are hunted with dogs and offered regularly on markets (4). Both the European mole and Russian desman have been extensively used in the past for fur production. Mole furs were already being worn in the Middle Ages and in 1930 more than 20 million mole skins were still marketed (4). Apart from fur desmans also provided musk for the perfume industry until over-exploitation reduced wild populations to non-utilisable levels in the 1920s. Potentially it is a species well worth managing for sustainable returns (4).

*Fruit bats* are preferred food items for indigenous people all over Australasia. In Micronesia and Polynesia they are considered a special delicacy, cooked and eaten whole with fur and viscera, and only served at social occasions. The strong demand for Fruit bats has had significantly detrimental effects on *Pteropus* populations in the Mariana islands since the Second World War (6). Consumption of Fruit bats on a commercial basis is now restricted to Guam, the southernmost island of the Marianas under USA-administration, which has been importing between 8,050 and 24,650 Fruit bats annually from 1975 to 1984 (6). In the forest zone of West Africa, Fruit bats (mainly *Eidolon* spp.) are also taken in large numbers and marketed in a smoked state, but more as a substitute for other types of bush meat (7). The bush meat trade also includes *pangolins* which are particularly relished by the local population.

In South America *armadillos* are very popular meat suppliers and killed for consumption wherever found. However, since it was discovered in the 1970s that armadillos might be a natural reservoir of *Mycobacterium leprae*, the causative agent of leprosy, they have also become important laboratory animals (8). It was found that this pathogen which had been extremely difficult to cultivate in the laboratory, could be grown in certain tissues of armadillos. Thus several medical centres started keeping colonies of armadillos for research into the causes and cures of leprosy. In 1981 ten such centres received support from the World Health Organisation; they use mainly *Dasypus novemcincuts*, but also *D. hybridus* and -*subanicola*. Most of the laboratory armadillos were caught in the Southern USA where *D. novemcinctus* is common, but efforts have also been initiated to breed armadillos in captivity (2).

*Mouse deer* are primitive, rabbit-size ruminants occurring in India and South-East Asia. They are widely sought by the natives for food. In the USA they are raised as laboratory animals for basic research on ungulates (1). *Water deer* are another primitive group of ruminants related to deer. They occur only in Southeastern China and Korea where they are hunted for their meat and skins. Early this century some Water deer escaped from Woburn Park in England and have since become well established as game animals in various localities. The species has also been reported from France (1).

*Giraffe* constitute a separate family of the ruminant artiodactyles and are quite common on some of the game ranches in Zimbabwe and South Africa where they are cropped together with antelopes. Like exclusively browsing antelopes, such as kudu and eland, giraffe are of particular ecological significance because they do not influence range condition. Their biomass is an addition to that of the otherwise utilised grazing stock and is thus of economic importance to the rancher (5). Limited numbers are also used as trophy animals.

## Literature references

1. ANONYMOUS (1991): *Microlivestock. Little-known Small Animals with a Promising Economic Future*. BOSTIK, National Research Council; National Academy Press, Washington, DC; p. 291-297 and 321-325
2. BARZDO, J., LUXMOORE, R., BROAD, S. and D. JONES (1984): *A survey of wildlife farming operations*. WCMC/IUCN, Cambridge (unpubl. report)
3. DE VOS, A. (1976): Game as food. A report on its significance in Africa and Latin America. *Unasylva*, 29(116), p. 2-12
4. HERTER, K. (1967): Bau und Leben der Insektenfresser. In: *Grzimeks Tierleben*, Vol. 10, p. 176, 220 and 226; Kindler, Zürich
5. JANSEN, D., BOND, I. and B. CHILD (1992): *Cattle, Wildlife, Both or Neither - Results of a Financial and Economic Survey of Commercial Ranches in Southern Zimbabwe*. WWF-Multispecies Animal Production Systems Project, Project Paper No. 27; WWF, Harare, Zimbabwe
6. PAYNE, N. (1986): The Trade in Pacific Fruit Bats. *TRAFFIC Bulletin*, 8(2), p. 25-27
7. ROTH, H.H. and B. HOPPE-DOMINIK (1996): Le marché de viande de brousse à Toumodi en Côte d'Ivoire (publication in preparation)
8. WALSH, G.P., STORRES, E.E., MEYERS, W. and C.H. BINFORD (1977): Naturally acquired leprosy-like disease in the nine-banded armadillo (*Dasypus novemcinctus*): recent epizootiologic findings. *Journal of the Reticuloendothelial Society*, 22(4), p. 363-367

# PART III: WILDLIFE PRODUCTS

Apart from the indirect and non-use values of wildlife referred to in Part I, wildlife resources produce a wide variety of important direct economic returns from services and products as shown in Part II. It is not possible in the framework of this book to describe in technical detail all of the wildlife-derived commodities. It is important, however, to generally relate resource utilisation to the direct economic benefits derived from the use of its products. Therefore, Part III first discusses the existing constraints on the use of wildlife products and then describes the customary uses, the practice of processing and marketing, and the trade in the most economically important wildlife products.

## 3.1 CONSTRAINTS ON THE USE OF WILDLIFE PRODUCTS

Wildlife product marketing is affected by the demand for the product, the nature of the product (including perishability), the level of technology (processing, storing, transport, etc.) and the cost of marketing. Although there may be a desire for the product, the processing or transport costs may be too high resulting in prices which depress demand. Thus it is possible to increase the value of wildlife products by investing in more advanced or reduced-cost processing and transport as well as by increasing demand.

Since, by their nature, most wild populations are located far from urban centres, wildlife product marketing is affected by distance, by poor infrastructure, and by small producers who are usually widely dispersed. It is also affected by the attitudes to wildlife which have often excluded it from playing its role in national economies. This has resulted in very little investment in research and development of wildlife products and their markets. In addition, the adoption of research from other countries which concentrates on exotic commodities diminishes the role of indigenous commodities including wildlife products. All these factors affect the development and market for wildlife products and together result in difficulties in achieving the economies of size which may be necessary for a viable industry.

The marketing possibilities are also affected by national and international regulations. These regulations may either prohibit trade or may inhibit it by affecting prices. Regulations increase prices directly through adding to the cost of processing and marketing and indirectly by creating barriers to entry which result in monopolies with their attendant higher prices and lower quantities being traded. Regulations can be useful where they can be effectively implemented, but where opportunities for black markets exist, prohibition can be more damaging than encouragement of legitimate markets. They drive the trade underground making it difficult to control. Where regulations are not designed to be prohibitive but rather facilitative, it is important to ensure that the benefits from imposing the regulations are greater than the direct and indirect costs.

## Processing and marketing of wildlife products

The diversity of wildlife products is such that it is not possible to develop a single marketing system or strategy; however, the goal is always the same; to increase returns from wildlife utilisation. A marketing system is efficient if the input-output ratios are maximised. Changes to processing or marketing must either increase consumer satisfaction, or decrease costs or both. A cheaper process which reduces consumer satisfaction may not be economically efficient. To develop products for marketing requires an understanding of the characteristics of participants to determine the degree and level of demand and potential demand for the different products.

Marketing encompasses all the activities involved in the flow of products from the primary producer to the final consumer. It includes not only physical transfer from one to another, but all the processes which add value to the product through processing, storing, transporting and pricing. The production and processing of the products are affected by:

- the nature of the product - whether it is perishable, free range, poisonous etc.;
- location and natural environment - climate, distance to market, etc.;
- technology - appropriate harvesting techniques, appropriate processing methods, sophisticated transport and storage, etc.

The marketing process needs to take into account the nature of the consumers and prospective consumers (their culture, superstitions, fashions, incomes, etc.). It needs to take account of available wild animal products and of the technologies available for using these products to meet consumer demands both locally and internationally. These marketing interactions are all governed by various norms and regulations that are designed either to protect the consumers or the wild animal populations or other animals, and sometimes, to help develop the production and marketing of wild-animal products.

The marketing of wildlife products has to overcome the negative and indifferent demands which exist for some of them, poorly developed infrastructure, technologies and institutions. Many wildlife products are luxuries and are therefore likely to be income elastic, which means that demand will increase as international income increases and it can be expected that the terms of trade will move increasingly in favour of the wildlife industry. This should encourage governments to invest in the marketing of wildlife products.

## Regulation of trade in wildlife products

In most countries there are legal provisions and administrative machinery regulating wildlife utilisation and the trade in wildlife products. Regulations and control mechanisms may pertain to:

- the conservation of the wildlife resources with a view to preventing over-exploitation and extermination of certain species or introduction of non-indigenous noxious species;
- the protection of consumers of wildlife products against fraud and health hazards, generally enforcing hygiene standards of production and marketing;

- the control of animal diseases by preventing transmission of pathogens from wild animals to domestic stock.

Apart from the private sector, marketing and trade in wildlife products may therefore involve wildlife, forestry, municipal, health, veterinary and customs authorities. However, there are great differences in official involvement depending on the local administrative efficiency or capability and the nature and scope of trade.

In many developing countries control of local or national trade in wildlife products is non-existent or inadequate. It is often regulated only by commercial licensing systems with no public health concern. Only in those countries where game farming or cropping operations are on a considerable commercial scale, as in New Zealand, Australia, South Africa, Zimbabwe and Namibia, have the local Veterinary Services become increasingly involved in the control of the slaughter, dressing, refrigeration and inspection of game carcasses and the management of game-processing plants. If, however, production from wildlife is destined for export, the trade always becomes subject to stringent conditions of control which are usually stipulated by the importing countries.

In respect of the protection of endangered species, the international trade is regulated by the Convention on International Traffic in Endangered Species of Wild Flora and Fauna (CITES) to which more than a hundred countries had adhered in 1993. Every country which is a party to CITES is required to designate a conservation authority charged with the execution of the provisions of CITES. These oblige CITES countries to permit importation of products from any of the listed endangered species only on the basis of a CITES export permit issued by the exporting country as discussed in Chapt. 1.5.

As regards the prevention of the spread of animal diseases and zoonoses, the main constraints world-wide are the veterinary regulations controlling the economically very important diseases of cloven-hooved animals, such as Foot and Mouth Disease, Tuberculosis and Brucellosis; in the case of wild pigs also African and Classical Swine Fever and Trichinosis. Most of these have little effect on the wild animal, but can be serious or even fatal when transmitted to domestic stock. The European Union, as the main importer of game meats, issued in 1992, a Council Directive[1)] which regulates the import of meat and other products of wild, cloven-hooved animals into member countries. This relates not only to marketing, but to all aspects of the killing of wild animals for meat production and the processing, inspection and preservation of derived game meats. The prescribed conditions seek to minimise possible public health and veterinary hazards, and being very stringent, they determine to a large extent the marketability and the economics of the trade in wildlife products with countries of the EU. Outside the European Union conditions of import are prescribed by the respective countries. These may vary from no conditions at all to most exacting requirements equivalent to those required for domestic meat hygiene and often modelled on the EU Directive. They are further discussed in Chapt. 3.2.1.

---

1) ANONYMOUS (1992): EC Directive 92/94/EEC of 16 June 1992 on public health and veterinary problems relating to the killing of game animals and marketing of game meat. EEC Bulletin No. L268/35 of 14 September, 1992; Brussels

## 3.2 COMMODITIES DERIVED FROM WILDLIFE RESOURCES

Subsistence societies have through the ages used virtually all parts of wild animals for various purposes. There is also a wide range and diversity of commercially utilised products. PAX and ARNDT[1)] undertook between 1925 and 1945 to compile a most comprehensive manual describing the chemical and physical properties of all animal products, their use, socio-economic importance and trade. However, this mammoth undertaking remained incomplete and is by now largely outdated. The following sections focus only on the most important commodities derived from reptiles, birds and mammals.

### 3.2.1 GAME MEAT

Whereas the term "venison" denotes only the edible flesh of deer or other wild animals taken by hunting, "game meat" encompasses any flesh of wild animals, including large reptiles, birds and small mammals. Of all products derived from wildlife, game meat has been historically the most important one and it is still of great significance as a valued food item all over the world.

#### Production and consumption

The **general role of wildlife as a source of food** and the difficulties of comparative assessment of game meat production and consumption have been discussed in Part I. The overall importance of game meat for the nutrition of man has decreased drastically in the 20th century as output of meat from domestic animals has increased rapidly. According to FAO's statistics (5,6) and a variety of supplementary sources studied, meat derived from wildlife for human nutrition comprises no more than about 1.0-1.5% of the world meat production. However, the meat of a wide range of wild animals is in many countries still a major food item in the diet of local populations (12). There is hardly any species of reptiles, birds or mammals which is not eaten by man, although dietary customs vary greatly and religious or spiritual beliefs often preclude certain species from consumptive use.

The ***developing countries***, with a share of slightly over 40% of global meat production, account for nearly 70% of the estimated global output of game meat, with Africa being the most important producing and consuming region. In tropical and Southern Africa, game contributes substantially to the total supply of animal products although prices of "bush meat" are often higher than those of other meats. In a number of western, central and southern African countries, estimated annual per capita supply of game meat was between 3 and 5 kg in the 1970s (20). In the meantime with fast population growth and urbanisation, per capita consumption has probably considerably decreased among the local population, whereas demand from foreign tourists has risen. Generally, in the developing countries, most of the game meat is consumed close to the place of hunting. Lean meat with relatively high protein and low fat and cholesterol content, such as venison, is preferred in the devel-

1) PAX, F. and W. ARNDT (eds.) (1929-1945): The primary resources of the animal kingdom (in German); 4 vols.; Gebr. Borntraeger, Berlin; 2235 pp.

oped countries, whereas consumers in many developing countries like fatty meat, obtained in particular from rodents.

In contrast to the developing countries, consumption of game meat in the ***industrialised countries*** has generally increased during the last decades, partly because of greater availability due to modern marketing possibilities and partly for social and gourmet reasons. Tab. 51 shows that the annual per capita consumption of game meat in European countries ranged in 1980-90 between 5-6 kg (in France) and almost nil (in the UK, Ireland and Netherlands). This was probably due to different availability within the country and long-lasting traditions. A sizeable proportion of the total consumption of game meat in European countries is derived from imports, which makes the EU the most important market for suppliers of game meat in international trade. In the USA, the largest recorded producer of game meat, average annual per capita consumption is estimated roughly at 1 kg (5). However, most of the game meat is consumed by the hunters and their families rather than marketed.

**Table 51:** Game meat consumption in European countries[1)]

| Year | F | I | B/L | D | DK | NL | UK | IRL |
|---|---|---|---|---|---|---|---|---|
| 1980 | 6.5 | 3.8 | 1.9 | 1.1 | 0.6 | 0.1 | 0.1 | < 0.1 |
| 1985 | 5.3 | 3.9 | 2.3 | 0.9 | 0.5 | 0.1 | 0.1 | < 0.1 |
| 1990 | 5.7 | 3.8 | 2.7 | 0.9 | 0.6 | 0.3 | 0.1 | 0.3 |

1) calculated from kg live weight per head of human population and year

Source: Bundesministerium für Ernährung, Landwirtschaft und Forsten, Bonn, Germany: Fleischwirtschaft in Zahlen (1986 und 1992)

The **medium term prospects for global meat production and consumption** must be viewed against a projected population increase from 4.0 billion in 1990 to 5.8 billion by the year 2010 in the developing countries alone (7). Therefore, these will have to account most for future growth in meat production. As output of meat from domestic animals increases, population pressures, expansion of agriculture, deforestation and desertification are expected to further reduce wildlife habitat. Thus the proportion of total meat supply that is game meat is likely to diminish further. To some extent, the decline of "bush meat" may be offset by increase of controlled wildlife utilisation and game farming, but much of this production is destined for export or foreign visitors. Hence, particularly among the rural population, per capita consumption of game meat is likely to decrease further. Notably in Africa this could be accompanied by a reduction in per capita consumption of total meat. In Asia and Latin America the economic outlook, and consequently prospects of consumer incomes and demand for meat, are better, probably leading to some improvement in average meat consumption levels.

## Consumer attitudes and nutritional value

In most countries of the world the meat of game animals is regarded as a particularly **healthy and valuable food**. Whereas traditional attitudes and spiritual beliefs play an important role in many developing countries in making "bush meat" or other locally obtained types of meat preferred food items, consumption of venison in the industrialised countries also seems to be more and more favoured for environmental reasons. It is produced under natural conditions, not subjected to any possible hormone treatment and often marketed directly to the consumer.

**Physiological and nutritional facts** seem to justify these consumer attitudes. In general meat of wild animals has a high content of biologically valuable ***protein*** and is rather lean. Furthermore it contains all nutritionally important water soluble B-vitamins and more iron than the meat of domestic animals. There are, however, differences of physical structure as well as chemical composition, not only in comparison with domestic animals but also between different wild animal groups as Tab. 52 shows.

**Table 52:** Composition of various game meats in comparison with domestic animals1)

| Species | Total protein | Fat<br>in g/100 g meat | Water | References |
|---|---|---|---|---|
| Fallow deer | 22.6 - 23.30 | 0,55 1.10 | 74.3 | (24) (14) |
| Red deer | 21.42 - 22.9 | 2.4 - 2.88 | 74.8 | (27) (2) |
| Roe deer | 21.4 | 1.25 - 3.55 | | (9) |
| Wild boar | 19.1 - 21.20 | 6.30 - 9.6 | 71.5 | (16) (27) |
| Hippopotamus2) | 24.20 | 0.92 | 73.6 | (22) |
| Hare | 24.35 | 5.45 | 69.2 | (27) |
| Rabbit | 20.8 - 21.0 | 7.60 - 8.0 | 69.6 | (24) (15) |
| Nutria | 19.40 | 9.40 | 69.4 | (25) |
| Cattle | 22.4 - 22.67 | 4.45 - 6.63 | 69.7 | (17) (27) |
| Dom. pig | 21.26 - 22.4 | 2.1 - 6.24 | 71.5 | (17) (27) |
| Wild duck | 17.4 - 19.9 | 4.2 - 15.2 | | (2) |
| Pheasant | 22.8 - 23.89 | 2.8 - 7.71 | 67.4 | (3) (27) |
| Ostrich 2) | 21.20 | 1.4 | 75.7 | (15) |

1) composition varies according to carcass part analysed; 2) meat from only one individual analysed

In wild animals the proportion of red muscle fibres is generally higher and their diameter is greater than in the domestic forms (26). ***Fat*** is predominantly laid down intracellularly rather than intercellularly between muscle fibres and in the connective tissue as in domestic animals. Nevertheless, the fat content of game meat can vary considerably according to the cut. Many wild animals build up specific fat deposits, for example in the subcutis or around the kidney. Therefore, the carcass as a whole may have a high fat content although the muscles remain lean. All game meat cuts except *Wild boar* give an energy value of about 100-130 kcal/100 g of raw meat. The protein content may fluctuate in the same species seasonally as a result of the change in food availability, or because of genetic factors. In *hares*, for example, the protein content of muscular tissue drops from 22.2 to 20.8% during winter time and fat deposits diminish likewise (10).

Some of the **differences between the meat of wild and domestic animals** are due to the fact that domestication may alter the anatomy and physiology of the digestive organs of the progenitor species. Allometric comparison of the *wild* and *domestic pig*, for example, shows that the small intestine of the latter is 39.1% longer in relation to body size than that of *Wild boar*, its surface is 43.2% larger and its weight 53.5% greater. This change has resulted in an increase of the associated lymphatic organs with consequent effects on digestive capacity (28). On the other hand the different composition of game meat may simply be caused by environmental influences. Wild *Fallow deer*, for example, have less fat and more protein in their muscular tissue than farmed deer (24).

The ***composition in amino acids*** of game meat is generally similar to that in domestic animals; some species have a greater proportion of essential amino acids, notably leucine and lysine, whereas tryptophan, hydroxiporlin and cysteine may be lower (11). The ***composition of fat*** varies with the tissue. Meat cuts which contain much intermuscular fat tissue contain mainly saturated triglycerides, whereas in lean cuts unsaturated phospholipids, like lecithin, predominate. The higher content of polyunsaturated fatty acids, such as linolic and linolenic acid, in *wild herbivores* reflects their consumption of vegetable oils rather than so-called "hard fats" associated with *domestic livestock*. Apart from increased essential amino acids the concentration of polyunsaturated fatty acids is the most important nutritive aspect of game meat (11). The ***vitamin content*** of game meat is similar to the meats of domestic animals, although *Red deer* venison was found to be particularly rich in B-complex vitamins (27). Game meat also has higher levels of myoglobin which makes it look redder than domestic meats. This accounts for increased ***quantitites of iron*** which is bound to the myoglobin molecules.

The typical ***flavour*** of game meat is caused in part by the different pattern of fatty acid composition but also by the different nutrition of the wild animals. The so-called "haut gout", however, is a result of undesirable bacterial processes after the animal has been killed. These may be enhanced by *ante mortem* stress which causes depletion of the muscle glycogen and consequently diminished formation of lactic acid. This acid is produced by anaerobic glycogen breakdown after death which causes the pH-value to fall from 7.0 during life to 5.5-5.8 *post mortem*. It gives flavour to the meat and the low pH-value delays bacterial putrefaction.

The **acceptance of game meat** and the **flavour preferences of consumers** vary greatly according to culture and tradition. In West African countries it is mainly the type and flavour of the gravies which accompany a dish of usually smoked bush meat that are most important. In Ghana, for example, rumen contents are widely used to produce the especially desired flavour. In European countries the consumer value of venison is largely determined by its structural delicacy. This is normally obtained by a process of maturation. Whereas agriculturally produced meat usually originates from relatively young animals, venison is derived from any age class. Therefore, venison of older animals (with a greater proportion of fibrous tissue) is usually soaked or prepared with specific marinades, usually of an acidic nature (19).

## Processing and preservation

Methods of processing and preservation of game meat vary widely. They depend on tradition and flavour preferences, climatic factors, technical and economical means, intended marketing, public health and veterinary concerns.

Simple bush meat processing facility at Kumasi in Ghana. Carcasses of mainly duikers, Cane rats and bushbuck are eviscerated and cut into portions for sale after the hair has been singed and scraped off. (Photo: T. HOFMANN)

As long as the game meat only serves domestic needs or local markets it is usually consumed fresh, smoked or dried. For this purpose carcasses are eviscerated and cut into portions of varying sizes. The skin is usually left on, either because this is eaten with the flesh or because it protects the carcass during transport against bacterial contamination. Only a few countries regulate the processing and preservation of meat destined for local or national markets. In Germany, for example, inspection of any game carcass intended for human consumption has been prescribed since 1981. It includes *ante mortem* observation of the game animal to be killed, but this responsibility is vested in the hunter.

The **suitability of game meat for human consumption** may be impaired by endogenic or exogenic factors. Amongst the first are bacterial and viral infections or infestations with parasites. These do not necessarily have to be macroscopically manifest. Their detection often requires laboratory facilities and the range of possible pathogens varies greatly according to climatic zones and geographical regions. Generally, wild animals enjoy good health and clinically manifest diseases are rare. However, depending on ecological and management conditions, levels of parasitic infestation can vary and wild animals may become latent carriers or reservoirs of pathogens which are sometimes transmissible to man or domestic animals. The converse may also occur where wild animals are closely associated with man and his animals. Exogenic factors include stress of the game animal as a result of adverse hunting or slaughtering methods, and secondary bacterial contamination in the course of processing, transport and storage of the game carcass. These factors must be carefully controlled in large-scale commercial production of game meat, particularly under hot and humid climatic conditions. Undue stress of the game animal before death results in a lack of lactic acid in its muscles and an increased *post mortem* pH-value in them. Such carcasses are prone to rapid putrefaction. To avoid secondary bacterial contamination it is essential to process the killed animals skilfully under the maximum possible hygienic conditions of either mobile or locally built small processing plants, or to transport them with minimum delay to special centralised game abattoirs. In all cases processed game carcasses must be cooled down as soon as possible, unless - as in Southern Africa - so-called "biltong," salted and sun-dried meat strips, or tinned meat (for pet food) are produced. For export purposes, game carcasses are usually deboned and deep frozen.

## Marketing and trade

Marketing practices for game meat vary widely according to the different systems of wildlife exploitation, ranging from subsistence hunting for "bush meat" to large-scale commercial harvesting of wild animals in order to supply urban and luxury markets. Although referred to as "subsistence hunting" the exploitation of wildlife by village hunters in the forested areas of West and Central Africa, South-East Asia and Northern South America, is often partly on a commercial basis. Meat which is not consumed locally is sold through ***traditionally developed marketing systems*** which remain uncontrolled and thus unrecorded. More recent studies in West and Central Africa (1, 29, 13, 23) have shown the considerable volume and socio-economic importance of such local trade in game meat.

The large-scale commercial production of game meat is usually predominantly intended for export. However, statistical information on the resulting ***international trade*** in game meats is scanty. FAO puts world exports at some 45,000 t, close to US$ 200 mill. in value terms, in recent years. This is less than 0.5% of global trade in total meat, though prices of most types of game meat are higher than those of meat from domestic animals. However, as in the case of production, trade data for individual countries are often incomplete (21). For the calendar year 1990, the European Statistical Office (EUROSTAT) reported total imports of 38,110 t into the European Union (EU), of which as much as 12,733 t was trade amongst the twelve member states. Germany (13,815 t), France (10,551 t) and Italy (5,608 t) were the biggest individual importing countries. EU trade statistics differentiate between meat of non-domestic rabbits and hares and other game meat. Of the total import of 12,936 t of meat of *non-domestic rabbits* and *hares* in 1990, most came from Argentina (7,134 t), with smaller shipments from Chile, Uruguay and Australia. The category "other game meat" comprises meat from *deer, Wild boar, wild birds* and *several other species*. East European countries, New Zealand and Austria were the main suppliers in 1990. Smaller quantities came from South Africa, Namibia and China. EU exports to Third World countries were small, leaving a net import of 25,414 t of game meat in 1990. This was the last year for which EUROSTAT data were conventionally published. Subsequent information can only be obtained by electronic evaluation of the computerised data. For Germany, the largest importer within EU and world-wide, imports up to 1992 are shown in Tab. 53. In Germany and other European countries, demand for game meat, which had been curbed by the consequences of the accident at the Chernobyl nuclear power plant in the former USSR in April 1986, seems to have strengthened again in the 1990s.

**Table 53:** Imports of game meat into Germany[1)]

| Countries exporting meat into Germany | 1988 metr. tons | 1990 metr. tons | 1991 metr. tons | 1992 metr. tons |
|---|---|---|---|---|
| European Union | 3.037 | 2,011 | 1,918 | 2,301 |
| East Europe | 5,129 | 5,082 | 7,021 | 6,655 |
| Argentina | 4,217[2)] | 3,911[2)] | 4,993[2)] | 3,917[2)] |
| New Zealand | 1,128 | 1,600 | 3,199 | 4,984 |
| Total imports | 15,025 | 13,815 | 17,828 | 18,820 |

1) united economy as of 1990; 2) mostly meat of hares

Source: Statistisches Bundesamt, Wiesbaden, Germany

Other importers of some significance are the USA and Sweden. More recently Japan, the world's largest net importer of total meat, has begun also to buy some game meat in the international market.

The main suppliers of game meat and the different animal groups from which this is derived are shown in Tab. 54. Argentina remains the largest exporter of hares and New Zealand more than doubled its export of venison in the early 1990s to a record level of 13,000 t (8).

On many game farms in Southern Africa game meat is cut into small strips for salting and sun-drying as "biltong". (Photo: HENDERSON BROS.)

This success has been facilitated by less competition from East European countries, previously the most important venison suppliers. Apart from large quantities of *kangaroo* meat mainly for pet food, Australia also exports *deer* venison, *rabbits* and meat from *feral wild pigs, Water buffalo* and *horses*. South Africa and Namibia are the only African exporters of *antelope* meat. More recently, smaller quantities of such specialities like *ostrich* and *crocodile*-tail meat have been supplied by a number of African and other countries.

**Table 54:** Range of commercially produced game meats[1)]

| Types of game meat | Most important producer countries |
|---|---|
| Crocodile tails | South-East Asian countries, Zimbabwe, Colombia |
| Snakes | China |
| Turtles | China, South-East Asian countries, India |
| Ostrich meat | South Africa, Zimbabwe, Israel, USA, Australia |
| Quail, pheasant | Japan, France, Italy, UK |
| Kangaroo meat | Australia |
| Hares, rabbits | Argentina, Australia |
| Nutria | USA, Argentina |
| Capybara | Venezuela |
| Wild boar meat | Australia, East European countries, France, Germany |
| Deer venison | New Zealand, East European countries, Canada |
| Reindeer vension | former USSR, Canada, Fennoscandia |
| Antelope meat | South Africa, Namibia, Zimbabwe, former USSR |
| Bison meat | Canada |

1) only meats which are derived from farming or ranching or intensive commercialised hunting operations

In order to protect the consumer, but also to enhance the trade in game meat, the EU as the single most important importer of game meat issued in 1992 a Council Directive (4) which regulates all ***public health and veterinary aspects inherent in the trade,*** as mentioned in Chapt. 3.1. After this is enacted into the national legislation of the member states it will form the legal framework for both import and interstate trade in game meat. It applies to the carcasses or meat of any free-ranging terrestrial mammals and game birds. Imports will be permitted only from those countries which are considered to be free of any diseases which might pose a hazard to public health and/or the livestock industry of the importing country. Furthermore, game meat consignments will normally be accepted only from acknowledged game processing or packing plants in which production and marketing conditions conform with those described in detail in an Annexe to the above Directive and which are in force in the importing country. The onus to create, maintain and to prove such conditions lies with the exporting country and experts from the EU may inspect them from time to time.

The USA federal regulations designed to control poaching forbid the sale of edible parts of resident wildlife altogether. Veterinary regulations on the import of game meat into USA from any source are so stringent that foreign meat exporters usually seek less demanding markets. In Canada it is also illegal to sell products of wild animals except for some native groups. "Farmed" venison can be sold at "farm gate" sites without inspection, or through special registered and inspected slaughter houses. *Bison* have the status of "domestic animals" and their meat may thus be traded freely. New Zealand has also declared *farmed deer* to be "domestic animals"; but they must be slaughtered in special Deer Slaughter Premises (DSP) which have to comply with the standards of export abattoirs for domestic animals and also those of the EU. Australia has the great advantage of being free from Foot and Mouth Disease, African Swine Fever, Tuberculosis and Brucellosis. The meat of farmed wild ungulates is exported through approved slaughter houses. *Wild rabbits* are mainly exported to South-East Asia where no *ante mortem* inspection is demanded. *Kangaroos*, whose meat in the past has mainly been utilised for pet food manufacture, have since 1980 been legally classified as "game animals" and are thus "fit for human consumption". African countries are in the most disadvantaged position for the export of game meat due to the endemic presence of Foot and Mouth Disease, African Swine Fever, Tuberculosis, Brucellosis, and other transmissable diseases. At present only Botswana, Swaziland and Namibia are permitted to export meat of wild cloven-hooved animals to countries of the EU.

Generally, the stringent regulations enacted by the EU and others are likely to limit the international trade in game meat. The export business will be restricted to countries which are financially, administratively and technically capable of not only creating and maintaining European standards of meat hygiene, but which will generally control and manage their wildlife resources to ensure their sustainable utilisation. In this respect the developing countries will have to compete with the East European countries which are all desperate for foreign exchange earnings from game meat exports. World-wide, game farming and ranching is likely to see its share in total game meat supply rise. This will enable the production of larger quantities of game meat that qualifies for export into the European market. However, intensification of game production may also adversely affect the unique qualities of wild game meat and thus tarnish its image among consumers in the industrialised countries who hope to buy and consume a healthy product.

## Literature references

1. ANADU, P.A., ELAMAH, P.O. and J.F. OATES (1988): The Bushmeat Trade in Southwestern Nigeria: A Case Study. *Humcn Ecology*, 16(2), p. 199-208
2. ANONYMOUS (1990): *Food Processor II database* (Venison 3.22 F.T.); ESHA Research; Salem, Oreg., USA
3. ANONYMOUS (1991): *Micro Regal database* (Version 1.1); ed. FEINBERG, M.; Paris
4. ANONYMOUS (1992): *EC Directive 92/45/ECC of 16 June 1992 on public health and veterinary problems relating to the killing of game animals and marketing of game meat*. EEC Bulletin No. L268/35 of 14 September, 1992; Brussels
5. ANONYMOUS (1993a): *Statistics from the FAO Interlinked Computer System*. Food and Agriculture Organisation of the United Nations, Statistics Division; Rome
6. ANONYMOUS (1993b): *World meat situation and outlook*. Food and Agriculture Organisation of the United Nations, Rome; June 1993
7. ANONYMOUS (1993c): *Agriculture: Towards 2010*. Food and Agriculture Organisation of the United Nations, Rome; Nov. 1993

8. ANONYMOUS (1993d): *The New Zealand deer industry, an overview of production, marketing and trade.* FAS/USDA report; US Embassy, Wellington, New Zealand; June 1993
9. ANONYMOUS (1993e): *Prodi 4.2 plus database*; ed. KLUTHE, B and P. KASSEL. Wissenschaftiche Verlagsgesellschaft, Stuttgart, Germany
10. CAMBERO, M.I., DE LA HOZ, L., SANT, B. and J.A. ORDONEZ (1991): Fett- und chemische Zusammensetzung bei spanischen Wildhasen. *Fleischwirtschaft*, 71, p. 54-55
11. CRAWFORD, M. (1973): A reevaluation of the nutritional role of animal products. In: *Proceedings of the IIIrd World Conference on Animal Production*, ed. REID, R.L.; p. 21-35; University of Sydney, Sydney
12. DE VOS, A. (1978): Game as food. *Unasylva*, 29 (116), p. 2-12
13. Falconer, J. (1992); *Non-timber forest products in Southern Ghana. A Summary Report.* Overseas Development Administration; Forestry Series No. 2; London; 23 pp.
14. FREUDENREICH, P. and K. FISCHER (1989): Untersuchungen zur Fleischqualität von Damtieren. *Landwirtschaftliche Wildhaltung, Wildtiere in Gehegen*; Heft 5; p. 87
15. HERZOG, R. (1994): Fleischerzeugung mit Gehegewild und Kaninchen. *Fleischwirtschaft*, 74, p. 257
16. HONIKEL, K.O. (1995): *pers. comm.*
17. HONIKEL, K.O. and R. WELLHÄUSER (1993a): Zusammensetzung verbrauchergerecht zugeschnittener Rindfleischstücke. *Fleischwirtschaft*, 73, p. 1051
18. HONIKEL, K.O. and R. WELLHÄUSER (1993b): Zusammensetzung verbrauchergerecht zugeschnittener Schweinefleischstücke. *Fleischwirtschaft*, 73, p. 863
19. JANSEN, A. and C. RING (1993): Wirkung der Beize auf Schwarzwildbret. *Fleischwirtschaft*, 73, p.693-694
20. KROSTITZ, W. (1979): The new international market for game meat. *Unasylva*, 31(123), p. 32-36
21. LUXMOORE, R.A. (1989): International trade. In: *Wildlife production systems*, ed. HUDSON, R.J., DREW, K.R. and L.M. BASKIN; p. 28-49; Cambridge University Press, Cambridge
22. MANN, I. and B. GRZIMEK (1962): Vom Fleisch der Flusspferde und Antilopen: *Fleischwirtschaft*, 42, p. 502-504
23. ROTH, H.H. and B. HOPPE-DOMINIK (1996): Le marché de viande de brousse à Toumodi en Côte d'Ivoire (*publication in preparation*)
24. SCHWARK, H.J., BRÜGGEMANN, J. and M. GOLZE (1990): Der Schlachtkörper des Damwildes und seine Zusammensetzung. *Monatshefte für Veterinärmedizin*, 45, p. 504-506
25. SPERBER, E., LEYK, W. and E. GEHLE (1982): Zusammensetzung und organoleptische Eigenschaften des Fleisches vom Nutria. *Fleischwirtschaft*, 62, p. 1329
26. SZENTKUTI, L. and O. SCHLEGEL (1985): Genetische und funktionelle Einflüsse auf Fasertypenanteile und Faserdurchmesser im *M. longissimus dorsi* und *M. semitendinosus* von Schweinen. Untersuchungen an trainierten Haus- und immobil gehaltenen Wildschweinen. *Deutsche Tierärztliche Wochenschrift*, 92 p. 93-97
27. UHEROVÁ, R., BUCHTOVÁ, V. and M. TAKÁCSOVÁ (1992): Nährwertfaktoren im Wildfleisch. *Fleischwirtschaft*, 72, p. 1155-1156
28. UHR, G. (1994): *Vergleichende Untersuchung am Darmtrakt des Haus- und Wildschweins unter besonderer Berücksichtigung des Darmschleimhautimmunsystems.* Habilitationsschrift; Tierärztliche Hochschule, Hannover, Germany
29. WILSON, V.J. and B.L.P. (1991): *La chasse traditionelle et commerciale dans le sud ouest du Congo.* TURACO; Research Report No. 4, Chapt. 16, p. 279-389; Turaco Press, Brussels; in association with CONOCO OIL Co.

## 3.2.2 FATS AND OILS

Animal fats and oils (lipids) have always been a valued product. All animals contain different types of lipids, composed of varying proportions of different saturated and unsaturated fatty acids. However, only those which have a high percentage of lipids in most of the body tissues (such as oily fish), or in which the lipids are concentrated in a discrete tissue (such as whales, seals or dugongs) have been commercially exploited for oil. Some wild animals deposit fat seasonally in specific organs or parts of the body and these too have been widely utilised for fat extraction.

### Production and uses

Animal fats and oils provide high-energy food, fuel for the production of light and heat, and have in the past also been used in industrial applications and for medicinal purposes (see also Chapt. 3.2.8). Accordingly the range of wild animals exploited at one time or another for fat and oils is wide and comprises reptiles, birds and mammals. However, the great majority of animal oils produced as a prime product are of marine origin.

Today animal fats as food have been widely replaced by vegetable oils and mineral oil products have been introduced for fuel even in the remotest areas. Therefore, exploitation of wild animals for fat and oils is now largely limited to subsistence societies or to specific species and situations. Apart from the on-going limited harvesting of whales and seals, fats and oils are also derived for commercial purposes, but only to a small extent, from Sea birds, farmed emu and some rodents and carnivores. In all these cases, however, fats and oils are secondary products only.

**Whale oil** was the main product of the whaling industry (see Chapt. 2.3.5). Species exploited for their oil were mostly Baleen or Whalebone whales, initially *Right whales (Balaenidae)*; however, following their near extinction, the *rorquals (Balaenopteridae)* were the main source. Oil was also obtained from some Toothed whales, mainly *Sperm whales (Physeter catodon)*, but this oil is chemically different from that from Whalebone whales (see below). Most of a whale's oil is contained in the blubber and at first this was the sole source. However, substantial quantities are present in the bones and other tissues and later technology has allowed the recovery of this oil. Blubber was stripped from the whale, minced and cooked in open boilers, or later in pressure vessels or rotary cookers, until the oil separated. Oil from bones and meat was extracted by steam cooking under pressure and centrifugation of the resulting liquor (14). Whale oil was used initially for lighting, but by the 18th century it was also used for soft-soap production and for scouring wool in the textile trade (9). With the introduction in 1903 of the Normann process for converting liquid animal oils to solid fats by hydrogenation, whale oil became a major stock for the production of margarine and cooking fats. Glycerine, a by-product in the production of soap from whale oil, assumed great importance for the production of explosives in the First World War. A minor use of whale oil was in the production of drying oils for paints (4). Production of whale oil is at present negligible, since the whaling that is carried on is mainly for meat. However, when production was at its peak in the 1930s, more than half a million long tons of oil were being produced yearly (1). The maximum recorded production in one season was 584,225 long tons in 1937/38.

In the Sperm whale the head is greatly modified to house a complex organ consisting of a hollow case and a mass of oily tissue known as the junk (7). The waxy oil found in this is called spermaceti. It contains a greater proportion of waxes (mostly acetyl palmitate) than the blubber oil (12) to the extent that it solidifies on exposure to the air. Spermaceti was once extensively used as lamp oil, burning with a clearer and less odorous flame than whale oil. This property was the basis of the Yankee sperm whaling industry that flourished in the 18th and 19th centuries (4). With the introduction of mineral oil and other illuminants, sperm oil was principally used for its lubricating properties as the presence of waxes rendered it unsuitable for production of edible fats. Sperm oil retains its lubricating properties at high temperatures and pressures, and is thus an excellent cutting oil for machining metals.

**Seal oil** has been used by small fishing communities since prehistoric times. In the Hebrides, Orkney and Shetland, small-scale sealing continued, mainly for oil, until this century. The oil was used for lighting, medicinally, or as a leather dressing (11, 3). Oil was the original product from the hunting of *Harp seals* and the sole product from the *Elephant seal* industry (see Chapt. 2.3.4). Harp seals when killed were stripped of the skin and blubber together. This "sculp" was then deblubbered and originally the blubber minced and the oil allowed to separate spontaneously. This "cold drawn" oil represented only about 50-70% of the total, the remainder being extracted by boiling to produce a lower grade oil (6). More recently all the oil has been cooked out by steam pressure and purified by centrifugation. In South Georgia the skins of Elephant seals were minced together with the blubber and the mince dried out in pressure cookers (5). Oil obtained as a by-product of Fur seal industries was ordinarily used mainly for dressing the pelts. Following world fat shortages after the Second World War, much seal oil was hydrogenated and used in margarine and cooking fat production.

**Seabird oil** has been an important local subsistence resource for many small communities. Communal nesting birds such as *guillemots* (*Uria* spp.) and other alcids, or *shearwaters* (*Mutton birds - Puffinus* spp.) have been exploited from the Arctic to New Zealand. The only seabirds that have been commercially exploited for oil, however, appear to be penguins. In the Falkland Islands, *Rockhopper penguins (Eudyptes cristatus)* were boiled for oil in tripots similar to those used by sealers. Between 1864 and 1866 about 63,000 gallons of penguin oil were produced (13). A more substantial industry was carried on from about 1849 to 1920 in Macquarie Island where *Royal penguins (E. schlegeli)* and *King penguins (Aptenodytes patagonicus)* were utilised (8). Sea bird oil seems to have been used for similar purposes as whale and seal oils.

## Chemical composition

Chemically, whale oil consists of a complex mixture of the triglyceride esters of fatty acids, principally oleic acid (about a third) and palmitic acid (about an eighth). Monounsaturated and polyunsaturated acids, free fatty acids and phospholipids are also present (2). Sperm whales produce a different type of oil, consisting mainly of wax esters (esters in which higher fatty alcohols, principally acetyl alcohol, replaces the glycerol of the triglycerides, with only about a fifth triglyceride lipid (2).

Seal oil is generally very similar in composition to whale oil. The unsaturated oleic acid, constitutes about a third of the total mass. The main saturated acids are myristic and palmitic, which make up about 8-14% of the total (15,10,2).

## Trade

Oil from marine mammals, and to a lesser extent Sea birds, was once of considerable industrial importance. Today, however, it has been largely replaced by oils derived from industrial fishing and the vegetable oil industry. Shortages of catch material (particularly whales) and changes in public attitudes have led to the abandonment of nearly all the oil trade from warm-blooded marine species.

## Literature references

1. ANONYMOUS (annual): *International Whaling Statistics*. Bureau of International Whaling Statistics, Sandefjord, Norway
2. ACKMAN, R.G. and F. LAMOTHE (1989): Marine Mammals. In: *Marine biogenic lipids, fats and oils*, ed. ACKMAN, R.G.; vol. 2, p. 179-381, CRC Press, Boca Raton, Fl., USA
3. BERRY, R.J. (1985): *The natural history of Orkney*. Collins, London
4. BONNER, W.N. (1980): *Whales*. Blandford Press, Poole, Dorset, UK
5. BONNER, W.N. (1982): *Seals and Man: a study of interactions*. Washington University Press, Seattle, Wash., USA
6. BUSCH, B.C. (1982): *Seals and Man: The war against the seals: a history of the North American seal fishery*. McGill-Queen's University Press, Kingston, Canada
7. CLARKE, M.R. (1978): Physical properties of spermaceti oil in the sperm whale. *Journal of the Marine Biology Association of the United Kingdom*, 58, p.19-26
8. CUMPSTON, J.S. (1968): *Macquarie Island*. Publication No. 93, ANARE Scientific Report A (1); Antarctic Division, Department of External Affairs, Canberra, Australia
9. JACKSON, G. (1978): *The British whaling trade*. Adam and Charles Black, London
10. JANSGAARD, P.M. and P.J. KE (1968): Principal fatty acids of depot fat and milk lipids from Harp seal (*Pagophilus groenlandicus*) and Hooded seal (*Crystophora cristata*). *Journal of the Fisheries Research Board Canada*, 25, p. 2419-2432
11. LOCKLEY, R.M. (1966): *Grey seal, common seal*. André Deutsch, London
12. MORRIS, R.J. (1975): The lipid structure of the spermaceti organ of the sperm whale (*Physeter catodon*). *Deepsea Research*, 20, p. 911-916
13. STRANGE, I. (1972): *The Falkland Islands*. David and Charles, Newton Abbot, Devon, UK
14. TØNNESON, J.N. and A.O. JOHNSEN (1982): *The History of Modern Whaling*. C. Hurst and Co., London
15. WINTER, G. and W.J. NUNN (1950): The component fatty acids of Elephant seal oil. *Journal of Science, Food and Agriculture*, 1, p. 18-21

## 3.2.3 GAMESKINS

Depending on their ultimate use the larger-sized hides and relatively smaller skins of game animals may be utilised with or without the hair, wool or feathers still attached. They are accordingly referred to as gameskins, fur- and featherskins, or pelts (2). Most gameskins are derived as a bye-product (though often a valuable one) during the course of the killing of game animals for meat or other products such as trophies. Conversely, the production of fur- and featherskins is more often the primary objective of the hunting or trapping of certain animals. Gameskins in the narrow sense are mainly destined for leather manufacture, though they may also be used to prepare trophies, rugs, miscellaneous artefacts or collagen preparations. Furskins are specifically used for the production of furs, fur garments or derivatives, such as felt, and are discussed separately in Chapt. 3.2.3.

### Production and uses

Historically, most gameskins were derived from **hunting activities** but this now accounts for a relatively small proportion of the total production. Those that are produced in this way may be used for various purposes. For example, gameskins from sporting operations are

Salting of a zebra skin on a game farm in Zimbabwe for commercialisation as a floor rug. The skin of a zebra has a much greater value than its meat. (Photo: HENDERSON BROS.)

very likely to end up as *trophies* modelled by taxidermists, or made into decorative *rugs*. Those used in taxidermy must be exceptionally well prepared; there must be no scope for bacterial spoilage since even the loss of a few hairs may spoil the finished work. Moreover, the animal's hide or skin must be very carefully removed to facilitate its subsequent reconstruction. Gameskins destined for use as *rugs* must be prepared to the same high standards. When the hunting is being done to provide *food*, the hide or skin may be consumed along with the meat. The hair is usually burnt off and the skin scraped and cleaned in water prior to butchering the animal. In many countries, for example in West Africa, the cooked or roasted skin of mammals is considered to be a delicacy. If it is removed prior to consumption, the hide or skin may be reserved for making crude harnesses or ropes, to upholster seats, to cover musical instruments or to construct any of a large range of artefacts.

The most important sources of gameskins for economic use are **farming, ranching and culling activities**. Despite the subtle difference between these operations they all tend to provide relatively large numbers of hides and skins. Although working conditions may vary, they are usually good enough for commercial exploitation for subsequent use in a tannery.

The range of game animals from which hides or skins may be utilised for one purpose or another includes essentially any reptile, bird and mammal species.

## Composition and structure

Like any material of biological origin, gameskins exhibit considerable variation in their chemical composition and physical structure. These may be attributable to the type of animal from which the hide or skin came, but are often compounded by a variety of individual factors including age, sex, season, climate, nutritional status, environment, and so on. Fundamental similarities do exist, however, among all gameskins, as well as fish skins, feather- or furskins and the hides and skins of domestic animals. Indeed, because relatively little research has been completed on gameskins in particular, most current understanding is based on studies of hides and skins from domestic animals.

The **chemical composition** of most gameskins is typically 65% water, 30% protein, 4% fat and 1% ash; with the greatest variation occurring in the last two components. Many animals of course may exhibit heavy deposits of subcutaneous fat, but *ostriches* and *pigs*, for example, often contain particularly high levels inside the skin itself. Similarly, many *reptiles* may provide high levels of ash due to mineral deposits in the surface scales. Irrespective of the general chemical composition, the distribution of particular components is subject to topographic and stratigraphic variations associated with the physical structures of gameskins.

The most important **physical structures** of gameskins can be observed in a transverse section. The outermost *epidermal layer* is very thin, cellular and characterised by an abundance of the same protein (keratin) found in hair, fur, wool, feathers, nails and horns. In fact, these structures are outgrowths of the epidermis. Where such outgrowths occur, they also produce invaginations which may go deep into the underlying middle layer of the gameskin

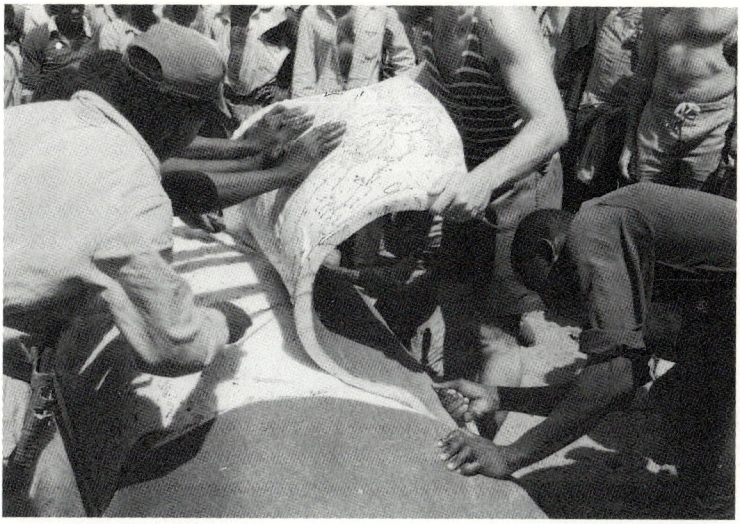

Early stages of the flaying of a hippopotamus. The hide is cut into smaller panels to facilitate handling. (Photo: I.B. LEACH)

which is called the *dermis*. It constitutes the bulk of the thickness of a hide or skin and is characterised by the presence of relatively few cells and an abundance of randomly interwoven collagen fibres. Underneath the dermis, and attaching it to other tissues of the animal's body, is the *hypodermis*. This diffuse layer varies considerably in thickness and consists of various proportions of fat, muscle and connective tissue.

Despite the fairly regular pattern of physical structure, there is a considerable variation in particular species. The thickness alone may vary from less than 1 mm in small *snake* skins to more than 35 mm in the case of a *hippopotamus*. These differences are accompanied by variations of the overall size as well; one of the biggest gameskins is the hide of an *elephant* which may exceed 210 kg in weight and 13.5 $m^2$ (6) in area. Furthermore, there are the important attributes of colour, surface texture and general aesthetic appeal. In some species, once the overlying hair has been removed together with any pigments in the epidermis, the residual dermis is indistinguishable from many other species as in the case of hides from *zebras* and *horses*. Conversely, the surface appearance of some skins is very distinctive as in the case of the irregular protuberances (know as "butts") on the part of an *ostrich* skin.

Only rarely would the inherent physical and chemical characteristics of a particular gameskin prevent its use in certain applications. However, any gameskin may be rendered completely unusable by bacterial spoilage. This may be avoided by using fresh gameskins (for whatever purpose) as quickly as possible. But, given the circumstances under which many gameskins must be protected by preservation pending storage and transportation to centralised facilities for additional processing.

## Preservation

There are three principle types of preservation procedures based on the application of chemicals (bacteristats and bactericides), temperature control (refrigeration and freezing) and moisture control (drying and salting).

The **use of chemicals** (such as disinfectants) to reduce or eliminate bacterial growth is widespread in domestic and commercial applications. However, chemical preservation is often impractical or hazardous in the isolated areas where gameskins are often produced, and the technique may present problems during the course of marketing. Accordingly, bacteristats and bactericides are rarely used alone to preserve gameskins, although they may be applied as adjuncts in other, more conventional, procedures. They may also be used to provide very short-term preservation when, for example, gameskins collected in the bush need to be protected until they can be delivered to the field station or camp for more thorough preservation.

**Refrigeration and freezing** are used routinely to protect all sorts of biological materials. Refrigeration has been applied commercially to high quality and high value furskins such as *mink*. However, the procedure is expensive in terms of capital and running costs, and is therefore not widely used. It should be appreciated though, that cooling in general is considered an essential element of any good preservation operation. For example, simply washing or rinsing gameskins in cool water may well reduce their initial temperature and remove some of the contaminating bacteria.

Long-established techniques based on the **control of moisture** have remained popular and effective. In the simplest drying procedure gameskins are suspended within vertical frames. Under normal circumstances the moisture content should decrease to the target of 10-20% within 1 to 2 days for thin skins, and 2 to 5 days for the thicker hides. Drying is not recommended for very thick hides where spoilage may begin to occur before the protracted drying is complete. Ideally, the drying should be completed in an open-sided shed, designed to provide shelter against direct sunlight and rain. Drying outside on the floor, should be avoided because of the deleterious effects of sunlight and the hazard of contamination. Contrary to popular belief, drying can be used successfully in most tropical and temperate countries. Problems due to the protracted drying in the wet season may be avoided by the application of very small quantities of short-term chemical preservatives (3). Suspension-dried gameskins sometimes exhibit problems of slow and incomplete rehydration during subsequent processing. However, this may be overcome during leather manufacture, for example, by changes in tanning procedures and the use of appropriate wetting agents or soaking aids.

The most popular method of preservation involves the use of salt. This preservative action is not due to its effect on the pH of tissue moisture, nor to bactericidal action, nor the fact that salt does achieve some degree of dehydration. Surprisingly perhaps, preservation with salt is also based on moisture control but unlike drying, where most of the water is removed, with salting the water is physically modified. Salt renders the water biochemically unavailable for metabolism by bacteria. In theory, the amount of salt required in preservation is equivalent to about 25% of the fresh weight of a gameskin. In practise though, it may be necessary to apply as much as 50 or 100% to the flesh surface. The time required for salt to diffuse through the full thickness of a hide or skin varies considerably from less than a day for many *domesticated animals*, to 5 days for moderately thick *elephant* hides and up to 10 days for the thickest *hippopotamus* hides. Hides, skins and gameskins preserved with salt are generally preferred by tanners.

## Tanning

Tanning for the manufacture of leather goods is the most important commercial application for gameskins. Many different species have already been used to make various types of leather (5) and it is likely that any hides or skins from game animals could be used in the tanning industry.

Fresh gameskins may be tanned immediately, but those which have been preserved must be soaked to reverse the effects of the drying or salting. The next operation usually involves the use of an alkaline solution to dissolve away the epidermis and plump-up the rest of the hide or skin to facilitate removal of residual hypodermal tissue. Eventually both the epidermis and hypodermia are removed so that only the dermis remains. The tanning procedure proper will vary according to the type of leather that is required. One of the commonest procedures is based on the use of chromium salts to provide a mineral type of tannage. Irrespective of which type of tanning method is used, the leathers are finally subjected to miscellaneous finishing operations including splitting (to reduce the thickness), dyeing, oiling and drying.

Gameskins which have been tanned using conventional, commercial procedures may be used like leathers made from the hides and skins of domestic animals. If the leather made from gameskins has no special features it may be used like *kangaroo*, to make shoes (8). Many gameskins, however, yield very distinctive leathers. The unique appearance of *ostrich*, for example, means it is considered very attractive in fashion garments. Almost all leathers made from *reptile* skins are very distinctive and may be used to make high-value, sought after articles such as belts, wallets, cases, shoes and boots.

Given the ancient origins of the tanning technique and its widespread application, it is often assumed that it must be very simply and easy. In principal, this is so, but to exploit the full physical and aesthetic appeal of fameskins requires considerable technical skill and access to specialised equipment not to mention some artistic flair and knowledge of current fashion trends. Since it is difficult and expensive to practice and develop these skills, they are often restricted to places with a tradition for processing different types of gameskins.

## Marketing and trade

The marketing of gameskin has (and still is) adversely affected by irregular supplies, variations in the quality of skins and unreliable trade information. Statistics on the availability of gameskins are not always produced and may not be collated or made available to those who might be able to use the data. Information is often inaccurate and inconsistent; and the trade statistics combined into general categories such as "reptiles" or "furskins". Similarly, production may be quoted in terms of "numbers", "pieces", "weight" or "unspecified" with no reference to the method of preservation, so that interpretation and comparisons are virtually impossible.

Under such circumstances, much of the marketing tends to be undertaken by individuals, based on personal experiences. Companies involved in the utilisation of gameskins tend to restrict themselves to known sources of raw materials, from established traders, and treat others with caution. Many organisations are sceptical of test batches of gameskin submitted for their assessment, because they may bear little resemblance to later supplies produced on a larger scale in the field. Neither is it safe to rely on grading standards, where they exist. The main problems with these is that the leather-making substance in the raw material (the dermis) cannot be observed through the overlying epidermis, on one side, and hypodermis on the other. Therefore, they may not provide a reliable measure of the potential value of the gameskins to a tanner.

In theory, any gameskin could be used to make leather; however, commercial exploitation has in the past been restricted to a relatively small number of species. The restrictions were rarely to do with the technical aspects of the tanning, and where they once were, the problems were overcome with a little research and development. Gameskins used in the tanning industry, tend to have been employed for one of two reasons; as a supplement or substitute for hides and skins from domestic species or as a speciality material. In the first availability, during shortages in conventional supplies. Examples of these would probably include *impala, wildebeest* and *wild pig* (including *peccaries*). The more significant category of gameskins is those used for speciality purposes, and includes all the *reptiles*, *birds* and some mammals such as *buffalo, elephant* and *hippopotamus*. Due mainly to their distinctive appearance, these materials are almost always in demand, especially for high-value fashion goods, and tend to attract high prices. Some, such as *ostrich* skins, may only exhibit distinctive patters on a small part of their area, and may be difficult and expensive to process. Despite, or because of all these disadvantages (from the tanner's point of view) leathers made from distinctive gameskins are considered especially attractive. Articles made from ostrich skins, for example, are among the most expensive leather goods.

With the exception of reptile and ostrich skins (see relevant chapters) there are no special marketing channels for gameskins destined for leather or rug production on the scale of those established for furskins. Most new suppliers of gameskins usually begin by using merchants already trading in hides and skins from domesticated animals which are established in almost all countries. Thereafter trade may concentrate on specialist tanners of gameskins some of which are listed in the International Leather Directory (7).

The economic importance of international trade in game skins is generally difficult to assess for reasons of the inadequate statistics referred to previously. The commercialisation of gameskins from any particular species is very susceptible to availability, local marketing facilities and the existence of relevant tanning skills. There are, in addition, the trade restrictions imposed by CITES and the introduction of more restrictive national and international legislation designed to protect the health and welfare of producers as well as consumers. These do, of course, take precedence over any of the technical and economic aspects of

Processing of an ostrich skin in an abattoir in Zimbabwe for use into highly priced specialty leather. (Photo: K. WILSON)

marketing discussed in Chapt. 3.1.1. Significant trade in gameskins for leather production or decorative purposes is limited to those animal groups which are utilised on a large scale for meat production, i.e. *kangaroos, capybara, wild pigs, deer* and *antelope* (7, 8). Rarely are gameskins of greater monetary value than the meat, like with *zebras* (4). The economic importance of the skin trade in these animal groups of animals is discussed in the relevant chapters of Part II.

## Literature references


1. ANONYMOUS (annual): *International Leather Directory*. Benn, London
2. ANONYMOUS (1975): *International glossary of leather terms*. International Council of Tanners, London
3. BARRETT, J.C., LEACH, I.B. and M.R. ASCOUGH (1984): The recycling of silicoflouride and boric acid liquors for hide preservation. *Journal of the Society of Leather Technologists and Chemists*, 68(5), p. 146-149
4. ELTRINGHAM, S.K. (1984): *Wildlife Resources and Economic Development*. John Wiley, New York
5. FUCHS, K.H.P (undated): *The chemistry and technology of novelty leathers*. Food and Agriculture Organisation of the United Nations, FAO, Rome
6. LEACH, I.B. (1988): Preservation of elephant and hippopotamus hides. *Leather*, April 1988, p. 79
7. LUXMOORE, R.A. (1989): International trade. In: *Wildlife production systems*, ed. HUDSON, R.J., DREW, K.R. and L.M. BASKIN; p. 28-50; Cambridge University Press, Cambridge
8. RAMSAY, B.J. (1994): *Commercial use of wild animals in Australia*. Bureau of Resource Sciences; Australian Government Publishing Service, Canberra.

### 3.2.4 FURSKINS AND HAIR

Gameskins and hides are utilised to produce leather and leather articles whereas furskins serve for the manufacture of garments, blankets or rugs. After a furskin has been removed, or "skinned", from a carcass, it is usually called a pelt. Contrary to leather production, the use of furskins implies careful conservation of all hair. Pelts are often processed into bleached or dyed furs and a wide array of articles of fur fashion. Next to game meat for human nutrition, furskins for clothing purposes constitute one of the oldest and most important raw materials derived from wildlife. Some furs are sheared and the hair used for various purposes; only rarely is hair collected directly from wild animals for weaving.

#### Production and uses

Pelts are produced from a wide range of mammals, but the great majority of commercially traded wild-animal furs originate from rodents and carnivores; Tabs. 26, 28 and 30 in Part II show the diversity of species utilised. Historically the range of species from which fur was derived might have been wider, and in fact the skins of all hunted animals were utilised for clothing or utility purposes, unless they were eaten with the flesh, as described in Chapt. 3.2.3. The production of pelts as objects of trade, as opposed to using them only for subsistence, developed very early on, as described in Chapt. 2.3.4. From the Middle Ages, trade in raw pelts and production of furs in Europe became more and more a fashion-dependent activity, culminating in an extravagant use of the furskin resource in the course of industrialisation in Europe and North America. Strong demand for specific types of fur resulted in the over-exploitation and near extermination of some of the wild furbearers and this fact combined with economic considerations gave rise to the ***production of furs from captive-bred stocks***. As described in Chapt. 2.3.4, fur farming developed rapidly during the first decades of the 20th century, and particularly the selective breeding of different colour mutants greatly influenced the fur market. Although fur production remained governed by the whims of fashion, it became limited to a few dominating species, *mink, foxes, beaver* and *seals*, later including more and more *nutria, muskrat* and *raccoon*. Farming of the almost exterminated *chinchilla* commenced in the 1920s and produced sufficient quantities of furskins from 1930 onwards. *Sable* were also successfully propagated in captivity in the former USSR from 1929 onwards, but farming still does not furnish more than about 10-20% of the traded sable pelts. Other species, such as the *raccoon*, the *American badger* and the *Striped skunk*, were farmed only as long as this was profitable; now furskins of these species are produced almost exclusively by trapping of wild populations. More recently *Pole cats* (or *fitch*) and *Raccoon dogs* (or *tanuki*) were also successfully farmed whereas attempts to breed the *American marten* (or *Canadian sable*), *River otter* and *lynx* have so far remained experimental or without economic importance. In the 1960s conservation oriented international trade restrictions, particularly in respect of *Spotted cats* and *seals*, futher reduced the diversity of the fur market, and since 1980 fur production has concentrated on mink farming. More than 60% and perhaps as much as 85% of all pelts traded are produced in farms, and after 50-80 years of selective breeding, mink, Silver and Arctic fox, chinchilla and sable may be considered semi-domesticated. The total output from fur farms in 1983 was approximately 33 million pelts as opposed to approximately 18 million from wild or feral populations of the three most important wild furbearing species, muskrat, nutria and raccoon (3).

The fur trade has a long tradition in Europe, and from the Middle Ages Germany, France, Italy, and also Greece, became centres for processing of raw pelts. *Squirrel* pelts were extensively traded by the Hanseatic merchants along the Baltic Sea coast as early as the 14th and 15th centuries (4). Coats made out of *ermine* fur were a symbol of royalty, and the wearing of hats and trimmings of *beaver* fur the privilege of rich merchants in the 18th and 19th centuries. The special craft of furriery developed and refined the production of high quality furs into an art, particularly in Leipzig. It took centuries of trial and error before special processes were developed that were compatible with the objective of producing such furs for the varying fashion demands. Only with the increasing cost of skilled labour after the Second World War did fur production gradually shift to countries with low levels of labour cost, such as Greece, Korea, China, Hong Kong, the Ukraine, and Argentina, some of which were also producers of raw pelts. Therefore the trade in raw or dressed furskins into Europe regressed and was largely replaced by importation of semi-finished products such as fur plates and linings, from these countries. The fur industry of Canada, as one of the world's greatest producers of raw pelts, has, however, retained its major importance (10) and is the leading supplier of fur apparel to the North American market (2).

Less valuable furskins may simply serve utility purposes in the form of overcoats and jackets, as lining in warm clothing, parkas, caps and boots, or as bedspreads. Most furskins, however, are processed for decorative or fashion purposes and utilised as trimmings, collars, stoles or prestigious capes and coats. Valuable and particularly conspicuous furs are often perceived as a status symbol of wealthy snobs and have thus at times provoked strong public resentment. This has in some cases brought about political pressures against the fur trade, although this is as legitimate and important as any other industry which utilises raw materials derived from wild animals.

Some furskins are plucked or shorn in the process of fur production. Hair from these and the pelts of *Wild rabbits* and *hares* (mainly from Australia) are used for felt production; the tail hair of some *mustelid species* may also serve for paint brushes. Some bovids and all camelids shed their wooly winter hair and this is locally collected from semi-domesticated species such as *yak, muskox, camel* and *dromedary* for weaving. The only hair, however, which is commercialised internationally because of its extreme fineness is *vicuña* wool and to a lesser extent also *guanaco* hair, as described in Chapt. 2.3.11.

## Pelt preparation and processing

The pelt preparation process includes the choice and application of most appropriate trapping methods, trap maintenance and frequent checking, cleaning and careful handling of the carcass, and skinning and processing of the raw pelts. Neglect at any of these stages may result in 25-50% loss of value to the trapper at grading (6). The skinning of wild animals for the purpose of furpelt production requires much greater care than that described for gameskins and hides in Chapt. 3.2.3. It is carried out in two ways, according to the different species trapped: open removal of the pelts by cutting the skin lengthwise down the belly (*beaver, badger, some wolves, bears*) or casing of pelts by cutting the skin from rear heel to heel across the anus and turning the inside out (*muskrat, nutria, squirrels, Wild mink, raccoon, oppossum*). The hairsides of cased pelts of some species (*coyote, foxes, lynxes, marten,* most *fisher* and some *badgers* and *wolves*) are turned out again during the drying process (6). The casing of pelts protects the hair during the removal of flesh and allows the drying of the pelts in a stretched natural form on species-sized boards or iron frames. Whereas raw skins destined for leather production are usually wet salted or preserved chemically, furpelts are only airdried. Most furskins are marketed in the form of such airdried raw pelts.

Production of fur garments from farm mink pelts. (Photo: Deutsches Pelz Institut)

During further industrial processing the pelts are first "dressed", i.e. they are softened by soaking in brine, washed, and their subcutaneous connective tissue is meticulously removed. The pelts are then "pickled", i.e. they are submerged into a solution of salt and acid. The pickled furskins are tanned with an addition of certain lipids which penetrate into the furskin and ensure its permanent elasticity. After a drying process the tanned furskins are placed into slowly turning drums filled with sawdust and fanned with warm air. Thereby the hair is loosened and superfluous fat extracted. Ultimately, the leather of the pelts has to be manually cut thinner and stretched into its original form.

The processed furskins are matched into bundles of similar looking pelts which requires very special experience (10). The bundles are then cut and joined into a variety of fur products. The usage of fur types and garment styles depends very much on fashion demand which often appears to be cyclic, i.e. a particular type of fur and its usage may again become fashionable after decades of other preferences (10). Generally the processing of furskins and production of high-quality fur garments involves much skilled hand labour which makes it costly in industrialised countries, as described above.

## Marketing and trade

The marketing of raw pelts for fur production was initially to a large extent a monopoly of a few large trading enterprises such as the Hudson Bay Company in London, and the Anglo-Russian Muskovy which exploited the Siberian furbearer resources through Arkhangelsk until the 17th century. Thereafter, the fair in Nishnij Novgorod became the most important centre for Russian fur sales until after the Russian Revolution Sojuzpushina, based in Leningrad and Moscow, was created as a State monopoly. The Hudson Bay Company introduced auctioning for the fast marketing of perishable raw skins. Later other trading companies like Eastwood and Holt (London), Trappers Union (North Bay) and Dalgety

(Dunedin), established themselves and competed in the marketing process. Since the 1960s, however, many producers have exported their raw pelts directly. The most important producer probably was the former USSR, but the USA, Canada and Australia are the most important exporters. With the above-described transfer of the processing technology some of the smaller producers, like Argentina and China, are now processing their raw pelts locally into semi-finished products such as plates and linings, which they then export. The most important exporters of finished fur apparel into the USA in 1993 were Canada with a total import value of US$ 43.8 mill., Hong Kong with US$ 38.4 mill., China with US$ 27.7 mill. and Greece with US$ 24.7 mill. (2).

**Table 55:** Relative commercial importance of furs from wild animal species

| Animal order (see Part II) | Relative commerical importance of furs[1)] | | | | | |
|---|---|---|---|---|---|---|
| | Great[2)] | CITES | Fair[2)] | CITES | Little[2)] | CITES |
| **Marsupialia** | | | oppossum<br>Brushtail possum | -<br>- | wallabies<br>kangaroos | -<br>- |
| **Insectivora** | | | | | mole | - |
| **Lagomorpha** | | | Wild rabbit | - | Snow hare | - |
| **Rodentia** | nutria[3)], muskrat<br>Can. beaver[3)]<br>chinchilla[4)] | -<br>-<br>- | Sib./Red squirrel<br>bobak, suslik | -<br>- | hamster, burunduk<br>chipmunk, viscacha | -<br>- |
| **Carnivora** | | | Amer. martens[5)] | - | zorrino | III |
| **-mustelids** | mink[3)], sable[3)] | - | fitch[6)], Striped | | Chin. weasel | III |
| | | | skunk | - | fisher[7)], Amer. | |
| | | | River otters | II | badger | - |
| **-raccoons/bears** | raccoon | - | | | Black bear | - |
| | | | | | Polar bear | II |
| **-dogs/foxes** | Red fox mutants[3)] | - | Wild Red fox and | | Patagonian fox | II |
| | Arctic fox[3)] | - | coyote | - | other foxes | - |
| | | | Raccoon dog[3)] | - | | |
| **-vivverids** | | | | | Ind. Civet cats | III |
| **-cats** | | | Can. lynx/bobcat | II | other lynxes | II |
| | | | Leopard cat[8)] | II | some "spotted cats" | II |
| **Pinnipedia** | | | Fur seals | II | seals | II |
| **Artiodactyla** | | - | reindeer[3)] | - | guanaquito | II |
| **Primates** | | | | | guereza[9)] | II |

CITES = trade restrictions in terms of the Convention on International Traffic in Endangered Species of Wild Fauna and Flora

1) in terms of both numbers traded and individual monetary value; 2) scientific names see overview tables in the respective chapters of Part II; 3) both wild and farmed; 4) only farmed; 5) *Martes americana* (Can. sable), *-foina* (Stone marten); 6) Pole cat; 7) *Martes pennanti*; 8) *Prionailurus bengalensis* sspp.; 9) *Colobus* spp.

In Tab. 55 an attempt is made to classify commonly marketed furskins according to their commercial importance, both in terms of trade volume and individual value. Of greatest economic importance is the trade in *mink* skins and Fig. 20 shows that almost 42 million were marketed in 1988. This production, however, exceeded demand and as a result prices crashed with production declining to about 20 million in 1993. Only about 430,000 mink pelts are derived from wild populations in the USA and Canada (7, 9), the remainder come from farms in 26 countries all over the world (3). These offer a large number of different colour varieties, from white to beige and grey; more than half the entire international fur business is now based on mink. Therefore, the annual auction prices for mink raw pelts more or less guide and regulate all trade in furs.

Next in quantitative importance are *muskrat, nutria* and *raccoon* pelts of which about 20 million are annually traded and most of these still originate from wild or feral populations (9). Fox pelts, of which about 6 million were marketed in 1987/88 (see Fig. 20) comprise mainly the blue colour phase of the *Arctic fox*, and the silver mutant of the *Red fox*. About 4.5 million of these were produced in farms in Scandanavia and about 1.0 million in Poland (1). A large number of Wild Red fox pelts are derived from the control of this species in the USA (445,000 in 1982/83) and Australia (351,000 in 1982/83) (8), and from hunting in Germany and Eastern Europe.

**Figure 20:** Origin and numbers of farm pelts offered on "western" markets in 1983-1993 (after ANONYMOUS (1))

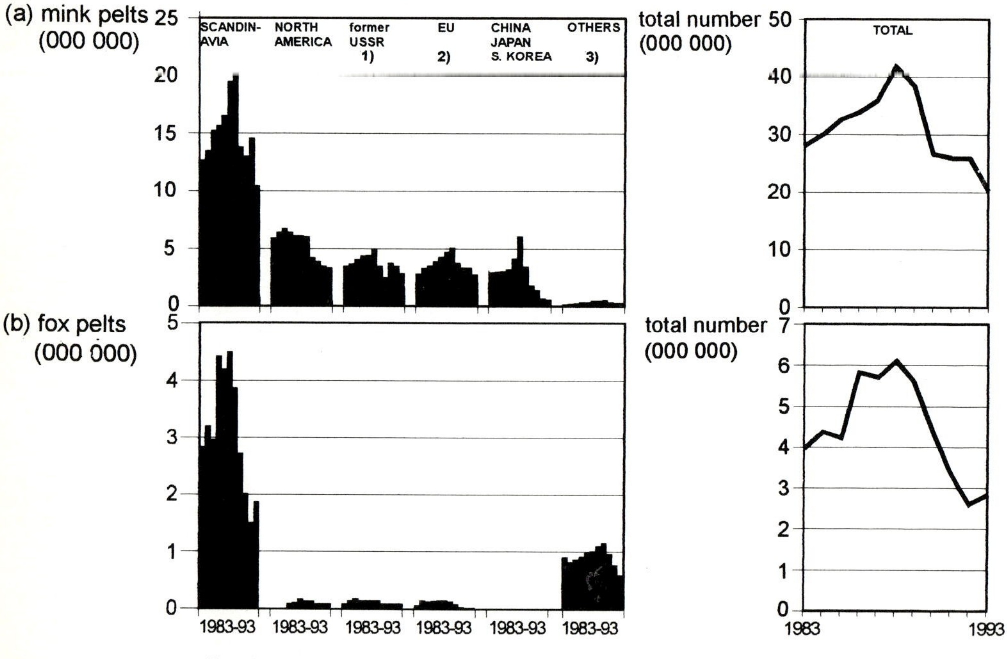


1) only exports to western markets according to auction data; internal production was much higher (see Table 29); 2) Germany, France, England, Ireland, Holland, Belgium, Italy, Spain (in order of importance); 3) Poland, Argentina and others.

*Opossum* pelts are also produced in large numbers in North and South America, about 1.8 million per annum (9, 5), but only a small percentage of these are traded internationally. Confused with the opossum are sometimes *Brushtail possums* which are hunted, and more recently also farmed, in Tasmania and New Zealand, as discussed in Chapt. 2.3.1. The export of Brushtail possum pelts has gained some economic importance with about 1.5-2.0 mill. exported per annum in recent years. The majority of these are processed in Korea and

Hong Kong into trimmings and linings but some are imported directly by EU countries amongst other uses for industrial applications.

Of particularly high value are *chinchilla* and *sable* pelts. Best quality chinchilla pelts may fetch US$ 50, but sable pelts of the black variety may have a value of up to US$ 400/skin. *American martens* are also traded at high prices as " Canadian sable". The most valuable pelts from North American trapping are *lynxes* which averaged Can.$ 205 (approx. US$ 166) per skin in 1982 (3). Trade in *beaver* skins had great economic importance until the second half of the 19th century, then nutria fur started to replace beaver as discussed in Chapt. 2.3.3. Now, only limited numbers, about 650,000/annum in 1983/84, are exported from the USA and Canada from strictly managed wild populations (9) and some beaver ranches (3) to highly specialised fur producers in Argentina, Germany and Italy. Similarly, the formerly important trade in *squirrel* skins has declined, and the Siberian and Canadian species have completely replaced the European squirrel.

The *Asian bobak marmot* is traded in large numbers from Mongolia, but is only used to simulate more expensive furs. Also the various *Asian ground squirrel* species are traded in significant numbers under the name of "suslik" but produce only second rate fur. Australia and New Zealand export large numbers of *Wild rabbit* and *wallaby* furskins. These are of lesser importance to the fur industry, because most of them are either used for felt or, like *reindeer* skins, for leather garment production. The trade in *seal* pelts was of considerable economic importance until the 1960s. Since then it has rapidly declined to insignificance mainly due to subjective pressures by animal welfare groups, as discussed in Chapt. 2.3.5.

Other furbearing wild animals have very limited economic importance. Many of the *Large* and *Spotted cats* are excluded from international trade, i.e. they are listed on Appendix I of the Convention on International Trade in Endangered Species of Wild Fauna and Flora (CITES). The market for Small cat furs boomed throughout the late 1950s and 1970s, and

**Figure 21:** Net trade in furskins of Latin American Small cats in 1976-1985 (according to CITES Country Reports from FITZGERALD ((12) in Chapt. 3.2.5))

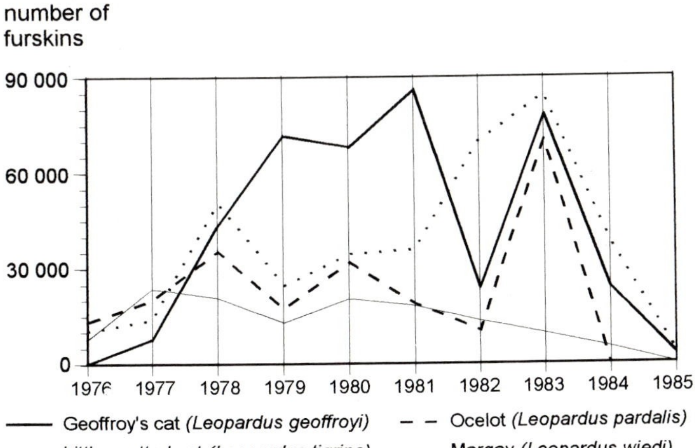

in 1980 European nations alone imported an estimated 430,000 skins per year (11). Fig. 21 shows the net trade in skins of the most important Latin American species in 1976-85 and how this collapsed as a result of the trade ban. All other fairly important furbearers, mainly *Asian Small cats*, such as the *Tiger cat*, the *North American lynxes*, and such species as the *River otters, Argentinian foxes* and *zorrino, Fur seals* and the *guanaco*, are listed in Appendix II, and their trade is therefore controlled by the countries of origin.

Some other species, such as the *ermines, Asian weasels,* and *Yellow-throated marten,* have been placed by India on Appendix III and require an export permit in that country, although China is the most important producer of *Siberian* or *Kolinsky weasel* fur plates. Generally the provisions of CITES have been further supplemented and detailed by national legislation in most of the importing countries, like the USA and the EU, curtailing the fur trade even further. It should be noted, however, that furbearing animals in many countries constitute an important part of their wildlife resources, and that the export of furskins may be an essential element of their national economy. In Argentina, as an example of a major furskin producer, the total value of hides, skins and fur products derived from wildlife in 1976-79 equalled 51% of the value of the beef export, and 92% of the horse meat, mutton and pork exports together (5).

## Literature references


1. ANONYMOUS (1993): *World production for offering at the "western" markets*. Official Auction Statistics; Oslo Fur Auctions, Oslo, Norway
2. ANONYMOUS (1994): *U.S. Fur apparel imports*. Statistics of the US Department of Commerce; Washington, DC
3. BARZDO, J., LUXMOORE, R., BROAD, S. and D. JONES (1984): *A survey of wildlife farming operations*. WCMC/IUCN; Cambridge (unpubl. report)
4. BRASS, E. (1925): *Aus dem Reich der Pelze*. Verlag Neue Pelzwaren - und Kürschner-Zeitung, Berlin
5. FUJITA, H.O. and J.O. CALVO (1981): Las exportaciones de productos y subproductos de la fauna silvestre en el quinquenio 1976/1980. *IDIA*; No. 397-400; Instituto Nacional de Technologia Agropecuaria (INTA), Buenos Aires
6. HALL, E.G. and M.E. OBBARD (1987): 55. Pelt Preparation. In: *Wild Furbearer Management and Conservation in North America*, ed. NOVAK, M., BAKER, J.R., OBBARD, M.E. and B. MALLOCH; p. 843-861; Ministry of Natural Resources, Ontario, Canada
7. NOVAK, M., OBBARD, M.E., JONES, J.G., NEWMAN, R., BOOTH, A., SATTERTHWAITE, A. and G. LINSCOMBE (1987): *Furbearer Harvests in North America*, 1600-1984. Supplement to *Wild Furbearer Management and Conservation in North America* (see (6))
8. RAMSEY, B.J. (1994): *Commercial use of wild animals in Australia.* Bureau of Resource Sciences, Australian Government Publishing Services; Canberra, Australia
9. SHIEFF, A. and J.A. BAKER (1987): *Marketing and International Fur Markets* In: *Wild Furbearer Management and Conservation in North America*, ed. NOVAK, M., BAKER, J.R., OBBARD, M.E. and B. MALLOCH; p. 862-877; Ministry of Natural Resources, Ontario, Canada
10. SCHIPPER, S.S. (1987): 57. Garment Manufacture. In: *Wild Furbearer Management and Conservation in North America*, ed. NOVAK, M., BAKER; J.R., OBBARD, M.E. and B. MALLOCH; p. 878-888; Ministry of Natural Resources, Ontario, Canada
11. WILLIAMS, T. (1985): Small cats: Forgotten, Exploited. *Audubon*, 87(6), p. 34-40

## 3.2.5 FEATHERS

Feathers are epidermal products of the birds' skin consisting predominantly of keratin. Similar to the hair of mammals, the shaft of a feather grows from an inversion of the epidermis into the dermis and becomes the axis of the feather vane. This is structured diversely according to the feather's function and may contain a variety of melanins and lipochromes which give the feather its specific coloration. Feathers which are particularly decorative are usually called plumes whereas the short woolly feathers which keep the bird's body warm are called dunes or down. Like hair, feathers are regularly shed and renewed.

### Production and uses

**Feathers for decorative or ornamental purposes** are usually obtained by hunting or trapping of the birds for food or sport, utilising their plumage only as a secondary product; this still applies to *birds of prey, galliforms* and *psittacines*. However, in cases of feathers with a particularly high monetary value their commercialisation in the past has been a primary reason for large-scale killing of certain bird species. In the 19th century this use contributed significantly to the extirpation or local diminution of some bird species, like, for example, *Wild ostriches* in Western Asia and North Africa, *Great White egrets* in South-East Europe and Yellow-billed egrets in South America, to name only a few. In fact, this abuse of birds and the drive to stop it led to the formation of the first major national and international bird conservation bodies in the 1890s (10, 9). Today, the commercial use of feathers is almost entirely based on the collection of moulted feathers from captive-held birds, the plucking of farmed species and the harvesting of down from eiderduck nests. In ornamental as well as utility usage, feathers of domestic fowl, often dyed, have widely replaced those of wild birds. However, some species are still hunted in numbers primarily for ornamental use of their feathers, such as for example, the *birds of paradise* (*Paradisaeidae*) in Papua New Guinea (8, 13), the *Scarlet ibis* (*Eudocimus ruber*) in the Guyanas (12), and *macaws* (*Anodorhynchus* and *Ara* spp.) in South America (21). With the reviving demand in the USA for specific feathers for ornamentation or tying of fish flies, as mentioned below, some feathers are unfortunately also procured through illegal hunting. In 1988, for example, the US Fish and Wildlife Service filed charges against 41 persons of a poaching network extending over nine states involved in the killing and sale of more than 600 protected birds of over 50 species (4).

It is interesting to note in this context that some native tribes in South America were always well aware of the need for sustainability of the feather use. They only immobilised their quarries and released them after plucking some of the desired feathers (17). Others kept the procurer bird species, such as *hokkos, harpyies, macaws* and *amazones*, in captivity to be able to repeatably harvest their feathers. The demand for *peacock* and *ostrich* feathers was undoubtedly the main reason for the early semi-domestication of these birds. Another sustainable form of utilisation developed in South America to produce *egret* plumes. Private landowners in Columbia and Venezuela leased out breeding colonies on their lands, so-called "garzeros", for collection of egret feathers only during the moulting season when these are naturally shed and drop down into the swamp (17).

Historically, feathers of almost any bird have been extensively used by man since earliest times for a great variety of purposes (17). In colder regions feathered birdskins were used like furs for clothing and bedding; in warmer regions feathers served mainly for decoration, but also for a variety of utility purposes. In some cultures, feathers, like those of the *quetzal* (*Pharomachrus mocino*) in Central America or of the *Paradise birds (Paradisaeidae)* in New Guinea, assumed the function of money. Many bird species were worshipped and their feathers used for ritual purposes, for example various *birds of prey, owls* and the *Sacred ibis* (*Threskiornis aethiopicus*) in Ancient Egypt. These and specimens of many other bird species were mummified and buried together with important persons. Some early cultures developed the decorative use of feathers to an artistic level, such as the Aztecs and Mayas in Mexico, some of the South American Indians and the Polynesians. In many cultures the use of feathers had symbolic importance, certain feathers indicating the status and achievements of its bearer. For the North American Indians the use of feathers also had intricate religious significance, birds being considered as the messengers between man and the Gods. Many of the customary and ritual uses of feathers in aboriginal societies have survived to-date.

However, also in Western Cultures the use of plumes for ornamental purposes has a long history. It was initially introduced from the orient where *peacock* feathers were widely used to indicate the status of the bearer. They were worn in bunches on hats or processed into decorative mosaics. The use of peacock feathers was adopted by the Romans and later on introduced to Western Europe by the crusading knights. All through the Middle Ages peacock feathers together with the white plumes of *egrets* or *herons* played an important role until they were replaced by the *ostrich* feather which could be more easily bleached and dyed. During the second half of the 19th century feathers became a general fashion article and their use developed into a veritable "ars plumaria" (16). A great variety of bird feathers mainly from *galliform, colubriform, falconiform, psittacine* and *passerine species* were commercialised and at the turn of the century the feather trade grew into an economically quite important activity.

The First World War drastically changed attitudes and fashion, and this resulted in an abrupt decline of the use of feathers for ornamental and fashion purposes. Demand for bird feathers diminished to such an extent that the well-developed ostrich farming industry in South Africa crashed, as described in Chapt. 2.2.1. As a result, the quite extensive European industry for crafting decorative feather fashion articles and gadgets, such as feather stoles, boas, fans and artificial flower ornaments made of feathers, declined to economic insignificance. The craft of the "plumassier" has been slowly dying out since because of lack of this specific fashion demand. Nevertheless, since the 1950s the market for feathers in general has picked up again, partly because of **new uses of feathers in industrial applications**; a large proportion of the farmed ostrich feathers, for example, are used as dusters mainly in the automobile and computer industries.

Partly, however, increasing demand is also a result of the **revival of specific human ornamentation habits**. In the USA the Pueblo Indians in Arizona and New Mexico, for example, have been using feathers of no less than 219 different bird species (24) including some which are now endangered. In the wake of the developing racial self-consciousness and pride, the Indian population is also reviving its traditional and religious use of feathers and this has given rise to the development of commercial production and trade in Indian feather artefacts and replicas. However, also in Europe numerous clubs have been formed which foster traditional ways of living and these have become a new market for ornamental feathers. Curio production for tourists is another area of increasing use of ornamental feathers.

Another traditional usage of feathers is the **tying or dressing of fish flies**, mainly for trout and salmon fishing. This has been a tradition in the United Kingdom since the Victorian era (14) and used to involve feathers from a great variety of exotic birds from all over the British Empire, the hackle feathers of the now rare *Gray Jungle fowl* (*Gallus sonnerati*) being amongst the most popular. Flies were "created" individually by highly specialised craftsmen as works of art and were given specific names, such as "Thunder and Lightning", "Silver Doctor", "Bronze Pirate" and the popular "Fairy King". These Victorian flies were not only dressed with feathers, often matched from different areas of the bird's body or dyed, but also contained various other materials such as silk, tinsel and pieces of fur. After the turn of the century simpler fly patterns and practicality took over, but traditional tying of Atlantic salmon flies has revived in the USA since the 1950s together with the increasing popularity of the sport of fly fishing. In the 1980s the number of professional Atlantic Salmon fly tyers was estimated at 500-750 persons and the Federation of Fly Fishers alone had over 10,000 members holding annual conclaves and producing several professional periodicals (20). However, fish flies are not only used for fishing but have become - as so many other items - the object of wealthy collectors. These do not only pay large sums for new artistic designs of flies mounted in shadow boxes for display, but often add complete stuffed specimens of the birds whose feathers have been used, to their collections. The fly tying has thus become a rather controversial subject, accused by conservationists of stimulating and supporting a black market in endangered bird skins and feathers (20). On the other hand, the demand for feathers for fly tying has induced the commercial breeding in captivity of various rare or endangered species, such as the *Red, Gray, Ceylon* and *Green Jungle fowl* (7, 20).

An entirely different use of birds' feathers is the **collection of down** for making very light and warm bed covers and lined garments. Down are shaftless feathers of a particularly fine structure which cover the bird's body underneath the contour feathers. Although domestic geese have always been the main source of bed down, historically, a wide range of *anseriform, lariform* and *other seabirds* were also utilised for this purpose (see Tab. 21). Down of the highest quality is obtained from the *Common eiderduck* (*Somateria mollissima*) which breeds in very large numbers on the coasts of Greenland, Iceland, Norway, Spitzbergen, and Northwestern Siberia. The females pluck their own belly down and stuff the nests with it, covering also the eggs with down when they leave the nest. Therefore, the use of eiders can be sustainable if collection is restricted to nests in which the young have already hatched. In this case the female bird is induced to renewed plucking and stuffing of the nest and the rearing of the young is not endangered. Unfortunately, collection of eider down in the past has not always taken account of the conservation aspect. Breeding colonies were raided several times a year, destroying nests and eggs, and this has greatly reduced eiderduck populations. Efforts to protect the eider breeding grounds and to regulate down collection were initiated in Northern Norway as early as the 1890s (18). Islands on which eiderduck were breeding were leased out by their owners and resident fisher families pegged out their territories for down collection. In some areas, like Iceland, artificial nests are offered and the eiders are virtually "farmed" (11). Although the use of eider down has greatly diminished, production in some areas has retained economic importance as a source of cash income for the local population. Collection, today, is controlled through the issue of licenses which in turn qualify for obtaining export permits. One nest yields only about 20 g of down and about 24 nests might produce 1 kg of raw down. However, about 6-6.5 kg of raw down are required to produce 1 kg of cleaned down (18).

## Marketing of raw feathers and processing

For the now limited commercial use of the feathers of smaller birds, such as the *Paradisaeidae*, these were skinned on the spot by the hunter, leaving all feathers and also the beak and skull on the skin. The feathered skins would be locally preserved by drying and/or chemical treatment. Large batches of usually stuffed feathered skins used to be collected by local, often Chinese, traders for shipment to Europe. For larger birds only wing, tail or neck feathers are plucked for sale, and down is collected from the nests into larger quantities of 1.5-2 kg. All feathers are locally dried, roughly cleaned, and graded before they are shipped in bags to the consumer countries. Before the First World War London and to a lesser extent also Paris were the main centres of the raw feather trade. Great quantities of feathers and feathered skins of an incredible *variety of birds* were annually auctioned, similar to the sale of furskins. Now, feathers are usually sold directly or through intermediary traders who buy feathers from individual producers, such as sportsmen and bird breeders, to the few remaining millinery factories and craftshops specialised in the processing and crafting of feathers for ornamental purposes or in the dressing of fish flies. Industrial consumers usually import directly from overseas producers such as ostrich farm and down-marketing cooperatives.

As was the case for the furrier trade **processing of feathers** used to be a special industry involving highly sophisticated skills. Just around Paris some 600 factories and workshops existed in 1914 for the processing and crafting of feathers which engaged about 50,000 female labourers (15). Various machines were developed for cleaning, washing, bleaching, dying and drying of feathers. Depending on the end product the feathers would be steam heated before being drawn into any fancy form. Only very few of these specialised factories have survived into the present time, usually as a sideline of the millinery industry. In Austria and Germany for example, formerly most important countries for feather processing, there are now less than a dozen establishments which supply various types of plumes for hats, but also ornamental feather products such as carnival and dancing accessories.

*Eider* and *swan* down have been more and more replaced by *geese* and *duck* down which have become the raw material base of a still economically important industry. However, only a very few establishments still process eider down, mainly because of its prohibitive price. This fluctuated in Germany, one of the main consumers, between DM 550 and 1,000 per kg (approx. US$ 370-700) for down imported from Canada, Iceland or Finland. Although, compared with geese, 60% less down is required for making super light eider down duvets or blankets, these are a luxury good which only a few can afford.

## Trade

Induced by the increasing affluence during the Victorian era and fashion demands, the feather trade reached an all-time **peak before the First World War**. Annual world consumption of feathers for ornamental purposes at that time was estimated at 300 million featherskins (19). France alone imported about 50,000 t of foreign bird feathers from 1890 to 1929, by which time the market had slumped to insignificance (13). At just one auction in London in 1910 as many as 600,000 featherskins of *smaller birds* were offered amongst which the *birds of paradise* featured strongly (22). Of greatest importance, however, were *ostrich* feathers, followed by those of *egrets* and *herons* which were called in the feather trade "aigrette", "crosse" or misleadingly "osprey". In 1913 South Africa's ostrich farms exported about 512 t of feathers which were worth almost £st 3 mill. at that time, as mentioned in Chapt. 2.2.1. Australia and Argentina also exported some ostrich feathers, the latter 27 t in 1908 (1, 15). Trade in *rhea* feathers, called "vauteur", predominantly from *Rhea americana*, was also of economic importance, although rhea feathers commanded lower prices than those from ostriches (1); this trade fluctuated at the turn of the century between 45 and 100 t per annum of which an estimated 40% originated from farmed birds in Argentina (17). France alone consumed about 45 t in 1912 (16). The much narrower *emu* feathers were considered inferior in quality and traded sometimes erroneously as *cassowary* feathers. The latter were mainly used by the Papuas for ornamental purposes and rarely traded internationally (see Chapt. 2.2.1).

After the **decline of the feather fashion** from 1914 onwards, as mentioned above, only few species continued to be traded in important numbers such as *birds of paradise*. Hunting of these for plumes in New Guinea remained to be deeply entrenched to-date as a social and economic activity of the indigenous people (8); thus the feathers of these beautiful birds are still traded although all of them were placed on Appendix II of CITES. The volume of **present-day traffic in feathers** is rather difficult to assess, but is generally rather limited. Most of the plumes still used for decorative or ornamentation purposes originate either from non-endangered species, captive bred or farmed bird stocks. A review of the international traffic in feathers of endangered species recorded under the provisions of CITES, shows that ***net world imports*** are negligible compared with the volume of the trade in live birds (5). During 1983-88 feathers from *galliform birds*, mainly captive bred *Gallus sonnerati*, the *American rhea, falconiform* and *strigiform birds* dominated, followed by a very limited trade in feathers from *psittacines, birds of paradise, flamingos* and *cranes*. In terms of weights, a total of 2,368 kg of rhea feathers were traded whereas in other groups totals did not exceed some hundreds of kilograms over a six year period. As regards the birds of paradise, 360 "bodies" or "specimens" were traded legally in addition to 366 unit-unspecified export/imports. Feathers of some particularly brightly coloured birds, such as *macaws* in Brazil and the *Scarlet ibis* in the Guyanas, may now be processed locally and crafted into flower-shaped ornaments for sale to tourists who then export these (25, 6). Just one art centre in French Guiana, for example, was found in 1987 to sell 20,000 such feather ornaments each year, including about 1,000 of which were made of the primary feathers of the Scarlet ibis. Considering that an ibis has only 8 of these feathers, it could take as many as 1,500 birds to just supply that single variety of ornament (3). In order to curb this clearly destructive over-exploitation, the species was placed in 1987 on Appendix II of CITES.

As with most other wildlife products there is, of course, ***illicit international traffic in ornamental feathers***, as reported for *birds of paradise* (13) and documented recently for *psittacines* (*Ara* spp., *Anodorhynchus hyacinthus*), *Harpy eagle* (*Harpia harpya*), *toucans* (*Ramphastos* spp.), and *Roseate spoonbills* (*Ajaia ajaja*) (6). Based on the local procurement of feathers, as mentioned above, there is also ***internal trade in ornamental feathers*** through intermediary dealers. One of these in the USA, for example, has been selling feathers worth about US$ 25,000 per annum, which were exclusively obtained from aviculturists against payments of up to US$ 15 per tail feather depending on the species. Well-dressed fish flies were sold in 1987 at US$50-60, but special creations for collectors may fetch hundreds or even thousands of dollars (20).

The ***ostrich feather trade*** had diminished by 1929 to some 38 t, worth only £st 43,000 (2). Only after the Second World War did it revive with other technical uses of ostrich feathers added, as mentioned above. In 1962 the value of feathers exported by the South African ostrich industry was 77% of its total production, reported at R 1.3 mill. (approx. US$ 1.8 mill.) (23). As described in Chapt. 2.2.1, ostrich and emu farming have since developed as a predominantly leather and meat producing industry. Although only a by-product, the offer of feathers from this industry is bound to greatly increase once this has reached full production level.

Mainly as a result of urbanisation in Europe ***trade in bed down*** generally trebled during the first decade of the century. In contrast to other feather trades it was little affected by the First World War and in 1925 had again reached the pre-war level. However, the use of feathers from wild birds for filling of pillows was completely, and that of down largely, replaced by domestic fowl. Therefore, but also because of the misuse of the eiderduck breeding grounds, discussed above, trade in eider down diminished to only a fraction of the important volumes commercialised around the turn of the century. Annual imports at that time from Greenland and Iceland alone, which produced the best down, into Denmark were about 3,000-4,000 kg, other producer countries being Great Britain, Norway and Russia (17). Present imports, mainly from Canada, Iceland and Finland into Germany, for example, fluctuate between just 600 and 800 kg/annum.

## Literature references


1. ANONYMOUS (1914): Straussenzucht in British Afrika, Amerika, Australien. *Deutsche Kolonialzeitung*, 48, p. 514
2. ANONYMOUS (1928): *Official Yearbook of the Union and of Basutoland, Bechuana Protectorate and Swaziland. No. 10, 1927/28* (cited after NEUNZIG 1929/33)
3. ANONYMOUS (1987): *Positions on proposals to amend the Appendices of CITES at the Sixth Meeting of the Conference of the Parties;* TRAFFIC Network (cited after FITZGERALD 1989)
4. ANONYMOUS (1988): News and announcements. *Ornithological Newsletter,* No. 61; American Ornithological Society, Columbus, Ohio, USA
5. ANONYMOUS (1991): *Review of significant trade in species listed on Appendix II of CITES, 1983-1988.* WCMC, Cambridge (unpubl. background data document)
6. ANONYMOUS (1995): Seizures and prosecutions: France. *TRAFFIC Bulletin,* 15(2 and 3), p. 92 and 116

7. BRAMELD, D. (1985): Return of the Jungle Cock: The un-endangered gray jungle fowl. *Fly fishing*, Jan. 1985, p. 30-32
8. DAWNES, M.C. (1969): *Report of the Wildlife Ecology Section 1968/69*. Department of Agriculture, Stock and Fisheries, Papua New Guinea; 12 pp. (unpubl. report)
9. DIAMOND, A.W. (1987): Uses of Birds: A global view. In: *The Value of Birds*, ed DIAMOND, A.W. and F.L. FILION; p. 103-104; ICBP Technical Publication, No. 6, Cambridge
10. DOUGHTY, R.W. (1975): *Feather Fashions and Bird Preservation*. University of California Press, Berkley, Ca., USA
11. DOUGHTY, R.W. (1979): Eider husbandry in the North Atlantic: trends and prospects. *Polar Record*, 19, p. 447-459
12. FITZGERALD, S. (1989): *International Wildlife Trade: Whose Business is it?* WWF (USA), Washington, DC
13. INSKIPP, T. and S. WELLS (1979): *International trade in wildlife*. Earthscan, London
14. KELSON, G.M. (1896): *The Salmon Fly*. London
15. KNAUER, E. (1914): *Vogelschutz und Federindustrie. Eine Streitfrage der Zeit.* W. Braumüller, Wien and Leipzig
16. LEFEVRE, E. (1914): *Le commerce et l'industrie de la plume pour pàrure*. Chez l'auteur, Paris
17. NEUNZIG, R. (1929-33): Vogelbälge und Federn. In: *Die Rohstoffe des Tierreichs*, ed. PAX, F. and W. ARNDT; Vol. 1, Part 1, p. 568-845; Gebr. Borntraeger, Berlin
18. PEARSON, H.J. and E. BIDWELL (1894): On birds'-nesting to the north of Norway in 1893. *The Ibis*, 6, p. 226-238
19. REINHARDT, L. (1912): *Die Erde und die Kultur*. Bd. III. Kulturgeschichte der Nutztiere. Ernst Reinhardt, München
20. SCHNEIDER, K. (1987): Director of Feather Way, Lakeside, Ca. USA - *in litt.*
21. SCHNEIDER, K. (1988): *The last feathers*. Feather Way, Lakeside, Ca., USA
22. SCHWARTZ, M. (1911): Sind internationale Vereinbarungen zum Schutz solcher Vögel erwünscht, deren Bestand dadurch gefährdet ist, dass man sie ihrer Schmuckfedern wegen verfolgt? In: *Proceedings of the Vth International Congress of Ornithologists*, Berlin 1910; p. 839-846
23. SMIT, D.G.v.Z. (1963): *Ostrich Farming in the Little Karoo*. Deparment of Agricultural Technical Services, Bulletin No. 358; Pretoria, South Africa
24. TYLER, H.A. (1982): Pueblo Birds and Myths (*cited after* SCHNEIDER 1988)
25. VILLALBA-MACIAS, J.S. (undated): pers. comm. (*cited after* FITZGERALD 1989)

## 3.2.6 TROPHIES AND CURIOS

A trophy, in the narrow sense of the word, is defined as "something won or given in victory, especially when preserved or mounted as memorial". For a hunter it is a treasured memento of daring moments of pursuit and overcoming of a wild beast, inspiring pride and joy at its beauty. Historically, however, animal trophies mainly constituted charms. They had a strong spiritual or magical significance, being displayed or carried either to please the deities or spirits into awarding hunting fortune, or to acquire some of the desired properties of the killed animal, such as power, velocity, courageousness and so on. Much of this original feeling about trophies has continued into modern times, becoming psychologically coupled with vanity and status symbolism. Therefore, trophies as wildlife products are here discussed together with the various types of "curios" commercially derived from wild animals.

### Development of the use of trophies

The oldest animal trophies known are the skulls of the *Cave bears* found at Stone Age sites of Central Europe, dating back at least 100,000 years (4). The custom of offering trophies to the deities continued in Europe into the early Mediaeval Age, but it is still practised in aboriginal societies in Africa, Australia and South America. As with the Germanic hunters who would string up skulls of the *aurochs* and *bison* as sacrifices and signs of hunting success, indigenous hunters in Africa may still do this with *buffalo* skulls. Similarly, decorative wearing of certain animal parts as token of status has been continued into modern times. Just as ancient hunters would wear *aurochs* horns, *boar* tusks or *bear* teeth, so modern hunters like to decorate themselves, in Europe with certain *bird* feathers and the "beards" of *chamois*, and in Africa with *leopard* skin bands around their hats and *elephant* hair bracelets. Specific *fur* robes were indicative of royalty in Europe until the 19th century, head decorations with *eagle* feathers a sign of chieftaincy amongst the North American Indians, and display of *elephant* tusks the prerogative of important chiefs and now of heads of state and ministers in Africa.

In Europe, Christianity probably suppressed the mystically oriented trophy cult until the scientific interest in wild animals and their trophies emerged during the 18th and 19th centuries. Naturally this focused on the indigenous game animals which had become abundant, and it became standard practice of hunters to collect and mount any *deer* antlers for trophies (5). The emerging interest in natural history also prompted the development of taxidermy, the "art of preparing, stuffing, and mounting of skins of animals in a life-like manner". The British Museum displayed its first, rather unpolished, mounts in 1753, and many other mounts were seen in Europe during the 19th century (6). Around 1850 the Rowland Ward studio was founded in London and became a focal point not only for taxidermy but generally for the collection and evaluation of trophies from *deer* and *horned bovids* world-wide. At the end of the 19th century it started publishing in regular intervals record measurements of big game trophies (7). With the development of sports hunting and the periodic convening of international trophy exhibitions since 1910, internationally recognised criteria and scoring formulae were determined for the evaluation of any ungulate trophy (4). This gave impetus to the development of a new, competitive trophy cult in which the possession of record or rare trophies became most important (2). Since the 1950s when hunting grounds all over the world became easily accessible, sports hunting has to a large extent been commercialised into trophy-hunting tourism, as described in Chapt. 1.7. In this context the processing and mounting, and also tourism-orientated production and sale, of trophies and wildlife curios has developed into a distinct secondary industry. Thus, there is often no clear distinction anymore between a trophy in the narrow sense and a wildlife curio. Both may be derived from a wide variety of game species, either hunted, ranched or farmed. Mostly, however they comprise the antlers and the horns of ungulates, often mounted with the head entire, stuffed game specimens or skulls bleached and mounted on shields; they may also include tusks of *elephants, hippos, warthogs* and *boars*, claws of *large carnivores*, particularly conspicuous furskins and feathers, feather ornaments or entire stuffed birds, particularly *raptors*.

Mainly in Africa, however, the taxidermy trade also produces a wide range of rather profane utility curios, such as bookends, ashtrays, bottle openers, ice buckets, dustbins, stools, all incorporating parts of hunted animals; or uses by-products, such as trimmings of tanned furskins for making table mats or handbag trims, tail hair for bracelets, or even pieces of dung to set in resin. All of these items have a flair of Africa and their sale to tourists helps to realise the full economic potential of the wildlife resource without endangering any species.

Curios carved out of ivory were widely offered before the trade in raw ivory was banned, but their production and marketing has a rather different background, discussed in Chapt. 3.2.6.

## Preparation and mounting

The preparation of an animal trophy in the form of a high-quality mount requires not only special skill but also considerable anatomical knowledge, sometimes special field research and some artistic aptitude. The objective is to transfer all species-specific details of the live animal to the mount. Gameskins intended for taxidermy have to be removed particularly carefully from the hunted animal. The pelts should be open, i.e. cuts are to be made from the lower lip to the anus and along the insides of each leg. Extra care is required to skin out feet, head, leaving claws, hooves, ears, nostrils, eyelids, lips and gums attached to the pelt. The flayed skins are usually preserved dry-salted until they reach the taxidermist. They are then treated much the same way as other game- or furskins, as described in Chapts. 3.2.3 and 3.2.4.

For full-body or shoulder mounts forms made out of polyurethane or fibre glass have to be cast from moulds which match exactly the type and body measurements of the particular animal. Horns or antlers are fitted onto the forms and secured with screws and adhesives. The head skin (cape) or full-mount skin is then drawn over the form and sewn into place. The shaping of finer details is possible by using special putties and adhesives. The dried mounts are painted, cleaned and touched off before final delivery to the client.

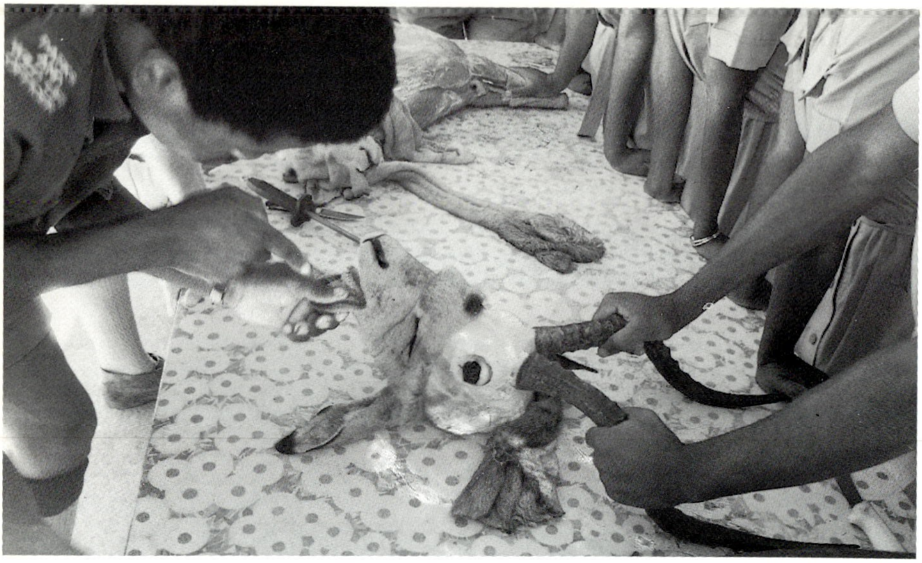

Head of an impala during the final stages of flaying by a taxidermist for the purpose of mounting it. (Photo: I.B. LEACH)

In Europe antlers or horns are usually mounted only with a piece of skull onto a wooden shield. The furskins of bears or large cats are usually prepared into rugs, sometimes with head mounts attached.

## Marketing and economic importance

A hunter always has the option either to take his trophies with him, to have them shipped home in a preserved state, or to have them processed locally before being shipped; he may also choose to sell the trophies. His choice will depend on the prices and quality of local taxidermy services. These, as well as curio processing are very labour intensive and therefore generally well placed in such developing countries which promote hunting tourism, provided the necessary skills and experience are imported from the industrialised countries.

Zimbabwe may serve as an example: in 1989 the taxidermy industry consisted of seven enterprises employing a total of about 100 partly skilled workers (6). The prime objective of the taxidermy enterprises is to cater for overseas trophy hunters, as a complement to the developing wildlife tourism industry. Despite high freight costs the average prices for mounts delivered to the overseas clients in 1988 were about 25% less than those charged by taxidermy studios in Europe or the USA (6). Preparation and mounting of trophies are generally of a high quality and these originate predominantly from *impalas, leopards, kudus, sable, reedbuck* and *buffalo*. Nevertheless, only about 15% of hunted trophies are fully processed in Zimbabwe. The majority were sent out in a raw state for processing in overseas studios and this is indicative of the economic potential for taxidermy services. On average each Plains Game safari in 1988 brought some US$ 7,500 into the country, on top of which a potential taxidermy order of about US$ 2,500 could be expected. A hunter may pay up to US$ 25,000 for an *elephant* hunt, but with other *plains game animals* included, may also leave a processing order for as much as additional US$ 8,000. In total, the taxidermy sector in Zimbabwe was estimated to have added in 1988 about 6% to the total net return from the hunting safari business (6).

A more detailed study in 1993 showed that taxidermy services in Zimbabwe earned on average US$ 365 from each of 1,301 analysed hunts, which represented 3.61% of the total gross value of hunting, calculated at US$ 13.16 mill. (3). In addition to these returns the taxidermy sector earns considerable amounts from the production and sale of wildlife curios to non-hunting tourists.

The world-wide importance of the processing and shipping of trophies and wildlife curio sales is difficult to assess. Only the traffic in trophies of animals which are listed on Appendix II of the Convention on International Trade in Endangered Species of Wild Fauna and Flora (CITES) and which therefore require special export/import permits, can be analysed to this end. A review of the Annual Reports by Parties of CITES in 1983-88 by the World Conservation Monitoring Centre (WCMC) (1) shows that during this period a total of 4,689 trophies from carnivores, 4,076 from Cloven-hoofed ungulates, 954 from Odd-toed ungulates and 725 from primates were exported from countries of origin; bird trophies originated mostly from the falconiforms (46). Amongst the carnivores *lions* and *bears* predominated and the most prominent group of Cloven hoofed trophy animals were *Lechwe, Sable* and a few *other antelope species*; the majority of Odd-toed ungulates were *White rhino.* Apart from *birds of prey* there were very few other bird species traded as "trophies", but some of the traffic in "feathers" or "bodies", referred to in Chapt. 3.2.5, may have originated from sports hunting.

These numbers are most certainly minimum figures considering that many trophies are probably declared as "skins", "skin pieces", "specimens" or not at all. The vast majority of trophies, however, originate no doubt from non-endangered species.

## Literature references


1. ANONYMOUS (1991): *Review of significant trade in species of animals listed on Appendix II of CITES, 1983-1988.* WCMC, Cambridge (unpubl. background document)
2. BUBENIK, A.B. (1989): Sport hunting in continental Europe. In: *Wildlife production systems,* ed. HUDSON, R.J., DREW, K.R. and L. M. BASKIN; p. 115-133; Cambridge University Press, Cambridge
3. FERGUSSON, R.A. (1994): *Sport hunting in Zimbabwe - 1993.* Zimbabwe Association of Tour and Safari Operators (ZATSO/ZPHGA); Harare, Zimbabwe
4. HALTENORTH, T. and W. TRENSE (1956): *Das Grosswild der Erde und seine Trophäen.* Bayerischer Landwirtschaftsverlag, München; 436 pp.
5. MÜLLER-USING, D. (1955): Von Ursprung und Sinn der Trophäen. *Österreichisches Weidwerk,* Jhg. 1955, Wien; p. 11-21
6. PEAK, R. (1989): Processing of hunted trophies. In: *Proceedings of the SADCC/GTZ Workshop on Processing and Marketing of Wildlife Products, 13-21 Nov. 1989, Bulawayo, Zimbabwe, Agenda point 2.3*; ed. FGU CONSULTING, Königstein, Germany
7. ROWLAND WARD (1910): *Records of Big Game.* 6th edition; London

## 3.2.7 IVORY

Ivory is the creamy-white substance that makes up the bulk of tusks from the elephant, hippopotamus, narwhal, walrus and warthog and also the fossil tusks of the mammoths from Siberia and North America. Vegetable ivory comes from a palm nut: the hard endosperm of a seed of the South American corozo palm. Elephant ivory, however, has always far outweighed the other ivory sources in terms of a commodity for man's use and is described in this chapter.

### Production and uses

Ivory may be a **primary or secondary product**. In the first case elephants are killed for the express purpose of taking their tusks for monetary gain. The rest of the animal is either left to rot or the meat is eaten by local people. In the second instance elephants are killed for other reasons such as the prevention of damage to agriculture or silviculture, protection of humans and their houses, regulation of the numbers of elephants in protected areas such as national parks, and also in the form of recreational and trophy hunting. The utilisation of their tusks together with meat and hides then becomes a secondary aspect and is unrelated to the motive of killing the elephant. Historically, the first situation was dominant, particularly in Africa where commercial hunting for ivory developed in the course of European colonisation. This type of indiscriminate "production" of ivory reduced elephant populations significantly in South and West Africa early on during the 19th century (3, 11). After the Second World War, with the rapid development of agriculture in Africa, the official control of elephants for the above-mentioned reasons became more and more important and furnished significant amounts of ivory for commercialisation. However, the illegal killing of elephants purely for their ivory has continued all over Africa and to a lesser extent also in Asia (12) to the present day.

Remains of an elephant rotting after removal of the tusks by commercialised poachers in Ivory Coast. (Photo: H.H. ROTH)

The **history of using ivory** as an art material for carvers dates back to the earliest times, as described in Chapt. 2.3.8. It has inspired work of beautiful detail and fineness in many countries of the world (4). After the art of ivory carving declined in the course of the collapse of the Roman Empire it reached new sophistication in Arab and Asian cultures. With the spread of Indian and Chinese cultural influences to Europe during the 18th and 19th centuries, the appeal of ivory grew in the West as well as in the Orient. The demand in Europe for ivory billiard balls, buttons, knife handles and piano keys then became great (8). It resulted in steadily increasing imports of ivory into Europe until the First World War, as described in Chapt. 2.3.8. Thereafter demand declined and remained relatively low until the early 1970s when world ivory trade increased to such an extent that the survival of the African elephant seemed to be at stake. This provoked a world-wide ban on ivory trade in 1989 which in turn has put most of the ivory carvers out of business.

The ivory of the mammoth also has been used for carving, especially in the Mediaeval period and later in the early 20th century, in Europe. It seems now to be replacing elephant ivory in some carving centres, especially in Germany.

The **main carving centres** for elephant ivory developed early in some Asian countries. From the 16th century until the end of the Moghul rule in 1858, *India* produced some of the finest ivory carvings in the world. Solid ivory furniture was made with beautiful lace-work tracery, and tables and ceilings were inlaid with ivory. Paintings on thin slices of ivory were commissioned as well as beautiful figure carvings. With the start of British rule, their greatest patrons, the Moghul rulers, were gone. Europeans brought in better tools and mass production started, much for export. Around the Second World War, India was importing 246 t of ivory annually, the world's major importer, but business declined in the 1960s when the average annual imports were only 37 t as Hong Kong and then Japan took over most of the trade. Nevertheless, there were 7,200 ivory craftsmen in 1978 and over 2,000 in 1988, the largest number of carvers in the world, in the main centres of Delhi, Jaipur, Jodhpur, Trivandrum and Bangalore (10).

After *Japan* re-opened its borders in the 18th century, the ivory carving industry became famous for its netsukes (ornate toggles for the traditional kimono). These exquisite and unique miniature carvings became very popular with foreign collectors. Demand for other items such as piano keys and signature seals, sculptures, jewellery and accessories, as well as parts for traditional Japanese musical instruments increased within Japan right up until the 1989 international ivory ban. Annual ivory imports rose from 8.3 t in the mid-1880s to 255 t in the 1970s and Japan by the early 1980s became the world's greatest ivory consumer. It was in vogue to buy ivory in this increasingly affluent country. Cylindrical signature seals, required by all adults, consumed 55% of Japan's raw ivory by the early 1980s. The left-over chips were made into jewellery which became very fashionable in Japan (6).

Up until the CITES ban, *China* had the third-largest ivory carving industry after Japan and Hong Kong, employing about 1,400 craftsmen (7). Most of the work was exported wholesale or bought by foreign tourists. Government factories, all in Eastern China, still produced specialised carvings for the tastes to which they catered hundreds of years ago. The best carvers were in Beijing while other main centres were in Guangzhou, Hangzhou, Nanjing, Shanghai and Suzhou. They carved people, buddhas, flowers, birds, animals and large detailed landscapes, as well as ivory fans and magic balls with up to 45 hollow balls, intricately carved, one inside the other.

During China's 1949 Revolution, many ivory carvers fled to *Hong Kong* and by the late 1950s, it had suddenly become the main raw ivory importer and largest ivory manufacturing centre. Due to skilled, cheap labour and long working hours, ivory craftsmen could produce all ivory items very competitively for the foreign market.

When international tourism to *Africa* became a major industry in the 1960s, ivory jewellery carved in Hong Kong and India was on sale in many curio shops in African countries. It was not until the 1970s that entrepreneurs started commercial ivory carving industries in Zimbabwe, Zambia, Botswana and South Africa. In Malawi, however, carving had already been introduced in the 1920s by Sri Lankans. African animal carvings became the most popular in Southern Africa, and leftover ivory was used to make jewellery. By 1983, Zimbabwe had 182 ivory carvers and the largest industry in Southern Africa (5). In Zambia, many of the ivory carvers were in fact born in Zaire, a country with a long tradition in carving ivory for local demand. West Africa also had a long tradition in ivory carving; until recently considerable quantities of ivory were carved in Senegal, Nigeria, Cameroon, and particularly the Ivory Coast whose raw ivory imports as a non-CITES country rose from 0.7 t in 1972 to almost 46 t in 1980 (11).

Netsukes, miniature ivory sculptures, are used by Japanese men to anchor lacquer boxes to their Kimono sashes. Source: Metropolitan Museum of Art; gift of Mrs. Russell Sage, 1910; (by courtesy of the New York Zoological Society)

## Composition and physical properties

Ivory consists of dentine, the hard dense tissue forming the main part of teeth. Elephant ivory can be divided into three categories: Asian which is the hardest, Central and West African from the forest subspecies which is also hard, and thirdly the Eastern and Southern African soft ivory from the savanna subspecies. Hard ivory is heavier with a finer grain and is more translucent, whereas soft ivory is whiter and more porous. Mammoth ivory has the advantage of its remarkable size, but frequently it is twisted and much of it is cracked and brown and often has a pungent smell. It is a very hard ivory and therefore more brittle. Other ivories are too small and too hard for most carving, but hippo ivory has been more commonly used to make netsukes.

## Processing and ivory preferences

While craftsmen in India, China and Hong Kong preferred soft ivory as it is easier to carve and is less likely to crack, the Japanese preferred hard tusks for making signature seals, parts for traditional Japanese musical instruments, sculptures and netsukes. Of course, large 20 kg tusks were in the greatest demand everywhere, but particularly in Japan for seals, and to make piano keys for which soft ivory was used. The Japanese master carvers needed to serve a ten-year apprenticeship to carve the more difficult hard ivory into the finest, most detailed and intricate work in the world. These masters would use about 100 hand tools: hammers, knives, chisels and files to produce their works of art. They would handsketch a pencil design on a piece of ivory and then meticulously start shaping it, taking about 10 days for a 10 cm sculpture.

For most ivory production electric tools including band saws and dentist drills have been used in recent years, being much faster. Yet Chinese and Indian master carvers also continued to produce more individual work with hand tools, as did the traditional carvers in Southern India and Malawi up until the ban. Normally several people would be involved in producing an ivory carving, each person concentrating on a different stage: tusk cutting, the design, initial carving, detailed carving and polishing.

## International trade

In 1975 the Convention on International Trade and Endangered Species of Wild Fauna and Flora (CITES) came into force and this has had a significant effect on the ivory trade. The Asian elephant was placed right away on Appendix I of CITES, i.e. commercial trade of ivory from this species was no longer allowed. The African elephant, however, was listed only as of 1977 on Appendix II, i.e. controlled trade in ivory from Africa was allowed.

Exports of raw ivory from Africa were estimated to have risen by that time from about 200 t in the mid-1950s to about 600 to 900 t in the mid-1970s, but despite the CITES listing increased further to about 1,000 t in the early 1980s before declining as of 1985, as discussed in Chapt. 2.3.8. Whereas the price of ivory between the two World Wars ranged from US$ 2.50 to US$ 9.00/kg, it started to increase rapidly in the early 1970s to US$ 150/kg in 1988 (1) and even to an average of US$ 290/kg in Japan in 1989 (2). Tab. 56 shows that total imports from 1979 to 1988 followed the export trend, i.e. they diminished generally from 1985 onwards. Japan and Hong Kong together imported in 1979-84 between 600 and 800 t net per annum and in 1985-88 between 300 and 160 t net. This represented an average annual consumption of about 75 to 50% of all raw ivory imported by CITES-countries (1).

From 1979 to 1982, Hong Kong alone was importing 668 t per year, but some of this was directly re-exported to China and Japan. India also used to import much of its raw ivory from Hong Kong in recent years, rather than direct from Africa. Traders in Asia preferred to buy through an entrepôt despite the 10% mark-up as they could select smaller or larger tusks and buy smaller quantities, whereas at the ivory auctions in Africa, tusks were often auctioned in large lots and sometimes unsorted. European ivory dealers also bought African tusks to sell to Europe and Asia.

**Table 56:** Imports of raw ivory by major consumers 1979-1988[1)]
(according to BARBIER and coll. (1))

| Importers | 1979 t | 1980 t | 1981 t | 1982 t | 1983 t | 1984 t | 1985 t | 1986 t[2)] | 1987 t[2)] | 1988 t[2)] |
|---|---|---|---|---|---|---|---|---|---|---|
| Hong Kong | **366** | **376** | **427** | **322** | **442** | **270** | **109** | **129** | **150** | **133** |
| Japan | **279** | **240** | **286** | **257** | **425** | **384** | **206** | **29** | **103** | **75** |
| EU[3)] | 179 | 203 | 39 | 39 | 56 | 24 | 49 | 12 | 13 | 4 |
| India | 17 | 19 | 19 | 24 | 23 | 30 | 21 | 8 | 6 | 4 |
| China | 7 | 10 | 10 | 54 | 20 | 7 | 7 | 19 | 39 | 50 |
| USA | 6 | 23 | 11 | 7 | 20 | 55 | 24 | 17 | 21 | 9 |
| Others[4)] | 125 | 96 | 103 | 188 | 32 | -60 | 333 | 386 | 38 | -122[5)] |
| Total | **979** | **967** | **895** | **891** | **1,018** | **710** | **749** | **600** | **370** | **153** |

1) net imports based on customs trade statistics and CITES documentation through WCMC, Cambridge, compiled by the London Environmental Centre; 2) because of adoption of the CITES Quota System there may have been an incentive to evade customs and CITES systems in some countries; 3) only Germany, France, Belgium and UK; 4) difference to total net import of all countries (min. estimate), including particularly imports by Thailand, Singapore, Macao and Taiwan which were at times significant; 5) in some years some importers, like Belgium, UK and Singapore, became net exporters, most likely through destocking

Illicit dealers constantly re-routed their ivory from Africa via different entrepôts in the 1980s, Taiwan, Singapore and Macao, always a step ahead of those trying to stop the illegal trade. Finally, from 1987 to mid-1989 Dubai was the central entrepôt for ivory without proper CITES documentation, where it would be partly worked and then could be imported legally into countries in Eastern Asia as carved ivory.

In 1986, CITES tried to enforce an ivory quota system in Africa whereby each country could sell tusks proportional to its number of elephants. However, elephant poaching continued rampantly in most of Africa. In mid-1989 North America, Western Europe and Japan brought in domestic legislation banning elephant ivory imports. Finally, at the October 1989 CITES meeting, a world-wide ban was agreed by a more than two-thirds majority of representatives of 103 states, and their ruling was put into effect on 18 January 1990. All commercial, international trade in ivory was prohibited between all CITES member states with the exception of China, Zambia, Zimbabwe, Malawi, Botswana and South Africa and later Namibia who took out reservations, and Hong Kong, for whom Britain took out a reservation for six months to help the traders sell their stocks. The domestic ivory bans caused immediate changes in the world ivory trade (2). Consumer demand in the USA and in Europe declined due to the risk of buying an illegal item and because it fell out of fashion. Consequently prices for ivory generally dropped. Some traders started to sell ivory at much reduced prices but others stored it remaining optimistic about the future value of ivory. The long-term trend of the price of raw ivory will be essentially dependent on the effectiveness of the ban, the establishment of illegal trading and the opening of new markets (1, 2). Since the ban, those governments which took out reservations to allow them controlled trade have not exported any raw ivory. Generally the ban has been more effectively upheld than expected. It has, however, largely destroyed the legal carving industries, especially in India, Hong Kong, China and Southern Africa (9).

## Literature References


1. BARBIER, E.B., BURGESS, J.C., SWANSON, T.M. and D.W. PEARCE (1990): *Elephants, economics and ivory*. Earthscan, London
2. CALDWELL, J.R. and R.A. LUXMOORE (1990): Recent changes in world ivory trade. *TRAFFIC Bulletin*, 11(4), p. 50-58
3. CATTRICK, A. (1959): *Spoor of Blood*. Howard Timmins, Cape Town
4. CLUTTON-BROCK, J. (1987): *A Natural History of Domesticated Mammals*. Cambridge University Press and British Museum (Natural History), Cambridge and London; p. 116-120
5. MARTIN, E.B. (1984): Zimbabwe's ivory carving industry. *TRAFFIC Bulletin*, VI (2), p. 33-38
6. MARTIN, E.B. (1985): *The Japanese ivory industry*. WWF Japan, Tokyo
7. MARTIN, E.B. (1988): China's ivory carving industry. *TRAFFIC Bulletin*, 10 (1/2), p. 4-7
8. MARTIN, E.B. (1990a): Ivory billiard balls. *BBC Wildlife*, 8 (9), p. 622-623
9. MARTIN, E.B. (1990b): After the ivory bans. *Wildlife Conservation*, 93 (6), p. 28-31
10. MARTIN, E.B. and L. VIGNE (1989): The decline and fall of India's ivory industry. *Pachyderm*, 12, p. 4-21
11. ROTH, H.H. and I. DOUGLAS-HAMILTON (1991): Distribution and status of elephants in West Africa, *Mammalia*, 55 (5), p. 489-527
12. SUKUMAR, R. (1989): *The Asian elephant: ecology and management*. Cambridge University Press, Cambridge; p. 165-173

## 3.2.8 OTHER PRODUCTS

Apart from the above-discussed commodities, wild animals produce a great variety of other useful raw materials. Of particular importance are all those substances to which medicinal properties are attributed. Many of these were also formerly used in Europe, but have by now been replaced by synthetic medicines. In other continents, however, a wide range of wildlife products are still in great demand for medical purposes. Amongst these are the horns of the *rhinoceros* and *Saiga antelope*, the velvet and bone of *deer* antlers, the fat of a number of species, such as *Spiny-tailed lizards, dugongs,* certain *rodents* and *carnivores*, the venom of *snakes*, the bile of *bears* and also snakes, the milk of *eland* and *moose*, the bones, blood, specific organs and even droppings of *some animals*. Likewise, substances which have a strong scent have been used since early times as perfumes. Certain glandular secretions of penetrating odour from *crocodiles, beavers, muskrats, Civet cats* and *Musk deer* are still widely used in the perfume industry.

Some of these products have already been mentioned in previous chapters and others are only of non-commercial interest. A few, however, are of outstanding economic importance because of their particularly high commercial value, and have significant implications for conservation; only these will be discussed below.

## RHINOCEROS HORN

In contrast to the bovid ungulates, rhinos grow horns which consist of fibrous keratin with no supporting bone structure. It has a unique consistency of great beauty when carved, and is considered to be a superior fever-reducing remedy in traditional Chinese medicine.

### Production and uses

Rhinos in Asia and Africa have been killed for thousands of years for their horn and other products. In the 1970s, with the surge of automatic weapons, marauding soldiers and corrupt government officials, the African rhinoceros species produced large supplies of horn and became the major source for the trade in rhino horn. Poaching became the most important cause for a dramatic decline of the African species, as described in Chapt. 2.3.9.

The Chinese started using rhino horn as a medicine over a thousand years ago, and traditional Chinese doctors all over Eastern Asia, as well as certain other Asians, still believe it to be very efficacious. Chinese throughout Eastern Asia prefer *Asian rhino* horn, because, being much smaller than African horn, its healing powers are considered to be more concentrated. The misunderstanding in the West that the horn was used by the Chinese as an aphrodisiac probably arose from Gujarati traders in Zanzibar meeting European travellers. In the state of Gujarat and parts of Rajasthan in Western India, rhino horn was believed by a few to be an aphrodisiac, but nowhere else.

The pharmacist in an Oriental medicine clinic in Seoul examines a Korean medicine book in order to prescribe African rhino horn to a customer. (Photo: E. MARTIN)

During the last ten years, the main consumers of Asian rhino horn have been residents of Hong Kong, Macao, Singapore, Thailand and Taiwan. Pharmacists take shavings from a complete horn on the counter, then sometimes mix them with other ingredients, usually as a remedy for high fever, especially in children. The patient then infuses the substance in hot water and drinks it like a tea.

*African rhino* horns, being one-tenth the import price of Asian, were more commonly imported into China and South Korea to be made into packaged medicines. Government factories in China would produce millions of tablets and medicinal balls each year, containing many ingredients including rhino horn, for export all over Eastern Asia. In South Korea, one specific medicine with 30 ingredients, including rhino horn, was commonly manufactured: the "chung sim hwan" ball. This was locally prescribed to cure high blood pressure and nose bleeds, rather than exported. Although medicines with rhino horn had also been made in Japan in the past, the country stopped using the horn soon after joining the Convention on International Trade in Endangered Species of Wild Fauna and Flora (CITES) in 1980, and little horn has been smuggled in since then.

The Chinese had another use for rhino horn. During the Ming and Chin dynasties, artisans carved exquisite cups and ornaments out of rhino horn for the Imperial Court. On a smaller scale rhino horn was carved into a variety of items in many other parts of the world including Europe, before it became rare. It was initially the demand for rhino horn in North Yemen that caused massive poaching of rhinos in East and Central Africa after 1970. Traditional daggers, called jambiyyas, were worn daily by many Yemeni men, and those with rhino horn handles were, and still are, considered to be the most beautiful and prestigious. When the country overthrew the ruling Imams and escaped from its feudal system in the late 1960s, many workers went to Saudi Arabia during the oil boom in the early 1970s and sent home large amounts of money (3). More and more people could afford jambiyyas with rhino horn handles as opposed to the ubiquitous Water buffalo horn handles. Yemeni traders wanted large African horns as more handles could be made out of these. The jambiyya industry grew, especially in Sanaa, and craftsmen in their small workshops cut, filed and polished pieces of rhino horn into dagger handles, usually adding gold coin-like decoration, in a manner unchanged for generations. The left-over scraps and shavings would be collected and sold to China and South Korea to be made into medicines.

## Marketing and trade

Due to the great economic growth in both Eastern Asia and Yemen the demand for rhino horn increased in the 1970s, prices rose and the supply out of Africa steadily grew through that decade.

Yemen imported no less than 3,000 kg of rhino horn a year, 40% of all the horn on the world market, but from 1985 to 1990 imports fell to about 300 kg a year (4). This was not only because rhinos had become so scarce, but also because craftsmen were producing more handles from Water buffalo horn and a new substitute, plastic. Furthermore, the import price for rhino horn had greatly increased, from US$ 37 in 1974 to over US$ 1,000 per kg in the mid-1980s when affluent Taiwan became its main importer via corrupt Taiwanese businessmen in South Africa. In 1990, African horn was selling retail in Taipei for US$ 4,221 a kg in the medicine shops while Asian horn was an astronomical US$ 54,040 per kg (2). Some traders were also buying it as an investment in Taiwan, realising the species could become extinct.

By 1977 international trade in products of all five rhino species was completely banned in terms of CITES. Nevertheless rhinos continued to be illegally killed. The number of *Black rhino*, for example, fell from an estimated 14,785 in 1980 (5) to 3,452 in 1991 (1). The majority of these have been poached for illegal trade of their horns, which must have realised millions of dollars. Gradually, however, consumer countries started to introduce internal trade bans. The domestic markets are now essentially closed in Brunei, Hong Kong, India, Indonesia, Japan, Malaysia, Nepal and the Phillipines. By 1992, China, South Korea, Taiwan and Yemen remained the main consumers, but these countries in mid-1993 made a significant effort to enforce new legislation to stop the internal trade.

## ANTLER VELVET

In contrast to the permanent horns of the bovid ungulates, male deer grow new antlers every year which they shed at the end of their annual sexual cycle. Antlers consist of bone substance which is produced by and under a strongly vascularised skin, called velvet. Once the antlers are fully developed the velvet dries and is stripped off by the stag. Growing antlers covered with well-vascularised velvet as well as fully grown velvetless antlers are both commodities widely used in East Asian countries for medical purposes.

### Production and uses

The production of antler velvet was the primary motivation in East Asia for keeping deer captive in enclosures and has become a most important economic aspect of modern deer farming, as described in Chapt. 2.3.12.

All deer species which grow large multiple-point antlers may be used for velvet production. However, there are distinct consumer preferences and generally *Fallow* and *Reindeer* velvet is considered inferior to that derived from *Red, Sika* and *Sambar deer.* In China, *Roe deer* velvet is also commercialised (8). Velvet is harvested by sawing off the antler during its growth stage when it is still covered with well-vascularised velvet. Shed antlers are also collected but do not serve velvet production; they may, however, also be used in pulverised form for medicinal purposes.

Velvet has probably been used for thousands of years in Chinese medicine (7). It is considered a general tonic and body strengthener and thus to be useful for a variety of ailments. It is not specifically used as aphrodisiac, as often believed, but may, similarly to Ginseng, also indirectly improve sex drive and potency.

Production and processing are geared to consumer expectations which differ between countries (13). ***Korea***, the largest export market, demands large four- to five-point antlers preferably from Siberian and Manchurian Red deer, cut before there is any sign of calcification and dried with the blood retained. This is achieved by heat-sealing of the cut. On the ***Hong Kong*** market and in ***China*** two-point velvet antlers from sika, between 15-30 cm long from which the blood is drained out, usually by application of a vacuum, are preferred. Such velvet has a yellow to white colour and a particularly fine texture. In both cases the antlers are dipped several times into near boiling water and then well dried. The blood-in antlers are either sectioned or packaged in toto or into boxes. The bloodless product is usually sliced very finely into pieces of only 25-50 μm thickness which are marketed in packets. Velvet slices are usually mixed with a variety of herbs depending on the ailment. The mix is cooked and the resulting decoction drunk hot. Velvet slices may also be steamed with chicken, duck or pork and be consumed as a soup. Sometimes, velvet is made into a powder and dispensed as such. In ***Taiwan*** velvet consumption patterns are closely related to the local deer production. Sambar antlers between 15 and 25 cm length from which velvet is cut fresh for direct sale to the consumer are preferred. The sliced velvet and also the highly priced blood caught during velvetting, are preserved in drinkable alcohol such as whisky or rice wine, and mixed with medicinal herbs. The tincture which results from such mixes is drunk by the patient. Only if the demand exceeds local production is some imported deep-frozen velvet utilised.

In some of the Russian deer farms sika velvet is processed directly into so-called Pantocrin, for domestic consumption or export (6). This is a dilute medicinal liquid which contains the physiologically effective substances of velvet, including sex hormones and minerals, in a concentrated form.

## Marketing and trade

The main producers of antler velvet today are China with an estimated 115 t/annum (12) and the GUS countries (former USSR) which produce about 34-35 t of conserved velvet per annum (equivalent to about 100 t fresh antlers) (9). In the 1980s about 53% of this had originated from the Central and East Asian *Red deer subspecies* (6). In 1990 an additional 40 t of antler velvet were obtained from reindeer of which 25 t were exported (9). China is not only a producer but also the main consumer of velvet; domestic use in 1981 was estimated at 80 t (6) and total revenue from antlers at US$ 35 mill. (7). International exports from China and the former USSR countries in 1977-82 was about 40-50 t/annum and 12-14 t/annum respectively but has steadily risen since (10). The majority of this trade was routed through Singapore and Hong Kong to Taiwan, Thailand, South Korea and Hong Kong itself. The next important exporters were New Zealand, the USA, Germany and Canada. In 1981-84 New Zealand sold 83% by value of its produced velvet to the above-mentioned two countries (11).

Display of traditional Chinese medicines including antler velvet and rhino horn, in a Chinese medicine shop in South-East Asia in 1980. (Photo: E. MARTIN)

The price for dried velvet in South Korea fluctuated in 1980-85 from US$ 276 to US$ 463 per kg, whereas in Taiwan only between US$ 85 and US$ 185 per kg was paid probably for fresh-frozen velvet (11). Using the price for dried velvet the total value of internationally traded velvet during these years approximated to US$ 18-22 mill.

Apart from the velvet, large quantities of hard antlers are also traded but only fetch much lower prices. Other deer parts used in Oriental medicine, such as penises, tails, tendons and embryos, are traded in low quantities but realise the highest unit prices (11).

## MUSKS

Many mammals, but also some reptiles secrete substances of penetrating and persistent odour from differently located skin glands usually for territorial or sexual marking. These are generally called musk and some of them constitute important raw materials for the perfume industry and are also used in Oriental medicine.

### Production and uses

Musk is commercially obtained mainly from *Civet cats* and *Musk deer*. Efforts are also made to utilise farmed *crocodiles, beavers* and *muskrats* for the production of musk which, however, is of inferior quality.

**Civet musk** originates from special pouched glands situated in the male civet between the testicles and the praeputium of the penis and in females between the vulva and anus. It is commercially produced mainly from the *African Civet cat* in Ethiopia, but to a much lesser extent is also collected from *Asian civets*. Early missionaries and travellers reported in detail how Civet cats in Ethiopia were kept in small wooden cages for the purpose of regular extraction of the highly valued civet musk (20, 24). This practice has not changed since, and despite efforts to introduce civet farming to other African countries, Ethiopia has remained the main civet producer, as described in Chapter 2.3.4.

The musk, also called "civet", is harvested by opening the pouch of the male gland with thumb and forefinger and scooping it out with a special horn "spoon", at intervals of 9-12 days during which 8-10 g collects in the pouch (25). Females are less productive because their civet is washed out or diluted by passing urine and therefore rarely kept. After extraction the pouch is filled with an oily mixture of rancid butter and low quality civet which is said to prevent the gland from drying up. Some civet may also be collected from places where the cats mechanically remove their accumulated secretion (25). Harvested civet is stored in ox horns which must be sealed to prevent its oxidation.

Civet musk is a yellowish, fouwl-smelling oily substance which has been analysed to contain mainly a fatty-acid-related ketone, called civetone. Despite production of synthetic civetone under the name of Civetine, natural civet musk has retained its place in the perfume industry. It has powerful holding properties for other scents and is mainly used as a fixative in the manufacture of expensive perfumes (27). To a lesser degree it is also used as a scent, substituting for the more expensive musk from deer. Arabs also use civet for medical purposes and in Japan it is used as a preservative of antibiotics (16)

**Musk of the Musk deer** originates from the male animal's preputial glands which form a sac on the ventral side of the penis, the so-called "pod". It is commercially produced mainly from the *Siberian* and *Dwarf Musk deer* in Eastern Siberia and China, and to a lesser degree also from the *Himalayan species* in India, Nepal, Bhutan and Tibet. This musk has been used for thousands of years in Chinese perfumery and medicine, but has also been introduced to the Western European perfume industry (27). Traditionally deer are hunted and the entire musk pod excised for commercialisation in toto. The serious decline of Musk deer in the wild has induced their farming in captivity since 1958 in China, as described in Chapt 2.3.12.

The musk is harvested in captive males once a year by inserting a small spatulum into the gland-pod's orifice and scooping the secretion out onto paper or into a small dish. After extraction an antiseptic is inserted into the gland to prevent infection (28). A maximum of 18g (10 g dry weight) of musk may be harvested per annum which is comparable to yields obtained from males killed in the wild (22). The musk is usually dried in a desiccator, then weighed and stored in an airtight container.

Musk from Musk deer is a milky-yellow substance which is secreted from the epithelial cells of the glands. In the gland's sac it matures into a granular, red-brown substance (29, 18). It is mainly used as a scent or fixative for high-quality perfumes (27). In China it is also used for medical purposes (29). There are different synthetic aromatic nitro-musk compounds available, e.g. musk xylene, musk ketone and musk ambrette. These are to some extent used as a cheap substitute for the deer musk in perfumery, but cannot replace this in its application in Oriental medicine. The nitro-musks are being added extensively to wash powders, soaps and cosmetics, and have been more recently identified as dangerous water pollutants, concentrating in fatty tissues of fish and mussels, with geno- and neuro-toxic effects (21).

## Marketing and trade

**Civet musk** is marketed by the small farmers via roving tradesmen; only the bigger farmers sell directly to the exporters in Addis Ababa. Like other musk, civet is often adulterated, with butter, oil, mashed bananas etc. (26). Therefore, the Ethiopian government introduced a system of quality control through the Pasteur Institut in Addis Ababa based on the method of BENNET and SEABORS, used in perfumery since 1929. Any civet intended for export would be controlled for the percentage of civetone it contains. If this was less than 40%, export would not be allowed (25).

Trade of civet musk from Ethiopia to the Near East and India through Arab merchants existed in the Middle Ages, and ever since Ethiopia has retained its near monopoly in this trade, producing almost 90% of total world supplies (16). In Ethiopia most of the civet originates from the southwestern provinces. It was formerly traded to Europe via Nubia and Egypt, but after the construction of the railways towards the end of the 19th century it was shipped out through Djibouti. In 1959 Ethiopia exported 1,194 kg of civet valued at US$ 226,890 (i.e. about US$ 190/kg), which made it one of the major export items (25). Exports in 1975-78 averaged 1,943 kg/annum (17), and in 1981 the price paid for a kilogram of civet in the UK was as much as £st 1,060 (approx. US$ 2,150) (16). Main importers of civet are France and the USA, lesser quantities being purchased by other European countries, Japan and Saudi Arabia (25).

**Musk of Musk deer** is usually marketed in a granular or powdered form in bottles. As adulteration is frequent, marketing of musk for medical purposes in China is controlled by the State Pharmaceutical Corporation (19). The trade in musk pods from Siberia to Eastern Asian countries has a long tradition and reached a peak of 18,200 pods in 1855. Early in the 20th century hunting of Musk deer in Russia was fortunately banned (15). It was resumed in a controlled manner in the 1950s and Russia is again the most important producer, followed by China, Nepal and India. Musk from farmed animals is commercially produced in only very limited quantities in China (23). The greatest importer is Japan having annually purchased between 109 and 727 kg from 1960 to 1983. Imports even rose in 1988 to an astounding 810 kg and this was seen as stockpiling by Japanese dealers in anticipation of Japan dropping its CITES Reservation agains musk import (14). Japan is followed by Korea, which imported between 56 and 132 kg/annum and Taiwan which imported between 2 and 34 kg/annum during 1980-85 (11). Japan's import represents well over 50% of the internationally traded musk, but is only a fraction of what used to be exported by China and the Indian continent at the turn of this century (23).

Musk is by far the most valuable wildlife product fetching prices of up to US$ 24,000/kg, and even US$ 45,000/kg for the granular product, in Japan in 1978 (22).

## Literature references

### Rhino horn

1. BROOKS, M. (1992): Chairman's Report: IUCN/SSC African Rhino Specialist Group. *Pachyderm,* 15, p. 6
2. MARTIN, C.B. and E.B. (1991): Profligate spending exploits wildlife in Taiwan. *Oryx,* 25 (1), p. 19
3. MARTIN, E.B. and C.B. (1982): *Run Rhino Run*. Chatto and Windus, London; p. 56
4. VIGNE, L. and E.B. MARTIN (1993): Yemen's rhino horn trade increases. *Oryx*, 27 (2), p. 94
5. WESTERN, D. and L. VIGNE, (1985): The deteriorating status of African rhinos. *Oryx*, 19, p. 216

### Antler velvet

6. DREW, K.R., BAI, Q. and E.V. FADEEV (1989): Deer farming in Asia. In: *Wildlife production systems*, ed. HUDSON R.J., DREW K.R. and L.M. BASKIN; p. 334-346; Cambridge University Press, Cambridge
7. FLETCHER, T.J. (1984): 16. Other deer. In: *Evolution of domesticated animals*, ed. MASON, I.L.; p. 138-145; Longman Group, London and New York
8. GREER, E.C. and R.W. DOUGHTY (1976): Wildlife Utilisation in China. *Environmental Conservation*, 3(3), p. 200-208
9. JUDIN, A.M. (1993): *Velvet antlers as drug resource* (in Russian). Nauka, Novosibirsk, Russia; 120 pp.

10. LEE, C.H. and T.S. CHA'NG (1985): Marketing and utilisation of deer products in Asia (*cited after* LUXMOORE 1989)
11. LUXMOORE, R.A. (1989): International trade. In: *Wildlife production systems*, ed. HUDSON, R.J., DREW, K.R. and L.M. BASKIN; p. 28-49; Cambridge University Press, Cambridge
12. PINNEY, B. (1981): Delegation to China (*cited after* DREW and coll. 1989)
13. RENNIE, N. (1982): Antler velvet. In: *The Farming of Deer: World Trends and Modern Techniques*; ed. YEREX, D.; p. 141-144; Agricultural Promotion Associates; Wellington, New Zealand

**Musks**

14. ANONYMOUS (1988): Japan Stockpiles Musk. *TRAFFIC (USA)*, 8(1), p. 17
15. BANNIKOV, A.G., USTINOV, S.K. and LOBANOV, P.N. (1980): *The Musk Deer, Moschus moschiferus, in the USSR*. IUCN, Gland, Switzerland (unpubl. report)
16. BARZDO, J., LUXMOORE, R., BROAD, S. and D. JONES (1984): *A survey of wildlife farming operations*. WCMC/IUCN, Cambridge (unpubl. report)
17. BERHANU, A.L. (1984): 29. Civets. In: *Evolution of domesticated animals*, ed. MASON I.L.; p. 237-238; Longman Group, London and New York
18. BI, S.Z., YAN, Y.H., QUEING, Z.X., SHENG, P.T., WU, Y.M., CHEN, C.F., XU, H. J., YANG, G.K., YIN, T.B. and Y.L. LU (1980): Dissection and analysis of the musk gland of *M. moschiferus* and a preliminary investigation into its histology (*cited after* GREEN 1989)
19. BISTA, R.B., SHRESTA, M.N. and B. KATTEL (1979): Domestication of the dwarf musk deer (*Moschus berezovskii*) in China (*cited after* GREEN 1989)
20. CECCHI, A. (1888): *Fünf Jahre in Ostafrika*. F. Brockhaus, Leipzig; p. 444-446
21. GEYER, H.J., RIMKUS, G., WOLF, M., ATTAR, A., STEINBERG, C. and A. KETTRUP (1994): Synthetische Nitromoschus-Duftstoffe und Bromocyclen - Neue Umweltchemikalien in Fischen und Muscheln bzw. Muttermilch und Humanfett. *Zeitschrift für Umweltchemie und Ökotoxikologie*, 6(1), p. 9-17
22. GREEN, M.J.B. (1988): The musk trade, with particular reference to its impact on the Himalayan population of *Moschus chrysogaster* (*cited after* GREEN 1989)
23. GREEN, M.J.B. (1989): Musk production from musk deer. In: *Wildlife production systems*, ed. HUDSON, R.J., DREW, K.R. and L.M. BASKIN; p. 401-409; Cambridge University Press, Cambridge
24. GUIBOURT, N.Y.B. (1869/70): *Histoire naturelle des drogues simples* (6ième edition). Vol. 4; Baillière, Paris
25. KRUG, W. (1969): *Civet farming in Ethiopia*. FAO Animal Health Project, Jimma, Ethiopia; (unpubl. report)
26. PARRY, E.J. (1925): *Parry's Cyclopedia of Perfumes*. Vol. 2; p. 473-486; J. and A. Churchill, London
27. POUCHER, W.A. (1936): *Perfumes, Cosmetics and Soaps*. Vol. 1; p. 123-124; Chapman and Hall, London
28. ZHANG, B.L. (1983): Musk deer: their capture, domestication and care according to Chinese experience and methods. *Unasylva*, 35, p. 16-24
29. ZHANG, B.L., DANG, F.M. and B.S. LI, (1979): The Farming of Musk Deer (*cited after* GREEN 1989)

# INDEX OF ANIMAL NAMES